MC68HC12

An Introduction:

Software and Hardware Interfacing

MC68HC12

An Introduction:

Software and Hardware Interfacing

Han-Way Huang
Minnesota State University • Mankato

THOMSON

DELMAR LEARNING

Australia • Canada • Mexico • Singapore • Spain • United Kingdom • United States

MC68HC12 An Introduction: Software and Hardware Interfacing
Han-Way Huang

Business Unit Director:
Alar Elken

Executive Editor:
Sandy Clark

Senior Acquisitions Editor:
Gregory L. Clayton

Senior Development Editor:
Michelle Ruelos Cannistraci

Editorial Assistant:
Jennifer Luck

Executive Marketing Manager:
Maura Theriault

Channel Manager:
Fair Huntoon

Marketing Coordinator:
Karen Smith

Executive Production Manager:
Mary Ellen Black

Production Manager:
Larry Main

Production Editor:
Stacy Masucci

Art/Design Coordinator:
David Arsenault

Technology Project Manager:
David Porush

Technology Project Specialist:
Kevin Smith

Library of Congress
Cataloging-in-Publication Data:

ISBN 0-7668-83448-4

Contents

Chapter 2   68HC12 Assembly Programming   29

Chapter 3   68HC12 Members & Hardware & Software Development Tools   75

Chapter 4   Advanced Assembly Programming   109

Chapter 5   C Language Programming   165

Chapter 6   Interrupts, Resets & Operation Modes   199

Chapter 7   Parallel Ports   229

Chapter 8   Timer Functions   289

Chapter 9   Serial Interface – SCI & SPI   359

Chapter 10   Analog-to-Digital Converter   427

Appendixes

Glossary   701

Index   711

Preface

The 16-bit 68HC12 was designed as an upgrade to the popular 8-bit 68HC11 microcontroller. If the 68HC11 is considered as the first-generation microcontroller from Motorola, then the 68HC12 should be considered as the second-generation microcontroller. The design of the 68HC12 has adopted many features from several 8-bit and 16-bit microcontrollers:

1. *Full feature timer system.* The 68HC12 timer system provides input-capture, output-compare, pulse-accumulator, pulse-width-modulation (PWM), real-time interrupt (RTI), and computer-operate-properly (COP) system. The enhanced timer system of the 68HC12 has the capability to capture two edges before interrupting the CPU. This feature allows the 68HC12 to handle high frequency signals even when the CPU is very busy.

2. *In system programming* (ISP) capability. Most 68HC12 members provide on-chip flash memory and allow the software to be upgraded in the system.

3. *Background debug mode* (BDM). This BDM circuit provides a one-line interface for accessing the internal resources of the 68HC12 and hence allows a low-cost source-level debugger to be designed.

4. *Multiple serial interfaces.* The 68HC12 supports industrial standard SPI, I^2C, and CAN bus. The SPI and I^2C allow the 68HC12 to interface with numerous peripheral devices with serial interface such as LED driver, LCD, A/D converter, D/A converter, real-time clock, SRAM, EEPROM, and phase-locked loop. The CAN bus was proposed as a communication bus for automotive applications. However, this bus is also widely used in automation control applications.

5. *Fuzzy logic support.* The 68HC12 provides a group of instruction to support the fuzzy logic operations. These instructions should facilitate the programming of fuzzy logic applications.

These features appear to be the features most desired by the end user. With these features the 68HC12 is perfect for those who want to learn about modern microcontroller interfacing and applications.

Intended Audience

This book is written for two groups of readers:

1. Students in electrical and computer engineering and technology who are taking an introductory course of microprocessor interfacing and applications.

For this group of readers this book provides a broad and systematic introduction to the microprocessors and microcontrollers.

2. Senior electrical engineering and computer engineering students and working engineers who want to learn about the 68HC12 and use it in a design project. For this group of readers, this book provides numerous more complicated examples to explore the functions and applications of the 68HC12.

Prerequisites

This book has been written with the assumption that the reader has taken a course on digital logic design and has been exposed to high-level language programming. The knowledge in digital logic will greatly facilitate the learning of the 68HC12. The knowledge in assembly language programming is not required because one of the writing goals of this book is to teach the 68HC12 assembly language programming.

Approach

During the past few years, we have seen a new trend in teaching microprocessor applications:

Using the microcontroller along with a high-level language such as C to teach microprocessor interfacing and programming.

This book follows this trend. The 68HC12 is the latest 16-bit microcontroller developed by Motorola that combines all of the useful features from many other 8- and 16-bit microcontrollers. According to the survey posted in the website www.microcontroller.com, more than 80% of the applications are written in the C language while 75% of the applications are written in assembly language. It is not difficult to tell from these statistics that many applications use both the assembly and C languages. The code generated by a C compiler is still much larger than its equivalent in assembly language. Many time-critical applications are still written in assembly language. Both the assembly and C languages are used in illustrating the programming of the 68HC12 in this text.

The writing of this text has put microcontroller learning as its focus. Each subject is started with background issues followed by the specific implementation in the 68HC12. Numerous simple and complex examples are provided in appropriate places to illustrate the programming and interfacing of the 68HC12.

Textbook Organization

Chapter 1 presents the basic concepts about computer hardware and software, microcontroller applications, the 68HC12 addressing modes, and a subset of the 68HC12 instructions.

Chapter 2 introduces basic assembly programming skills and the 68HC12 instructions.

Chapter 3 provides a brief summary on the features of the 68HC12 members, and hardware and software development tools. A tutorial on using the Axiom CME-12BC demo boards and the use of D-Bug12 monitor commands are also provided in this chapter.

Chapter 4 illustrates more advanced assembly programming skills and subroutine calls.

Chapter 5 provides a brief tutorial on the C language syntax and the use of the ImageCraft ICC12 C compiler.

Chapter 6 discusses the concepts and programming of interrupts and resets.

Chapter 7 introduces the basic concepts about parallel I/O. This chapter also covers the interfacing and programming of simple I/O devices, including DIP switches, keypad scanning and debouncing, LEDs, LCDs, D/A controllers, and stepper motor control.

Chapter 8 explores the operation and applications of the timer system including input capture, output compare, real time interrupt, pulse accumulator, and pulse width modulation.

Chapter 9 deals with serial communication interface, including SCI (asynchronous) and SPI (synchronous). Chapter 10 introduces the A/D converter and its applications in temperature, humidity, and barometric pressure measurement.

Chapter 11 provides a detailed explanation of the operation of the BDM mode and also a tutorial on the use of a BDM-based source-level debugger designed by Axiom.

Chapter 12 presents the CAN 2.0 protocol and the 68HC12 CAN module. Several examples on the programming of the CAN module are provided.

Chapter 13 describes the 68HC12 internal SRAM, EEPROM, and flash memory. This chapter also explores the issues related to external memory expansion: address space assignment, decoder design, and timing analysis.

Pedagogical Features

Each chapter opens with a list of *Objectives*. Every subject is presented in a step-by-step manner. Background issues are presented before the specifics related to each 68HC12 function are discussed. Numerous *Examples* are then presented to demonstrate the use of each 68HC12 I/O function. Procedural steps and flowcharts are used to help the reader understand the program logic in most examples. Each chapter concludes with a *Summary* and numerous *Exercises* and *Lab Assignments*.

Demo Boards

A demo board is essential for learning the 68HC12. Although general programs can be debugged by using a simulator, the testing of peripheral functions cannot be done without a piece of hardware demo board. The Axiom CME-12BC is chosen as the demo board for testing the programs developed for this book. The CME-12BC supports all of the 68HC12 features with the exception of the enhanced timer function. This demo board uses the Motorola *D-Bug12* as its debugging monitor. This monitor provides functions similar to those available in the Buffalo for the 68HC11. Users can use commands to display and change memory and register contents, set breakpoints, and download programs for execution on the CME-12BC demo board.

How to Use the CD

This text includes a complementary CD for all users. The CD contains:

1. Programs in the text. A CD icon is placed on the text pages beside those programs that are in the CD.

2. PDF files of datasheets and manuals of the 68HC12 members and peripheral chips that are covered in this text. These files allow the user to refer to details of these chips at their fingertips.

3. Freeware cross assembler: MiniIDE (from Mgtek).

4. Demo version of ICC12 (good for one month). The demo version of the ICC12 can also be downloaded from the Imagecraft website. For best results, a standard or professional compiler should be used.

Supplements

A CD dedicated to instructors who adopt this text is also available from the publisher. This CD contains solutions to all exercise problems. In addition, a set of Powerpoint lecture notes provides a starting-point for instructors who teach the 68HC12 microcontroller.
ISBN: 0-7668-3449-2

Acknowledgements

This book would not be possible without the help of a number of people, and I would like to express my gratitude to all of them. I would like to thank Tony Plutino of Motorola (he just retired from Motorola) for his help over the years. He sent me samples of microcontroller chips, demo boards, and software to support my writing of this book. I would like to thank Greg Clayton, Senior Acquisition Editor of Delmar Thomson Learning, and Michelle Ruelos Cannistraci, Senior Development Editor, for their enthusiastic support during the preparation of this book. I also appreciate the outstanding work of the production staff, including Stacy Masucci, David Arsenault and the staff of Phoenix Creative Graphics, led by John Shanley.

I would like to express my thanks for the many useful comments and suggestions provided by colleagues who reviewed this text during the course of its development, especially to:

Norm Grossman, DeVry University, Long Beach, CA
Gilbert Seah, DeVry University, Scarborough, Ontario
Roman Stemprok, University of North Texas, Denton, TX
Ralph Tanner, Western Michigan University, Kalamazoo, MI
Asad Yousuf, Savannah State University, Savannah, GA
Jamie Zipay, Oregon Institute of Technology, Portland, OR

I am grateful to my wife, Su-Jane and my sons, Craig and Derek, for their encouragement, tolerance, and support during the entire preparation of this book. Finally, I want to dedicate this book to the memory of my parents. They gave me the education and they taught me how to be a good man.

Han-Way Huang
June, 2002

Avenue of Feedback

The Author has taught microcontrollers for more than 15 years and has written several textbooks for other microcontrollers including the Z80, the 8051, and the 68HC11. The Author welcomes the report of errors and suggestions for improvement. Your input will be greatly appreciated. Error reports and suggestions can be sent directly to the author (at han-way.huang@mnsu.edu) or the publisher.

About the Author

Dr. Han-Way Huang received his M.S. and Ph.D., degrees in Computer Engineering from Iowa State University. He has taught microprocessor and microcontroller applications extensively for 15 years. Before teaching at Minnesota State University, Mankato, Dr. Huang worked for four years in the computer industry. In addition to this book, Dr. Huang has also authored books on the Zilog Z80 (in Chinese), the Motorola 68HC11 (*An Introduction to the 68HC11* from Delmar Learning) and the 8051 microcontroller, (*Using the MCS-51 Microcontroller* from Oxford University Press).

1

Introduction to the
68HC12 Microcontroller

1.1 Objectives

After completing this chapter you should be able to:

- define or explain the following terms: computer, processor, microprocessor, microcontroller, embedded system, hardware, software, cross assembler, cross compiler, RAM, DRAM, SRAM, ROM, PROM, EPROM, EEPROM, flash memory, byte, word, nibble, bus, KB, MB, mnemonic, opcode, and operand

- explain the differences between the inherent, immediate, direct, extended, relative, indexed, and indirect indexed addressing modes

- write a sequence of arithmetic and data transfer instructions to perform simple operations

1.2 Basic Computer Concepts

A computer is made up of hardware and software. The hardware consists of the following major components:

- *The processor.* The processor is the brain of a computer system. All computation operations are performed in the processor. A computer system may have one or multiple processors. A computer system that consists of multiple processors is called *multiprocessor* computer system.

- *The memory.* The memory is the place where software programs and data are stored. A memory system can be made of semiconductor chips, magnetic material, and/or optical discs.

- *Input device.* Input devices allow the computer user to enter data and programs into the computer so that computation can be performed. Switches, keypads, keyboards, mice, microphones, and thumb wheels are examples of input devices.

- *Output device.* The results of computation are displayed via output devices so that the user can read them and equipment can be controlled. Examples of output devices include CRT displays, LEDs, seven-segment displays, liquid-crystal-displays (LCD), printers, plotters, and so on.

As shown in Figure 1.1, the processor communicates with other components through a *common bus.* A *bus* is simply a group of conducting wires that allow signals to travel from one point to another. The common bus actually consists of three buses: address bus, data bus, and control bus. Each bus is a group of signals of the same nature.

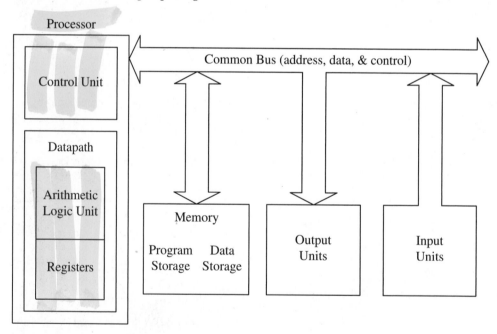

Figure 1.1 ■ Computer organization

1.2.1 The Processor

The processor, which is also called the central processing unit (CPU), can be further divided into two major parts:

Datapath. The datapath consists of a *register file* and an *arithmetic logic unit* (ALU). The register file consists of one or more registers. A register is a storage location in the CPU. It is used to hold data or a memory address during the execution of an instruction. Because the register file is small and close to the ALU, accessing data in registers is much faster than accessing data in memory outside the CPU. The register file makes program execution more efficient. The number of registers varies from computer to computer. All of the arithmetic computations and logic evaluations are performed in the ALU. The ALU receives data from main memory and/or the register file, performs a computation, and, if necessary, writes the result back to main memory or registers.

Control unit. The control unit contains the hardware instruction logic. It decodes and monitors the execution of instructions and also acts as an arbiter as various portions of the computer system compete for resources of the CPU. The system clock synchronizes the activities of the CPU. All CPU activities are measured by clock cycles. At the time of this writing, the highest clock rate of a PC has reached more than 2.0 GHz, where

1 GHz = 1 billion ticks (or cycles) per second

The control unit maintains a register called the *program counter* (PC), which controls the memory address of the next instruction to be executed. During the execution of an instruction, the presence of overflow, an addition carry, a subtraction borrow, and so forth, is flagged by the system and stored in another register called a *status register*. The resultant flags are then used by programmers for program control and decision-making.

WHAT IS A MICROPROCESSOR?

The processor in a very large computer is built from a number of integrated circuits. A *microprocessor* is a processor fabricated on a single integrated circuit. A *microcomputer* is a computer that uses a microprocessor as its CPU. Early microcomputers were quite simple and slow. However, many of today's desktop or notebook computers have become very sophisticated and even faster than many larger computers were only a few years ago.

One way to classify microprocessors is to use the number of bits (referred to as *word length*) that a microprocessor can manipulate in one operation. A microprocessor is 8-bit if it can only work on 8 bits of data in one operation. Microprocessors in use today include 4-bit, 8-bit, 16-bit, 32-bit, and 64-bit. The first 4-bit microprocessor is the Intel 4004, which was introduced in 1971. Since then many companies have joined in the design and manufacturing of microprocessors. Today 8-bit microprocessors are the most widely used among all microprocessors, whereas 64-bit microprocessors emerged only a few years ago and are mainly used in high performance workstations and servers.

Many 32-bit and 64-bit microprocessors also incorporate on-chip memory to enhance their performance. Microprocessors must interface to input/output devices in many applications. However, the characteristics and speed of the microprocessor and input/output devices are quite different. Peripheral chips are required to interface input/output devices with the microprocessor. For example, the integrated circuit i8255 is designed to interface a parallel device such as a printer or seven-segment display with the Intel 8-bit microprocessor 8085.

Microprocessors have been widely used since their invention. It is not an exaggeration to say that the invention of microprocessors has revolutionized the electronics industry. However, the following limitations of microprocessors have led to the development of microcontrollers:

- A microprocessor requires external memory to execute programs.
- A microprocessor cannot directly interface with I/O devices. Peripheral chips are needed.
- Glue logic (such as address decoders and buffers) is needed to interconnect external memory and peripheral interface chips with the microprocessor.

Because of these limitations, a microprocessor-based design cannot be made as small as might be desired. The development of microcontrollers not only eliminated most of these problems but also simplified the hardware design of microprocessor-based products.

WHAT IS A MICROCONTROLLER?

A microcontroller is a computer implemented on a single very large-scale integration (VLSI) chip. A microcontroller contains everything a microprocessor contains plus one or more of the following components:

- *Memory*. Memory is used to store data or programs. A microcontroller may incorporate certain amounts of SRAM, ROM, EEPROM, EPROM, or flash memory.
- *Timer*. Timer function is among the most useful and complicated functions in most microcontrollers. The timer function of most 8-bit and 16-bit microcontrollers consists of input capture, output compare, counter, pulse accumulator, and pulse width modulation (PWM) modules. They can be used to measure frequency, period, pulse width, and duty cycle. They can also be used to create time delay and generate waveforms.
- *Analog-to-digital converter* (ADC). The ADC is often used with a sensor. A sensor can convert certain non-electric quantities into an electric voltage. The ADC can convert a voltage into a digital value. Therefore, by combining an appropriate sensor and the ADC, a data acquisition system that measures temperature, weight, humidity, pressure, or airflow can be designed.
- *Digital-to-analog converter* (DAC). A DAC can convert a digital value into a voltage output. This function can be used in controlling DC motor speed, adjusting brightness of the light bulb, the fluid level, and so on.
- *Direct memory access* (DMA) controller. DMA is a data transfer method that requires the CPU to perform initial setup but does not require the CPU to execute instructions to control the data transfer. The initial setup for a DMA transfer includes the setup of source and destination addresses and transfer byte count. This mechanism can speed up data transfer by several times.
- *Parallel I/O interface* (often called parallel port). Parallel ports are often used to drive parallel I/O devices such as seven-segment displays, LCDs, or parallel printers.
- *Asynchronous serial I/O interface*. This transfer method transmits data bits serially without using a clock signal to synchronize the sender and receiver.
- *Synchronous serial I/O interface*. This interface transfers data bits serially and uses a clock signal to synchronize the sender and receiver.
- *Memory component interface circuitry*. All memory devices require certain control signals to operate. This interface generates appropriate control signals to be used by the memory chips.
- *Digital signal processing (DSP) feature*. Some microcontrollers (for example, the Motorola HC16) provide features such as the multiply-accumulate instruction that can execute in one clock cycle to support DSP computation.

The Motorola 68HC12 family of microcontrollers were introduced in 1996 as an upgrade for the 68HC11 microcontrollers. Members in this family implement the same instruction set and addressing modes but differ in the amount of on-chip memory and peripheral functions. The 68HC12 has the following features:

- 16-bit CPU
- Supports a standard 64KB address space
- Some members support a paged memory expansion scheme that increases the standard memory space by means of predefined windows in address space
- 768 bytes to 4KB of on-chip EEPROM
- 1KB to 12KB of on-chip SRAM
- 8-bit or 10-bit A/D converter
- 32KB to 128KB of on-chip flash or ROM memory
- Timer module that includes input capture, output compare, and pulse accumulator functions. This is the most complicated module in the 68HC12 microcontroller.
- Pulse-width modulation (PWM)
- Synchronous peripheral interface (SPI)
- Asynchronous serial communication interface (SCI)
- Byte data link communication (BDLC)
- Controller area network (CAN)
- Computer operating properly (COP) watchdog timer
- Single-wire background debug mode (BDM)
- Instructions for supporting fuzzy logic

A 68HC12 member may have from 80 to 112 pins. These signal pins are needed to support the I/O functions. A block diagram of the 912BC32 is shown in Figure 1.2.

For the time being, you may feel lost when seeing these names if you are new to microcontrollers. However, all of them will be explained in detail in the appropriate chapters.

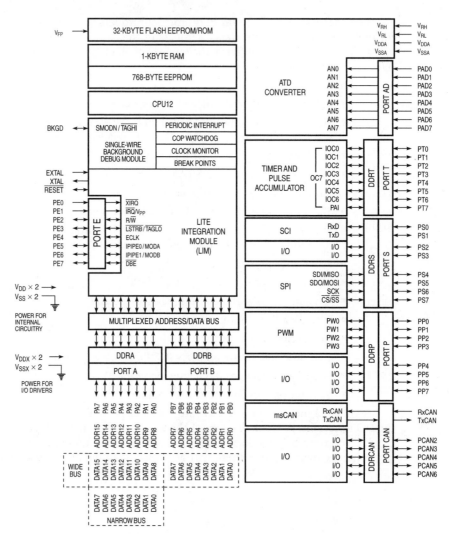

Figure 1.2 ■ Block diagram and pin assignment of the 68HC912BC32
(Redrawn with permission of Motorola)

EMBEDDED SYSTEMS

A microcontroller is designed to be used in a product to provide certain control functions. End users are interested in the features and functions of the product rather than the microcontroller. A product that incorporates one or more microcontrollers and has this nature is called an *embedded product* or *embedded system*. A cell phone is a good example of an embedded system. The cell phone contains a microcontroller that is responsible for making the phone call, accepting incoming calls, accessing Internet Web sites, displaying Web pages, handling all user input and output, keeping track of time, and so on.

Today's automobiles are also good examples of embedded systems. Most of today's new cars incorporate several microcontrollers. One microcontroller is responsible for controlling the instrument display. In order to carry out this function, the display microcontroller needs to collect information using different sensors. These sensors are used to measure temperature, speed, distance, fuel level, and so on. Other microcontrollers are responsible for fuel injection control, cruise control, global positioning, giving warning when the car is too close to other cars or objects, and many other safety features available in high-priced vehicles. These microcontrollers may need to exchange information or even coordinate their operations. The Controller Area Network (CAN) bus is designed for this purpose and this bus will be discussed in detail in Chapter 12.

Another example of an embedded system is the home security system. A typical home security system consists of a microcontroller along with several sensors and actuators. Among the list of sensors are the temperature sensor, smoke detector, humidity sensor, motion detectors, and so on. When the security system detects the smoke, it may activate the alarm or even call the fire department. When the temperature is lower than a certain degree, the system may start the heater to warm the house. When the temperature is higher than a certain degree, it will start the air conditioner. When the house owner is away from home, the security system can be armed to call the neighbor or even the police department if the motion detector detects people trying to open the door. A home security system can be as sophisticated as you want.

1.2.2 Memory

Memory is the place where software programs and data are stored. A computer may contain semiconductor, magnetic, and/or optical memory. Only semiconductor memory will be discussed in this chapter. Semiconductor memory may be further classified into two major types: *random access memory*, or RAM, and *read-only memory*, or ROM.

RANDOM ACCESS MEMORY

Random access memory allows the processor to read from and write into any location on the memory chip. It takes about the same amount of time to perform a read or a write operation to any memory location. Random access memory is *volatile* in the sense that it cannot retain data without power.

There are two types of RAM technology: *dynamic RAM* (DRAM) and *static RAM* (SRAM). Dynamic random access memories are memory devices that require periodic refresh operations in order to maintain the stored information. *Refresh* is the process of restoring binary data stored in a particular memory location. The dynamic RAM uses one transistor and one capacitor to hold one bit of information. The information is stored in the capacitor in the form of electric charges. The charges stored in the capacitor will leak away over time; so periodic refresh operations are needed to maintain the contents of the DRAM. The time interval over which each memory location of a DRAM chip must be refreshed at least once in order to maintain its contents is called the *refresh period*. Refresh periods typically range from a few milliseconds to over a hundred milliseconds for today's high-density DRAMs.

Static random access memories are designed to store binary information without the need for periodic refreshes. Four to six transistors are used to store one bit of information. As long as power is stable, the information stored in the SRAM will not be degraded.

RAM is mainly used to store dynamic programs or data. A computer user often wants to run different programs on the same computer, and these programs usually operate on different sets of data. The programs and data must therefore be loaded into RAM from the hard disk or other secondary storage, and for this reason they are called dynamic.

READ-ONLY MEMORY (ROM)

ROM is *nonvolatile*. When power is removed from ROM and then reapplied, the original data will still be there. However, as its name implies, ROM data can only be read. If the processor attempts to write data to a ROM location, ROM will not accept the data, and the data in the addressed ROM memory location will not be changed.

Mask-programmed read-only memory (MROM) is a type of ROM that is programmed when it is manufactured. The semiconductor manufacturer places binary data in the memory according to the request of the customer. To be cost effective, many thousands of MROM memory units, each consisting of a copy of the same data (or program), must be sold. MROM is the major memory technology used to hold microcontroller application programs and constant data. Most people simply refer to MROM as ROM.

Programmable read-only memory (PROM) is a type of programmable read-only memory that can be programmed in the field (often by the end user) using a device called a PROM programmer or PROM "burner". Once a PROM has been programmed, its contents cannot be changed. Because of this, PROMs are also called one-time programmable ROM (OTP). PROMs are fuse-based; in other words, end users program the fuses to configure the contents of the memory.

Erasable programmable read-only memory (EPROM) is a type of read-only memory that can be erased by subjecting it to strong ultraviolet light. The circuit design of EPROM requires us to erase the contents of a location before writing a new value into it. A quartz window on top of the EPROM integrated circuit permits ultraviolet light to be shone directly on the silicon chip inside. Once the chip is programmed, the window should be covered with dark tape to prevent gradual erasure of the data. If no window is provided, the EPROM chip becomes one-time programmable (OTP) only. Many microcontrollers incorporate on-chip one-time programmable EPROM to save cost for those users who do not need to reprogram the EPROM. EPROM is often used in prototype computers, where the software may be revised many times until it is perfected. EPROM does not allow erasure of the contents of an individual location. The only way to make changes is to erase the entire EPROM chip and reprogram it. The programming of an EPROM chip is done electrically using a device called an EPROM programmer. Today, most programmers are universal in the sense that they can program many types of devices including EPROM, EEPROM, flash memory, and *programmable logic devices.*

Electrically erasable programmable read-only memory (EEPROM) is a type of nonvolatile memory that can be erased by electrical signals and reprogrammed. Like EPROM, the circuit design of EEPROM also requires users to erase the contents of a memory location before a new value can be written into it. EEPROM allows each individual location to be erased and reprogrammed. Unlike EPROM, EEPROM can be erased and programmed using the same programmer. However, EEPROM pays the price for being very flexible in its erasability. The cost of an EEPROM chip is much higher than that of an EPROM chip of comparable density.

Flash memory was invented to incorporate the advantages and avoid the drawbacks of both the EPROM and EEPROM technologies. Flash memory can be erased and reprogrammed in the system without using a dedicated programmer. It achieves the density of EPROM, but it does not require a window for erasure. Like EEPROM, flash memory can be programmed and erased electrically. However, it does not allow the erasure of an individual memory location—the user can only erase a block or the entire chip. Today, more and more microcontrollers are incorporating on-chip flash memory for storing programs and static data.

1.3 The Computer's Software

Programs are known as *software.* A *program* is a set of instructions that the computer hardware can execute. A program is stored in the computer's memory in the form of binary numbers called *machine instructions.* For example, the 68HC12 machine instruction

0001 1000 0000 0110

adds the contents of accumulator B and accumulator A together and leaves the sum in accumulator A. The machine instruction

0100 0011

decrements the contents of accumulator A by 1. The machine instruction

1000 0110 0000 0110

places the value 6 in accumulator A.

Writing programs in machine language is extremely difficult and inefficient:

1. *Program entering.* The programmer will need to memorize the binary pattern of every machine instruction, which can be very challenging because a microprocessor may have several hundred machine instructions, and each machine instruction can have different length. Constant table lookup will be necessary during the program entering process. On the other hand, programmers are forced to work on program logic at a very low level because each machine instruction implements only a very primitive operation.

2. *Program debugging.* Whenever there are errors, it is extremely difficult to trace the program because the program consists of only sequences of 0s and 1s. A programmer will need to identify each machine instruction and then think about what operation is performed by that instruction. This is not an easy job.

3. *Program maintenance.* Most programs will need to be maintained in the long run. A programmer who did not write the program will have a hard time reading the program and following the program logic.

Assembly language was invented to simplify the programming job. An *assembly program* consists of assembly instructions. An *assembly instruction* is the mnemonic representation of a machine instruction. For example, in the 68HC12:

ABA stands for "add the contents of accumulator B to accumulator A." The corresponding machine instruction is 00011000 00000110.

DECA stands for "decrements the contents of accumulator A by 1." The corresponding machine instruction is 0100 0011.

A programmer no longer needs to scan through 0s and 1s in order to identify what instructions are in the program. This is a significant improvement over machine language programming.

The assembly program that the programmer enters is called *source program* or *source code.* A software program called an *assembler* is then invoked to translate the program written in assembly language into machine instructions. The output of the assembly process is called *object code.* It is a common practice to use a *cross assembler* to translate assembly programs. A cross assembler is an assembler that runs on one computer but generates machine instructions to be run on a different computer with a totally different instruction set. In contrast, a *native assembler* runs on a computer and generates machine instructions to be executed by machines that have the same instruction set. The freeware **as12** is a cross assembler that runs on an IBM PC and generates machine code that can be downloaded into a 68HC12-based computer for execution. The list file generated by an assembler shows both the source code and machine code (encoded in hexadecimal format). Here is an example:

```
line    addr.   machine code              source code
─────────────────────────────────────────────────────────────
1:              =00001000                 org     $1000
2:      1000    B6 0800                   ldaa    $800
3:      1003    BB 0801                   adda    $801
4:      1006    BB 0802                   adda    $802
5:      1009    7A 0900                   staa    $900
6:                                        end
─────────────────────────────────────────────────────────────
```

There are several drawbacks to programming in assembly language:

- The programmer must be very familiar with the hardware organization of the microcontroller on which the program is to be executed.

- A program (especially a long one) written in assembly language is extremely difficult to understand for anyone other than the author.

- Programming productivity is not satisfactory for large programming projects because the programmer needs to work on the program logic at a very low level.

For these reasons, high-level languages such as Fortran, COBOL, PASCAL, C, C++, and JAVA were invented to avoid the drawbacks of assembly language programming. High-level languages are very close to plain English and hence a program written in high-level language becomes easier to understand. A statement in high-level language often needs to be implemented by tens or even hundreds of assembly instructions. The programmer can now work on the program logic at a much higher level and achieve higher productivity. A program written in a high-level language is also called a *source program*, and it requires a software program called a *compiler* to translate it into machine instructions. A compiler compiles a program into *object code*. Just as there are cross assemblers, there are *cross compilers* that run on one computer but translate programs into machine instructions to be executed on a computer with a different instruction set.

High-level languages have their own problems. One of their problems is that the resulting machine code cannot run as fast as its equivalent in assembly language due to the inefficiency of the compilation process. For this reason, many time-critical programs are still written in assembly language.

The C language has been used extensively in microcontroller programming in the industry. The Web site at *Microcontroller.com* reported that C has been used to program 80% of the embedded systems, whereas assembly language is used in 75% of embedded systems. Both C and assembly language will be used throughout this text. The C programs in this book will be compiled using the ImageCraft ICC12 compiler and tested on Axiom evaluation boards.

1.4 The 68HC12 CPU Registers

The 68HC12 microcontroller has many registers. These registers can be classified into two categories: CPU registers and I/O registers. CPU registers are used solely to perform general-purpose operations such as arithmetic, logic, and program flow control. I/O registers are mainly used to configure the operations of peripheral functions, to hold data transferred in and out of the peripheral subsystem, and to record the status of I/O operations. The I/O registers in a microcontroller can further be classified into *data*, *data direction*, *control*, and *status registers*. These registers are treated as memory locations when they are accessed. CPU registers do not occupy the 68HC12 memory space.

The CPU registers of the 68HC12 are shown in Figure 1.3 and are listed below. Some of the registers are 8-bit and others are 16-bit.

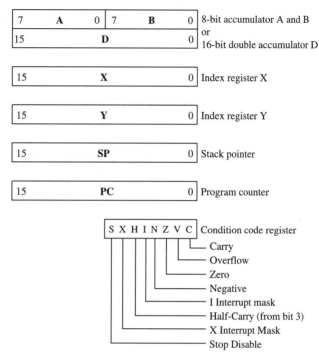

Figure 1.3 ■ MC68HC12 CPU registers

General-purpose accumulator A and B. Both A and B are 8-bit registers. Most arithmetic functions are performed on these two registers. These two accumulators can also be concatenated to form a single 16-bit accumulator that is referred to as the D accumulator.

Index registers X and Y. These two registers are used mainly in forming operand addresses during the instruction execution process. However, they are also used in several arithmetic operations.

Stack pointer (SP). A stack is a first-in-first-out data structure. The 68HC12 has a 16-bit stack pointer that points to the top byte of the stack (shown in Figure 1.4). The stack grows toward lower addresses. The use of stack will be discussed in Chapter 4.

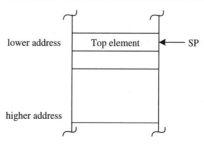

Figure 1.4 ■ 68HC12 stack structure

Program counter (PC). The 16-bit PC holds the address of the next instruction to be executed. After the execution of an instruction, the PC is incremented by the number of bytes of the executed instruction.

Condition code register (CCR). This 8-bit register is used to keep track of the program execution status, control the execution of conditional instructions, and enable/disable the interrupt handling. The contents of the CCR register are shown in Figure 1.3. The function of each condition code bit will be explained in later sections and chapters.

The 68HC12 supports the following types of data:

- Bits
- 5-bit signed integers
- 8-bit signed and unsigned integers
- 8-bit, 2-digit binary-coded-decimal numbers
- 9-bit signed integers
- 16-bit signed and unsigned integers
- 16-bit effective addresses
- 32-bit signed and unsigned integers

Negative numbers are represented in two's complement format. Five-bit and nine-bit signed integers are formed during addressing mode computations. Sixteen-bit effective addresses are formed during addressing mode computations. Thirty-two-bit integer dividends are used by extended division instructions. Extended multiply and extended multiply-and-accumulate instructions produce 32-bit products.

A multi-byte integer (16-bit or 32-bit) is stored in memory from most significant to least significant bytes starting from low to higher addresses. A number can be represented in binary, octal, decimal, or hexadecimal format. An appropriate prefix (shown in Table 1.1) is added in front of the number to indicate its base:

Base	Prefix	Example
binary	%	%10001010
octal	@	@1234567
decimal		12345678
hexadecimal (shorthand hex)	$	$5678

Table 1.1 ■ Prefixes for number bases

1.5 Memory Addressing

Memory consists of a sequence of directly addressable "locations." A memory location is referred to as an *information unit*. A memory location can be used to store information such as data, instructions, and the status of peripheral devices. An information unit has two components: its *address* and its *contents*, as shown in Figure 1.5.

Address ⟶ Contents

Figure 1.5 ■ The components of a memory location

Each location in memory has an address that must be supplied before its contents can be accessed. The CPU communicates with memory by first identifying the location's address and then passing this address on the address bus. This is similar to when a UPS delivery person needs an address to deliver a parcel. The data are transferred between memory and the CPU along the data bus (see Figure 1.6). The number of bits that can be transferred on the data bus at once is called the *data bus width* of the processor.

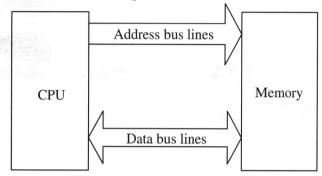

Figure 1.6 ■ Transferring data between CPU and memory

The size of memory is measured in bytes. A byte consists of 8 bits. A 4-bit quantity is called a *nibble*. A 16-bit quantity is called a *word*. To simplify the quantification of memory, the unit *kilobyte* (KB) is often used. K is given by the following formula:

$K = 2^{10} = 1024$

Another frequently used unit is *megabyte* (MB), which is given by the following formula:

$M = K^2 = 2^{20} = 1024 \times 1024 = 1048576$

The 68HC12 has a 16-bit address bus and a 16-bit data bus. The 16-bit data bus allows the 68HC12 to access 16-bit data in one operation. The 16-bit address bus enables the 68HC12 to address directly up to 2^{16} (65,536) different memory locations. Certain 68HC12 members use paging techniques to allow user programs to access more than 64KB.

In this book, we will use the notation *m[addr]* and *[reg]* to refer to the contents of a memory location at *addr* and the contents of the register *reg*, respectively. For example,

m[$20]

refers to the contents of the memory location at $20 and *[A]* refers to the contents of accumulator A.

1.6 68HC12 Addressing Modes

A 68HC12 instruction consists of one or two bytes of opcode and zero to five bytes of operand addressing information. The opcode byte(s) specifies the operation to be performed and the addressing modes to be used to access the operand(s). The first byte of a two-byte opcode is $18.

Addressing modes determines how the CPU accesses memory locations to be operated upon. Addressing modes supported by the 68HC12 are summarized in Table 1.2.

Addressing mode	Source format	Abbre.	Description
Inherent	INST (no externally supplied operands)	INH	Operands (if any) are in CPU registers
Immediate	INST #opr8i or INST #opr16i	IMM	Operand is included in instruction stream. 8- or 16-bit size implied by context
Direct	INST opr8a	DIR	Operand is the lower 8 bits of an address in the range $0000-$00FF
Extended	INST opr16a	EXT	Operand is a 16-bit address
Relative	INST rel8 or INST rel16	REL	An 8-bit or 16-bit relative offset from the current PC is supplied in the instruction
Indexed (5-bit offset)	INST oprx5,xysp	IDX	5-bit signed constant offset from x,y,sp, or pc
Indexed (pre-decrement)	INST oprx3,-xys	IDX	Auto pre-decrement x, y, or sp by 1 ~ 8
Indexed (pre-increment)	INST oprx3,+xys	IDX	Auto pre-increment x, y, or sp by 1 ~ 8
Indexed (post-decrement)	INST oprx3,xys-	IDX	Auto post-decrement x, y, or sp by 1 ~ 8
Indexed (post-increment)	INST oprx3,xys+	IDX	Auto post-increment x, y, or sp by 1 ~ 8
Indexed (accumulator offset)	INST abd,xysp	IDX	Indexed with 8-bit (A or B) or 16-bit (D) accumulator offset from x, y, sp, or pc
Indexed (9-bit offset)	INST oprx9,xysp	IDX1	9-bit signed constant offset from x, y, sp, or pc (lower 8-bits of offset in one extension byte)
Indexed (16-bit offset)	INST oprx16,xysp	IDX2	16-bit constant offset from x, y, sp, or pc (16-bit offset in two extension bytes)
Indexed-Indirect (16-bit offset)	INST [oprx16,xysp]	[IDX2]	Pointer to operand is found at 16-bit constant offset from (x, y, sp, or pc)
Indexed-Indirect (D accumulator offset)	INST [D,xysp]	[D,IDX]	Pointer to operand is found at x, y, sp, or pc plus the vlaue in D

Note
1. INST stands for the instruction mnemonic.
2. opr8i stands for 8-bit immediate value.
3. opr16i stands for 16-bit immediate value.

4. opr8a stands for 8-bit address
5. opr16a stands for 16-bit address
6. oprx3 stands for 3-bit value (amount to be incremented or decremented).

Table 1.2 ■ M68HC12 addressing mode summary

1.6.1 Inherent Mode

Instructions that use this addressing mode either have no operands or all operands are in internal CPU registers. In either case, the CPU does not need to access any memory locations to complete the instruction.

For example, the following two instructions use inherent mode:

```
NOP        ; this instruction has no operands
INX        ; operand is a CPU register
```

1.6.2 Immediate Mode

Operands for immediate mode instructions are included in the instruction stream. The CPU does not access memory when this type of instruction is executed. An immediate value can be 8-bit or 16-bit depending on the context of the instruction. An immediate value is preceded by a # character in the assembly instruction.

For example:

```
LDAA   #$55            ; A ← $55
```

places the value $55 (hexadecimal) in accumulator A when this instruction is executed.

```
LDX    #$2000          ; X ← $2000
```

places the hex value $2000 in index register X when this instruction is executed.

```
LDY    #$44            ; Y ← $0044
```

places the hex value $0044 in index register Y when this instruction is executed. Only an 8-bit value was supplied in this instruction. However, the assembler will generate the 16-bit value $0044 because the CPU expects a 16-bit value when this instruction is executed.

1.6.3 Direct Mode

This addressing mode is sometimes called zero-page addressing because it is used to access operands in the address range of $0000–$00FF. Since these addresses begin with $00, only the eight low-order bits of the address need to be included in the instruction, which saves program space and execution time.

For example:

```
LDAA   $20             ; A ← m[$20]
```

fetches the contents of the memory location at $0020 and puts it in accumulator A.

```
LDX    $20             ; X_H ← m[$20], X_L ← m[$21]
```

fetches the contents of memory locations at $0020 and $0021 and places them in the upper and lower bytes (X_H and X_L) of the index register X, respectively.

1.6.4 Extended Mode

In this addressing mode, the full 16-bit address of the memory location to be operated on is provided in the instruction. This addressing mode can be used to access any location in the 64-KB memory map.

For example:

```
LDAA   $2000           ; A ← m[$2000]
```

copies the contents of the memory location at $2000 into accumulator A.

1.6.5 Relative Mode

The relative addressing mode is used only by branch instructions. Short and long conditional branch instructions use relative addressing mode exclusively. Branching versions of bit manipulation instructions (BRSET and BRCLR) may also use the relative addressing mode to specify the branch target.

A short branch instruction consists of an 8-bit opcode and a signed 8-bit offset contained in the byte that follows the opcode. Long branch instructions consist of an 8-bit prebyte, an 8-bit opcode, and a signed 16-bit offset contained in two bytes that follow the opcode.

Each conditional branch instruction tests certain status bits in the condition code register. If the bits are in a specified state, the offset is added to the address of the next memory location after the offset to form an effective address, and execution continues at that address; if the bits are not in the specified state, execution continues with the instruction immediately following the branch instruction.

Both 8-bit and 16-bit offsets are signed two's complement numbers to support branching forward and backward in memory. The numeric range of short branch offset values is $80 (-128) to $7F (127). The numeric range of long branch offset values is $8000 (-32768) to $7FFF (32767). If the offset is zero, the CPU executes the instruction immediately following the branch instruction, regardless of the test result.

Branch offset is often specified using a label rather than a numeric value due to the difficulty of calculating the exact value of the offset. For example, in the following instruction segment:

```
minus  .
       .                        ; if N (of CCR) = 1
       .                        ;           PC ← PC + branch offset
       bmi      minus           ; else
       ...                      ;           PC ← PC
```

The instruction *bmi minus* will cause the 68HC12 to execute the instruction with the label *minus* if the N flag of the CCR register is set to 1.

The assembler will calculate the appropriate branch offset when the symbol that represents the branch target is encountered. Using a symbol to specify the branch target makes the programming task easier.

1.6.6 Indexed Modes

There are quite a few variations for the indexed addressing scheme. The indexed addressing scheme uses a postbyte plus 0, 1, or 2 extension bytes after the instruction opcode. The postbyte and extensions implement the following functions:

- Specify which index register is used.
- Determine whether a value in an accumulator is used as an offset.
- Enable automatic pre- or post-increment or pre- or post-decrement.
- Specify the size of increment or decrement.
- Specify the use of 5-, 9-, or 16-bit signed offsets.

The indexed addressing scheme allows:

- the stack pointer to be used as an index register in all indexed operations
- the program counter to be used as an index register in all but autoincrement and autodecrement modes
- the value in accumulator A, B, or D to be used as an offset
- automatic pre- or post-increment or pre- or post-decrement by –8 to +8
- a choice of 5-, 9-, or 16-bit signed constant offsets
- indirect indexing with 16-bit offset or accumulator D as the offset

Table 1.3 is a summary of indexed addressing mode capabilities and a description of postbyte encoding. The postbyte is noted as *xb* in instruction descriptions.

Postbyte code (xb)	Source code syntax	Comments rr: 00 = X, 01 = Y, 10 = SP, 11 = PC	
rr0nnnnn	r n, r -n, r	**5-bit constant offset** n = -16 to +15 r can be X, Y, SP or PC	
111rr0zs	n, r -n, r	**Constant offset** (9- or 16-bit signed) z- 0 = 9-bit wth sign in LSB of postbyte (s) 1 = 16-bit if z = s = 1, 16-bit offset indexed-indirect (see below) r can be X, Y, SP, or PC	-256 < n < 255 0 < n < 65,536
111rr011	[n, r]	**16-bit offset indexed-indirect** rr can be X, Y, SP, or PC	0 < n < 65,536

rr1pnnnn	n, -r
	n, +r
	n, r-
	n, r+
		Auto pre-decrement/increment or auto post-decrement/increment;
		p = pre-(0) or post-(1), n = -8 to -1, +1 to +8
		r can be X, Y, or SP (PC not a valid choice)
		        +8 = 0111
		        ...
		        +1 = 0000
		        -1 = 1111
		        ...
		        -8 = 1000	

111rr1aa	A, r
	B, r
	D, r
		Accumulator offset (unsigned 8-bit or 16-bit)
		aa- 00 = A
		        01 = B
		        10 = D (16-bit)
		        11 = see accumulator D offset indexed-indirect
		r can be X, Y, SP, or PC	

111rr111	[D, r]	**Accumulator D offset indexed-indirect** r can be X, Y, SP or PC	

Table 1.3 ■ Summary of indexed operations

5-BIT CONSTANT OFFSET INDEXED ADDRESSING

This indexed addressing mode adds a 5-bit signed offset that is included in the instruction postbyte to the base index register to form the effective address. The base index register can be X, Y, SP, or PC. The range of the offset is from –16 to +15. Although the range of the offset is short, it is the most often used offset. Using this indexed addressing mode can make the instruction shorter.

For example:

LDAA 0,X

loads the contents of the memory location pointed to by index register X (i.e., X contains the address of the memory location to be accessed) into accumulator A.

STAB -8, X

stores the contents of accumulator B in the memory location with the address, which is equal to the contents of index register X minus 8.

9-BIT CONSTANT OFFSET INDEXED ADDRESSING

This indexed addressing mode uses a 9-bit signed offset, which is added to the base index register (X, Y, SP, or PC) to form the effective address of the memory location affected by the instruction. This gives a range of –256 through +255 from the value in the base index register.

For example:

LDAA $FF, X

loads the contents of the memory location located at the address, which is equal to 255 plus the value in index register X.

LDAB   -20, Y

loads the contents of the memory location with the address, which is equal to the value in index register Y minus 20.

16-BIT CONSTANT OFFSET INDEXED ADDRESSING

This indexed addressing mode specifies a 16-bit offset to be added to the base index register (X, Y, SP, or PC) to form the effective address of the memory location affected by the instruction. This allows access to any location in the 64-KB address space. Since the address bus and the offset are both 16 bits, it does not matter whether the offset value is considered to be a signed or an unsigned value. The 16-bit offset is provided as two extension bytes after the instruction postbyte in the instruction flow.

16-BIT CONSTANT INDIRECT INDEXED ADDRESSING

This indexed addressing mode adds a 16-bit offset to the base index register to form the address of a memory location that contains a pointer to the memory location affected by the instruction. The square brackets distinguish this addressing mode from 16-bit constant offset indexing.

For example:

LDAA [10, X]

In this example, index register X holds the base address of a table of pointers. Assume that X has an initial value of $1000, and that $2000 is stored at addresses $100A and $100B. The instruction first adds 10 to the value in X to form the address $100A. Next, an address pointer ($2000) is fetched from memory at $100A. Then the value stored in $2000 is read and loaded into accumulator A.

AUTO PRE/POST DECREMENT/INCREMENT INDEXED ADDRESSING

This indexed addressing mode provides four ways to automatically change the value in a base index register as a part of instruction execution. The index register can be incremented or decremented by an integer value either before or after indexing takes place. The base index register may be X, Y, or SP.

Pre-decrement and pre-increment versions of the addressing mode adjust the value of the index register before accessing the memory location affected by the instruction—the index register retains the changed value after the instruction executes. Post-decrement and post-increment versions of the addressing mode use the initial value in the index register to access the memory location affected by the instruction, and then change the value of the index register.

The 68HC12 allows the index register to be incremented or decremented by any integer value in the ranges –8 through –1, or 1 through 8. The value need not be related to the size of the operand for the current instruction. These instructions can be used to incorporate an index adjustment into an existing instruction rather than using an additional instruction and increasing execution time. This addressing mode is also used to perform operations on a series of data structures in memory.

For example:

```
STAA  1, -SP
```

stores the contents of accumulator A at the memory location with the address equal to the value of stack pointer SP minus 1. After the store operation, the contents of SP are decremented by 1.

```
LDX    2, SP+
```

loads the contents of the memory locations pointed to by the stack pointer SP and also increments the value of SP by 2 after the instruction execution.

ACCUMULATOR OFFSET INDEXED ADDRESSING

In this indexed addressing mode, the effective address is the sum of the values in the base index register and an unsigned offset in one of the accumulators. The value in the base index register itself is not changed. The base register can be X, Y, SP, or PC, and the accumulator can be either of the 8-bit accumulators (A or B) or the 16-bit D accumulator.

For example, the instruction:

```
LDAA   B,X
```

loads the contents of the memory location into accumulator A with the address equal to the sum of the values of accumulator B and index register X. Both B and X are not changed after the instruction execution.

ACCUMULATOR D INDIRECT INDEXED ADDRESSING

This indexed addressing mode adds the value in accumulator D to the value in the base index register to form the address of a memory location that contains a pointer to the memory location affected by the instruction. The instruction operand does not point to the address of the memory location to be acted upon, but rather to the location of a pointer to the location to be acted upon. The square brackets distinguish this addressing mode from accumulator D offset indexing.

For example, the following instruction sequence implements a computed GOTO statement:

```
        JMP     [D, PC]
GO1  DC.W     target1    ; the keyword DC.W reserves two bytes to hold the
GO2  DC.W     target2    ; value of the symbol that follows
GO3  DC.W     target3    ;              "
        ...
target1...
        .
        .
target2...
        .
        .
target3...
        .
        .
```

In this instruction segment, the names (also called labels) *target1*, *target2*, and *target3* are labels that represent the addresses of the memory locations that the JMP instruction may jump to. The names GO1, GO2, and GO3 are also labels. They represent the memory locations that hold the values of the labels *target1*, *target2*, and *target3*, respectively.

The values beginning at GO1 are addresses of potential destinations of the jump instructions. At the time the *JMP [D, PC]* instruction is executed, PC points to the address GO1, and D holds one of the values $0000, $0002, or $0004 (determined by the program some time before the JMP).

Assume that the value in D is $0002. The JMP instruction adds the values in D and PC to form the address of GO2 and jumps to target2. The locations of target1 through target3 are known at the time of program assembly but the destination of the JMP depends on the value in D computed during program execution.

1.7 Addressing More than 64KB

Some 68HC12 devices (for example, 68HC912DG128) incorporate hardware that supports addressing a larger memory space than the standard 64KB. The expanded memory system is accessed by using the bank-switching scheme.

The devices with expanded memory treat the 16KB of memory space from $8000 to $BFFF as a program memory window. Expanded-memory devices also have an 8-bit program page register (PPAGE), which allows up to 256 16KB program memory pages to be switched into and out of the program memory window. This provides up to 4MB of paged program memory.

Accessing expanded memory will be discussed in later chapters.

1.8 68HC12 Instruction Examples

In this section we will examine several groups of instructions to develop a feel for the 68HC12 instruction set and learn some simple applications of these instructions.

1.8.1 The Load & Store Instructions

The load instruction copies the contents of a memory location or places an immediate value into an accumulator or a register. Memory contents are not changed.

Store instructions copy the contents of a CPU register into a memory location. The contents of the accumulator or CPU register are not changed. Store instructions automatically update the N and Z flags in the condition code register (CCR).

Table 1.4 shows a summary of load and store instructions.

All except for the relative addressing mode can be used to select the memory location or value to be loaded into an accumulator or a CPU register. All except for the relative and immediate addressing modes can be used to select the memory location to store the contents of a CPU register.

For example, the following instruction loads the contents of the memory location pointed to by index register X into accumulator A:

LDAA 0,X

The following instruction loads the contents of the memory location at $1004 into accumulator B:

LDAB $1004

The following instruction copies the contents of accumulator A into the memory location at $20:

STAA $20

Mnemonic	Function	Operation
LDAA	Load A	$(M) \Rightarrow A$
LDAB	Load B	$(M) \Rightarrow B$
LDD	Load D	$(M:M+1) \Rightarrow (A:B)$
LDS	Load SP	$(M:M+1) \Rightarrow SP$
LDX	Load index register X	$(M:M+1) \Rightarrow X$
LDY	Load index register Y	$(M:M+1) \Rightarrow X$
LEAS	Load effective address into SP	Effective address $\Rightarrow$ SP
LEAX	Load effective address into X	Effective address $\Rightarrow$ X
LEAY	Load effective address into Y	Effective address $\Rightarrow$ Y
Store Instructions		
Mnemonic	Function	Operation
STAA	Store A	$(A) \Rightarrow M$
STAB	Store B	$(B) \Rightarrow M$
STD	Store D	$(A) \Rightarrow M, (B) \Rightarrow M+1$
STS	Store SP	$(SP) \Rightarrow M, M+1$
STX	Store X	$(X) \Rightarrow M:M+1$
STY	Store Y	$(Y) \Rightarrow M:M+1$

Table 1.4 ■ Load and store instructions

The following instruction stores the contents of index register X in memory locations at $8000 and $8001:

```
STX    $8000
```

When dealing with a complex data structure such as a record, we often use an index register or the stack pointer to point to the beginning of the data structure and use the indexed addressing mode to access the elements of the data structure. For example, there is a record that contains the following four fields:

- ID number (unit *none*, size four bytes)
- height (unit *inch*, size one byte)
- weight (unit *pound*, size two bytes)
- age (unit *year*, size one byte)

Suppose this record is stored in memory starting at $6000. Then we can use the following instruction sequence to access the weight field:

```
LDX    #$6000          ; set X to point to the beginning of data structure
LDD    5, X            ; copy weight into D
```

1.8.2 Transfer & Exchange Instructions

A summary of transfer and exchange instructions are displayed in Table 1.5. Transfer instructions copy the contents of a register or accumulator into another register or accumulator. Source content is not changed by the operation. TFR is a universal transfer instruction, but other mnemonics are accepted for compatibility with the M68HC11. The TAB and TBA instructions affect the N, Z, and V condition code bits. The TFR instruction does not affect the condition code bits.

Transfer Instructions		
Mnemonic	**Function**	**Operation**
TAB	Transfer A to B	$(A) \Rightarrow B$
TAP	Transfer A to CCR	$(A) \Rightarrow CCR$
TBA	Transfer B to A	$(B) \Rightarrow A$
TFR	Transfer register to register	$(A, B, CCR, D, X, Y, \text{ or } SP) \Rightarrow A, B, CCR, D, X, Y, \text{ or } SP$
TPA	Transfer CCR to A	$(CCR) \Rightarrow A$
TSX	Transfer SP to X	$(SP) \Rightarrow X$
TSY	Transfer SP to Y	$(SP) \Rightarrow Y$
TXS	Transfer X to SP	$(X) \Rightarrow SP$
TYS	Transfer Y to SP	$(Y) \Rightarrow SP$
Exchange Instructions		
Mnemonic	**Function**	**Operation**
EXG	Exchange register to register	$(A, B, CCR, D, X, Y, \text{ or } SP) \Leftrightarrow A, B, CCR, D, X, Y, \text{ or } SP$
XGDX	Exchange D with X	$(D) \Leftrightarrow X$
XGDY	Exchange D with Y	$(D) \Leftrightarrow Y$
Sign Extension Instructions		
Mnemonic	**Function**	**Operation**
SEX	Sign extend 8-bit operand	$(A, B, CCR) \Rightarrow X, Y, \text{ or } SP$

Table 1.5 ■ Transfer and exchange instructions

Exchange instructions exchange the contents of pairs of registers or accumulators. For example:

    EXG   A, B

exchanges the contents of accumulator A and B.

    EXG   D,X

exchanges the contents of double accumulator D and index register X.

The SEX instruction is a special case of the universal transfer instruction that is used to sign-extend 8-bit two's complement numbers so that they can be used in

16-bit operations. The 8-bit number is copied from accumulator A, accumulator B, or the condition code register to accumulator D, index register X, index register Y, or the stack pointer. All the bits in the upper byte of the 16-bit result are given the value of the most significant bit of the 8-bit number. For example,

    SEX   A,X

copies the contents of accumulator A to the lower byte of X and duplicates the bit 7 of A to every bit of the upper byte of X.

    SEX   B,Y

copies the contents of accumulator B to the lower byte of Y and duplicates the bit 7 of B to every bit of the upper byte of Y.

Transfer instructions allow operands to be placed in the right register so that the desired operation can be performed. For example, if we want to compute the squared value of accumulator A, we can use the following instruction sequence:

    TAB   ; B ⇐ (A)
    MUL   ; A:B ⇐ (A) × (B)

Applications of other transfer and exchange instructions will be discussed in Chapters 2 and 4.

1.8.3 Move Instructions

A summary of move instructions is listed in Table 1.6. These instructions move data bytes or words from a source $(M1, M: M+1_1)$ to a destination $(M2, M: M+1_2)$ in memory. Six combinations of immediate, extended, and indexed addressing are allowed to specify source and destination addresses (IMM ⇒ EXT, IMM ⇒ IDX, EXT ⇒ EXT, EXT ⇒ IDX, IDX ⇒ EXT, IDX ⇒ IDX).

Transfer Instructions		
Mnemonic	**Function**	**Operation**
MOVB	Move byte (8-bit)	$(M1) \Rightarrow M2$
MOVW	Move word (16-bit)	$(M:M+1_1) \Rightarrow M:M+1_2$

Table 1.6 ■ Move instructions

For example, the following instruction copies the contents of the memory location at $1000 to the memory location at $2000:

    MOVB $1000, $2000

The following instruction copies the 16-bit word pointed to by X to the memory location pointed by Y:

    MOVW 0,X, 0,Y

1.8.4 Add & Subtract Instructions

Add and subtract instructions allow the 68HC12 to perform fundamental arithmetic operations. A summary of Add and Subtract instructions is shown in Table 1.7.

Add Instructions		
Mnemonic	**Function**	**Operation**
ABA	Add B to A	$(A) + (B) \Rightarrow A$
ABX	Add B to X	$(B) + (X) \Rightarrow X$
ABY	Add B to Y	$(B) + (Y) \Rightarrow Y$
ADCA	Add with carry to A	$(A) + (M) + C \Rightarrow A$
ADCB	Add with carry to B	$(B) + (M) + C \Rightarrow B$
ADDA	Add without carry to A	$(A) + (M) \Rightarrow A$
ADDB	Add without carry to B	$(B) + (M) \Rightarrow B$
ADDD	Add without carry to D	$(A:B) + (M:M+1) \Rightarrow A:B$
Subtract Instructions		
Mnemonic	**Function**	**Operation**
SBA	Subtract B from A	$(A) - (B) \Rightarrow A$
SBCA	Subtract with borrow from A	$(A) - (M) - C \Rightarrow A$
SBCB	Subtract with borrow from B	$(B) - (M) - C \Rightarrow B$
SUBA	Subtract memory from A	$(A) - (M) \Rightarrow A$
SUBB	Subtract memory from B	$(B) - (M) \Rightarrow B$
SUBD	Subtract memory from D	$(D) - (M:M+1) \Rightarrow D$

Table 1.7 ■ Add and subtract instructions

Example 1.1

Write an instruction sequence to add 3 to the memory locations at $10 and $15.

Solution: A memory cannot be the destination of an ADD instruction. Therefore, we need to copy the memory content into an accumulator, add 3 to it, and then store the sum back to the same memory locations.

```
LDAA   $10        ; copy the contents of $10 into accumulator A
ADDA   #3         ; add 3 to A
STAA   $10        ; store the sum back to $10
LDAA   $15        ; copy the contents of $15 into accumulator A
ADDA   #3         ; add 3 to A
STAA   $15        ; store the sum back to $15
```

Example 1.2

Write an instruction sequence to add the byte pointed to by index register X with the following byte, and place the sum at the memory location pointed to by index register Y.

Solution: The byte pointed to by index register X and the following byte can be accessed by using the indexed addressing mode.

```
LDAA  0,X            ; put the byte pointed to by X in A
ADDA  1,X            ; add the following byte to A
STAA  0,Y            ; store the sum at the location pointed to by Y
```

▲

Example 1.3

▼

Write an instruction sequence to add the numbers stored at $800 and $801 and store the sum at $804.

Solution: To add these two numbers, we need to put one of them in an accumulator:

```
LDAA  $800           ; copy the number in $800 to accumulator A
ADDA  $801           ; add the second number to A
STAA  $804           ; save the sum at $804
```

▲

1.9 Instruction Queue

The 68HC12 executes one instruction at a time and many instructions take several clock cycles to complete. When the CPU is executing the instruction, it does not need to access memory for the operand in every clock cycle. The design of the 68HC12 takes advantage of this fact to prefetch instruction bytes from the memory and put them in a queue to speed up the instruction execution process.

There are two 16-bit queue stages and one 16-bit buffer. Program information is fetched in aligned 16-bit words. Unless buffering is required, program information is first queued into stage 1, and then advanced to stage 2 for execution.

At least two words of program information are available to the CPU when execution begins. The first byte of object code is in either the even or odd half of the word in stage 2, and at least two more bytes of object code are in queue.

Queue logic manages the position of program information so that the CPU itself does not deal with alignment. As it is executed, each instruction initiates at least enough program word fetches to replace its own object code in the queue.

The buffer is used when a program word arrives before the queue can advance. This occurs during execution of single-byte and odd-aligned instructions. For instance, the queue cannot advance after an aligned, single-byte instruction is executed, because the first byte of the next instruction is also in stage 2. In these cases, information is latched into the buffer until the queue can advance.

1.10 Instruction Execution Cycle

In order to execute a program, the microprocessor or microcontroller must access memory to fetch instructions or operands. The process of accessing a memory location is called a *read cycle,* the process of storing a value in a memory location is called a *write cycle,* and the process of executing an instruction is called an *instruction execution cycle.*

When executing an instruction, the 68HC12 performs a combination of the following operations:

- One or multiple read cycles to fetch instruction opcode byte(s) and addressing information.
- One or more read cycles to fetch the memory operand(s) (optional).
- The operation specified by the opcode.
- One or more write cycles to write back the result to either a register or a memory location (optional).

1.11 Summary

The invention of the microprocessor in 1971 resulted in the revolution of the electronics industry. The first microprocessor, the Intel 4004, implemented a simplified CPU onto an integrated circuit (IC). Following the introduction of the 4-bit 4004, Intel introduced the 8-bit 8008, 8080, and 8085 over three years. The introduction of the 8085 was a big success. The key to its success lies in its programmability. Through this programmability, many products can be designed and constructed. Other companies also joined in the design and manufacturing of microprocessors. Zilog, Motorola, and Rockwell are among the more successful ones.

The earliest microprocessors still needed peripheral chips to interface with I/O devices such as seven-segment displays, printers, timers, and so on. Memory chips were also needed to hold the application program and dynamic data. Because of this, the products designed with microprocessors could not be made as small as desired. Then came the introduction of microcontrollers, which incorporated the CPU, some amount of memory, and peripheral functions such as parallel ports, timers, and serial interface functions onto one chip. The development of microcontrollers has had the following impacts:

- I/O interfacing issue is greatly simplified.
- External memory is no longer needed for many applications.
- System design time is greatly shortened.

A microcontroller is not designed to build a desktop computer. Instead, it is used as the controller of many products. End users of these products do not care about what microcontrollers are used in their appliances. They only care about the functionality of the product. A product that uses a certain microcontroller as a controller and has this characteristic is called an *embedded system.* Cell phones, automobiles, cable modems, HDTVs, and home security systems are well-known embedded systems.

Over the last 20 years, we can see clearly how a microcontroller needs to incorporate some or all of the following peripheral functions in order to be useful:

- Timer module that incorporates input capture, output compare, real-time interrupt, and counting capability
- Pulse-width modulation function for easy waveform generation
- Analog-to-digital converter
- Digital-to-analog converter
- Temperature sensor
- Direct memory access (DMA) controller

- Parallel I/O interface
- Serial I/O interface such as UART, SPI, Microwire, I²C, and CAN
- Memory component interface circuitry

The 68HC12 from Motorola implements most of the above peripheral modules and the CPU onto one VLSI chip.

Memory is where software programs and data are stored. Semiconductor memory chips can be classified into two major categories: random-access memory (RAM) and read-only memory (ROM). RAM technology includes DRAM and SRAM. MROM, PROM, EPROM, EEPROM, and flash memory are read-only memories.

Programs are known as *software*. A program is a set of instructions that the computer hardware can execute. In the past, system designers mainly use assembly language to write microcontroller application software. The nature of assembly language forces an assembly programmer to work on the program logic at a relatively low level. This hampers programming productivity. In the last 10 years, more and more people have turned to high-level programming languages to improve their programming productivity. C is the most widely used language for embedded system programming.

Although system designers use assembly or high-level languages to write their programs, the microcontroller can only execute machine instructions. Programs written in assembly or high-level languages must be translated into machine instructions before they can be executed. The program that performs the translation work is called an *assembler* or *compiler*, depending on the language to be translated.

A machine instruction consists of *opcode* and *addressing information* that specifies the operands. Addressing information is also called *addressing mode*. The 68HC12 implements a rich instruction set along with many addressing modes for specifying operands. This chapter examines the functions of a few groups of instructions. Examples are used to explore the implementation of simple operations using these instructions.

The execution of an instruction may take several clock cycles. Because the 68HC12 does not access memory in every clock cycle, it performs an instruction prefetch to speed up the instruction execution. A two-word (16-bit word) instruction prefetch queue and a 16-bit buffer are added to hold the prefetched instructions.

1.12 Exercises

E1.1 What is a processor?

E1.2 What is a microprocessor? What is a microcomputer?

E1.3 What makes a microcontroller different from a microprocessor?

E1.4 How many bits can the 68HC12 CPU manipulate in one operation?

E1.5 How many different memory locations can the 68HC12 access?

E1.6 Why must every computer have some amount of nonvolatile memory?

E1.7 Why must every computer have some amount of volatile memory?

E1.8 What is source code? What is object code?

E1.9 Convert 5K, 8K, and 13K to decimal representation.

E1.10 Write an instruction sequence to swap the contents of memory locations at $800 and $801.

E1.11 Write an instruction sequence to add 10 to memory locations at $800 and $801, respectively.

E1.12 Write an instruction sequence to set the contents of memory locations at $800, $810, and $820 to 10, 11, and 12 respectively.

E1.13 Write an instruction sequence to perform the operations equivalent to those performed by the following high-level language statements:

```
I := 11;
J := 33;
K = I + J – 5;
```

Assume variables I, J, and K are located at $800, $900, and $1000, respectively.

E1.14 Write an instruction sequence to subtract the number stored at $810 from that stored at $800 and store the difference at $805.

E1.15 Write an instruction sequence to add the contents of accumulator B to the 16-bit word stored at memory location $800 and $801. Treat the value stored in B as a signed number.

E1.16 Write an instruction sequence to copy four bytes starting from $800 to $900-$903.

E1.17 Write an instruction sequence to subtract the contents of accumulator B from the 16-bit word at $800-$801 and store the difference at $900-$901. Treat the value stored in B as a signed value.

E1.18 Write an instruction sequence to swap the 16-bit word stored at $800-$801 with the 16-bit word stored at $900-$901.

E1.19 Give an instruction that can store the contents of accumulator D at the memory location with an address larger than the contents of X by 8.

E1.20 Give an instruction that can store the contents of index register Y at the memory location with an address smaller than the contents of X by 10.

2

68HC12 Assembly Programming

2.1 Objectives

After completing this chapter you should be able to:

- use assembler directives to allocate memory blocks, define constants, and create a message to be output

- write assembly programs to perform simple arithmetic operations

- write program loops to perform repetitive operations

- use program loops to create time delays

- use Boolean and bit manipulation instructions to perform bit field manipulations

2.2 Assembly Language Program Structure

An assembly language program consists of a sequence of statements that tells the computer to perform the desired operations. From a global point of view, a 68HC12 assembly program consists of three sections. In some cases these sections can be mixed to provide better algorithm design. The three sections are:

- *Assembler directives.* Assembler directives instruct the assembler how to process subsequent assembly language instructions. Directives also provide a way to define program constants and reserve space for dynamic variables. Some directives may also set a location counter.
- *Assembly language instructions.* These instructions are 68HC12 instructions. Some instructions are defined with labels.
- *Comments.* There are two types of comments in an assembly program. The first type is used to explain the function of a single instruction or directive. The second type explains the function of a group of instructions or directives or a whole routine. Adding comments makes a program more readable.

Each line of a 68HC12 assembly program, excluding certain special constructs, is comprised of four distinct fields. Some of the fields may be empty. The order of these fields is:

1. Label
2. Operation
3. Operand
4. Comment

2.2.1   The Label Field

Labels are symbols defined by the user to identify memory locations in the programs and data areas of the assembly module. For most instructions and assembler directives, the label is optional. The rules for forming a label are as follows:

- A label must start at column one and begin with a letter *(A-Z, a-z)*, and the letter can be followed by letters, digits, or special symbols. Some assemblers permit special symbols to be used. For example, the assembler from IAR Inc. allows a symbol to start with a question mark *(?)*, at character *(@)*, and underscore *(_)*, in addition to letters. Digits and the dollar *($)* character can also be used after the first character in the IAR assembler.
- Most assemblers restrict the number of characters in a label name. The as12 assembler reference manual does not mention the limit. The IAR assembler allows a user-defined symbol to have up to 255 characters.
- The as12 assembler allows a label to be terminated by "*:*".

Example 2.1

▼

Valid and Invalid labels.

The following instructions contain valid labels:

```
begin  ldaa    #10      ; label begins in column 1
```

```
print:   jsr      hexout    ; label is terminated by a colon
         jmp      begin     ; instruction references the label begin
```

The following instructions contain invalid labels:

```
here is  adda     #5        ; a space is included in the label
  loop   deca               ; labels begins at column 2
```

2.2.2 The Operation Field

This field contains the mnemonic names for machine instructions and assembler directives. If a label is present, the opcode or directive must be separated from the label field by at least one space. If there is no label, the operation field must be at least one space from the left margin.

Example 2.2

Examples of operation fields

```
         adda     #$02      ; adda is the instruction mnemonic
  true   equ      1         ; equate directive equ occupies the operation field
```

2.2.3 The Operand Field

If an *operand field* is present, it follows the *operation field* and is separated from the operation field by at least one space. The operand field may contain operands for instructions or arguments for assembler directives. The following instructions include the operand field:

```
TCNT     equ      $0084     ; $0084 is the operand field
         adda     $0090     ; $0090 is the operand field
```

2.2.4 The Comment Field

The comment field is optional and is added mainly for documentation purposes. The comment field is ignored by the assembler. Here are the rules for comments:

- Any line beginning with an * is a comment.
- Any line beginning with a ; (semi-colon) is a comment. In this book, we will use ; to start a comment.
- You must have a ; (semi-colon) prefixing any comment on a line with mnemonics.

Examples of comments appear in the following instructions:

```
; this program computes the square root of N 8-bit integers.
org      $1000              ; set the location counter to $1000
dec      lp_cnt             ; decrement the loop count
```

In this chapter, we will use the Motorola Freeware cross-assembler *as12* as the standard to explain every aspects of the assembly programming.

2.3 Assembler Directives

Assembler directives look just like instructions in an assembly language program, but they tell the assembler to do something other than creating the machine code for an instruction. The available assembler directives vary with the assembler. Interested readers should refer to the user's manual of the specific assembler for details.

We will discuss assembler directives supported by the as12 in detail. In the following discussion, statements enclosed in brackets [] are optional. All directives and assembly instructions can be in either upper- or lowercase.

End

The end directive is used to end a program to be processed by the assembler. In general, an assembly program looks like this:

```
(your program)
end
```

The *end* directive indicates the logical end of the source program. Any statement following the end directive is ignored. A warning message will be raised if the end directive is missing from the source code; however, the program will still be assembled correctly.

org (origin)

The assembler uses a *location counter* to keep track of the memory location where the next machine code byte should be placed. If the programmer wants to force the program or data array to start from a certain memory location, then this directive can be used.

For example, the statement:

```
org    $1000
```

forces the location counter to be set to $1000.

The *org* directive is mainly used to force a data table or a segment of instructions to start with a certain address. As a general rule, this directive should be used as infrequently as possible. Using too many *orgs* will make your program less reusable.

db (define byte)
dc.b (define constant byte)
fcb (form constant byte)

These three directives define the value of a byte or bytes that will be placed at a given memory location. The *db* (or fcb, or dc.b) directive assigns the value of the expression to the memory location pointed to by the location counter. Then the location counter is incremented. Multiple bytes can be defined at a time by using commas to separate the arguments.

For example, the statement:

```
array   db        $11,$22,$33,$44,$55
```

initializes five bytes in memory to:

```
$11
$22
$33
$44
$55
```

and the assembler will use *array* as the symbolic address of the first byte whose initial value is $11. The program can also force these five bytes to a particular address by adding the *org* directive. For example, the sequence:

```
org    $800
array  db      $11,$22,$33,$44,$55
```

initializes the contents of memory locations at $800, $801, $802, $803, and $804, to $11, $22, $33, $44, and $55, respectively.

dw (define word)
dc.w (define constant word)
fdb (form double bytes)

These three directives define the value of a word or words that will be placed at a given address. The value can be specified by an integer or an expression. For example, the statement:

```
vect_tab        dw      $1234, $5678
```

initializes the two words starting from the current location counter to $1234 and $5678, respectively. After this statement, the location will be incremented by four.

fcc (form constant character)

This directive allows us to define a string of characters (a message). The first character in the string is used as the delimiter. The last character must be the same as the first character because it will be used as the delimiter. The delimiter must not appear in the string. The space character cannot be used as the delimiter. Each character is encoded by its corresponding ASCII code.

For example, the statement:

```
alpha  fcc      "def"
```

will generate the following values in memory:

```
$64
$65
$66
```

and the assembler will use the label *alpha* to refer to the address of the first letter, which is stored as the byte $64. A character string to be output to the LCD display is often defined using this directive.

fill (fill memory)

This directive allows a user to fill a certain number of memory locations with a given value. The syntax of this directive is as follows:

```
fill    value, count
```

where the number of bytes to be filled is indicated by *count* and the value to be filled is indicated by *value*.

For example, the statement:

```
space_line      fill        $20, 40
```

will fill 40 bytes with the value of $20 starting from the memory location referred to by the label *space_line*.

ds (define storage)
rmb (reserve memory byte)
ds.b (define storage bytes)

Each of these three directives reserves a number of bytes given as the arguments to the directive. The location counter will be incremented by the number that follows the directive mnemonic.

For example, the statement:

```
buffer  ds       100
```

reserves 100 bytes in memory starting from the location represented by the label *buffer*. After this directive, the location counter will be incremented by 100. The content(s) of the reserved memory location(s) are not defined.

ds.w (define storage word)
rmw (reserve memory word)

Each of these directives increments the location counter by the value indicated in the number-of-words argument multiplied by two. In other words, if the *ds.w* expression evaluates to *k* then the location counter is advanced by 2k. These directives are often used with a label. For example, the statement:

```
dbuf   ds.w      20
```

reserves 40 bytes starting from the memory location represented by the label *dbuf*. None of these 40 bytes are initialized.

equ (equate)

This directive assigns a value to a label. Using *equ* to define constants will make our program more readable.

For example, the statement:

```
loop_cnt        equ        40
```

informs the assembler that whenever the symbol *loop_cnt* is encountered, it should be replaced with the value of 40.

loc

This directive increments and produces an internal counter used in conjunction with the backward tick mark (`). By using the *loc* directive and the ` mark you can write program segments like the following example, without thinking up new labels:

```
            loc
            ldaa       #2
loop`       deca
            bne        loop`
            loc
loop`       brclr      0,x $55 loop`
```

This code segment will work perfectly fine because the first loop label will be seen as loop001, whereas the second loop label will be seen as loop002. The assembler actually sees this:

```
            loc
            ldaa       #2
loop001     deca
            bne        loop001
            loc
loop002     brclr      0,x $55 loop002
```

You can also set the *loc* directive with a valid expression or number by putting that expression or number in the operand field. The resultant number will be used to increment the suffix to the label.

2.4 Software Development Issues

A complete discussion of issues involved in software development is out of the scope of this text. However, we do need to take a serious look at some software development issues because embedded system designers must spend a significant amount of time on software development.

As we all know, software development starts with *problem definition*. The problem presented by the application must be fully understood before any program can be written. At the problem definition stage, the most critical thing is to get you, the programmer, and your end user to agree upon what needs to be done. To achieve this, asking questions is very important. For complex and expensive applications, a formal, written definition of the problem is formulated and agreed upon by all parties.

Once the problem is known, the programmer can begin to lay out an overall plan of how to solve the problem. The plan is also called an *algorithm*. Informally, an algorithm is any well-defined computational procedure that takes some value, or set of values, as input, and produces some value, or set of values, as output. An algorithm is thus a sequence of computational steps that transforms the input into the output. We can also view an algorithm as a tool for solving a well-specified computational problem. The statement of the problem specifies in general terms the desired input/output relationship. The algorithm describes a specific computational procedure for achieving that input/output relationship.

An algorithm is expressed in *pseudocode* that is very much like C or PASCAL. What separates pseudocode from "real" code is that in pseudocode, we employ whatever expressive method that is most clear and concise to specify a given algorithm. Sometimes the clearest method is English, so do not be surprised if you come across an English phrase or sentence embedded within a section of "real" code.

An algorithm provides not only the overall plan for solving the problem but also documentation to the software to be developed. In the rest of this book, all algorithms will be presented in the following format:

Step 1

....

Step 2

....

An earlier alternative for providing the overall plan for solving software problems was the use of flowcharts. A flowchart shows the way a program operates. It illustrates the logic flow of the program. Therefore, flowcharts can be a valuable aid in visualizing programs. Flowcharts are not only used in computer programming, they are also used in many other fields, such as business and construction planning.

The flowchart symbols used in this book are shown in Figure 2.1. The *terminal symbol* is used at the beginning and end of each program. When it is used at the beginning of a program, the word *Start* is written inside it. When it is used at the end of a program, it contains the word *Stop*.

The *process box* indicates what must be done at this point in the program execution. The operation specified by the process box could be shifting the contents of one general purpose register to a peripheral register, decrementing a loop count, and so on.

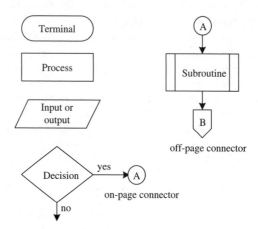

Figure 2.1 ■ Flowchart symbols used in this book

The *input/output box* is used to represent data that are either read or displayed by the computer.

The *decision box* contains a question that can be answered either yes or no. A decision box has two exits, also marked yes or no. The computer will take one action if the answer is yes and will take a different action if the answer is no.

The *on-page connector* indicates that the flowchart continues elsewhere on the same page. The place where it is continued will have the same label as the on-page connector. The *off-page connector* indicates that the flowchart continues on another page. To determine where the flowchart continues, you need to look at the following pages of the flowchart to find the matching off-page connector.

Normal flow on a flowchart is from top to bottom and from left to right. Any line that does not follow this normal flow should have an arrowhead on it.

When the program gets complicated, the flowchart that documents the logic flow of the program also becomes difficult to follow. This is the limitation of the flowchart. In this book, we will use both the flowchart and the algorithm procedure to describe the solution to a problem.

After you are satisfied with the algorithm or the flowchart, *convert it to source code in one of the assembly or high-level languages*. Each statement in your algorithm (or each block of your flowchart) will be converted into one or multiple assembly instructions or high-level language statements. If you find an algorithmic step (or a block in the flowchart) requires many assembly instructions or high-level language statements to implement, then it might be beneficial to either (1) convert this step (or block) into a subroutine and just call the subroutine, or (2) further divide the algorithmic step (or flowchart block) into smaller steps (or blocks) so that it can be coded with just a few assembly instructions or high-level language statements.

The next major step is to *test your program*. Testing a program means testing for anomalies. Here you will first test for normal inputs that you always expect. If the result is what you expected then you go on to test the borderline inputs. Test for the maximum and minimum values of the input. When your program passes this test, then test for illegal input values. If your algorithm includes several branches, then you must use enough values to exercise all the possible branches. This is to make sure that your program will operate correctly under all possible circumstances.

In the rest of this book, most of the examples are well defined. Therefore, our focus is on how to design the algorithm that solves the specified problem as well as how to convert the algorithm into source code.

2.5 Writing Programs to do Arithmetic

In this section, we will use small programs that perform simple computations to demonstrate how a program is written.

Example 2.3

▼

Write a program to add the numbers stored at memory locations $800, $801, and $802, and store the sum at memory location $900.

Solution: This problem can be solved by the following steps:

Step 1
Load the contents of the memory location at $800 into accumulator A.

Step 2
Add the contents of the memory location at $801 into accumulator A.

Step 3
Add the contents of the memory location at $802 into accumulator A.

Step 4
Store the contents of accumulator A at memory location $900.

These steps can be translated into the as12 assembly program as follows:

```
org      $1000   ; starting address of the program
ldaa     $800    ; place the contents of the memory location $800 into A
adda     $801    ; add the contents of the memory location $801 into A
adda     $802    ; add the contents of the memory location $802 into A
staa     $900    ; store the sum at the memory location $900
end
```

▲

Example 2.4

▼

Write a program to subtract the contents of the memory location at $805 from the sum of the memory locations at $800 and $802, and store the result at the memory location $900.

Solution: The logic flow of this program is illustrated in Figure 2.2. The assembly program is as follows:

```
org      $1000   ; starting address of the program
ldaa     $800    ; copy the contents of the memory location at $800 to A
adda     $802    ; add the contents of memory location at $802 to A
suba     $805    ; subtract the contents of memory location at $805 from A
staa     $900    ; store the contents of accumulator A to $805
end
```

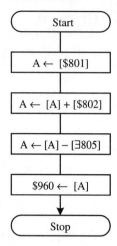

Figure 2.2 ■ Logic flow of program 2.4

Example 2.5

Write a program to subtract five from four memory locations at $800, $801, $802, and $803.

Solution: In the 68HC12, a memory location cannot be the destination of an ADD or SUB instruction. Therefore, three steps must be followed to add or subtract a number to or from a memory location:

Step 1
Load the memory contents into an accumulator.

Step 2
Add (or subtract) the number to (from) the accumulator.

Step 3
Store the result at the specified memory location.

The program is as follows:

```
org   $1000
ldaa  $800      ; copy the contents of memory location $800 to A
suba  #5        ; subtract 5 from A
staa  $800      ; store the result back to memory location $800
ldaa  $801
suba  #5
staa  $801
ldaa  $802
suba  #5
staa  $802
```

```
ldaa    $803
suba    #5
staa    $803
end
```

▲

Example 2.6

▼

Write a program to add two 16-bit numbers that are stored at $800~$801 and $802~$803, and store the sum at $900~$901.

Solution: This program is very straightforward:

```
org     $1000
ldd     $800    ; place the 16-bit number at $800~$801 in D
addd    $802    ; add the 16-bit number at $802~$803 to D
std     $900    ; save the sum at $900~$901
end
```

▲

2.5.1 Carry/Borrow Flag

The 68HC12 can add and subtract either 8-bit or 16-bit numbers and place the result in either 8-bit accumulators, A or B, or the double accumulator D. The 8-bit number stored in accumulator B can also be added to index register X. However, programs can also be written to add numbers larger than 16 bits. Arithmetic performed in a 16-bit microprocessor/microcontroller on numbers that are larger than 16 bits is called *multiprecision arithmetic*. Multiprecision arithmetic makes use of the carry flag (C flag) of the condition code register (CCR).

Bit 0 of the CCR register is the C flag. It can be thought of as a temporary 9th bit that is appended to any 8-bit register or 17th bit that is appended to any 16-bit register. The C flag allows us to write programs to add and subtract hex numbers that are larger than 16-bit. For example, consider the following two instructions:

```
ldd     #$8645
addd    #$9978
```

These two instructions add the numbers $8645 and $9978.

```
 $8645
+$9978
$11FBD
```

The result is $11FBD, a 17-bit number, which is too large to fit into the 16-bit double accumulator D. When the 68HC12 executes these two instructions, the lower sixteen bits of the answer, $1FBD, are placed in double accumulator D. This part of the answer is called the *sum*. The leftmost bit is called a *carry*. A carry of 1 following an addition instruction sets the C flag of the CCR register to 1. A carry of 0 following an addition clears the C flag to 0. This applies to both 8-bit and 16-bit additions for the 68HC12. For example, execution of the following two instructions:

```
ldd     #$1245
addd    #$4581
```

will clear the C flag to 0 because the carry resulting from this addition is 0. In summary:

- If the addition produces a carry of 1, the carry flag is set to 1.
- If the addition produces a carry of 0, the carry flag is cleared to 0.

2.5.2 Multiprecision Addition

For a 16-bit microcontroller like the 68HC12, multiprecision addition is the addition of numbers that are larger than 16 bits. To add the hex number $1A598183 to $76548290, the 68HC12 has to perform multiprecision addition.

```
   1   1 1
 $1A598183
+$76548290
 $90AE0413
```

Multiprecision addition is performed one byte at a time, beginning with the least significant byte. The 68HC12 does allow us to add 16-bit numbers at a time because it has the ADDD instruction. The following two instructions can be used to add the least significant 16-bit numbers together:

```
ldd    #$8183
addd   #$8290
```

Since the sum of the most significant digit has a sum greater than 16, it generates a carry that must be added to the next more significant digit, causing the C flag to be set to 1. The contents of double accumulator D must be saved before the higher bytes are added. Let's save these two bytes at $802-$803:

```
std    $802
```

When the second-to-most significant bytes are added, the carry from the lower byte must be added in order to obtain the correct sum. In other word, we need an "add with carry" instruction. There are two versions of this instruction: the ADCA instruction for accumulator A and the ADCB instruction for accumulator B. The instructions for adding the second-to-most-significant bytes are:

```
ldaa   #$59
adca   #$54
```

We also need to save the *second-to-most-significant* byte of the result at $801 with the following instruction:

```
staa   $801
```

The most significant bytes can be added using similar instructions, and the complete program with comments appears as follows:

```
ldd    #$8183   ; place the lowest two bytes of the first number in D
addd   #$8290   ; add the lowest two bytes of the second number to D
std    $802     ; store the lowest two bytes of the sum at $802-$803
ldaa   #$59     ; place the second-to-most significant byte of the first number in A
adca   #$54     ; add the second-to-most-significant byte of the
                ; second number and carry to A
staa   $801     ; store the second-to-most-significant byte of the sum at $801
ldaa   #$1A     ; place the most-significant byte of the first number in A
adca   #$76     ; add the most-significant byte of the second number and carry to A
staa   $800     ; store the most significant byte of the sum
end
```

Note that the LOAD and STORE instructions do not affect the value of the C flag (otherwise, the program would not work). The 68HC12 does not have a 16-bit instruction with the carry flag as an operand. Whenever the carry needs to be added, we must use the 8-bit instruction ADCA or ADCB. This is shown in the previous program.

Example 2.7

▼

Write a program to add two 4-byte numbers that are stored at $800~$803 and $804~$807, and store the sum at $810~$813.

Solution: The addition should start from the least significant byte and proceed to the most significant byte. The program is as follows:

```
org   $1000   ; starting address of the program
ldd   $802    ; place the lowest two bytes of the first operand in D
addd  $806    ; add the lowest two bytes
std   $812    ; save the sum of the lowest two bytes
ldaa  $801    ; place the second-to-most-significant byte of the
              ; first operand in A
adca  $805    ; add the second-to-most-significant byte of the
              ; second operand and carry to A
staa  $811    ; save the sum of the second-to-most significant bytes
ldaa  $800    ; place the most-significant byte of the first operand in A
adca  $804    ; add the most-significant byte of the second
              ; operand and carry to A
staa  $810    ; save the sum of the most-significant bytes
end
```

▲

2.5.3 Subtraction & the C Flag

The C flag also enables the 68HC12 to borrow from the high byte to the low byte during a multiprecision subtraction. Consider the following subtraction problem:

$39
−$74

We are attempting to subtract a larger number from a smaller one. Subtracting $4 from $9 is not a problem:

$39
−$74
    5

Now we need to subtract $7 from $3. To do this, we need to borrow from somewhere. The 68HC12 borrows from the C flag, thus setting the C flag. When we borrow from the next higher digit of a hex number, the borrow has a value of decimal 16. After the borrow from the C flag, the problem can be completed:

$39
−$74
$C5

When the 68HC12 executes a subtract instruction, it always borrows from the C flag. The borrow is either 1 or 0. The C flag operates as follows during a subtraction:

- If the 68HC12 borrows a 1 from the C flag during a subtraction, the C flag is set to 1.
- If the 68HC12 borrows a 0 from the C flag during a subtraction, the C flag is set to 0.

2.5.4 Multiprecision Subtraction

For a 16-bit microcontroller, multiprecision subtraction is the subtraction of numbers that are larger than 16 bits. To subtract the hex number $16753284 from $98765432, the 68HC12 has to perform multiprecision subtraction:

```
 $98765432
-$16757284
```

Like multiprecision addition, multiprecision subtraction is performed one byte at a time, beginning with the least significant byte. The 68HC12 does allow us to subtract two bytes at a time because it has the SUBD instruction. The following two instructions can be used to subtract the least significant two bytes of the subtrahend from the minuend:

```
ldd    #$5432
subd   #$7284
```

Since a larger number is subtracted from a smaller one, there is a need to borrow from the higher byte, causing the C flag to be set to 1. The contents of double accumulator D should be saved before the higher bytes are subtracted. Let's save these two bytes at $802~$803:

```
std    $802
```

When the second-to-most-significant bytes are subtracted, the borrow 1 has to be subtracted from the second-to-most-significant byte of the result. In other words, we need a "subtract with borrow" instruction. There is such an instruction, but it is called *subtract with carry*. There are two versions: the SBCA instruction for accumulator A, and the SBCB instruction for accumulator B. The instructions to subtract the second-to-most-significant bytes are:

```
ldaa   #$76
sbca   #$75
```

We also need to save the second-to-most-significant byte of the result at $801 with the following instruction:

```
staa   $801
```

The most significant bytes can be subtracted using similar instructions, and the complete program with comments is as follows:

```
org    $1000    ; starting address of the program
ldd    #$5432   ; place the lower two bytes of the minuend in D
subd   #$7284   ; subtract the lower bytes of the subtrahend from D
std    $802     ; save the lower two bytes of the difference
ldaa   #$76     ; place the second-to-most-significant byte of the minuend in A
sbca   #$75     ; subtract the second-to-most-significant byte of the
                ; subtrahend and the borrow from A
staa   $801     ; save the second-to-most-significant byte of the difference
ldaa   #$98     ; put the most-significant-byte of the minuend in A
sbca   #$16     ; subtract the most-significant-byte of the
                ; subtrahend and the borrow from A
staa   $800     ; save the most-significant-byte of the difference
end
```

Example 2.8

Write a program to subtract the hex numbers stored at $804~$807 from the hex number stored at $800~$803, and save the difference at $900~$903.

Solution: We will perform the subtraction from the least significant byte towards the most significant byte as follows:

```
org    $1000    ; starting address of the program
ldd    $802     ; place the lowest two bytes of the minuend in D
subd   $806     ; subtract the lowest two bytes of the subtrahend from D
std    $902     ; save the lowest two bytes of the difference
ldaa   $801     ; put the second-to-most-significant byte of the minuend in A
sbca   $805     ; subtract the second-to-most-significant byte of the
                ; subtrahend and the borrow from A
staa   $901     ; save the second-to-most-significant byte of the difference
ldaa   $800     ; put the most significant byte of the minuend in A
sbca   $804     ; subtract the most significant byte of the subtrahend
                ; and the borrow from A
staa   $900     ; save the most significant byte of the difference
end
```

2.5.5 Binary-Coded-Decimal (BCD) Addition

Although virtually all computers work internally with binary numbers, the input and output equipment generally uses decimal numbers. Since most logic circuits only accept two-valued signals, the decimal numbers must be coded in terms of binary signals. In the simplest form of binary code, each decimal digit is represented by its binary equivalent. For example, 2,538 is represented by:

0010 0101 0011 1000

This representation is called *binary-coded-decimal.* If the BCD format is used, it must be preserved during arithmetic processing.

The principal advantage of the BCD encoding method is the simplicity of input/output conversion; its major disadvantage is the complexity of arithmetic processing. The choice between binary and BCD depends on the type of problems the system will be handling.

The 68HC12 microcontroller can add only binary numbers, not decimal numbers. The following instruction sequence appears to cause the 68HC12 to add the decimal numbers 25 and 31 and store the sum at the memory location $800:

```
ldaa   #$25
adda   #$31
staa   $800
```

This instruction sequence performs the following addition:

```
 $25
+$31
 $56
```

When the 68HC12 executes this instruction sequence, it adds the numbers according to the rules of binary addition and produces the sum $56. This is the correct BCD answer, because the result represents the decimal sum of 25 + 31 = 56. In this example, the 68HC12 gives the appearance of performing decimal addition. However, a problem occurs when the 68HC12 adds two BCD digits and generates a sum greater than nine. Then the sum is incorrect in the decimal number system, as the following three examples illustrate:

```
   $ 1 8        $ 3 5        $ 1 9
 +$ 4 7       +$ 4 7       +$ 4 7
   $ 5 F        $ 7 C        $ 6 0
```

The answers to the first two problems are obviously wrong in the decimal number system because the hex digits F and C are not between zero and nine. The answer to the third example appears to contain valid BCD digits, but in the decimal system 19 plus 47 equals 66, not 60; this example involves a carry from the lower nibble to the higher nibble.

In summary, a sum in the BCD format is incorrect if it is greater than $9 or if there is a carry to the next higher nibble. Incorrect BCD sums can be adjusted by adding $6 to them. To correct the examples, do the following:

1. Add $6 to every sum digit greater than nine.

2. Add $6 to every sum digit that had a carry of one to the next higher digit.

Here are the problems with their sums adjusted:

```
   $ 1 8        $ 3 5        $ 1 9
 +$ 4 7       +$ 4 7       +$ 4 7
   $ 5 F        $ 7 C        $ 6 0
 +$   6       +$   6       +$   6
   $ 6 5        $ 8 2        $ 6 6
```

The fifth bit of the condition code register is the *half-carry*, or H flag. A carry from the lower nibble to the higher nibble during the addition operation is a half-carry. A half-carry of one during addition sets the H flag to one, and a half-carry of zero during addition clears it to zero. If there is a carry from the high nibble during addition, the C flag is set to one, which indicates that the high nibble is incorrect. A $6 must be added to the high nibble to adjust it to the correct BCD sum.

Fortunately, we don't need to write instructions to detect the illegal BCD sum following a BCD addition. The 68HC12 provides a *decimal adjust accumulator A* instruction, DAA, which takes care of all these detailed detection and correction operations. The DAA instruction monitors the sums of BCD additions and the C and H flags and automatically adds $6 to any nibble that requires it. The rules for using the DAA instruction are as follows:

1. The DAA instruction can only be used for BCD addition. It does not work for subtraction or hex arithmetic.

2. The DAA instruction must be used immediately after one of the three instructions that leave their sum in accumulator A. (These three instructions are ADDA, ADCA, ABA.)

3. The numbers added must be legal BCD numbers to begin with.

Example 2.9
▼

Write an instruction sequence to add the BCD numbers stored at memory locations $800 and $801, and store the sum at $810.

Solution:

```
ldaa  $800    ; load the first BCD number in A
adda  $801    ; perform addition
daa           ; decimal adjust the sum in A
staa  $810    ; save the sum
```

2.5.6 Multiplication & Division

The 68HC12 provides three multiply and five divide instructions. A brief description of these instructions appears in Table 2.1.

Mnemonic	Function	Operation	
EMUL	unsigned 16 by 16 multiply	$(D) \times (Y) \rightarrow$	Y:D
EMULS	signed 16 by 16 multiply	$(D) \times (Y) \rightarrow$	Y:D
MUL	unsigned 8 by 8 multiply	$(A) \times (B) \rightarrow$	A:B
EDIV	unsigned 32 by 16 divide	$(Y:D) \div (X)$ quotient $\rightarrow$ remainder $\rightarrow$	Y D
EDIVS	signed 32 by 16 divide	$(Y:D) \div (X)$ quotient $\rightarrow$ remainder $\rightarrow$	Y D
FDIV	16 by 16 fractional divide	$(D) \div (X) \rightarrow$ remainder $\rightarrow$	X D
IDIV	unsigned 16 by 16 integer divide	$(D) \div (X) \rightarrow$ remainder $\rightarrow$	X D
IDIVS	signed 16 by 16 integer divide	$(D) \div (X) \rightarrow$ remainder $\rightarrow$	X D

Table 2.1 ■ Summary of 68HC12 multiply and divide instructions

The *EMUL* instruction multiplies the 16-bit unsigned integers stored in accumulator D and index register Y and leaves the product in these two registers. The upper 16 bits of the product are in Y whereas the lower 16 bits are in D.

The *EMULS* instruction multiplies the 16-bit signed integers stored in accumulator D and index register Y and leaves the product in these two registers. The upper 16 bits of the product are in Y whereas the lower 16 bits are in D.

The *MUL* instruction multiplies the 8-bit unsigned integer in accumulator A by the 8-bit unsigned integer in accumulator B to obtain a 16-bit unsigned result in double accumulator D. The upper byte of the product is in accumulator A whereas the lower byte of the product is in B.

The *EDIV* instruction performs an unsigned 32-bit by 16-bit division. The dividend is the register pair Y and D with Y as the upper 16-bit of the dividend. Index register X is the divisor. After division, the quotient and the remainder are placed in Y and D, respectively.

The *EDIVS* instruction performs a signed 32-bit by 16-bit division using the same operands as the EDIV instruction does. After division, the quotient and the remainder are placed in Y and D, respectively.

The *FDIV* instruction divides an unsigned 16-bit dividend in double accumulator D by an unsigned 16-bit divisor in index register X, producing an unsigned 16-bit quotient in X and an unsigned 16-bit remainder in D. The dividend must be less than the divisor. The radix point of the quotient is to the left of the bit 15. In the case of overflow (the denominator is less than or equal to the nominator) or division-by-zero, the quotient is set to $FFFF, and the remainder is indeterminate.

The *IDIV* instruction divides an unsigned 16-bit dividend in double accumulator D by the unsigned 16-bit divisor in index register X, producing an unsigned 16-bit quotient in X, and an unsigned 16-bit remainder in D. If both the divisor and the dividend are assumed to have radix points in the same positions (to the right of bit 0), the radix point of the quotient is to the right of bit 0. In the case of division by zero, the quotient is set to $FFFF, and the remainder is indeterminate.

The *IDIVS* instruction divides the signed 16-bit dividend in double accumulator D by the signed 16-bit divisor in index register X, producing a signed 16-bit quotient in X, and a signed 16-bit remainder in D. If division-by-zero is attempted, the values in D and X are not changed, but the values of the N, Z, and V status bits are undefined.

Example 2.10

Write an instruction sequence to multiply the contents of index register X and double accumulator D, and store the product at memory locations $800~$803.

Solution: There is no instruction to multiply the contents of double accumulator D and index register X. However, we can transfer the contents of index register X to index register Y and execute the EMUL instruction. If index register Y holds useful information, then we need to save it before the data transfer.

```
sty    $810        ; save Y in a temporary location
tfr    x,y         ; transfer the contents of X to Y
emul               ; perform the multiplication
sty    $800        ; save the upper 16 bits of the product
std    $802        ; save the lower 16 bits of the product
ldy    $810        ; restore the value of Y
```

Example 2.11

Write an instruction sequence to divide the signed 16-bit number stored at memory locations $805~$806 by the 16-bit unsigned number stored at memory locations $820~$821, and store the quotient and remainder at $900~$901 and $902~$903, respectively.

Solution: Before we can perform the division, we need to place the dividend and divisor in D and X, respectively.

```
ldd    $805        ; place the dividend in D
ldx    $820        ; place the divisor in X
idivs              ; perform the signed division
stx    $900        ; save the quotient
std    $902        ; save the remainder
```

Because most arithmetic operations can be performed only on accumulators, we need to transfer the contents of index register X to D so that further division on the quotient can be performed. The 68HC12 provides two exchange instructions in addition to the TFR instruction for this purpose:

- The XGDX instruction exchanges the contents of accumulator D and index register X.
- The XGDY instruction exchanges the contents of accumulator D and index register Y.

The 68HC12 provides instructions for performing unsigned 8-bit by 8-bit and both signed and unsigned 16-bit by 16-bit multiplications. Since the 68HC12 is a 16-bit microcontroller, we expect that it will be used to perform complicated operations in many sophisticated applications. Performing 32-bit by 32-bit multiplication will be one of them.

Since there is no 32-bit by 32-bit multiplication instructions, we will have to break a 32-bit number into two 16-bit halves and use the 16-bit by 16-bit multiply instruction to synthesize the operation. Assume M and N are the multiplicand and the multiplier, respectively. These two numbers can be broken down as follows:

$M = M_H M_L$
$N = N_H N_L$

where, M_H and N_H are the upper 16 bits and M_L and N_L are the lower 16 bits of M and N, respectively. Four 16-bit by 16-bit multiplications are performed, and then their partial products are added together as shown in Figure 2.3.

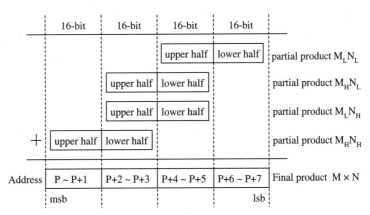

Figure 2.3 ■ Unsigned 32-bit by 32-bit multiplication

The procedure is as follows:

Step 1
Allocate eight bytes to hold the product. Assume these eight bytes are located at P, P+1, ..., and P+7.

Step 2
Generate the partial product $M_L N_L$ (in Y:D) and save it at locations P+4 ~ P+7.

Step 3
Generate the partial product $M_H N_H$ (in Y:D) and save it at locations P ~ P+3.

Step 4

Generate the partial product $M_H N_L$ (in Y:D) and add it to memory locations P+2 ~ P+5. The C flag may be set to 1 after this addition.

Step 5

Add the C flag to memory location P+1 using the ADCA (or ADCB) instruction. This addition may also set the C flag to 1. So, again, add the C flag to memory location P.

Step 6

Generate the partial product $M_L N_H$ (in Y:D) and add it to memory locations P+2 ~ P+5. The carry flag may be set to 1. So add the C flag to memory location P+1 and then add it to memory location P.

Example 2.12

Write a program to multiply the 32-bit unsigned integers stored at M~M+3 and N~N+3, respectively and store the product at memory locations P~P+7.

Solution: The following program is a direct translation of the previous multiplication algorithm.

```
        org     $800
M       rmb     4               ; reserved to hold the multiplicand
N       rmb     4               ; reserved to hold the multiplier
P       rmb     8               ; reserved to hold the product
        org     $1000
        ldd     M+2             ; place ML in D
        ldy     N+2             ; place NL in Y
        emul                    ; compute MLNL
        sty     P+4             ; save the upper 16 bits of the partial product MLNL
        std     P+6             ; save the lower 16 bits of the partial product MLNL
        ldd     M               ; place MH in D
        ldy     N               ; place NH in Y
        emul                    ; compute MHNH
        sty     P               ; save the upper 16 bits of the partial product MHNH
        std     P+2             ; save the lower 16 bits of the partial product MHNH
        ldd     M               ; place MH in D
        ldy     N+2             ; place NL in Y
        emul                    ; compute MHNL
; the following seven instructions add MHNL to memory locations P+2~P+5
        addd    P+4             ; add the lower half of MHNL to P+4~P+5
        std     P+4             ;          "
        tfr     y,d             ; transfer Y to D
        adcb    P+3
        stab    P+3
        adca    P+2
        staa    P+2
; the following six instructions propagate carry to the most significant byte
        ldaa    P+1
        adca    #0              ; add C flag to location P+1
        staa    P+1             ;          "
        ldaa    P
        adca    #0              ; add C flag to location P
        staa    P
```

```
; the following three instructions compute MLNH
        ldd     M+2     ; place ML in D
        ldy     N       ; place NH in Y
        emul            ; compute MLNH
; the following seven instructions add MLNH to memory locations P+2~P+5
        addd    P+4     ; add the lower half of MLNH to P+4~P+5
        std     P+4     ;           "
        tfr     y,d     ; transfer Y to D
        adcb    P+3
        stab    P+3
        adca    P+2
        staa    P+2
; the following six instructions propagate carry to the most significant byte
        ldaa    P+1
        adca    #0      ; add C flag to location P+1
        staa    P+1
        ldaa    P
        adca    #0      ; add C flag to location P
        staa    P
        end
```

▲

Example 2.13

▼

Write a program to convert the 16-bit binary number stored at $800~$801 to BCD format and store the result at $900~$904. Convert each BCD digit into its ASCII code and store it in one byte.

Solution: A binary number can be converted to BCD format using repeated division-by-10. The largest 16-bit binary number corresponds to the 5-digit decimal number 65535. The first division by 10 computes the least significant digit and should be stored in the memory location $904, the second division-by-10 operation computes the ten's digit, and so on. The ASCII code of a BCD digit can be obtained by adding $30 to each BCD digit. The program is as follows:

```
        org     $800
data    fdb     12345   ; place a number for testing
        org     $900
result  rmb     5       ; reserve five bytes to store the result
        org     $1000
        ldd     data    ; make a copy of the number to be converted
        ldy     #result
        ldx     #10     ; divide the number by 10
        idiv            ;           "
        addb    #$30    ; convert to ASCII code
        stab    4,Y     ; save the least significant digit
        xgdx            ; swap the quotient to D
        ldx     #10
        idiv
        addb    #$30    ; convert to ASCII code
        stab    3,Y     ; save the second-to-least-significant digit
        xgdx
        ldx     #10
```

```
idiv
addb    #$30
stab    2,Y      ; save the middle digit
xgdx
ldx     #10
idiv             ; separate the most-significant and second-to-most-
                 ; significant digits
addb    #$30
stab    1,Y      ; save the second-to-most-significant digit
xgdx             ; swap the most significant digit to B
addb    #$30     ; convert to ASCII code
stab    0,Y      ; save the most significant digit
end
```

2.6 Program Loops

Many applications require repetitive operations. We can write programs to tell computers to perform the same operation over and over. A *finite loop* is a sequence of instructions that will be executed by the computer for a finite number of times, while an *endless loop* is a sequence of instructions that the computer will execute forever.

There are four major loop constructs:

Do statement S forever

This is an endless loop in which statement *S* is repeated forever. In some applications, we might add the statement "If C then exit" to leave the infinite loop. An infinite loop is shown in Figure 2.4.

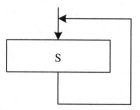

Figure 2.4 ■ An infinite loop

For i = n1 to n2 do S or For i = n2 downto n1 do S

Here, the variable *i* is the *loop counter*, which keeps track of the number of remaining times statement *S* is to be executed. The loop counter can be incremented (the first case) or decremented (the second case). Statement *S* is repeated $n2 - n1 + 1$ times. The value of $n2$ is assumed to be no smaller than $n1$. If there is concern that the relationship $n1 \leq n2$ may not hold, then it must be checked at the beginning of the loop. Four steps are required to implement a FOR loop:

Step 1
Initialize the loop counter and other variables.

Step 2
Compare the loop counter with the limit to see if it is within bounds. If it is, then perform the specified operations. Otherwise, exit the loop.

Step 3
Increment (or decrement) the loop counter.

Step 4
Go to step 2.

A *For-loop* is illustrated in Figure 2.5.

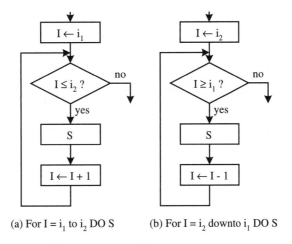

(a) For I = i_1 to i_2 DO S (b) For I = i_2 downto i_1 DO S

Figure 2.5 ■ For looping construct

While C Do S

Whenever a *While* construct is executed, the logical expression C is evaluated first. If it yields a false value, statement S will not be executed. The action of a While construct is illustrated in Figure 2.6. Four steps are required to implement a While loop:

Step 1
Initialize the logical expression C.

Step 2
Evaluate the logical expression C.

Step 3
Perform the specified operations if the logical expression C evaluates to true. Update the logical expression C and go to step 2. (Note: The logical expression C may be updated by external conditions or by an interrupt service routine.)

Step 4
Exit the loop.

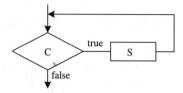

Figure 2.6 ■ The While ... Do looping construct

Repeat S Until C

Statement *S* is first executed then the logical expression *C* is evaluated. If *C* is false, the next statement will be executed. Otherwise, statement *S* will be executed again. The action of this construct is illustrated in Figure 2.7. Statement *S* will be executed at least once. Three steps are required to implement this construct:

Step 1
Initialize the logical expression *C*.

Step 2
Execute statement S.

Step 3
Go to Step 2 if the logical expression *C* evaluates to true. Otherwise, exit.

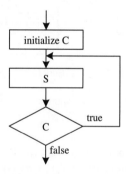

Figure 2.7 ■ The Repeat ... Until looping construct

To implement one of the looping constructs, we need to use unconditional branch or one of the conditional instructions. When executing conditional branch instructions, the 68HC12 checks the condition flags in the CCR register.

2.6.1 Condition Code Register

The contents of the condition code register are shown in Figure 2.8. The shaded characters are condition flags that reflect the status of an operation. The meanings of these condition flags are as follows:

7	6	5	4	3	2	1	0
S	X	H	I	N	Z	V	C

Figure 2.8 ■ Condition code register

- **C: the carry flag**
  Whenever a carry is generated as the result of an operation, this flag will be set to 1. Otherwise, it will be cleared to 0.
- **V: the overflow flag**
  Whenever the result of a two's complement arithmetic operation is out of range, this flag will be set to 1. Otherwise, it will be set to 0. The V flag is set to 1 when the carry from the most significant bit and the second most significant bit differ as the result of an arithmetic operation.
- **Z: the zero flag**
  Whenever the result of an operation is zero, this flag will be set to 1. Otherwise, it will be set to 0.
- **N: the negative flag**
  Whenever the most significant bit of the result of an operation is 1, this flag will be set to 1. Otherwise, it will be set to 0. This flag indicates that the result of an operation is negative.
- **H: the half-carry flag**
  Whenever there is a carry from the lower four bits to the upper four bits as the result of an operation, this flag will be set to 1. Otherwise, it will be set to 0.

2.6.2 Branch Instructions

Branch instructions cause program flow to change when specific conditions exist. The 68HC12 has three kinds of branch instructions including *short branches*, *long branches*, and *bit-conditional branches*.

Branch instructions can also be classified by the type of condition that must be satisfied in order for a branch to be taken. Some instructions belong to more than one category.

- **Unary (unconditional) branch** instructions always execute.
- **Simple branches** are taken when a specific bit in the CCR register is in a specific state as a result of a previous operation.
- **Unsigned branches** are taken when a comparison or test of unsigned quantities results in a specific combination of condition code register bits.
- **Signed branches** are taken when a comparison or test of signed quantities results in a specific combination of condition code register bits.

When a short-branch instruction is executed, a signed 8-bit offset is added to the value in the program counter when a specified condition is met. Program execution continues at the new address. The numeric range of the short branch offset value is $80 (-128) to $7F (127) from the address of the instruction immediately following the branch instruction. A summary of the short branch instructions is shown in Table 2.2.

When a long-branch instruction is executed, a signed 16-bit offset is added to the value in the program counter when a specified condition is met. Program execution continues at the new address. Long branch instructions are used when large displacements between decision-making steps are necessary.

Unary Branches		
Mnemonic	**Function**	**Equation or Operation**
BRA	Branch always	$1 = 1$
BRN	Branch never	$1 = 0$

Simple Branches		
Mnemonic	**Function**	**Equation or Operation**
BCC	Branch if carry clear	$C = 0$
BCS	Branch if carry set	$C = 1$
BEQ	Branch if equal	$Z = 1$
BMI	Branch if minus	$N = 1$
BNE	Branch if not equal	$Z = 0$
BPL	Branch if plus	$N = 0$
BVC	Branch if overflow clear	$V = 0$
BVS	Branch if overflow set	$V = 1$

Unsigned Branches		
Mnemonic	**Function**	**Equation or Operation**
BHI	Branch if higher	$C + Z = 0$
BHS	Branch if higher or same	$C = 0$
BLO	Branch if lower	$C = 1$
BLS	Branch if lower or same	$C + Z = 1$

Signed Branches		
Mnemonic	**Function**	**Equation or Operation**
BGE	Branch if greater than or equal	$N \oplus V = 0$
BGT	Branch if greater than	$Z + (N \oplus V) = 0$
BLE	Branch if less than or equal	$Z + (N \oplus V) = 1$
BLT	Branch if less than	$N \oplus V = 1$

Table 2.2 ■ Summary of short branch instructions

The numeric range of long-branch offset values is $8000 (-32768) to $7FFF (32767) from the instruction immediately after the branch instruction. This permits branching from any location in the standard 64-KB address map to any other location in the map. A summary of the long-branch instructions appears in Table 2.3.

Unary Branches		
Mnemonic	**Function**	**Equation or Operation**
LBRA	Long branch always	$1 = 1$
LBRN	Long branch never	$1 = 0$
Simple Branches		
Mnemonic	**Function**	**Equation or Operation**
LBCC	Long branch if carry clear	$C = 0$
LBCS	Long branch if carry set	$C = 1$
LBEQ	Long branch if equal	$Z = 1$
LBMI	Long branch if minus	$N = 1$
LBNE	Long branch if not equal	$Z = 0$
LBPL	Long branch if plus	$N = 0$
LBVC	Long branch if overflow is clear	$V = 0$
LBVS	Long branch if overflow set	$V = 1$
Unsigned Branches		
Mnemonic	**Function**	**Equation or Operation**
LBHI	Long branch if higher	$C + Z = 0$
LBHS	Long branch if higher or same	$C = 0$
LBLO	Long branch if lower	$C = 1$
LBLS	Long branch if lower or same	$C + Z = 1$
Signed Branches		
Mnemonic	**Function**	**Equation or Operation**
LBGE	Long branch if greater than or equal	$N \oplus V = 0$
LBGT	Long branch if greater than	$Z + (N \oplus V) = 0$
LBLE	Long branch if less than or equal	$Z + (N \oplus V) = 1$
LBLT	Long branch if less than	$N \oplus V = 1$

Table 2.3 ■ Summary of long branch instructions

Although there are many possibilities in writing a program loop, the following one is a common format:

```
loop:  .
       .
       .
       Bcc (or LBcc)  loop
```

Where cc is one of the condition codes (CC, CS, EQ, MI, NE, PL, VC, VS, HI, HS, LO, LS, GE, GT, LS, and LT).

Usually there will be a comparison or arithmetic instruction to set up the condition code for use by the conditional branch instruction.

2.6.3 Compare & Test Instructions

The 68HC12 has a set of compare instructions that are dedicated to the setting of condition flags. The compare and test instructions perform subtraction between a pair of registers or between a register and a memory location. The result is not stored, but condition codes are set by the operation. In the 68HC12, most instructions update condition code flags automatically, so it is often unnecessary to include a separate test or compare instruction. Table 2.4 is a summary of compare and test instructions.

Compare instructions		
Mnemonic	**Function**	**Operation**
CBA	Compare A to B	(A) - (B)
CMPA	Compare A to memory	(A) - (M)
CMPB	Compare B to memory	(B) - (M)
CPD	Compare D to memory	(D) - (M:M+1)
CPS	Compare SP to memory	(SP) - (M:M+1)
CPX	Compare X to memory	(X) - (M:M+1)
CPY	Compare Y to memory	(Y) - (M:M+1)
Test instructions		
Mnemonic	**Function**	**Operation**
TST	Test memory for zero or minus	(M) - $00
TSTA	Test A for zero or minus	(A) - $00
TSTB	Test B for zero or minus	(B) - $00

Table 2.4 ■ Summary of compare and test instructions

2.6.4 Loop Primitive Instructions

A lot of the program loops are implemented by incrementing or decrementing a loop count. The branch is taken when either the loop count is equal to zero or not equal to zero depending on the applications. The 68HC12 provides a set of loop primitive instructions for implementing this type of looping mechanism. These instructions test a counter value in a register or accumulator (A, B, D, X, Y, or SP) for zero or nonzero values as a branch condition. There are predecrement, preincrement, and test-only versions of these instructions.

The range of the branch is from $80 (-128) to $7F (127) from the instruction immediately following the loop primitive instruction. Table 2.5 shows a summary of the loop primitive instructions.

Mnemonic	Function	Equation or Operation
DBEQ cntr, rel	Decrement counter and branch if = 0 (counter = A, B, D, X, Y, or SP)	counter ← (counter) - 1 If (counter) = 0, then branch else continue to next instruction
DBNE cntr, rel	Decrement counter and branch if ≠ 0 (counter = A, B, D, X, Y, or SP)	counter ← (counter) - 1 If (counter) ≠ 0, then branch else continue to next instruction
IBEQ cntr, rel	Increment counter and branch if = 0 (counter = A, B, D, X, Y, or SP)	counter ← (counter) + 1 If (counter) = 0, then branch else continue to next instruction
IBNE cntr, rel	Increment counter and branch if ≠ 0 (counter = A, B, D, X, Y, or SP)	counter ← (counter) + 1 If (counter) ≠ 0, then branch else continue to next instruction
TBEQ cntr, rel	Test counter and branch if = 0 (counter = A, B, D, X, Y, or SP)	If (counter) = 0, then branch else continue to next instruction
TBNE cntr, rel	Test counter and branch if ≠ 0 (counter = A, B, D, X, Y, or SP)	If (counter) ≠ 0, then branch else continue to next instruction

Note. 1. **cntr** is the loop counter and can be accumulator A, B, or D and register X, Y, or SP.
2. **rel** is the relative branch offset and is usually a label

Table 2.5 ■ Summary of loop primitive instructions

Example 2.14

Write a program to add an array of N 8-bit numbers and store the sum at memory location $800~$801. Use the **For** $i = n1$ **to** $n2$ **do** looping construct.

Solution: We will use variable i as the array index. This variable can also be used to keep track of the number of iterations remained to be performed. We will use a two-byte variable *sum* to hold the sum of array elements. The logic flow of the program is illustrated in Figure 2.9.

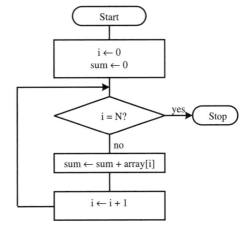

Figure 2.9 ■ Logic flow of example 2.14

The program is a direct translation of the flowchart shown in Figure 2.9.

```
N        equ     20              ; array count
         org     $800            ; starting address of on-chip SRAM
sum      rmb     2               ; array sum
i        rmb     1               ; array index
         org     $1000           ; starting address of the program
         ldaa    #0
         staa    i               ; initialize loop (array) index to 0
         staa    sum             ; initialize sum to 0
         staa    sum+1           ;     "

loop     ldab    i
         cmpb    #N              ; is i = N?
         beq     done            ; if done, then branch
         ldx     #array          ; use index register X as a pointer to the array
         abx                     ; compute the address of array[i]
         ldab    0,x             ; place array[i] in B
         ldy     sum             ; place sum in Y
         aby                     ; compute sum <- sum + array[i]
         sty     sum             ; update sum
         inc     i               ; increment the loop count by 1
         bra     loop
done     swi                     ; return to D-Bug12 monitor
; the array is defined in the following statement
array    db      1,2,3,4,5,6,7,8,9,10,11,12,13,14,15,16,17,18,19,20
         end
```

It is a common mistake for an assembly language programmer to forget to update the variable in memory. For example, we will not get the correct value for *sum* if we did not add the instruction *sty sum* in the program shown in Example 2.14.

Loop primitive instructions are especially suitable for implementing the *repeat S until C* looping construct as demonstrated in the following example.

Example 2.15

Write a program to find the maximum element from an array of N 8-bit elements using the *repeat S until C* looping construct.

Solution: We will use the variable **i** as the array index and also as the loop count. The variable *max_val* will be used to hold the array maximum. The logic flow of the program is shown in Figure 2.10.

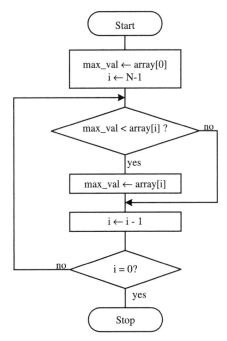

Figure 2.10 ■ Logic flow of example 2.15

The program is as follows:

```
N           equ     20                          ; array count
            org     $800                        ; starting address of on-chip SRAM
max_val     rmb     1                           ; memory location to hold array max
            org     $1000                       ; starting address of program
            ldaa    array                       ; set array[0] as the temporary array max
            staa    max_val                     ;         "
            ldx     #array+N-1                   ; start from the end of the array
            ldab    #N-1                        ; use B to hold variable i and initialize it to N-1
loop        ldaa    max_val
            cmpa    0,x                         ; compare max_val with array[i]
            bge     chk_end                     ; no update if max_val is larger
            ldaa    0,x                         ; update max_val
            staa    max_val                     ;         "
chk_end     dex                                 ; move the array pointer
            dbne    b,loop                      ; decrement the loop count, branch if not zero yet.
forever     bra     forever
array       db      1,3,5,6,19,41,53,28,13,42,76,14,20,54,64,74,29,33,41,45
            end
```

2.6.5 Decrementing & Incrementing Instructions

We often need to add or subtract one from a variable in our program. Although we can use one of the ADD or SUB instructions to achieve this, it would be more efficient to use a single instruction. The 68HC12 has a few instructions for us to increment or decrement a variable by one. A summary of decrement and increment instructions is listed in Table 2.6.

Decrement instructions		
Mnemonic	**Function**	**Operation**
DEC	Decrement memory by 1	M ← (M) - $01
DECA	Decrement A by 1	A ← (A) - $01
DECB	Decrement B by 1	B ← (B) - $01
DES	Decrement SP by 1	SP ← (SP) - $01
DEX	Decrement X by 1	X ← (X) - $01
DEY	Decrement Y by 1	Y ← (Y) - $01
Increment instructions		
Mnemonic	**Function**	**Operation**
INC	Increment memory by 1	M ← (M) + $01
INCA	Increment A by 1	A ← (A) + $01
INCB	Increment B by 1	B ← (B) + $01
INS	Increment SP by 1	SP ← (SP) + $01
INX	Increment X by 1	X ← (X) + $01
INY	Increment Y by 1	Y ← (Y) + $01

Table 2.6 ■ Summary of decrement and increment instructions

Example 2.16

▼

Use an appropriate increment or decrement instruction to replace the following instruction sequence:

```
ldaa    i
adda    #1
staa    i
```

Solution: The above three instructions can be replaced by the following instruction:

```
inc     I
```

▲

2.6.6 Bit Condition Branch Instructions

In certain applications, we need to make branch decisions based on the value of a few bits. The 68HC12 provides two special conditional branch instructions for this purpose. The syntax of the first special conditional branch instruction is:

```
[<label>]       BRCLR opr, msk, rel
```

where:

opr   specifies the memory location to be checked and can be specified using direct, extended, and all indexed addressing modes.

msk is an 8-bit mask that specifies the bits of the memory location to be checked. The bits to be checked correspond to those bit positions that are ones in the mask.

rel is the branch offset and is specified in 8-bit relative mode.

This instruction tells the 68HC12 to perform bitwise logical AND on the contents of the specified memory location and the mask supplied with the instruction, then branch if the result is zero.

For example, for the instruction sequence:

```
here    brclr    $66,$80,here
        ldd      $70
```

the 68HC12 will continue to execute the first instruction if the most significant bit of the memory location at $66 is 0. Otherwise, the next instruction will be executed.

The syntax of the second special conditional branch instruction is:

```
[<label>]        BRSET opr, msk, rel
```

where:

opr specifies the memory location to be checked and can be specified using direct, extended, and all indexed addressing modes.

msk is an 8-bit mask that specifies the bits of the memory location to be checked. The bits to be checked correspond to those bit positions that are ones in the mask

rel is the branch offset and is specified in 8-bit relative mode

This instruction tells the 68HC12 to perform the logical AND of the contents of the specified memory location inverted and the mask supplied with the instruction, then branch if the result is zero (this occurs only when all bits corresponding to ones in the mask byte are ones in the tested byte).

For example, for the following instruction sequence:

```
loop    inc      count
        ...
        brset    $66,$e0,loop
        ...
```

the branch will be taken if the most significant three bits of the memory location at $66 are all ones.

Example 2.17

Write a program to count the number of elements that are divisible by 4 in an array of N 8-bit numbers. Use the *repeat S until C* looping construct.

Solution: The lowest two bits of a number divisible by four are 00. By checking the lowest two bits of a number, we can determine if a number is divisible by four. The program is as follows:

```
N       equ      20
        org      $800
total   rmb      1
        org      $1000          ; starting address of the program
        ldaa     #0
        staa     total          ; initialize total to 0
```

```
            ldx      #array          ; use index register X as the array pointer
            ldab     #N              ; use accumulator B as the loop count
loop        brclr    0,x,$03,yes
            bra      chkend
yes         inc      total           ; add 1 to the total
chkend      inx                      ; move the array pointer
            dbne     b,loop
forever     bra      forever
array       db       2,3,4,8,12,13,19,24,33,32,20,18,53,52,80,82,90,94,100,102
            end
```

2.6.7 Instructions for Variable Initialization

We often need to initialize a variable to zero when writing a program. The 68HC12 has three instructions for this purpose. They are:

    [<label>]   clr   opr

where *opr* is a memory location specified using the extended mode and all indexed addressing (direct and indirect) modes. The memory location is initialized to zero by this instruction.

    [<label>]   clra

Accumulator A is cleared to 0 by this instruction.

    [<label>]   clrb

Accumulator B is cleared to 0 by this instruction.

2.7 Shift & Rotate Instructions

Shift and *rotate instructions* are useful for bit field manipulation. They can be used to speed up the integer multiply and divide operations if one of the operands is a power of two. A shift/rotate instruction shifts/rotates the operand by one bit. The 68HC12 has shift instructions that can operate on accumulators A, B, and D, or on a memory location. A memory operand must be specified using the extended or indexed (direct or indirect) addressing modes. A summary of shift and rotate instructions is shown in Table 2.7.

Logical shift instructions		
Mnemonic	**Function**	**Operation**
LSL <opr>	Logical shift left memory	
LSLA	Logical shift left A	
LSLB	Logical shift left B	
LSLD	Logical shift left D	
LSR <opr>	Logical shift right memory	
LSRA	Logical shift right A	
LSRB	Logical shift right B	
LSRD	Logical shift right D	
Arithmetic shift instructions		
Mnemonic	**Function**	**Operation**
ASL <opr>	Arithmetic shift left memory	
ASLA	Arithmetic shift left A	
ASLB	Arithmetic shift left B	
ASLD	Arithmetic shift left D	
ASR <opr>	Arithmetic shift right memory	
ASRA	Arithmetic shift right A	
ASRB	Arithmetic shift right B	
Rotate instructions		
Mnemonic	**Function**	**Operation**
ROL <opr>	Rotate left memory thru carry	
ROLA	Rotate left A through carry	
ROLB	Rotate left B through carry	
ROR <opr>	Rotate right memory thru carry	
RORA	Rotate right A through carry	
RORB	Rotate right B through carry	

Table 2.7 ■ Summary of shift and rotate instructions

Example 2.18

What are the values of accumulator A and the C flag after executing the ASLA instruction? Assume that originally A contains $95 and the C flag is 1.

Solution: The operation of this instruction is shown in Figure 2.11a.

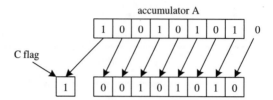

Figure 2.11a ■ Operation of the ASLA instruction

The result is shown in Figure 2.11b.

Original value	New value
[A] = 10010101 C = 1	[A] = 00101010 C = 1

Figure 2.11b ■ Execution result of the ASLA instruction

Example 2.19

What are the new values of the memory location at $800 and the C flag after executing the instruction ASR $800? Assume that the memory location $800 originally contains the value of $ED and the C flag is 0.

Solution: The operation of this instruction is shown in Figure 2.12a.

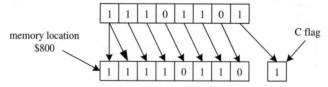

Figure 2.12a ■ Operation of the ASR $800 instruction

The result is shown in Figure 2.12b.

Original value	New value
[$800] = 11101101 C = 0	[$800] = 11110110 C = 1

Figure 2.12b ■ Result of the ASR $800 instruction

Example 2.20

What are the new values of the memory location at $800 and the C flag after executing the instruction LSR $800? Assume the memory location $800 originally contains $E7 and the C flag is 1.

Solution: The operation of this instruction is illustrated in Figure 2.13a.

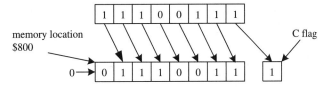

Figure 2.13a ■ Operation of the LSR $800 instruction

The result is shown in Figure 2.13b.

Original value	New value
[$800] = 11100111 C = 1	[$800] = 01110011 C = 1

Figure 2.13b ■ Execution result of LSR $800

Example 2.21

Compute the new values of accumulator B and the C flag after executing the instruction ROLB. Assume the original value of B is $BD and C flag is 1.

Solution: The operation of this instruction is illustrated in Figure 2.14a.

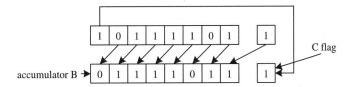

Figure 2.14a ■ Operation of the instruction ROLB

The result is shown in Figure 2.14b.

Original value	New value
[B] = 10111101 C = 1	[B] = 01111011 C = 1

Figure 2.14b ■ Execution result of ROLB

Example 2.22

What are the values of accumulator A and the C flag after executing the instruction RORA? Assume the original value of A is $BE and C = 1.

Solution: The operation of this instruction is illustrated in Figure 2.15a.

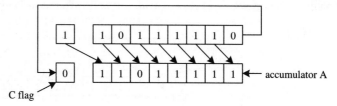

C flag

Figure 2.15a ■ Operation of the instruction RORA

The result is shown in Figure 2.15b.:

Original value	New value
[A] = 10111110 C = 1	[A] = 11011111 C = 0

Figure 2.15b ■ Execution result of RORA

Example 2.23

Write a program to count the number of zeros contained in memory locations $800~$801 and save the result at memory location $805.

Solution: The logical shift right instruction is available for double accumulator D. We can load this 16-bit value into D and shift it to the right sixteen times or until it becomes zero. The algorithm of this program is as follows:

Step 1
Initialize the loop count to 16 and the zero count to 0.

Step 2
Place the 16-bit value in D.

Step 3
Shift D to the right one place.

Step 4
If the C flag is 0, increment zero count by 1.

Step 5
Decrement loop count by 1.

Step 6
If loop count is zero, then stop. Otherwise, go to step 3.

The program is as follows:

```
                org     $800
                db      $23,$55
                org     $805
zero_cnt        rmb     1
lp_cnt          rmb     1
                org     $1000
                clr     zero_cnt        ; initialize the zero count to 0
                ldaa    #16
                staa    lp_cnt          ; initialize loop count to 16
                ldd     $800            ; place the 16-bit number in D
again           lsrd
                bcs     chk_end         ; branch if the lsb is a 1
                inc     zero_cnt
chk_end         dec     lp_cnt
                bne     again           ; have we tested all 16 bits yet?
forever         bra     forever
                end
```

Sometimes we need to shift a number larger than 16 bits. However, the 68HC12 does not have an instruction that does this. Suppose the number has k bytes and the most significant byte is located at *loc*. The remaining k −1 bytes are located at loc+1, loc+2, ..., loc+k-1, as shown in Figure 2.16.

Figure 2.16 ■ k bytes to be shifted

The logical shift-one-bit-to-the-right operation is shown in Figure 2.17.

Figure 2.17 ■ Shift-one-to-the-right operation

As shown in Figure 2.17:

- The bit seven of each byte will receive the bit zero of the byte on its immediate left with the exception of the most significant byte, which will receive a zero.
- Each byte will be shifted to the right by one bit. The bit zero of the least significant byte will be shifted out and lost.

The operation can therefore be implemented as follows:

Step 1
Shift the byte at *loc* to the right one place (using the LSR <opr> instruction).

Step 2
Rotate the byte at *loc+1* to the right one place (using the ROR <opr> instruction).

Step 3
Repeat step 2 for the remaining bytes.

By repeating this procedure, the given k-byte number can be shifted to the right as many bits as desired. The operation to shift a multi-byte number to the left should start from the least significant byte and rotate the remaining bytes toward the most significant byte.

Example 2.24

▼

Write a program to shift the 32-bit number stored at $800~$803 to the right four places.

Solution: The most significant to the least significant bytes are stored at $800~$803. The following instruction sequence implements the algorithm that we just described:

```
        ldab    #4          ; set up the loop count
        ldx     #$800
again   lsr     0,x
        ror     1,x
        ror     2,x
        ror     3,x
        dbne    b,again
```

2.8 Boolean Logic Instructions

When dealing with input and output port pins, we often need to change the values of a few bits. For these types of applications, Boolean logic instructions come in handy. A summary of the 68HC12 Boolean logic instructions is described in Table 2.8.

Mnemonic	Function	Operation
ANDA <opr>	AND A with memory	$A \leftarrow (A) \bullet (M)$
ANDB <opr>	AND B with memory	$B \leftarrow (B) \bullet (M)$
ANDCC <opr>	AND CCR with memory (clear CCR bits)	$CCR \leftarrow (CCR) \bullet (M)$
EORA <opr>	Exclusive OR A with memroy	$A \leftarrow (A) \oplus (M)$
EORB <opr>	Exclusive OR B with memory	$B \leftarrow (B) \oplus (M)$
ORAA <opr>	OR A with memory	$A \leftarrow (A) + (M)$
ORAB <opr>	OR B with memory	$B \leftarrow (B) + (M)$
ORCC <opr>	OR CCR with memory	$CCR \leftarrow (CCR) + (M)$
CLC	Clear C bit in CCR	$C \leftarrow 0$
CLI	Clear I bit in CCR	$I \leftarrow 0$
CLV	Clear V bit in CCR	$V \leftarrow 0$
COM <opr>	One's complement memory	$M \leftarrow \$FF - (M)$
COMA	One's complement A	$A \leftarrow \$FF - (A)$
COMB	One's complement B	$B \leftarrow \$FF - (B)$
NEG <opr>	Two's complement memory	$M \leftarrow \$00 - (M)$
NEGA	Two's complement A	$A \leftarrow \$00 - (A)$
NEGB	Two's complement B	$B \leftarrow \$00 - (B)$

Table 2.8 ■ Summary of Boolean logic instructions

The operand *opr* can be specified using all except the relative addressing modes. Usually, one would use the AND instruction to clear one or a few bits and use the OR instruction to set one or a few bits. The exclusive OR instruction can be used to toggle (change from 0 to 1 and from 1 to 0) one or a few bits.

For example, the instruction sequence:

```
ldaa    $56
anda    #$0F
staa    $56
```

clears the upper four pins of the I/O port located at $56.

The instruction sequence:

```
ldaa    $56
oraa    #$01
staa    $56
```

sets the bit 0 of the I/O port at $56.

The instructions sequence:

```
ldaa    $56
eora    #$0F
staa    $56
```

toggles the lower four bits of the I/O port at $56. The instructions (COMA and COMB) that perform one's complementing can be used if all of the port pins need to be toggled.

2.9 Bit Test & Manipulate Instruction

These instructions use a mask value to test or change the value of individual bits in an accumulator or in a memory location. BITA and BITB provide a convenient means of testing bits without altering the value of either operand. Table 2.9 shows a summary of bit test and manipulation instructions.

Mnemonic	Function	Operation
BCLR <opr>[2], msk8	Clear bits in memory	$M \leftarrow (M) \bullet (\overline{mm})$
BITA <opr>[1]	Bit test A	$(A) \bullet (M)$
BITB <opr>[1]	Bit test B	$(B) \bullet (M)$
BSET <opr>[2], msk8	Set bits in memory	$M \leftarrow (M) + (mm)$
Note.	1. <opr> can be specified using all except relative addressing modes for BITA and BITB. 2. <opr> can be specified using direct, extended, and indexed (exclude indiriect) addressing modes. 3. msk8 is an 8-bit value.	

Table 2.9 ■ Summary of bit test and manipulation instructions

For example, the instruction:

```
bclr    0,x,$81
```

clears the most significant and least significant bits of the memory location pointed to by index register X.

The instruction:

```
        bita    #$44
```

tests the bit six and bit two of accumulator A and updates Z and N flags of CCR register accordingly. The V flag in CCR register is cleared.

The instruction:

```
        bitb    #$22
```

tests the bit five and bit one of accumulator B and updates the Z and N flags of CCR register accordingly. The V flag in CCR register is cleared.

Finally, the instruction:

```
        bset    0,y,$33
```

sets the bits five, four, one, and zero of the memory location pointed to by index register Y.

2.10 Program Execution Time

The 68HC12 uses the ECLK (we will call it *E clock* from now on) signal as a timing reference. The frequency of the E clock is equal to one-half of the frequency of the crystal oscillator out of reset. The execution times of instructions are also measured in E cycles.

There are many applications that require the generation of time delays. Program loops are often used to create some amount of delay unless the time delay needs to be very accurate.

The creation of a time delay involves two steps:

1. Select a sequence of instructions that takes a certain amount of time to execute.

2. Repeat the instruction sequence for the appropriate number of times.

For example, the following instruction sequence takes 40 E clock cycles to execute:

```
loop    psha                    ; 2 E cycles
        pula                    ; 3 E cycles
        psha
        pula
        psha
        pula
        psha
        pula
        psha
        pula
        psha
        pula
        psha
        pula
        nop                     ; 1 E cycle
        nop                     ; 1 E cycle
        dbne    x,loop          ; 3 E cycles
```

If the 68HC12 runs under the control of a 16-MHz crystal oscillator, then the frequency and the period of the E clock signal are 8 MHz and 125 ns, respectively. The above instruction sequence will take 5 μs to execute.

Example 2.25

Write an instruction sequence to create a 100-ms time delay.

Solution: In order to create a 100-ms time delay, we need to repeat the above instruction sequence 20,000 times (100 ms ÷ 5 μs = 20,000). The following instruction sequence will create the desired delay:

```
            ldx        #20000      ; 2 E cycles
loop        psha                   ; 2 E cycles
            pula                   ; 3 E cycles
            psha                   ; 2 E cycles
            pula                   ; 3 E cycles
            psha                   ; 2 E cycles
            pula                   ; 3 E cycles
            psha                   ; 2 E cycles
            pula                   ; 3 E cycles
            psha                   ; 2 E cycles
            pula                   ; 3 E cycles
            psha                   ; 2 E cycles
            pula                   ; 3 E cycles
            psha                   ; 2 E cycles
            pula                   ; 3 E cycles
            nop                    ; 1 E cycle
            nop                    ; 1 E cycle
            dbne       x,loop      ; 3 E cycles
```

Example 2.26

Write an instruction sequence to create a delay of 10 seconds.

Solution: The instruction sequence in Example 2.25 can create no more than 327 ms delay. In order to create a longer time delay, we need to use a two-layer loop. For example, the following instruction sequence will create a 10-second delay:

```
            ldab       #100        ; 1 E cycle
out_loop    ldx        #20000      ; 2 E cycles
inner_loop  psha
            pula
            psha
            pula
            psha
            pula
            psha
            pula
            psha
            pula
            psha
            pula
            psha
            pula
```

```
        psha
        pula
        psha
        pula
        nop
        nop
        dbne            x,inner_loop
        dbne            b,out_loop          ; 3 E cycles
```

The time delay created by using program loops is not accurate. Some overhead is required to set up the loop count. For example, the one-layer loop has a 2-E-cycle overhead while the two-layer loop has much more overhead:

overhead = 1 E cycle (caused by the ldab #100 instruction)
           + 100 x 2 E cycles (caused by the out_loop ldx #20000 instruction)
           + 100 x 3 E cycles (caused by the dbne b,out_loop instruction)
           = 501 E cycles = 62.625 μs (at 8 MHz E clock)

This overhead can be reduced by placing a larger value in index register X and a smaller value in accumulator B. For example, by placing 50,000 in X and 40 in B, the overhead can be reduced to 25.125 μs. If higher accuracy is required then you should use one of the timer functions to create the desired time delay.

2.11 Summary

An assembly language program consists of three major parts: *assembler directives, assembly language instructions,* and *comments.* A statement of an assembly language program consists of four fields: *label, operation code, operand,* and *comment.* Assembly directives supported by the freeware *as12* are all discussed in this chapter.

The 68HC12 instructions are explained category by category. Simple program examples are used to demonstrate the applications of different instructions. The 68HC12 is a 16-bit microcontroller. Therefore, it can perform 16-bit arithmetic. Numbers greater than 16 bits must be manipulated using multiprecision arithmetic.

Microcontrollers are designed to perform repetitive operations. Repetitive operations are implemented by program loops. There are two types of program loops: *infinite loop* and *finite loop.* There are four major variants of the looping constructs:

- **Do** statement S **forever**
- **For** $i = n_1$ **to** n_2 **do** S or **For** $i = n_2$ **downto** n_1 **do** S
- **While** C **do** S
- **Repeat** S **until** C

In general, the implementation of program loops requires:

- the initialization of a loop counter (or condition)
- performing the specified operation
- comparing the loop count with the loop limit (or evaluating the condition)
- making a decision regarding whether the program loop should be continued

The 68HC12 provides instructions to support the initialization of a loop counter, decrementing (or incrementing) the loop counter, and deciding whether looping should be continued.

The shifting and rotating instructions are useful for bit field operations. Integer multiplication by a power of two and division by a power of two can be sped up by using the shifting instructions.

The 68HC12 also provides many Boolean logical instructions that can be very useful for setting, clearing, and toggling the I/O port pins.

2.12 Exercises

E2.1 Find the valid and invalid labels in the following statements, and explain why the invalid labels are invalid.

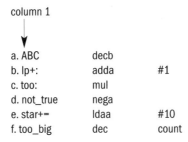

```
column 1

  a. ABC        decb
  b. lp+:       adda       #1
  c. too:       mul
  d. not_true   nega
  e. star+=     ldaa       #10
  f. too_big    dec        count
```

E2.2 Identify the four fields of the following instructions:

```
  a.          bne       not_done
  b. loop     brclr     0,x,$01,loop    ; wait until the least significant bit is set
  c. here:    dec       lp_cnt          ; decrement the variable lp_cnt
```

E2.3 Write a sequence of assembler directives to reserve 10 bytes starting from $800.

E2.4 Write a sequence of assembler directives to build a table of ASCII codes of lower-case letters a-z. The table should start from memory location $2000.

E2.5 Write a sequence of assembler directives to store the message "welcome to the robot demonstration!" starting from the memory location at $1050.

E2.6 Write an instruction sequence to add the two 24-bit numbers stored at $810~$812 and $813~$815, and save the sum at $900~$902.

E2.7 Write an instruction sequence to subtract the 6-byte number stored at $800~$805 from the 6-byte number stored at $810~$805, and save the result at $900~$905.

E2.8 Write a sequence of instructions to add the BCD numbers stored at $800 and $801 and store the sum at $803.

E2.9 Write an instruction sequence to add the 4-digit BCD numbers stored at $800~$801 and $802~$803, and store the sum at $900~$901.

E2.10 Write a program to compute the average of an array of N 8-bit numbers and store the result at $900. The array is stored at memory locations starting from $800. N is no larger than 255.

E2.11 Write a program to multiply two 3-byte numbers that are stored at $800~$802 and $803~$805, and save the product at $900~$905.

E2.12 Write a program to compute the average of the square of all elements of an array with thirty-two 8-bit unsigned numbers. The array is stored at $800~$81F. Store the result at $900~$901.

E2.13 Write a program to count the number of even elements of an array of N 16-bit elements. The array is stored at memory locations starting from $900.

E2.14 Write an instruction sequence to shift the 32-bit number to the left four places. The 32-bit number is located at $800~$803.

E2.15 Write a program to count the number of elements in an array that are smaller than 16. The array is stored at memory locations starting from $800. The array has N 8-bit unsigned elements.

E2.16 Write an instruction sequence to swap the upper four bits and the lower four bits of accumulator A (swap bit seven with bit three, bit six with bit two, and so on).

E2.17 Write a program to count the number of elements in an array whose bits three, four, and seven are zeroes. The array has N 8-bit elements and is stored in memory locations starting from $800.

E2.18 Write an instruction sequence to set bits three, two, one, and zero to one and clear the upper four bits.

E2.19 Find the values of condition flags N, Z, V, and C in the CCR register after the execution of each of the following instructions, given that [A] = $50 and the condition flags are N = 0, Z = 1, V = 0, and C = 1.

(a) SUBA #40	(b) TESTA	(c) ADDA #$50
(d) LSRA	(e) ROLA	(f) LSLA

E2.20 Find the values of condition flags N, Z, V, and C in the CCR register after executing each of the following instructions independently, given that [A] = $00 and the initial condition codes are N = 0, C = 0, Z = 1, and V = 0.

(a) TSTA	(b) ADDA #$40	(c) SUBA #$78
(d) LSLA	(e) ROLA	(f) ADDA #$CF

E2.21 Write an instruction sequence to toggle the odd number bits and clear the even number bits of the memory location at $66.

E2.22 Write a program to shift the 8-byte number located at $800-$807 to the left four places.

E2.23 Write a program to shift the 6-byte number located at $900-$905 to the right three places.

E2.24 Write a program to create a time delay of 100 seconds by using program loops.

E2.25 Write a program to create a time delay of five seconds using program loops.

3

68HC12 Members &
Hardware & Software
Development Tools

3.1 Objectives

After completing this chapter you should be able to:

- explain the differences between different 68HC12 members
- know the peripheral functions available at different 68HC12 members
- understand the types of hardware and software development tools available
- explain the functions of a source-level debugger
- know the features of the 68HC12 demo boards manufactured by Motorola
- know the features of the 68HC12 demo boards manufactured by Axiom
- use the D-Bug12 commands to view and change the contents of memory locations and CPU registers
- use the D-Bug12 commands to set breakpoints and trace program execution on the demo board
- use the MiniIDE program to enter, assemble, and download programs onto the CME-12BC demo board for execution

3.2 68HC12 Members

The 68HC12 was designed as an upgrade to the 8-bit 68HC11 microcontroller. However, the 68HC12 instructions are encoded differently from those of the 68HC11. The 68HC12 and 68HC11 are compatible on the source level only. All programs written for the 68HC11 can be reassembled or recompiled and run on the 68HC12 without modification, but the machine code for the 68HC11 cannot run on the 68HC12 hardware.

Like any other microcontroller family, the main differences among the 68HC12 members are in their on-chip peripheral functions. A summary of peripheral functions implemented by different 68HC12 members is listed in Table 3.1.

Using flash memory to hold application programs has become the trend of microcontroller applications. Most 68HC12 members incorporate a certain amount of flash memory.

Timer functions are very important in microcontroller applications. All members have incorporated *input-capture* (IC) and *output-compare* (OC) functions. The input-capture function allows the microcontroller to capture the arrival time of a signal edge. Pulse-width, period, duty cycle, and phase shift can be measured by using this function. The output compare function allows us to create time delays, trigger pin actions (toggle, pull high or low), generate waveforms, and so on. The pulse accumulation function can be used to count the number of events occurred within a certain time interval, measure frequency, and so forth.

Pulse width modulation (PWM) can be used to create a digital waveform with duty cycle ranges from zero to one hundred percent. Certain control applications require the voltage magnitude to be varied in order to perform their control operation. These applications do not respond quickly enough to the instantaneous change of the applied voltage. The variation of duty cycles becomes equivalent to the change of the average magnitude of the applied voltage. DC motor speed control is such an example.

All 68HC12 members except the 68HC812A4 implement one or two A/D converters with 10-bit resolution. By using an appropriate transducer to convert a non-electric quantity to a voltage, data acquisition applications can be implemented with the 68HC12. The 68HC812A4 implements an A/D converter with 8-bit resolution.

The 68HC12 members provide several serial interface functions to serve the needs of different applications. The *serial communication interface* (SCI) is designed to support asynchronous serial communication. This interface is mainly used for data communications that utilize the EIA232 (or RS232) standard. Most microprocessor demo boards use this interface to communicate with a PC.

The *serial peripheral interface* (SPI) is a synchronous serial interface that requires a clock signal to synchronize the data transfer between two devices. This interface is mainly used to interface with peripheral chips such as shift registers, seven-segment display and LCD drivers, A/D converters, D/A converters, SRAM and EEPROM, phase-lock-loop chips, and so on. These chips must also implement the SPI function in order to communicate with the 68HC12.

The *Inter-Integrated Circuit* (I²C) is a serial interface standard proposed by Philips. This interface has been very popular in the 8051/8052 microcontrollers. Like the SPI, the I²C interface was proposed to allow microcontrollers and peripheral chips to exchange data. The 912DG128, the 912DT128, and the 9S12DP256 support the I²C interface.

The *byte data link communication module* (BDLC) provides access to an external serial communication multiplex bus, and operates according to the SAE J1850 protocol. The SAE J1850 is proposed for low-speed (≤ 125Kbps) data communications in automotive applications. Only the 912B32, the 912BE32, and the 9S12DP256 support this protocol.

Several 68HC12 members implement the CAN 2.0 A/B protocol which is defined in the specification from Robert Bosch GmbH dated September 1991. The CAN protocol was primar-

Member	ROM (KB)	RAM (KB)	EEPROM (bytes)	Flash (KB)	Timer	I/O	Serial	A/D	PWM
68HC812A4	0	1	4K	0	8-ch 16-bit (IC or OC) RTI, pulse accumulator	up to 91	dual SCI, SPI	8-ch 8-bit	none
68HC912B32	0	1	768	32	8-ch 16-bit (IC or OC) RTI, pulse accumulator	up to 63	SCI, SPI, BDLC	8-ch 10-bit	4-ch 8-bit or 2-ch 16-bit
68HC912BC32	0	1	768	32	8-ch 16-bit (IC or OC) RTI, pulse accumulator	up to 63	SCI, SPI, CAN	8-ch 10-bit	4-ch 8-bit or 2-ch 16-bit
68HC912BE32	32	1	768	0	8-ch 16-bit (IC or OC) RTI, pulse accumulator	up to 63	SCI, SPI, BDLC	8-ch 10-bit	4-ch 8-bit or 2-ch 16-bit
68HC12D60	60	2	1K	0	8-ch 16-bit (IC or OC) RTI, pulse accumulator	up to 63 & 18 I	dual SCI, SPI, CAN	16-ch 10-bit	4-ch 8-bit or 2-ch 16-bit
68HC912D60	0	2	1K	60	8-ch 16-bit (IC or OC) RTI, pulse accumulator	up to 68 & 18 I	dual SCI, SPI, CAN	16-ch 10-bit	4-ch 8-bit or 2-ch 16-bit
68HC912DG128	0	8	2K	128	8-ch 16-bit (IC or OC) RTI, pulse accumulator	up to 68 & 18 I	dual SCI, SPI, I2C, 2 CAN	16-ch 10-bit	4-ch 8-bit or 2-ch 16-bit
68HC912DT128	0	8	2K	128	8-ch 16-bit (IC or OC) RTI, pulse accumulator	up to 68 & 18 I	dual SCI, SPI, I2C, 3 CAN	16-ch 10-bit	4-ch 8-bit or 2-ch 16-bit
MC9S12DP256	0	12	4K	256	8 channels of IC/OC	29	SCI, SPI, 5 CAN, I2C, BDLC	2 8-ch 10-bit	8-ch 8-bit or 4-ch 16-bit

Table 3.1 ■ Features of the 68HC12 family of products

ily designed to be used as a vehicle serial data bus, meeting the specific requirements of this field such as:

- Real-time processing
- Reliable operation in the electromagnetic interference (EMI) environment of a vehicle
- Cost effectiveness
- Required high bandwidth

The last member, 9SDP256, shown in Table 3.1, is the first member of the Star12 family (denoted as HCS12). This family shares the same instruction set and addressing modes with the 68HC12 family and implements the same peripheral functions. The main difference between them is the HCS12 family members use a more advanced semiconductor technology (0.25 µm) to implement the chip and run at higher bus frequency (25MHz compared to the 8MHz of the 68HC12 family). The HCS12 is still under active development at the time of this writing. The main focus of this text is the original 68HC12 family members.

3.3   Development Tools for the 68HC12

Development tools are essential for those who want to learn the 68HC12 or develop products based on the 68HC12. There are software and hardware development tools available to help us learn the 68HC12 and develop products that use the 68HC12.

Function generators, oscilloscopes, in-circuit emulators, logic analyzers, and demo boards are the most important hardware development tools. In this text we will discuss the use of demo boards only. In-circuit emulators and logic analyzers are very useful for the hardware development. However, in-circuit emulator and logic analyzers are relatively expensive debugging tools. The 68HC12 microcontroller has a *background debug mode* (BDM) that allows us to trace instruction executions on the target hardware (a 68HC12 microcontroller) from a PC or another 68HC12 running appropriate software. When performing debug activities in the BDM mode, the PC or the host 68HC12 communicates with the target 68HC12 via the BDM serial interface. This approach allows debug activities to be performed less intrusively. The BDM mode will be discussed in Chapter 11.

Software tools include text editors, terminal programs, cross assemblers, cross compilers, simulators, debuggers, and an integrated development environment. Text editors are used to enter and edit the programs. Any text editor can be used. A serious developer, however, will consider using a programmer's editor to improve the program entering productivity because a programmer's editor allows the user to perform operations such as keyword completion, syntax checking, parenthesis matching, keyword highlighting, and all the editing functions available in an ordinary text editor.

A *communication program* (also called a *terminal* program) allows the user to communicate a PC with the demo board and download the program onto a demo board for execution. Some demo boards require special software to communicate with the PC. However, the demo boards that we will deal with can communicate with the PC with a general-purpose terminal program such as the HyperTerminal bundled with the Windows 95/98/ME/NT operating system, Kermit, or ProComm.

Cross assemblers and *compilers* translate source programs into executable machine instructions. Cross assemblers and cross compilers generate executable code to be placed in the ROM, the EPROM, or the EEPROM of a 68HC12-based product. There are several freeware cross assemblers for the 68HC12. According to a survey posted in the Web site www.micro-controller.com more than 80% of the embedded applications are written in the C language. Other high-level languages pale by comparison. A tutorial on C will be given in Chapter 5. We will use the C compiler (ICC12) from ImageCraft to test all the C programs in this book.

A *simulator* allows the user to execute microcontroller programs without having the actual hardware. It uses computer memory to represent microcontroller registers and memory locations. The simulator interprets each microcontroller instruction by performing the operation required by the instruction, and then saves execution results in the computer memory. The simulator also allows the user to set the contents of memory locations and registers before the simulation run starts.

The *source-level debugger* is a program that allows you to find problems in your code at the high-level language (such as C) or assembly language level. A debugger may have the option to run your program on the demo board or use a simulator. Like a simulator, a debugger can display the contents of registers and memory (internal and external), and program code in separate windows. With a debugger, all debugging activities are done at the source level. You can see the value change of a variable after a statement has been executed. You can set a breakpoint at a statement in a high-level language. A source-level debugger requires a lot of computation. A debugger may run slowly if it needs to simulate the microcontroller instruction set instead of using the actual hardware to run the program. A source-level debugger needs to communicate with the monitor program on the demo board in order to display the contents of CPU registers and memory locations, set or delete breakpoints, trace program execution, and so on. Since the monitor programs on different evaluation boards may not be the same, a source-level debugger may be used only with one type of demo board. The BDM mode of the 68HC12 offers an alternative for implementing the source-level debugger.

Ideally, *integrated development environment* (IDE) software would provide an environment that combines the following software tools so that the user could perform all development activities without needing to exit any program:

- a text editor
- a cross assembler and/or compiler
- a simulator
- a source-level debugger
- a communication program

A full-blown IDE is certainly very useful but can be extremely expensive. However, many of today's cross compiler and assembler vendors add a text editor, project manager, and a terminal program into their cross software to make their product more competitive. The ICC12 from ImageCraft is an example. There is also freeware of this type available. The *MiniIDE* from Mgtek combines an assembler, text editor, and a terminal program. The *Ax12w* utility program bundled with the CME-12BC demo board also has this flavor.

3.4 68HC12 Demo & Evaluation Boards

A *demo board* is designed for end users to become familiar with certain microcontrollers. A demo board is also called an *evaluation board*. Whatever it is called, a demo board usually contains a small monitor program that allows the user to:

- display the values of registers and memory locations
- set values to registers and memory locations
- set breakpoints
- trace program execution
- enter assembler programs directly to the demo board
- disassemble the machine code on the demo board
- program the on-chip EPROM, EEPROM, or flash memory
- execute the program downloaded or entered to the demo board

A demo board with an on-board monitor is well suited for learning a microcontroller. We will review demo boards from Motorola and Axiom Manufacturing Inc.

3.4.1 68HC12 Demo Boards from Motorola

Motorola has three evaluation boards that have the D-Bug12 monitor on them. The HC12A4EVB is the only demo board that has external SRAM for downloading user programs for execution.

The *HC12A4EVB* is designed for the M68HC812A4 microcontroller. The features of the HC12A4EVB are as follows:

- Single-supply +3 to +5V DC power input
- Two RS-232 interfaces
- Two memory sockets populated with two 32-KB EPROMs, containing the D-Bug12 monitor program
- Two memory sockets populated with two 8-KB SRAMs
- Support for up to 1 MB of program memory space and 512 KB of data space using optional memory configurations
- 16-MHz crystal-controlled clock oscillator
- MCU mode control
- Alternate execution from on-chip EEPROM
- Header connectors for access to the 68HC812A4 I/O pins
- Connector for background debug mode (BDM) interface pins
- Prototype expansion area for customized interfacing with the MCU

Compared to other demo boards made by Motorola, this demo board has:

1. 16KB external SRAM available for downloading dynamic programs
2. 8-channel, 8-bit resolution A/D converter (other demo boards have 10-bit resolution)
3. no pulse width modulation function

The *M68EVB912B32* demo board has the following features:

- Single-supply +3 to +5V DC power input
- RS-232 interface
- BDM mode interface for debugging a user's target system
- 16MHz crystal for 8-MHz bus operation
- Four 2×20 header connectors for access to I/O lines
- V_{PP}/V_{DD} selection
- D-Bug12 monitor is resident on the on-chip flash memory of the 68HC912B32 microcontroller
- Prototype expansion area for customized interfacing with the MCU
- Can be jumper-configured to:
  a. run a program directly out of EEPROM
  b. control a remote "pod" MCU via the background debug mode interface
  c. reprogram EEPROM on either the host EVB or the "pod"
- 512-bytes of on-chip SRAM of the 68HC912B32 can be used to download user program for direct execution
- On-chip 32KB flash memory can be erased and reprogrammed to hold the user's program (you will lose the D-Bug12 monitor by doing this)

The *M68EVB912BC32* demo board has the following features:

- Single-supply +3 to +5V DC power input
- RS-232 interface
- BDM mode interface for debugging a user's target system
- 16MHz crystal for 8-MHz bus operation
- Four 2×20 header connectors for access to I/O lines
- V_{PP}/V_{DD} selection
- D-Bug12 monitor is resident on the on-chip flash memory of the 68HC912B32 microcontroller
- Prototype expansion area for customized interfacing with the MCU
- Can be jumper-configured to:
  a. run a program directly out of EEPROM
  b. control a remote "pod" MCU via the background debug mode interface
  c. reprogram EEPROM on either the host EVB or the "pod"
- 512-bytes of on-chip SRAM of the 68HC912B32 can be used to download user program for direct execution
- On-chip 32KB flash memory can be erased and reprogrammed to hold user program (you will lose the D-Bug12 monitor by doing this)
- CAN interface

The *M68EVB912D60* demo board has the following features:

- Single 5V power supply
- RS-232 interface
- BDM connector providing interface to the background debug mode
- 16MHz crystal oscillator
- Prototyping expansion area
- CAN physical interface
- Charge pump for the supply of the flash programming voltage
- Bar graph LED to assist with debugging
- 8-way DIP switch to assist with debugging

The *M68EVB912DG128* demo board has the following features:

- Single 5V power supply
- RS-232 interface
- BDM In connector providing interface to background debug mode
- 16MHz crystal oscillator
- Prototyping expansion area
- CAN physical interface
- Charge pump for the supply of the flash programming voltage
- Bar graph LED to assist with debugging
- 8-way DIP switch to assist with debugging

A photograph of the 68EVB912B32 demo board is shown in Figure 3.1.

Figure 3.1 ■ Motorola 68EVB912B32 evaluation board

3.4.2 Demo Boards from Axiom Manufacturing

Axiom designs and manufactures several demo boards for the 68HC12. The common features of these demo boards are:

- D-Bug12 Monitor
- Serial cable
- Wall plug power adaptor with 9V/300 mA output
- LCD interface (a connector is provided)
- Keypad interface (a connector is provided)
- BDM background debug mode interface
- SPI serial interface port
- 16MHz crystal oscillator operation

There are differences among these demo boards. The major differences are in the microcontroller unit that they include and the amount of external SRAM and EPROM that they provide. The incorporated microcontroller unit dictates the peripheral functions and the amount of internal memories available in each demo board. A summary of the differences of these demo boards is shown in Table 3.2.

demo board	MCU	External SRAM	External EEPROM	data bus width
CMD-12A4	HC812A4	64KB	64KB	16-bit
CME-12A4	HC812A4	32KB	32KB	8-bit
CME-12B32	HC912B32	64KB	32KB	16-bit
CME-12BC32	HC912BC32	64KB	32KB	16-bit
CME-12D60	HC912D60	64KB	32KB	16-bit
CMD912-xx*	selectable	256KB	16KB	16-bit

Note. CMD912-xx allows the user to choose from PM12*DP256 / MC9S12DP256 MCU or PM12DG128 / MC68HC912DG128 MCU or PM12D60/MC68HC912D60 MCU

Table 3.2 ■ A summary of differences of Axiom demo boards

Photographs of the CME-12BC32 and CME-12D60 demo boards are shown in Figure 3.2 and Figure 3.3, respectively.

Figure 3.2 ■ CME-12BC demo board (reprinted with permission of Axiom Manufacturing Inc.)

Figure 3.3 ■ CME-D60 demo board (reprinted with permission of Axiom Manufacturing Inc.)

3.5 BDM-based Debuggers

All 68HC12 microcontrollers incorporate the *background debug mode* (BDM) that enables less intrusive debugging activities of software. The BDM hardware communicates with external devices serially, via the BKGD pin. It usually uses CPU idle bus cycles to execute debugging commands while the CPU is operating normally, but can steal cycles from the CPU when necessary.

There are many vendors that provide the BDM-based source-level debugger. We will discuss the BDM mode and BDM-based debuggers and provide a tutorial on using the Axiom AX-BDM12 debugger in Chapter 11. If you are eager to use it, feel free to read that tutorial and try it out any time. Using the AX-BDM12 debugger does not require the understanding of the internal operation of the BDM module.

3.6 The D-Bug12 Monitor

The D-Bug12 is a monitor program designed for the 68HC12 microcontrollers. This monitor is used in several 68HC12 demo boards from Motorola and Axiom Manufacturing. It facilitates the writing, evaluation, and debugging of user programs.

The D-Bug12 monitor provides a set of commands that allow the user to display and modify memory and register contents, download programs to the demo board for execution, set breakpoints, trace program execution, and so on. A summary of the D-Bug12 command set is shown in Table 3.3.

Command	Description
ASM <address>	Single line assembler/disassembler
<CR>	Disassemble next instruction
<.>	Exit assembly/disassembly
BAUD <baudrate>	Set communications rate for the terminal
BF <StartAddress> <EndAddress> [<data>]	Fill memory with data
BR [<Address>]	Set/Display user breakpoints
BULK	Erase entire on-chip EEPROM contents
CALL [<address>]	Execute a user subroutine; return to D-Bug12 when finished
DEVICE [see description]	Select/define a new target MCU device
EEBASE <Address>	Inform D-Bug12 of the Target's EEPROM base address
FBULK	Erase the target processor's on-chip flash EEPROM
FLOAD <AddressOffset>	Program the target processor's on-chip Flash EEPROM from S-records
G [<Address>]	Go-begin execution of user program
GT <Address>	Go-Till—set a temporary breakpoint and begin execution of user program
HELP	Display D-Bug12 command set and command syntax
LOAD [<AddressOffset>]	Load user program in S-record
MD <StartAddress> [<EndAddress>]	Memory display—display memory contents in hex bytes/ASCII format
MDW <StartAddress> [<EndAddress>]	Memory display word—display memory contents in hex words/ASCII format
MM <Address> [<data>]	Memory modify—interactively examine/change memory contents
MMW <address> [<data>]	Memory modify word—interactively examine/change memory contents
MOVE <StartAddress> <EndAddress> <DestAddress>	Move a block of memory
NOBR [<Address> <Address>...]	Remove individual user breakpoints
RD	Register display—display the CPU registers
REGBASE	Inform D-Bug12 of the target I/O register's base address
RESET	Reset the target CPU
RM	Register modify—interactively examine/change CPU register contents
STOP	Stop execution of user code on the target processor and place it in background
T [<count>]	Trace <count> instructions
UPLOAD <StartAddress><EndAddress>	S_Record Memory display
USEHBR	Use Hardware Breakpoints
VERF [<AddressOffset>]	Verify S-Records against memory contents
<Register Name><Register Value>	Set register contents
Register Names:	PC, SP, X, Y, A, B, D
CCR Status Bits:	S, XM, H, IM, N, Z, V, C

Table 3.3 ■ D-Bug12 command-set summary

3.7 Using the Axiom CME-12BC Demo Board

We will need to have at least the following software programs in order to develop assembly programs to be downloaded onto the CME-12BC demo board for execution:

1. A text editor
2. A 68HC12 cross assembler
3. A terminal program

We prefer using an IDE program in developing assembly programs. The MiniIDE from Mgtek is a well-designed IDE for developing assembly programs for the 68HC12 and 68HC11 microcontrollers.

The freeware MiniIDE allows you to enter, assemble, and download the S-record file onto a demo board for execution without quitting any one of them. MiniIDE can be downloaded from the Web site at www.mgtek.com. It expires at the end of each year. Users must download it once a year. By doing this, you get an update of MiniIDE.

3.7.1 Starting the MiniIDE

The MiniIDE can be started by clicking on its icon. The screen should look like Figure 3.4 after the MiniIDE is started. The lower half in Figure 3.4 is the status window.

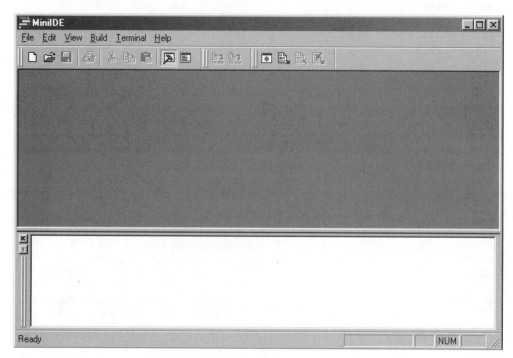

Figure 3.4 ■ MiniIDE startup screen

3.7.2 Communicating with the CME-12BC Demo Board

To communicate with the CME-12BC demo board, press the **Terminal** menu (as shown in Figure 3.5) and select **Show terminal window.** The terminal window will appear. Press the reset button (red button) of the CME-12BC demo board and the screen will change to what you see in Figure 3.6.

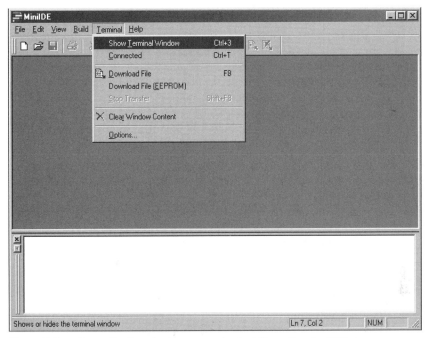

Figure 3.5 ■ Press the Terminal menu and select Show Terminal Window

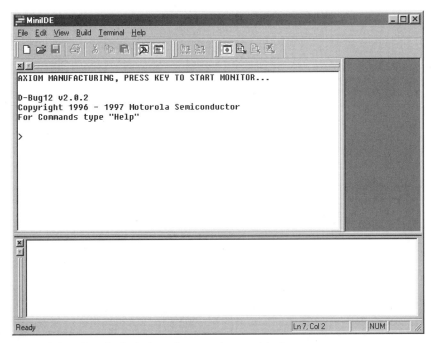

Figure 3.6 ■ Terminal window after pressing reset button

3.7.3 Using the D-Bug12 Commands

The D-Bug12 commands are provided to help the program-debugging process. Some of these commands can be used to set and display the contents of memory locations. In this section, character strings entered by the user appear in boldface and optional fields are enclosed in brackets [ ]. In the following section, the syntax of a command is presented and then examples are given.

BF <StartAddress> <EndAddress> [<Data>]

The *BF* command is used to fill a block of memory locations with the same value. For example, the following command will clear the internal memory locations from $800 to $8FF to zero:

>BF 800 8FF 0

The data field is optional. If we did not specify the data to be filled, then zero will be filled in the specified memory locations.

MD <StartAddress> [<EndAddress>]

The *MD* command (*memory display*) is used to display memory contents. This command displays memory contents as both hexadecimal bytes and ASCII characters, 16 bytes on each line. The *<StartAddress>* parameter must be supplied; the *<EndAddress>* parameter is optional. When the <EndAddress> parameter is not specified, a single line is displayed.

The number supplied as the <StartAddress> parameter is rounded down to the next lower multiple of 16, while the number supplied as the <EndAddress> parameter is rounded up to the next higher multiple of 16 minus 1. This causes each line to display memory in the range of $xxx0 to $xxxF. For example, if **$805** is entered as the start address and **$820** as the ending address, the actual memory range displayed would be $800 through $82F. This command allows us to examine the program execution result.

Example 3.1
▼

```
>md 800

0800   00 00 00 00 - 00 00 00 00 - 00 00 00 00 - 00 00 00 00      ................

>md 805 820

0800   00 00 00 00 - 00 00 00 00 - 00 00 00 00 - 00 00 00 00      ................
0810   00 00 00 00 - 00 00 00 00 - 00 00 00 00 - 00 00 00 00      ................
0820   00 00 00 00 - 00 00 00 00 - 00 00 00 00 - 00 00 00 00      ................
```

▲

MDW <StartAddress> [<EndAddress>]

This command (*memory display words*) displays the contents of memory locations as hexadecimal words and ASCII characters, 16 bytes on each line. The *<StartAddress>* parameter must be supplied; the *<EndAddress>* parameter is optional. When the <EndAddress> parameter is not supplied, a single line is displayed.

The number supplied as the <StartAddress> parameter is rounded down to the next lower multiple of 16, while the number supplied as the <EndAddress> parameter is rounded up to the next higher multiple of 16 minus 1. This causes each line to display memory in the range of $xxx0 through $xxxF. This command allows us to examine the program execution result.

Example 3.2

▼

```
>mdw 1000

1000  FC08 00CD - 0900 CE00 - 0A18 10CB - 306B 44B7     ............0kD.
>mdw 1000 1020

1000  FC08 00CD - 0900 CE00 - 0A18 10CB - 306B 44B7     ............0kD.
1010  C5CE 000A - 1810 CB30 - 6B43 B7C5 - CE00 0A18     .......0kC......
1020  10CB 306B - 42B7 C5CE - 000A 1810 - CB30 6B41     ..0kB........0kA
>
```

▲

MM <Address> [<Data>]

This command (*memory modify*) allows us to examine and modify the contents of memory locations one byte at a time. If the 8-bit data parameter is present on the command line, the byte at memory location *<Address>* is replaced with *<Data>* and the command is terminated. If no optional data is provided, then D-Bug12 enters the *interactive memory modify mode*. In the interactive mode, each byte is displayed on a separate line following the address of data. Once the memory modify command has been entered, single-character sub-commands are used for the modification and verification of memory contents. These sub-commands have the following format:

[<Data>] <CR>	Optionally update current location and display the next location.
[<Data>] </> or <=>	Optionally update current location and redisplay the same location.
[<Data>] <^> or <->	Optionally update current location and display the previous location.
[<Data>] <.>	Optionally update current location and exit Memory Modify.

With the exception of the carriage return (CR), the sub-command must be separated from any entered data with at least one space character. If an invalid sub-command character is entered, an appropriate error message is issued and the contents of the current memory location are redisplayed.

Example 3.3

▼

In this example, each line is terminated with a carriage return character. However, the character return character is non-displayable and hence is not shown.

```
>mm 800
0800 00
0801 00 FF
0802 00 ^
0801 FF
0802 00
0803 00 55 /
0803 55 .
>
```

▲

MMW <Address> [<Data>]

This command (*memory modify, word*) allows the contents of memory to be examined and/or modified as 16-bit hex data. If the 16-bit data is present on the command line, the word at memory location <Address> is replaced with <Data> and the command is terminated. If not, D-Bug12 enters the interactive memory modify mode. In the interactive mode, each word is displayed on a separate line following the address of the data. Once the memory modify command has been entered, single-character sub-commands are used for the modification and verification of memory contents. These sub-commands have the following format:

[<Data>] <CR>	Optionally update current location and display the next location.
[<Data>] </> or <=>	Optionally update current location and redisplay the current location.
[<Data>] <^> or <->	Optionally update current location and display the previous location.
[<Data>] <.>	Optionally update current location and exit Memory Modify.

With the exception of the carriage return (CR), the sub-command must be separated from any entered data with at least one space character. If an invalid sub-command character is entered, an appropriate error message is issued and the contents of the current memory location are redisplayed.

Example3.4

▼

In this example, each line is terminated with a carriage return character. However, the carriage-return character is non-displayable and hence is not shown.

```
>mmw 900
0900 00F0
0902 AA55 0008
0904 0000 ^
0902 0008 aabb
0904 0000
0906 0000 .
>
```

▲

Move <StartAddress> <EndAddress> <DestAddress>

This command (*move memory block*) is used to move a block of memory from one location to another, one byte at a time. Addresses are specified in 16-bit hex values. The number of bytes moved is one more than <EndAddress> minus <StartAddress>. The block of memory beginning at the destination address may overlap the memory block defined by <StartAddress> and <EndAddress>.

One of the uses of the *move* command might be to copy a program from RAM into the on-chip EEPROM.

Example 3.5

```
>move 800 8ff 900
>
```

RD

This command (*register display*) is used to display the 68HC12 CPU registers.

Example 3.6

```
>rd

 PC    SP    X     Y     D = A:B   CCR = SXHI NZVC
0000  0A00  0000  0000    00:00          1001 0000
>
```

RM

This command (*register modify*) is used to examine or modify the contents of the CPU12 registers interactively. As each register and its contents are displayed, D-Bug12 allows the user to enter a new value for the register in hex. If modification of the displayed register is not desired, entering a carriage return will cause the next CPU register and its contents to be displayed on the next line. When the last of the CPU registers has been examined or modified, the RM command displays the first register, giving the user an opportunity to make additional modifications to the CPU register contents. Typing a period as the first non-space character on the line will exit the interactive mode of the register modify command and return you to the D-Bug12 command prompt. The registers are displayed in the following order, one register per line: PC, SP, X, Y, A, B, CCR.

Example 3.7

```
>rm

PC=0000 1000
SP=0A00
IX=0000 0100
IY=0000
A=00
B=00 ff
CCR=90 d1
PC=1000 .
>
```

<RegisterName> <RegisterValue>

This command allows us to change the value of any CPU register (PC, SP, X, Y, A, B, D, or CCR). Each of the fields in the CCR may be modified by using the bit names shown in Table 3.4.

CCR bit name	Description	Legal Values
S	STOP enable	0 or 1
H	Half carry	0 or 1
N	Negative flag	0 or 1
Z	Zero flag	0 or 1
V	Two's complement over flg	0 or 1
C	Carry flag	0 or 1
IM	IRQ interrupt mask	0 or 1
XM	XIRQ interrupt mask	0 or 1

Table 3.4 ■ Condition code register bits

Examples 3.8

▼

```
>pc 2000

PC    SP    X     Y     D = A:B   CCR = SXHI NZVC
2000  0A00  0100  0000    00:FF         1101 0001
>x 800

PC    SP    X     Y     D = A:B   CCR = SXHI NZVC
2000  0A00  0800  0000    00:FF         1101 0001
>c 0

PC    SP    X     Y     D = A:B   CCR = SXHI NZVC
2000  0A00  0800  0000    00:FF         1101 0000
>z 1

PC    SP    X     Y     D = A:B   CCR = SXHI NZVC
2000  0A00  0800  0000    00:FF         1101 0100
>d 2010

PC    SP    X     Y     D = A:B   CCR = SXHI NZVC
2000  0A00  0800  0000    20:10         1101 0100
>
```

▲

ASM <Address>

This command invokes the one-line assembler/disassembler. It allows memory contents to be viewed and altered using assembly language mnemonics. Each entered source line is translated into object code and placed into memory at the time of entry. When displaying memory contents, each instruction is disassembled into its source mnemonic form and displayed along with the hex object code and any instruction operands.

Assembly mnemonics and operands may be entered in any mix of upper- and lower-case letters. Any number of spaces may appear between the assembler prompt and the instruction mnemonic or between the instruction mnemonic and the operand. Numeric values appearing in the operand field are interpreted as *signed* decimal numbers. Placing a $ in front of any number will cause the number to be interpreted as a hex number.

When an instruction is disassembled and displayed, the D-Bug12 prompt is displayed following the disassembled instruction. If a carriage return is the first non-space character entered following the prompt, the next instruction in memory is disassembled and displayed on the next line.

If a 68HC12 instruction is entered following the prompt, the entered instruction is assembled and placed into memory. The line containing the new entry is erased and the new instruction is disassembled and displayed on the same line. The next instruction location is then disassembled and displayed on the screen.

When entering branch instructions, the number placed in the operand field should be the absolute destination address of the instruction. The assembler calculates the two's complement offset of the branch and places the offset in memory with the instruction.

The assembly/disassembly process may be terminated by entering a period as the first non-space character following the assembler prompt.

The following example displays the assembly instructions from memory location $1000 ~ $1011. The carriage return character is entered at the > prompt of each line and the period character is entered at the last line.

```
>asm 1000
1000  FC0800      LDD    $0800              >
1003  CD0900      LDY    #$0900             >
1006  CE000A      LDX    #$000A             >
1009  1810        IDIV                      >
100B  CB30        ADDB   #$30               >
100D  6B44        STAB   4,Y                >
100F  B7C5        XGDX                       >
1011  CE000A      LDX    #$000A             >.
>
```

The following example enters a short program that consists of three instructions starting from memory location $1100:

```
>asm 1100
1100  FC0800      LDD    $0800
1103  F30802      ADDD   $0802
1106  7C0900      STD    $0900
1109  E78C        TST    12,SP              >.
>
```

BR [<Address> <Address> ...]

This command (*breakpoint set*) sets a breakpoint at a specified address or displays any previously set breakpoints. The function of a breakpoint is to halt user program execution when the program reaches the breakpoint address. When a breakpoint address is encountered, D-Bug12 disassembles the instruction at the breakpoint address, prints the CPU register contents, and waits for a D-Bug12 command to be entered by the user.

Breakpoints are set by typing the breakpoint command followed by one or more breakpoint addresses. Entering the breakpoint command without any breakpoint addresses will display all the currently set breakpoints.

A maximum of ten user breakpoints may be set at one time. Whenever our program is not working correctly and we suspect that the instruction at certain memory location is incorrect, we can set a breakpoint at that location and check the execution result by looking at the contents of the CPU registers or memory locations.

Example 3.9

▼

```
>br 1020 1040 1050                          ; set three breakpoints
Breakpoints: 1020   1040   1050
>br                                         ; display current breakpoints
Breakpoints: 1020   1040   1050
>
```

▲

NOBR [<Address> <Address>]

This command removes one or more previously entered breakpoints. If the NOBR command is entered without any argument, all user breakpoints are removed from the breakpoint table.

Example 3.10

▼

```
>br 1000 1010 1020 1040 1090
Breakpoints: 1000   1010   1020   1040   1090
>nobr 1000 1010
Breakpoints: 1020   1040   1090
>nobr
All Breakpoints Removed
>
```

▲

G [<Address>]

This command is used to begin execution of user code in real time. Before beginning the execution of user code, any breakpoints that were set with the BR command are placed in memory. Execution of the user program continues until a user breakpoint is encountered, a CPU exception occurs, the STOP or RESET command is entered, or the demo board's reset switch is pressed.

When the user code halts for any of these reasons and control is returned to D-Bug12, a message is displayed explaining the reason for user program termination. In addition, D-Bug12 disassembles the instruction at the current PC address, prints the CPU register contents, and waits for the next D-Bug12 command to be entered by the user.

If the starting address is not supplied in the command line parameter, program execution will begin at the address defined by the current value of the program counter.

Example 3.11

▼

```
>g 1000
User Breakpoint Encountered

 PC    SP    X     Y     D = A:B   CCR = SXHI NZVC
1012  0A00  1030  0000    01:13          1001 1001
1012  A600        LDAA  0,X

>
```

▲

GT <Address>

This command (*go till*) is similar to the G command except that a temporary breakpoint is placed at the address supplied at the command line. Any breakpoints that were set by the use of the BR command are *not* placed in the user code before program execution begins. Program execution begins at the address defined by the current value of the program counter. When user code reaches the temporary breakpoint and control is returned to D-Bug12, a message is displayed explaining the reason for user program termination. In addition, D-Bug12 disassembles the instruction at the current PC address, prints the CPU register contents, and waits for a command to be entered by the user.

Example 3.12

▼

```
>br 1010
Breakpoints: 1010
>pc 1000

PC    SP    X     Y      D = A:B    CCR = SXHI NZVC
1000  0A00  0032  0900      00:31         1001 0000
>gt 1036
Temporary Breakpoint Encountered

PC    SP    X     Y      D = A:B    CCR = SXHI NZVC
1036  0A00  0032  0900      00:31         1001 0000
1036  3F             SWI

>
```

▲

T [<count>]

This command (*trace*) is used to execute one or more user program instructions beginning at the current program counter location. As each program instruction is executed, the CPU register contents and the next instruction to be executed are displayed. A single instruction may be executed by entering the trace command immediately followed by a carriage return.

Because of the method used to execute a single instruction, branch instructions (Bcc, LBcc, BRSET, BRCLR, DBEQ/NE, IBEQ/NE, TBEQ/NE) that contain an offset that branches back to the instruction opcode do not execute. D-Bug12 appears to become stuck at the branch instruction and does not execute the instruction even if the condition for the branch instruction is satisfied. This limitation can be overcome by using the GT command to set a temporary breakpoint at the instruction following the branch instruction.

Example 3.13

▼

```
>pc 1000

PC    SP    X     Y      D = A:B    CCR = SXHI NZVC
1000  0A00  04D2  0900      00:05         1001 0000
>t                                                           ; trace one instruction

PC    SP    X     Y      D = A:B    CCR = SXHI NZVC
```

```
1003   0A00   04D2   0900      30:39      1001  0000
1003   CD0900         LDY    #$0900
>t 2                                              ; trace two instructions

 PC    SP     X      Y     D = A:B    CCR = SXHI NZVC
1006   0A00   04D2   0900      30:39      1001  0000
1006   CE000A         LDX    #$000A

 PC    SP     X      Y     D = A:B    CCR = SXHI NZVC
1009   0A00   000A   0900      30:39      1001  0000
1009   1810           IDIV
>
```

The first command in this example sets the program counter to $1000 so that we know where the program execution starts. This command is normally needed when tracing a program.

CALL [<Address>]

This command is used to execute a subroutine and returns to the D-Bug12 monitor program when the final RTS of the subroutine is executed. All CPU registers contain the values at the time the final RTS instruction was executed, with the exception of the program counter. The program counter contains the starting address of the subroutine. If a subroutine address is not supplied on the command line, the current value of the program counter is used as the starting address.

No user breakpoints are placed in memory before execution is transferred to subroutine. If the called subroutine modifies the value of the stack pointer during its execution, it must restore the stack pointer's original value before executing the RTS instruction. This restriction is required because a return address is placed on the user's stack that returns to D-Bug12 when the final RTS of the subroutine is executed. Obviously, any subroutine must obey this restriction in order to execute properly.

Example 3.14

```
>call 1000
Subroutine Call Returned

 PC    SP     X      Y     D = A:B    CCR = SXHI NZVC
1000   0A00   0032   0900      00:31      1001  0000
1000   FC0800         LDD    $0800


>
```

This command is useful for testing a subroutine without writing a testing program.

LOAD [<AddressOffset>]

This command is used to load S-record objects into memory from an external device. The address offset, if supplied, is added to the load address of each S-record before its data bytes are placed in memory. Providing an address offset other than zero allows object code or data to be loaded into memory at a location other than that for which it was assembled. During the loading process, the S-record data is not echoed to the control console. However, for each ten S-records that are successfully loaded, an ASCII asterisk character (*) is sent to the control console. When an S-record file has been successfully loaded, control returns to the D-Bug12 prompt.

The *load* command is terminated when D-Bug12 receives an S9 end-of-file record. If the object being loaded does not contain an S9 record, D-Bug12 does not return its prompt and continues to wait for the end-of-file record. Pressing the reset switch returns D-Bug12 to its command line prompt.

Example 3.15

▼

```
>load
*
>
```

This example downloads a file that is equal to or shorter than ten S-records, and hence only one * character is displayed on the screen. In addition to entering the **load** command followed by a carriage return, we also need to go back to the terminal program to specify the file to be downloaded. An example for selecting the **Download File** command in MiniIDE is shown in Figure 3.7.

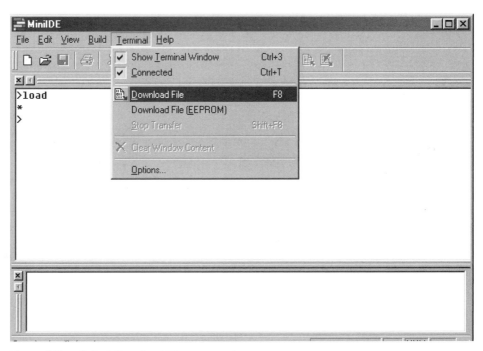

Figure 3.7 ■ Select Download File command

After the **Download File** command is selected, a popup window will appear, as shown in Figure 3.8. This window allows you to specify the file to be downloaded. Click on the **Open** button on the pop-up window and the file will be transferred to the demo board.

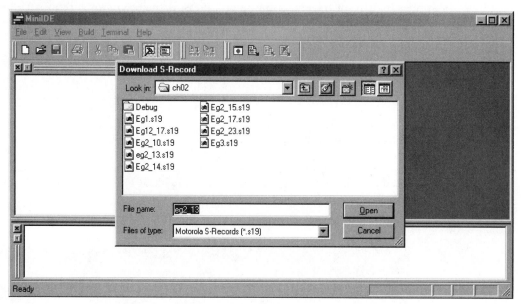

Figure 3.8 ■ Select an S-record file to be downloaded

3.7.4 Entering an Assembly Program

When a program is first created, we need to open a new file to hold it. To do that, we press the **File** menu from the MiniIDE window and select **New**, as shown in Figure 3.9. After we select the **New** command from the menu, the screen will change to what is displayed in Figure 3.10. In Figure 3.10, both the terminal (at the left-hand side) and editor windows (at the right-hand side) are open and probably not very convenient to work with. The terminal window can be closed by clicking on the × symbol on the upper-left corner of the terminal window. You may want to adjust the size of the text editor window before entering the program.

After adjusting the window size, we enter the program that converts a hex number into BCD digits in Example 2.13. This program performs the repeated divide-by-ten operation to the given number and adds hex number $30 to each remainder to convert it to its corresponding ASCII code. The result is shown in Figure 3.11.

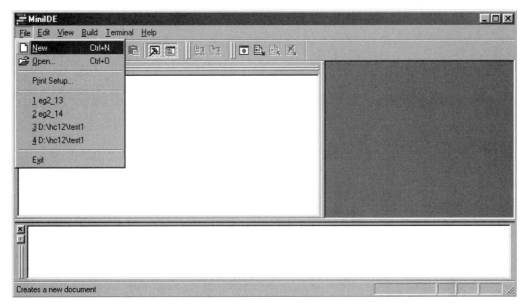

Figure 3.9 ■ Select New from the File menu to create a new file

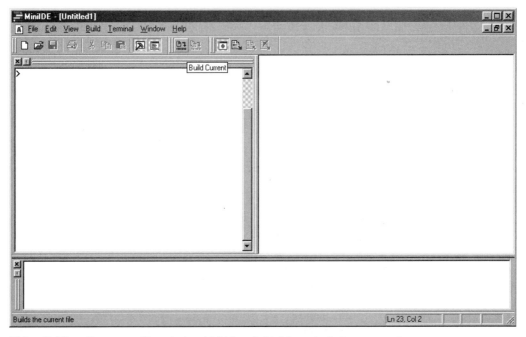

Figure 3.10 ■ Open an editor window (right hand side) for entering a new program

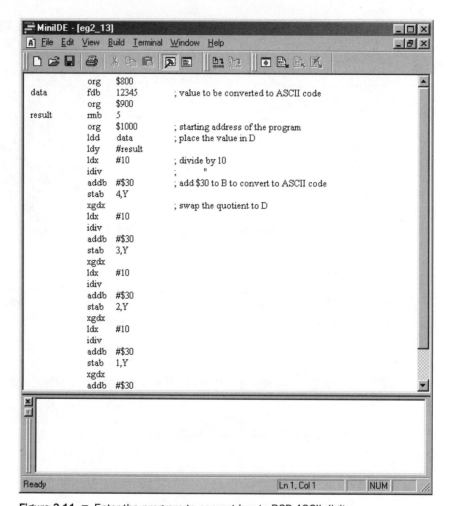

Figure 3.11 ■ Enter the program to convert hex to BCD ASCII digits

3.7.5 Assembling the Program

To assemble an assembly program, press the **Build** menu and select **Build eg2_13.asm** as shown in Figure 3.12. If the program is assembled successfully, the status window will display the corresponding message as shown in Figure 3.13. Two warnings appear in the message window. These two warnings concern the fact that the *end* directive is not supported by the cross assembler used in the MiniIDE program.

After the program is assembled successfully, you are ready to run and debug the program. The output of the assembly process is an S-record file, *eg3_13.s19*. The file name has a suffix **s19**. The S-record format is a common file format defined by Motorola to allow tools from different vendors to work on the same project.

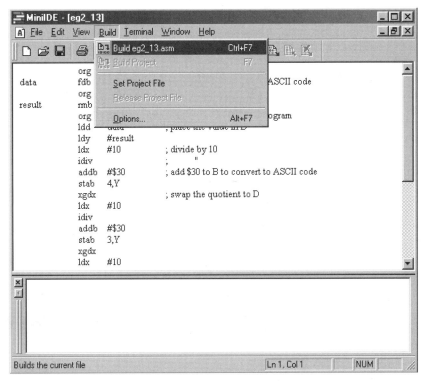

Figure 3.12 ■ Prepare to assemble the program eg2_13.asm

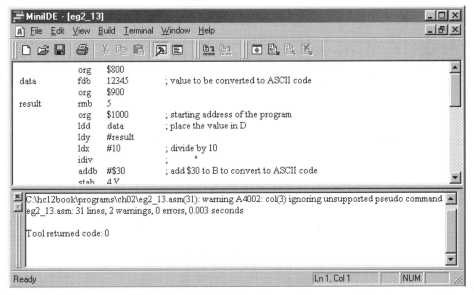

Figure 3.13 ■ Status window message shows that the previous compilation is successful

3.7.6 Running & Debugging the Program

We need to reopen the terminal window and download the program onto the demo board before it can be run and debugged. The screen after the program download will look like Figure 3.14.

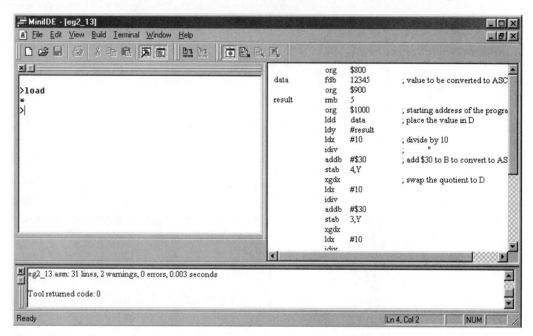

Figure 3.14 ■ Screen after downloading the program eg2_13 onto the CME-12BC demo board

Before we run the program we should verify that the program data is downloaded correctly into the memory. In this program, our program data is the decimal number 12,345, which corresponds to the hex number $3039. We can use the command *md 800* to verify it. The contents of the memory locations $800 and $801 displayed by the D-Bug12 monitor should be $3039. Otherwise, some error might have occurred.

We should also make sure that our program has been downloaded into the right memory area. This can be verified by using the *asm 1000* command (the eg2_13 program starts at $1000). The first few lines should look like this:

```
>asm 1000
1000   FC0800      LDD    $0800              >
1003   CD0900      LDY    #$0900             >
1006   CE000A      LDX    #$000A             >
1009   1810        IDIV                      >
100B   CB30        ADDB   #$30               >
100D   6B44        STAB   4,Y
```

After making sure that the program has been downloaded correctly, we will execute the program without setting any breakpoints and check the result to see if it is correct. The screen should appear like Figure 3.15.

The second line of the terminal window that contains the statement *User Breakpoint Encountered* is caused by the *software interrupt* (SWI) instruction. The SWI instruction caused

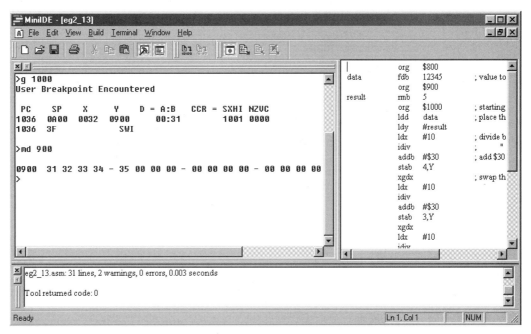

Figure 3.15 ■ Screen after running the eg2_13 program and display the contents at $900-$90F

the program control to be returned to the D-Bug12 monitor. The SWI instruction is often used as the last instruction of a program to be executed on the CME-12BC demo board.

The last line of the terminal window displays the contents of the memory locations $900 to $904. These five bytes represent the ASCII codes of one, two, three, four, and five. Therefore, the program executes correctly.

If our program does not work correctly, we can set breakpoints at locations that we have suspicions about or even trace the execution of some instructions to pinpoint the error.

Suppose the execution result (we take out the first *xgdx* instruction) appears as follows:

```
>md 900

0900   30 30 35 33 - 35 00 00 00 - 00 00 00 00 - 00 00 00 00    00535...........
>
```

We discovered that the first four digits are incorrect. Since this program is short, we can trace through it. One approach is like this:

Step 1
Trace through the first six instructions and check to see if the quotient (in index register X) and the memory contents at $904 are correct. The last seven lines of the terminal window are:

```
PC  SP  X   Y   D = A:B  CCR = SXHI NZVC
100F 0A00 04D2 0900   00:35     1001 0000
100F CE000A    LDX #$000A
>md 900

0900 00 00 00 00 - 35 00 00 00 - 00 00 00 00 - 00 00 00 00   ....5...........
>
```

These six lines tell us that:

- The index register contains hex value $04D2 (equal to decimal 1234) and is the correct quotient.
- The memory location at $904 contains hex value $35 and is the ASCII code of five.
- In the next division, the number to be divided by ten will be 1234.
- The next instruction to be executed is *LDX #$0A*.

Step 2

Trace the next two instructions. Oops! We find one error. The double accumulator D does not contain the value 1234 (it contains $35 instead) before the IDIV instruction is executed. We forgot to swap the value in index register X with the double accumulator D before performing the second division!

```
>t

PC  SP  X   Y   D = A:B  CCR = SXHI NZVC
1012 0A00 000A 0900   00:35    1001 0000
1012 1810      IDIV
>t

PC  SP  X   Y   D = A:B  CCR = SXHI NZVC
1014 0A00 0005 0900   00:03    1001 0000
1014 CB30      ADDB #$30
>
```

Step 3

Fix the error by inserting the *xgdx* instruction before the *ldx #$0a* instruction and re-run the program. After re-running the program, check the contents at $900 to $904 again. The last few lines on the screen of the terminal window will appear as follows:

```
>load
*
>g 1000
User Breakpoint Encountered

PC  SP  X   Y   D = A:B  CCR = SXHI NZVC
1036 0A00 0032 0900   00:31    1001 0000
1036 3F       SWI

>md 900

0900  31 32 33 34 - 35 00 00 00 - 00 00 00 00 - 00 00 00 00   12345..........
>
```

The memory locations from $900 to $904 contain correct values. So, we have fixed the error. Most debug sessions are similar to this example. Of course, longer programs will take more time to resolve and require us to use more commands to try different things before we can fix the bugs.

3.7.7 The CME-12BC Memory Map

You need to understand the memory map of the CME-12BC demo board in order to be able to download your program to the proper memory space for testing. The memory map of the CME-12BC demo board is shown in Figure 3.16.

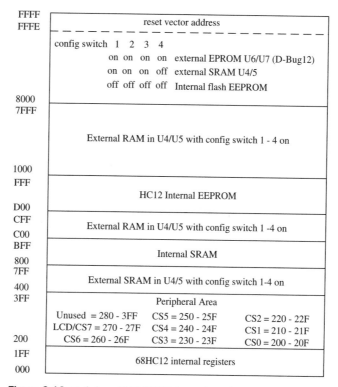

Figure 3.16 ■ Axiom CME-12BC demo board memory map

The 68HC912BC32 has several operation modes (discussed in Chapter 6). At this point, it is enough to know that after power on reset the CME-12BC demo board will be configured (with config switch 1 – 4 all on) to operate in *normal expanded wide* mode. In this mode, the memory space from $8000 to $FFFF (32KB) is assigned to external EPROM. The D-Bug12 monitor is located in this area. The CME-12BC has external SRAM chips that are configured to occupy the memory space from $0000 to $7FFF. However, several regions have been occupied by other devices:

1. $D00-$FFF: occupied by HC12 internal EEPROM. The 68HC912BC32 has 768 bytes of EEPROM located in this area. This EEPROM can be enabled or disabled by setting or clearing the EEON bit of the INITEE register. When this EEPROM is enabled, the memory space $D00-$FFF is assigned to EEPROM and the memory locations of the external SRAM in the same range are inaccessible.

2. $800-$BFF: occupied by internal SRAM. Internal SRAM is always enabled and hence the external SRAM locations in this area are invisible to the user.

3. $200-$3FF: This region is designed to be used by peripheral memory mapped devices. Users of the CME-12BC demo board can add peripheral devices as memory-mapped devices to the board. Eight chip-select signals have been provided.

4. $000-$1FF: This region is occupied by the internal peripheral registers and hence is not available for external SRAM. Any access to addresses in this range will be directed to the peripheral registers.

After learning the memory map of the CME-12BC demo board, you should have some idea of how to use the CME-12BC memory space in the normal expanded mode. A good choice would be:

1. Use the space from $1000 to $7FFF to hold your application programs. Since the 68HC12 doesn't distinguish between program memory and data memory, this area can also be used for data storage. However, you must be careful not to overlap your data with your program.

2. Use other SRAM areas for dynamic data. This includes three areas: $400-$7FF, $800-$9FF, and $C00-$CFF. The internal area from $A00 to $BFF has been used by the D-Bug12 monitor and is not available for application data use.

3.8 Summary

The 68HC12 microcontroller is developed as an upgrade for the 8-bit 68HC11 microcontroller family. All members implement the same architecture; in other words, they have the same instruction set and addressing modes but differ in the amount of peripheral functions that they implement.

The 68HC812A4 is the earliest member of the family and implements an 8-bit A/D converter. Other 68HC12 members implement a 10-bit A/D converter. The 68HC12 microcontroller implements a complicated timer system:

- The *input-capture* (IC) function latches the arrival time of a signal edge. This capability allows us to measure the frequency, period, and duty cycle of an unknown signal.

- The *output compare* (OC) function allows us to make a copy of the main timer, add a delay to this copy, and store the sum to an output compare register. This capability can be used to create a delay, generate a digital waveform, trigger an action on a signal pin, and so on.

- The *pulse accumulation* function allows us to count the events that arrived within an interval.

- The *pulse width modulation* module (PWM) enables us to generate digital waveforms of a certain frequency with a duty cycle ranging from 0 to 100 percent. This capability is useful in DC motor control.

The 68HC12 microcontroller also provides a wide variety of serial interface functions to appeal to different applications:

- The *SCI* function implements the EIA232 standard. A microcontroller demo board would use this interface to communicate with a PC.

- The *SPI* function is a synchronous interface that requires a clock signal to synchronize the data transfer between two devices. This interface is mainly used to interface with peripheral chips such as shift registers, seven-segment displays and LCD drivers, A/D and D/A converters with serial interface, SRAM and EEPROM with serial interface, phase-lock loop chips with serial interface, and so on.

- The *Inter-Integrated Circuit* (I²C) is a serial interface standard proposed by Phillips. This interface standard allows microcontrollers and peripheral devices to exchange data.

- The *Byte Data Link Communication* (BDLC) module is proposed for low-speed data communication in automotive applications. It provides access to an external serial communication multiplex bus that operates according to the SAE J1850 protocol.

- The *Controller Area Network* (CAN) was proposed to be used mainly as a vehicle serial data bus to provide reliable operation in the EMI environment and achieve the high bandwidth required in that environment.

Motorola and Axiom are two major vendors of demo boards for the 68HC12 microcontrollers. The features of these demo boards are reviewed in detail in this chapter.

Software development tools include the text editor, the terminal program, the cross assembler, the cross compiler, the simulator, the source-level debugger, and the *integrated development environment* (IDE). A sophisticated IDE should contain a text editor, a terminal program, a cross compiler, a cross assembler, and a source-level debugger. It allows us to perform all the development work without leaving any program.

In this text, we use MiniIDE to enter and test all the assembly programs. A tutorial on how to use MiniIDE is provided. The debugging activities on a CME-12BC demo board are facilitated by the command set provided by the D-Bug12 monitor.

3.9 Lab Exercises & Assignments

L3.1 Turn on the PC and start the MiniIDE program to connect to the CME-12BC or other demo boards from Axiom or Motorola. Then perform the following operations:

  a. Enter a command to set the contents of the memory locations from $800 to $8FF to 0.
  b. Display the contents of the memory locations from $800 to $8FF.
  c. Set the contents of the memory locations $800~$803 to 1, 2, 3, and 4, respectively.
  d. Verify that the contents of memory locations $800~$803 have been set correctly.

L3.2 Enter monitor commands to display the breakpoints, and set new breakpoints at $1020, $1050, and $1100. Delete breakpoints at $1050 and redisplay the breakpoints.

L3.3 Enter commands to place $10 and $0 in accumulators A and B.

L3.4 Use appropriate D-Bug12 commands to perform the following operations:

  a. set the contents of the memory location at $800 to 3.
  b. set the contents of the memory location at $801 to 4 and redisplay the location $800.
  c. set the contents of the memory location at $802 to 5 and redisplay the same location.
  d. set the contents of the memory location at $803 to 6 and return to the D-Bug12 command prompt.

L3.5 Invoke the one-line assembler to enter the following instructions to the demo board. Starting from address $1000, trace through the program and examine the contents of the memory locations at $800 and $801:

```
LDD     #$0000
STD     $0800
```

```
LDAB      #$00
INCB
LDX       $0800
ABX
STX       $0800
CMPB      #$14
BNE       $1008
SWI
```

L3.6 Use the text editor of MiniIDE to enter the following assembly program as a file with the file name *example6.a*:

```
              org         $800
sum           rmb         1
arcnt         rmb         1
              org         $1000
              ldaa        #20
              staa        arcnt
              ldx         #array
              ldaa        #0
              staa        sum
again         ldaa        0,X
              lsra
              bcs         next
              ldaa        sum
              adda        0,X
              staa        sum
next          inx
              dec         arcnt
              bne         again
              swi
              org         $2000
array         fcb         1,3,5,7,2,4,6,8,9,11,13,10,12,14,15,17,19,16,18,20
              end
```

After entering the program, perform the following operation:

a. Assemble the program.
b. Download the S-record file (file name *example6.s19*) to the demo board.
c. Display the contents of memory locations from $800 to $80F and $2000 to $202F.
d. Execute the program.
e. Display the contents of the memory location at $800.

This program adds all even numbers in the given array and stores the sum at $800.

L3.7 Write a program to count the number of elements in an array that are divisible by eight. The array has thirty 8-bit elements and is stored immediately after your program. Use the *repeat S until C* looping construct. Leave the result at $800.

L3.8 Write a program to swap the last element of an array with the first element, the second element with the second-to-last element, and so on. The array has thirty 8-bit elements.

L3.9 Write a program to find the *greatest common divisor* (gcd) of two 16-bit numbers stored at $800~$801 and $802~$803, and store the result at $810~$811.

4

Advanced Assembly Programming

4.1 Objectives

After completing this chapter you should be able to:

- access parameters stored in the stack and manipulate stack data structure
- access and manipulate arrays, vectors, and strings
- perform a binary-to-ASCII string and ASCII string–to-binary conversion
- write subroutines to perform certain functions
- make subroutine calls
- perform terminal I/O operations

4.2 Introduction

A program consists of two parts: *data structures* and *algorithms.* Data structures deal with how information is organized so that it can be processed efficiently. Algorithms deal with systematic methods for information processing. Common operations applied to data structures include adding and deleting elements, traversing and searching a data structure, balancing and sorting a data structure, and so on. There are many data structures in use today. However, we will deal with only the following data structures:

- *Stacks.* A stack is a data structure with a top and a bottom element. Elements can only be added or deleted from the top of the stack. From this standpoint, a stack is a first-in-last-out (LIFO) data structure.

- *Arrays.* An array is an ordered set of elements of the same type. The elements of the array are arranged so that there is a zeroth, first, second, ..., and nth element. An array may be one-, two-, or multidimensional. A vector is a one-dimensional array. A matrix is a two-dimensional array.

- *Strings.* A string is a sequence of characters terminated by a special character such as a NULL (ASCII code 0) or EOT (ASCII code 4). A computer system often needs to output a string to inform users of something or obtain some input from the user.

Algorithms are often represented in procedural steps or flowcharts. We will use both approaches in this book.

4.3 Stack

Conceptually, a stack is a list of data items whose elements can be accessed from only one end. A stack element has a top and a bottom. The operation that adds a new item to the top is called *push.* The top element can be removed by performing an operation called *pop* or *pull.* Physically, a stack is a reserved RAM area in main memory where programs agree to perform only push and pull operations. The structure of a stack is shown in Figure 4.1. Depending on the microcontroller, a stack has a pointer that points to the top element or to the memory location above the top element.

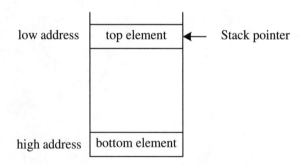

Figure 4.1 ■ Diagram of the 68HC12 stack

The 68HC12 has a 16-bit *stack pointer* (SP) to facilitate the implementation of the stack data structure. The stack pointer points to the top byte of the stack. Many companies follow the convention that the stack grows from higher addresses toward lower addresses. An area in main memory (which must be in RAM) is allocated for use as the stack area. If the stack grows into memory locations with addresses lower than the stack buffer, a *stack overflow* error occurs; if the stack is pulled too many times, a *stack underflow* occurs. We must check the stack overflow and underflow in order to make sure that the program won't crash. Of course, this checking adds overhead to the stack access.

The 68HC12 provides instructions for pushing and pulling all CPU registers except the stack pointer SP. A push instruction writes data from the source to the stack after decrementing the stack pointer. There are six push instructions: PSHA, PSHB, PSHC, PSHD, PSHX, and PSHY. A pull instruction loads data from the top of the stack to a register and then increments the stack pointer. There are also six pull instructions: PULA, PULB, PULC, PULD, PULX, and PULY. These instructions have equivalent store and load instructions combined with predecrement and postincrement index addressing modes as shown in Table 4.1.

Mnemonic	Function	Equivalent instruction
psha	push A into the stack	staa 1, -SP
pshb	push B into the stack	stab 1, -SP
pshc	push CCR into the stack	none
pshd	push D into stack	std 2, -SP
pshx	push X into the stack	stx 2, -SP
pshy	push Y into the stack	sty 2, -SP
pula	pull A from the stack	ldaa 1, SP+
pulb	pull B from the stack	ldab 1, SP+
pulc	pull CCR from the stack	none
puld	pull D from the stack	ldd 2, SP+
pulx	pull X from the stack	ldx 2, SP+
puly	pull Y from the stack	ldy 2, SP+

Table 4.1 ■ 68HC12 push and pull instructions and their equivalent load and store instructions

Example 4.1

▼

Assuming we have the following instruction sequence to be executed by the 68HC12, what would be the contents of the stack after the execution of these instructions?

```
lds      #$C00
ldaa     #$20
staa     1, -SP
ldab     #40
staa     1, -SP
ldx      #0
stx      2, -SP
```

Solution: The first instruction initializes the stack pointer to $C00. The second and third instructions together push the 8-bit value $20 into the stack. The fourth and fifth instructions push the 8-bit value 40 (hex $28) into the stack. The sixth and seventh instructions push the value 0 to the top two bytes of the stack. The contents of the 68HC12 stack are shown in Figure 4.2.

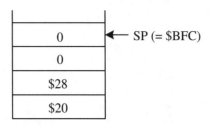

Figure 4.2 ■ The contents of the 68HC12 stack

One of the main uses of the stack is saving the return address for a subroutine call. Before the stack can be used, we need to set up the stack pointer. Since the stack can only be implemented in the RAM, we must make sure not to violate this requirement. Take the Axiom CME-12BC as an example: We can use either the on-chip SRAM or the external SRAM to implement the stack. The on-chip SRAM of the 68HC912BC32 occupies the address space from $800 to $BFF. However, the D-Bug12 monitor has used the area from $A00 to $BFF. Only the area from $800 to $9FF can be used for the stack area. A good initial setting for the stack pointer would be $A00 (when the stack is empty). We can also use external SRAM to implement the stack. The contiguous area from $1000 to $7FFF is available for downloading the user program and implementing the stack. Since the 68HC12 stack grows from higher addresses toward lower addresses, we can initialize the stack pointer to $8000.

▲

4.4 Indexable Data Structures

Vectors and matrices are indexable data structures. A vector is a sequence of elements in which each element is associated with an index i that can be used to access it. Conceptually, a vector is a one-dimensional data structure. To make address calculation easy, the first element is associated with the index zero and each successive element with the next integer. However, you can change the index origin of the array to one or some other value if you are willing to modify the routines. The elements of a vector or array have the same precision.

The assembler directives *db*, *dc.b*, and *fcb* can be used to define arrays with 8-bit elements. Directives *dw*, *dc.w*, and *fdb* can be used to define arrays with 16-bit elements. Suppose we have an array *vec_x* with six elements 11, 12, 13, 14, 15, and 16. It can be defined as follows:

```
vec_x  db      11,12,13,14,15,16
```

The assembler directives *ds*, *rmb*, and *ds.b* can be used to reserve memory locations for arrays with 8-bit elements. Directives *ds.w* and *rmw* can be used to reserve memory space for arrays with 16-bit elements.

Searching an array is a common operation for microcontroller applications. If the array is not sorted, the only way to find out whether a value is in the array is by performing a *sequential search*. A sequential search algorithm compares the value to be searched with every element in the array.

Example 4.2

Write a program to find out if the array *vec_x* contains a value *key*. The array has *n* 16-bit elements and is not sorted. Store the address of the element, which is equal to the key. Otherwise, store a –1 in it.

Solution: The algorithm for a sequential search is shown in Figure 4.3. To implement the sequential search algorithm, we will:

- use the double accumulator D to hold the search key
- use the index register X as a pointer to the array
- use the index register Y to hold the loop count

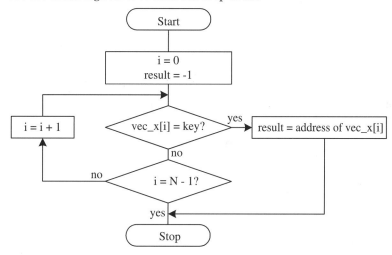

Figure 4.3 ■ Flowchart for sequential search

The program is as follows:

```
N           equ     30              ; array count
notfound    equ     -1
key         equ     190             ; define the searching key
            org     $800
result      rmw     1               ; reserve a word for result
            org     $1000           ; starting point of the program
            ldy     #N              ; set up loop count
            ldd     #notfound
            std     result          ; initialize the search result
            ldd     #key
            ldx     #vec_x          ; place the starting address of vec_x in X
loop        cpd     2,X+            ; compare the key with array element and
                                    ; increment the array pointer after comparison
            beq     found
            dbne    Y,loop          ; have we gone through the whole array?
            bra     done
found       dex                     ; need to restore the value of X to point to the
            dex                     ; matched element
            stx     result
done        swi
```

```
vec_x      dw      13,15,320,980,42,86,130,319,430,4
           dw      90,20,18,55,30,51,37,89,550,39
           dw      78,1023,897,930,880,810,650,710,300,190
           end
```

If an array (for example, a lookup table) needs to be searched frequently, then a sequential search is very inefficient. A better approach would be to sort the array and use the *binary search* algorithm. Suppose the sorted array (sorted in *ascending* order) has *n* elements and is stored at memory locations starting at the label *arr*. Let *max* and *min* represent the highest and lowest range of array indices to be searched and the variable *mean* represent the average of *max* and *min*. The idea of a binary search algorithm is to divide the sorted array into three parts:

- The portion of the array with indices ranging from *mean+1* to *max*.
- The element with index equal to *mean* (middle element).
- The portion of the array with indices ranging from *min* to *mean-1*.

The binary search algorithm compares the key with the middle element and takes one of the following actions based on the comparison result:

- If the key equals the middle element then stop.
- If the key is larger than the middle element, then the key can be found only in the portion of the array with larger indices. The search will be continued in the upper half.
- If the key is smaller than the middle element, then the key can be found only in the portion of the array with smaller indices. The search will be continued in the lower half.

The binary search algorithm can be formulated as follows:

Step 1
Initialize variables *max* and *min* to *n – 1* and *0*, respectively.

Step 2
If *max < min*, then stop. No element matches the key.

Step 3
Let *mean = (max + min)/2*.

Step 4
If *key* equals *arr*[mean], then key is found in the array, exit.

Step 5
If *key < arr*[mean], then set *max* to *mean – 1* and go to Step 2.

Step 6
If *key > arr*[mean], then set *min* to *mean + 1* and go to Step 2.

The above algorithm works for arrays sorted in ascending order. This algorithm can be modified to work for arrays sorted in descending order.

Example 4.3

Write a program to implement the binary search algorithm and a sequence of instructions to test it. Use an array of *n* 8-bit elements for implementation.

Solution: We will store a one as the result if the key is found in the array. Otherwise, a zero is stored as the search result. The program is as follows:

```
n           equ       30              ; array count
key         equ       69              ; key to be searched
            org       $800
max         rmb       1               ; maximum index value for comparison
min         rmb       1               ; minimum index value for comparison
mean        rmb       1               ; the average of max and min
result      rmb       1               ; search result
            org       $1000
            clra
            staa      min             ; initialize min to 0
            staa      result          ; initialize result to 0
            ldaa      #n-1
            staa      max             ; initialize max to n-1
            ldx       #arr            ; use X as the pointer to the array
loop        ldab      min
            cmpb      max
            lbhi      notfound        ; if min > max, then not found (unsigned comparison)
            addb      max             ; compute mean
            lsrb                      ;      "
            stab      mean            ; save mean
            ldaa      b,x             ; get a copy of the element arr[mean]
            cmpa      #key
            beq       found
            bhi       search_lo
            ldaa      mean
            inca
            staa      min             ; place mean+1 in min to continue
            bra       loop
search_lo   ldaa      mean
            deca
            staa      max
            bra       loop
found       ldaa      #1
            staa      result
notfound    swi
arr         db        1,3,6,9,11,20,30,45,48,60
            db        61,63,64,65,67,69,72,74,76,79
            db        80,83,85,88,90,110,113,114,120,123
            end
```

4.5 Strings

A string is a sequence of characters terminated by a NULL (ASCII code 0) or other special character such as EOT (ASCII code 4). In this book, we will use the NULL character to terminate a string. Common operations applied to strings include string concatenation, character and word counting, string matching, substring insertion and deletion, and so on.

Strings are also heavily used in input and output operations. Before a number (in binary format in memory) is output, it must be converted to ASCII code because most output devices only accept ASCII code. The number can be output in BCD or hex format. We looked at one example in Chapter 2 that converts a 16-bit binary number into BCD format by performing repeated divide-by-10 operations and adding $30 to each remainder (between zero and nine).

4.5.1 Data Conversion

In this section, we will deal with how data is converted from binary to ASCII and from ASCII to binary.

Example 4.4

Write a program to convert the unsigned 8-bit binary number in accumulator A into BCD digits terminated by a NULL character. Each digit is represented in ASCII code. The resulting string is stored in memory.

Solution: Since an 8-bit number can be from 0 to 255, we need 4 bytes to hold the result. By using the repeated division method, the conversion can be performed. The program is as follows:

```
test_dat    equ     220
            org     $800
out_buf     rmb     4
temp        rmb     2
            org     $1000       ; starting address of the program
            ldaa    #test_dat   ; load the test data
            ldy     #out_buf
            tab                 ; transfer the 8-bit value in B
; check to see if the number has only 1 digit
            cmpb    #9
            bhi     chk_99      ;
            addb    #$30        ; convert the digit into ASCII code
            stab    0,y         ; save the code and increment the pointer
            clr     1,y         ; terminated the string with NULL
            jmp     done
chk_99      clra
; check to see if the number has 2 digits
            cmpb    #99         ; is the number greater than 99?
            bhi     three_dig   ; if yes, the string has three digits
            ldx     #10
            idiv
            addb    #$30        ; convert the lower digit
            stab    1,y         ; store the lowest digit
            xgdx
            addb    #$30
            stab    0,y         ; save the upper digit
            clr     2,y         ; terminated the string with NULL
            bra     done
three_dig   ldx     #10
            idiv
            addb    #$30
            stab    2,y         ; save the least significant digit
            xgdx                ; swap the quotient to D
            ldx     #10
            idiv
            addb    #$30
            stab    1,y         ; save the middle digit
```

```
                 xgdx
                 addb    #$30
                 stab    0,y              ; save the ASCII code of the highest digit
                 clr     3,y              ; terminate the string with NULL
        done     swi
                 end
```

Example 4.5

Write a program to convert the 16-bit signed integer stored in the double accumulator D into a string of BCD digits and save the conversion result in a memory buffer starting with the label *out_buf*. The string must be terminated by a NULL character.

Solution: Because a string must be terminated by a NULL character, a signed 16-bit integer (in the range of –32768 to +32767) may need 2, 3, 4, 5, 6, or 7 bytes to store all of its digits in BCD format. If the number is negative, we need to add the minus sign as the first character of the converted string. A minus integer must be complemented. After performing these preprocessing steps, a repeated divide-by-10 operation is performed.

The program is as follows:

```
                 org     $800
        out_buf  rmb     7
        temp     rmb     2
        test_dat equ     -4300
                 org     $1000
                 ldd     #test_dat        ; load a test data
                 ldy     #out_buf         ; use Y as the pointer to the output buffer
                 cpd     #0
                 lbpl    chk_9            ; if plus, no complement
                 lbeq    zero             ; is the given number a 0?
; the following three instructions compute the magnitude of the given number in D
                 coma
                 comb
                 addd    #1
                 std     temp
                 ldaa    #$2D             ; place a negative sum
                 staa    0,y              ; store the sign
                 iny
                 ldd     temp
        chk_9    cpd     #9               ; does the number have only 1 digit?
                 lbhi    chk_99
                 addb    #$30             ; convert the single digit to ASCII
                 stab    0,y              ;            "
                 clr     1,y              ; terminate the string with NULL
                 lbra    done
        chk_99   cpd     #99              ; does the number have only 2 digits?
                 lbhi    chk_999          ; branch if the number has more than 2 digits
                 ldx     #10
                 idiv
                 addb    #$30             ; convert the ones digit to BCD ASCII
                 stab    1,y              ; save the ones digit
                 xgdx
```

```
            addb    #$30
            stab    0,y             ; save the tens digit
            clr     2,y             ; add a NULL character
            lbra    done
chk_999     cpd     #999            ; does the number have only 3 digits?
            lbhi    chk_9999        ; branch if the number has more than 3 digits
            ldx     #10
            idiv
            addb    #$30            ; convert the ones digit
            stab    2,y             ; save the ones digit
            xgdx
            ldx     #10
            idiv
            addb    #$30
            stab    1,y             ; save the tens digit
            xgdx
            addb    #$30            ; convert the hundreds digit
            stab    0,y
            clr     3,y             ; add a NULL character
            lbra    done
chk_9999    cpd     #9999           ; does the number have only 4 digits?
            lbhi    five_digit      ; branch if the number has 5 digits
            ldx     #10
            idiv
            addb    #$30
            stab    3,y             ; save the ones digit
            xgdx
            ldx     #10
            idiv
            addb    #$30
            stab    2,y
            xgdx
            ldx     #10
            idiv
            addb    #$30
            stab    1,y
            xgdx
            addb    #$30
            stab    0,y
            clr     4,y             ; add a NULL character
            lbra    done
five_digit  ldx     #10
            idiv
            addb    #$30
            stab    4,y             ; save the ones digit
            xgdx
            ldx     #10
            idiv
            addb    #$30
            stab    3,y             ; save the tens digit
            xgdx
            ldx     #10
            idiv
            addb    #$30
```

```
                    stab      2,y             ; save the hundreds digit
                    xgdx
                    ldx       #10
                    idiv
                    addb      #$30
                    stab      1,y             ; save the thousands digit
                    xgdx
                    addb      #$30
                    stab      0,y             ; save the ten-thousands digit
                    clr       5,y             ; add a NULL character
done                swi
zero                ldaa      #$30
                    staa      0,y
                    clr       1,y
                    swi
                    end
```

When the user enters a number from a keyboard, it is often in the form of a NULL-terminated ASCII string. We need to convert this into binary before it can be processed. Suppose *in_ptr* is the pointer to the input ASCII string, then the algorithm to convert an ASCII string into a binary number would be as follows:

Step 1
> sign ← 0
> error ← 0
> number ← 0

Step 2
If the character pointed to by *in_ptr* is the minus sign, then:
> sign ← 1
> in_ptr ← in_ptr + 1

Step 3
If the character pointed to by *in_ptr* is the NULL character,
then go to Step 4,
else if the character is not a BCD digit (i.e., m[in_ptr] > $39 or m[in_ptr] < $30), then:
> error ← 1;
> go to Step 4;

else:
> number ← number × 10 + m[in_ptr] - $30;
> in_ptr ← in_ptr + 1;
> go to Step 3;

Step 4
If sign = 1 and error = 0, then:
> number ← two's complement of number;

else
> stop;

The following example converts an ASCII string into a binary number.

Example 4.6

Write a program to convert a BCD ASCII string to a binary number and leave the result in double accumulator D. The ASCII string represents a number in the range of $-2^{15} \sim 2^{15} - 1$.

Solution: The following program is a direct translation of the above algorithm:

```
minus       equ     $2D             ; ASCII code of minus sign
            org     $800
in_buf      fcc     "9889"          ; input ASCII to be converted
            fcb     0               ; NULL character
out_buf     rmb     2
; buf1 is used to hold the current digit value and buf2 will be cleared to 0
buf2        rmb     1
buf1        rmb     1
sign        rmb     1               ; holds the sign of the number
error       rmb     1               ; indicates the occurrence of illegal character
            org     $1000
            clr     sign
            clr     error
            clr     out_buf
            clr     out_buf+1
            clr     buf2
            ldx     #in_buf
            ldaa    0,x
            cmpa    #minus          ; is the first character a minus sign
            bne     continue        ; branch if not minus
            inc     sign            ; set the sign to 1
            inx                     ; move the pointer
continue    ldaa    1,x+            ; is the current character a NULL character?
            lbeq    done            ; yes, we reach the end of the string
            cmpa    #$30            ; is the character not between 0 to 9?
            lblo    in_error        ;   "
            cmpa    #$39            ;   "
            lbhi    in_error        ;   "
            suba    #$30            ; convert to the BCD digit value
            staa    buf1            ; save the digit temporarily
            ldd     out_buf
            ldy     #10
            emul                    ; perform 16-bit by 16-bit multiplication
            addd    buf2            ; add the current digit value
            std     out_buf
            bra     continue
in_error    ldaa    #1
            staa    error
done        ldaa    sign            ; check to see if the original number is negative
            beq     positive
            ldaa    out_buf         ; if negative, compute its two's complement
            ldab    out_buf+1       ;   "
            coma                    ;   "
            comb                    ;   "
```

```
              addd        #1              ;   "
              std         out_buf
positive      swi
              end
```

4.5.2 Character & Word Counting

People probably will disagree on whether the white space characters such as the space, carriage return, line feed, and form feed characters should be included in the character count. We will include them in the character count because white space characters occupy disk space when a string is saved.

The procedure for finding the character count and the word count of a string is shown in Figure 4.4. In Figure 4.4, every non-NULL character causes the character count to be incre-

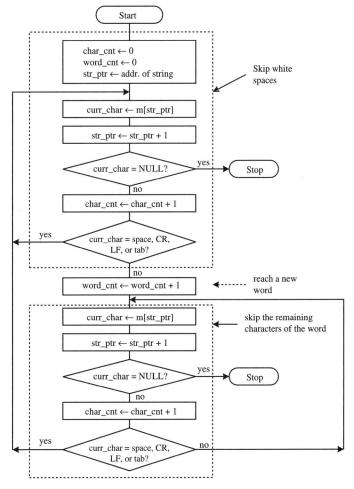

Figure 4.4 ■ Logic flow for counting characters and words in a string

mented by one. To look for a new word, we need to skip the white space characters during the scanning process. A new word is identified by the occurrence of a nonwhite character during the scanning process. Whenever a new word is encountered, we need to scan through the remaining nonwhite characters until the next white character is encountered.

The variables used in Figure 4.4 are:

- char_cnt: character count
- word_cnt: word count
- curr_char: current character
- char_ptr: character pointer

Example 4.7

Write a program to count the number of characters and words contained in a given string.

Solution: The following program is a direct translation of the algorithm in Figure 4.4:

```
tab          equ       $09
sp           equ       $20
cr           equ       $0D
lf           equ       $0A
             org       $800
char_cnt     rmb       1
word_cnt     rmb       1
string_x     fcc       "this is a strange test string to count chars and words."
             fcb       0
             org       $1000
             ldx       #string_x
             clr       char_cnt
             clr       word_cnt
string_lp    ldab      1,x+        ; get one character and move string pointer
             lbeq      done        ; is this the end of the string?
             inc       char_cnt
; the following 8 instructions skip white space characters between words
             cmpb      #sp
             beq       string_lp
             cmpb      #tab
             beq       string_lp
             cmpb      #cr
             beq       string_lp
             cmpb      #lf
             beq       string_lp
; a non-white character is the start of a new word
             inc       word_cnt
wd_loop      ldab      1,x+        ; get one character and move pointer
             beq       done
             inc       char_cnt
; the following 8 instructions check the end of a word
             cmpb      #sp
             lbeq      string_lp
             cmpb      #tab
             lbeq      string_lp
```

```
                    cmpb      #cr
                    lbeq      string_lp
                    cmpb      #lf
                    lbeq      string_lp
                    bra       wd_loop
        done        swi
                    end
```

▲

4.5.3 Word Matching

Searching certain words from a string is a common operation. The searching process starts from the beginning of the given string. The algorithm will look for the next new word from the given string and compare it with the given word. The comparison is performed character by character until two words are found to be either matched or unmatched. If two words are not equal, then the remaining characters of the word in the string must be skipped so that the same process can be repeated. The comparison must be performed one character beyond the last character of the given word if the comparison is matched, and there are three possible outcomes:

Case 1

The matched word is not the last word in the string. Comparison of the last characters will yield "*not equal*".

Case 2

The matched word is the last word in the given string, but there is one or a few white space characters between the last word and the NULL character. Comparison of the last characters will again give the result "*not equal*".

Case 3

The matched word is the last word of the given string, and it is followed by the NULL character. In this case, the comparison result for the last character is "*equal*".

A detailed flowchart for the word-matching algorithm is shown in Figure 4.5. In our convention, a punctuation mark that follows a word is not considered part of a word. The variables used in Figure 4.5 are:

- str_ptr: string pointer
- wd_ptr: word pointer
- match_flag: flag to indicate search result
- m[str_ptr]:   the character pointed to by str_ptr
- m[wd_ptr]:   the character pointed to by wd_ptr

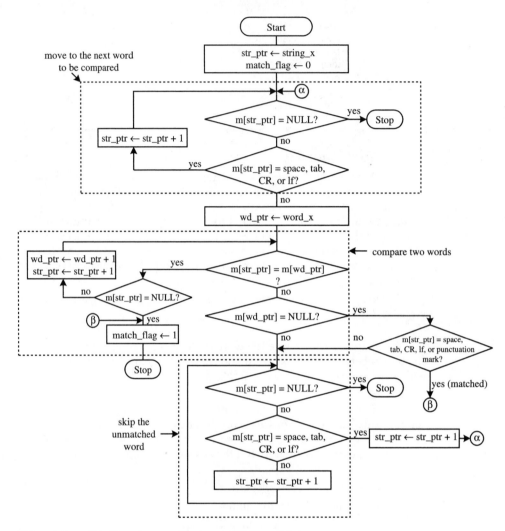

Figure 4.5 ■ Algorithm for searching a word from a string

Example 4.8

Write a program to search a given word from a string.

Solution: The following program is a direct implementation of the above algorithm:

```
tab          equ          $09          ; ASCII code of tab
sp           equ          $20          ; ASCII code of space character
cr           equ          $0D          ; ASCII code of carriage return
lf           equ          $0A          ; ASCII code of line feed
period       equ          $2E          ; ASCII code of period
comma        equ          $2C          ; ASCII code of comma
semicolon    equ          $3B          ; ASCII code of semicolon
exclamation  equ          $21          ; ASCII code of exclamation
```

```
null            equ         $0                  ; ASCII code of NULL character
                org         $800
match           rmb         1
                org         $1000
                clr         match
                ldx         #string_x
loop            ldab        1,x+
; the following 10 instructions skip white spaces to look for the next word in string_x
                tstb
                beq         done
                cmpb        #sp
                beq         loop
                cmpb        #tab
                beq         loop
                cmpb        #cr
                beq         loop
                cmpb        #lf
                beq         loop
; the first nonwhite character is the beginning of a new word to be compared
                ldy         #word_x
                ldaa        1,y+
next_ch         cba
                bne         end_of_wd
                cmpa        #null               ; check to see if the end of word is reached
                beq         matched             ;     "
                ldaa        1,y+                ; get the next character from the word
                ldab        1,x+                ; get the next character from the string
                bra         next_ch
; the following 10 instructions check to see if the end of the given word is reached
end_of_wd       cmpa        #null
                bne         next_wd
                cmpb        #cr
                beq         matched
                cmpb        #lf
                beq         matched
                cmpb        #tab
                beq         matched
                cmpb        #sp
                beq         matched
                cmpb        #period
                beq         matched
                cmpb        #comma
                beq         matched
                cmpb        #semicolon
                beq         matched
                cmpb        #exclamation
                beq         matched
; the following 11 instructions skip the remaining characters in the unmatched word
next_wd         ldab        1,x+
                beq         done
                cmpb        #cr
                lbeq        loop
                cmpb        #lf
                lbeq        loop
```

```
                    cmpb       #tab
                    lbeq       loop
                    cmpb       #sp
                    lbeq       loop
                    bra        next_wd
matched             ldab       #1
                    stab       match
done                swi
string_x            fcc        "This string contains certain number of words to be matched."
                    fcb        0
word_x              fcc        "This"
                    fcb        0
                    end
```

4.5.4 String Insertion

Inserting a substring into a string is another common operation. The insertion point will be given for such an operation. To insert a substring into a string, we need to make room for the substring to be inserted. All of the characters starting from the inserting point until the end of the string must be moved by the distance equal to the length of the substring. The main steps of the algorithm (also illustrated in Figure 4.6) are as follows:

Step 1
Count the number of characters that need to be moved (the NULL character is included because the resultant new string must be terminated by a NULL character).

Step 2
Count the number of characters in the substring to be inserted (excluding the NULL character).

Step 3
Move the characters in the string starting from the insertion point until the end of the string (move the last character first).

Step 4
Insert the substring (the first character of the substring is inserted first).

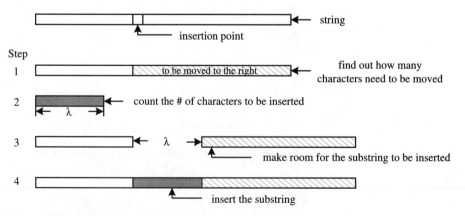

Figure 4.6 ■ Major steps of substring insertion

The program that implements the substring insertion is shown in Example 4.9.

Example 4.9

Write a program to implement the string insertion algorithm.

Solution: The following program is a direct translation of the above algorithm.

```
                    org         $800
ch_moved    rmb         1
char_cnt    rmb         1
sub_strg    fcc         "the first and most famous "
            fcb         0
string_x    fcc         "Yellowstone is national park."
            fcb         0
offset      equ         15
ins_pos     equ         string_x+offset    ; insertion point
            org         $1000
; the next 7 instructions count the number of characters to be moved
            ldaa        #1
            staa        ch_moved
            ldx         #ins_pos           ; use x to point to the insertion point
cnt_moved   ldaa        1,x+
            beq         cnt_chars
            inc         ch_moved
            bra         cnt_moved
cnt_chars   dex                            ; subtract 1 from x so it points to the NULL character
            ldy         #sub_strg          ; use y as a pointer to the substring
            clr         char_cnt
; the following 3 instructions count the move distance
char_loop   ldab        1,y+
            beq         mov_loop
            inc         char_cnt
            bra         char_loop
mov_loop    tfr         x,y                ; make a copy of x in y
            ldab        char_cnt
            aby                            ; compute the copy destination
            ldab        ch_moved           ; place the number of characters to be moved in B
again       movb        1,x-,1,y-
            dbne        b,again            ; make room for insertion
            ldx         #ins_pos           ; set up pointers to prepare insertion
            ldy         #sub_strg          ;              "
            ldab        char_cnt
insert_lp   movb        1,y+,1,x+
            dbne        b,insert_lp
            swi
            end
```

4.6 Subroutines

Good program design is based on the concept of *modularity*—the partitioning of a large program into subroutines. A main program contains the logical structure of the algorithm whereas smaller program units (called *subroutines*) execute many of the details.

The principles of program design in high-level language apply even more to the design of assembly language programs. Begin with a simple main program whose steps clearly outline the logical flow of the algorithm, and then assign the execution details to subroutines. Of course, subroutines may call other subroutines themselves. The structure of a modular program is illustrated in Figure 4.7.

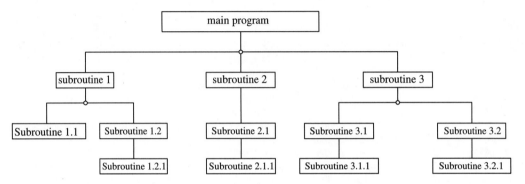

Figure 4.7 ▪ A structured program

A *subroutine* is a sequence of instructions that can be called from many different places in a program. A key issue in a subroutine call is to make sure that the program execution returns to the point immediately after the subroutine call (this address is called *return address*) when the subroutine completes its computation. This is normally achieved by saving and retrieving the return address in and from the stack. The program flow change involved in a subroutine call is illustrated in Figure 4.8.

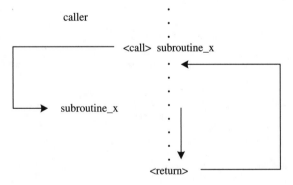

Figure 4.8 ▪ Program flow during a subroutine call

A microcontroller has dedicated instructions for making the subroutine call. The subroutine call instruction will save the return address, which is the address of the instruction immediately following the subroutine call instruction, in the system stack. When completing the computation task, the subroutine will return to the instruction immediately following the instruction that makes the subroutine call. This is achieved by executing a return instruction, which will retrieve the return address from the stack and transfer CPU control to it.

The 68HC12 provides three instructions (BSR, JSR, and CALL) for making subroutine calls and two instructions for returning from a subroutine. The syntax and operations performed by these instructions are listed and described below.

BSR <rel>   ; branch to subroutine

This instruction allows us to use only the *relative addressing mode* to specify the subroutine to be called. When this instruction is executed, the stack pointer is decremented by two, the return address is saved in the stack, and then the offset in the instruction is added to the current PC value and instruction execution is continued from there. In assembly language, the relative address is specified by using a label and the assembler will figure out the relative offset and place it in the program memory. For example, the following instruction calls the subroutine *bubble*:

```
bsr     bubble
```

JSR <opr>   ; jump to subroutine

This instruction allows us to use *direct, extended, indexed,* and *indexed indirect* addressing modes to specify the subroutine to be called. The subroutine can be located anywhere within 64KB. Like the BSR instruction, the 68HC12 first saves the return address in the stack and then jumps to execute the subroutine. Examples of the JSR instruction are as follows:

```
jsr     $ff                     ; call the subroutine located at $ff
jsr     sq_root                 ; call the subroutine sq_root
jsr     0,x                     ; call a subroutine pointed to by index register X
```

CALL <opr>   ; call a subroutine

This instruction is designed to work with expanded memory (larger than 64KB) supported by some 68HC12 members. Members with expanded memory treat the 16KB memory space from $8000 to $BFFF as a program memory window. An 8-bit program page register (PPAGE) is added to select one of the 256 16KB program memory pages to be accessed. To support subroutine calls in expanded memory, the *call* instruction pushes the current value of the PPAGE register along with the return address onto the stack and then transfers program control to the subroutine (three bytes are pushed into the stack). The <opr> field in the call instruction specifies the page number and the starting address of the subroutine within that page. The new page number will be loaded into the PPAGE register when the call instruction is executed. Extended, indexed, and indexed indirect addressing modes can be used to specify the subroutine address within a page.

Writing assembly programs to be run in expanded memory requires an assembler that supports this feature. Neither the Motorola freeware *as12* assembler nor the *MiniIDE* by Mgtek supports expanded memory.

RTS   ; return from subroutine

This instruction loads the program counter with a 16-bit value pulled from the stack and increments the stack pointer by two. Program execution continues at the address restored from the stack.

RTC   ; return from call

This instruction terminates subroutines in expanded memory invoked by the call instruction. The program page register (PPAGE) and the return address are restored from the stack; program execution continues at the restored address. For code compatibility, CALL and RTC are also executed correctly by 68HC12 members that do not have expanded memory capability.

4.7 Issues in Subroutine Calls

The program unit that makes the subroutine call is referred to as *caller* and the subroutine called by other program unit is referred to as *callee*. A subroutine is *entered* when it is being executed. There are three issues involved in subroutine calls.

- *Parameter passing.* The caller usually wants the subroutine to perform computations using the parameters passed to it. There are several methods available for passing parameters to the subroutine:

  1. Use registers. In this method, parameters are placed in CPU registers before the subroutine is called. This method is very convenient when there are only a few parameters to be passed.

  2. *Use the stack.* In this method, parameters are pushed into the stack before the subroutine is called. The stack must be cleaned up after the computation is completed. This can be done by either the caller or the callee.

  3. *Use the global memory.* Global memory is accessible to both the caller and the callee. As long as the caller places parameters in global memory before it calls the subroutine, the callee will be able to access them.

- *Result returning.* The result of a computation performed by the subroutine can be returned to the caller using three methods:

  1. *Use registers.* This method is most convenient when there are only a few bytes to be returned to the caller.

  2. *Use the stack.* The caller creates a hole of certain size before making the subroutine call. The callee places the computation result in the hole before returning to the caller.

  3. *Use global memory.* The callee simply places the value in the global memory and the caller will be able to access them.

- *Allocation of local variables.* In addition to the parameters passed to it, a subroutine may need memory locations to hold temporary variables and results. Temporary variables are called *local variables* because they only exist when the subroutine is entered. Local variables are always allocated in the stack so that they are not accessible to any other program units.

Although there are several methods for allocating local variables, the most efficient one is using the *LEAS* instruction. This instruction loads the stack pointer with an effective address specified by the program. The effective address can be any indexed addressing mode except an indirect address. For example, to deallocate 10 bytes, we can use the indexed addressing mode with SP as the base register and –10 as the offset:

```
leas     -10,sp     ; allocate ten bytes in the stack
```

This instruction simply subtracts 10 from SP and puts the difference back to SP. The general format for allocating space to local variables is:

```
leas     -n,sp     ; allocate n bytes in the stack
```

where n is the number of bytes to be allocated.

Before the subroutine returns to the caller, the space allocated to local variables must be deallocated. Deallocation is the reverse of allocation and can be achieved by the following instruction:

```
leas     n,sp     ; deallocate n bytes from the stack
```

4.8 The Stack Frame

The stack is used heavily during a subroutine call: the caller may pass parameters to the callee, and the callee may need to save registers and allocate local variables in the stack. The region in the stack that holds incoming parameters, return addresses, saved registers, and local variables is referred to as the *stack frame*. Some microprocessors have a dedicated register for managing the stack frame—the register is referred to as the *frame pointer*. The 68HC12, however, does not have a register dedicated to the function of the frame pointer. Since the stack frame is created during a subroutine call, it is also called the *activation record* of the subroutine. The stack frame exists as long as the subroutine is not exited. The structure of a stack frame is shown in Figure 4.9.

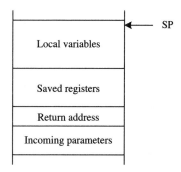

Figure 4.9 ■ Structure of the 68HC12 stack frame

The reason for having a dedicated frame pointer is due to the fact that the stack pointer may change during the lifetime of a subroutine. Once the stack pointer changes value, there can be problems in accessing the variables stored in the stack frame. The frame pointer is added to point to a fixed location in the stack and can avoid this problem. Since the 68HC12 does not have a dedicated frame pointer, we will not use the term frame pointer in the following discussion.

Example 4.10

▼

Draw the stack frame for the following program segment after the last *leas −10,sp* instruction is executed:

```
              ldd      #$1234
              pshd
              ldx      #$4000
              pshx
              jsr      sub_xyz
              ...
sub_xyz       pshd
              pshx
              psy
              leas     -10,sp
              ...
```

Solution: The caller pushes two 16-bit words into the stack. The subroutine *sub_xyz* allocated 10 bytes in the stack and saved 3 16-bit registers into the stack. The resultant stack frame is shown in Figure 4.10.

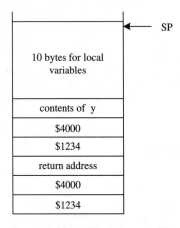

Figure 4.10 ■ Stack frame of Example 4.10

If a subroutine saves registers when it is entered, it must restore them in the reverse order (i.e., *last-in-first-out*) before returning to the caller. For example, if a subroutine has the following instructions to save the CPU registers when it is entered:

```
pshd
pshx
pshy
pshc
```

then it must have the following instructions to restore these registers:

```
pulc
puly
pulx
puld
```

4.9 Examples of Subroutine Calls

The examples in this section illustrate parameter passing, local variable allocation, and result returning.

4.9.1 Finding the Greatest Common Divisor

In this example, we will use CPU registers to pass parameters and return the result to the caller. The algorithm for finding the *gcd* (greatest common divisor) of two integers *m* and *n* is as follows:

Step 1
If m = n then
    gcd ← m;
    return;

Step 2
If n < m then swap m and n.

Step 3
gcd ← 1.
If m = 1 or n = 1 then return.

Step 4
For i = 2 to m do
    if (m % i = 0 and n % i = 0)
        gcd ← i;

Example 4.11

Write a program to compute the greatest common divisor of two 16-bit unsigned integers.

Solution: To implement the above algorithm, we will need the following local variables:

gcd_local: used as temporary gcd
i_local: used as a value to test divide m and n
m_local: used to hold the incoming m value
n_local: used to hold the incoming n value

Index register Y will be used and hence must be saved in the stack. The stack frame is shown in Figure 4.11. The program is as follows:

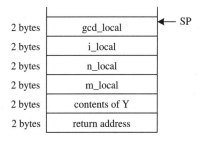

Figure 4.11 ■ Stack frame of Example 4.11

```
gcd_local    equ      0             ; the distance of variable gcd_local from the stack top
i_local      equ      2             ; the distance of variable i_local from the stack top
m_local      equ      4             ; the distance of variable m_local from the stack top
n_local      equ      6             ; the distance of variable n_local from the stack top
local_var    equ      8             ; number of bytes to be allocated to local variables
             org      $800
m            dw       375           ; the first operand
n            dw       1250          ; the second operand
gcd          rmw      1
             org      $1000
             ldx      m
             ldd      n
             jsr      find_gcd
             std      gcd           ; save the gcd in memory
             swi
find_gcd     pshy
```

```
                    leas      -local_var,sp       ; allocate space for local variables
                    stx       n_local,sp
                    std       m_local,sp
                    ldy       #1
                    sty       gcd_local,sp        ; initialize gcd to 1
                    cpd       n_local,sp          ; compare m with n
                    beq       m_equ_n             ; gcd = m if m = n
                    blo       m_less_n            ; it is fine if m < n
                    exg       d,x                 ; swap m and n
                    std       n_local,sp          ; also make sure the stack frame copy of
                    stx       m_local,sp          ; m and n are swapped
m_less_n            cpd       #1
                    beq       done                ; if m = 1, then gcd = 1
                    ldx       #2
                    stx       i_local,sp          ; initialize i to 2
loop                ldx       i_local,sp
                    cpx       m_local,sp
                    bhi       done
                    ldd       m_local,sp
                    idiv                          ; divide m by i
                    cpd       #0
                    bne       next_i
                    ldd       n_local,sp
                    ldx       i_local,sp
                    idiv                          ; divide n by i
                    cpd       #0
                    bne       next_i
                    ldd       i_local,sp
                    std       gcd_local,sp        ; set i as the current gcd
next_i              ldx       i_local,sp
                    inx
                    stx       i_local,sp          ; increment i
                    jmp       loop
m_equ_n             ldd       m_local,sp
                    bra       exit
done                ldd       gcd_local,sp
exit                leas      local_var,sp        ; deallocate local variables
                    puly
                    rts
                    end
```

4.9.2 Multiple Byte Division

The 68HC12 provides instructions for 16-bit by 16-bit signed and unsigned divisions, and also for 32-bit by 16-bit signed and unsigned divisions. However, the 68HC12 does not have instructions for performing higher precision divisions; say, for example, 32-bit by 32-bit division. We will need to synthesize such operations.

The most popular method for performing high precision divide operations is the repeated subtraction method. The conceptual hardware for implementing the repeated subtraction algorithm is shown in Figure 4.12, and is detailed in the following steps.

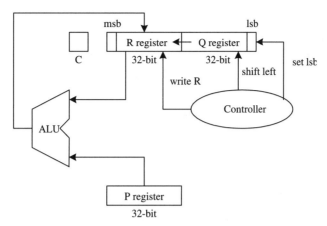

Figure 4.12 ■ Conceptual hardware for implementing repeated subtraction method

Step 1
Shift the register pair (R, Q) one bit left.

Step 2
Subtract register P from register R, put the result back to R if the result is non-negative.

Step 3
If the result of step 2 is negative, then set the least significant bit of Q to 0. Otherwise, set the least significant bit of Q to 1.

Perform this procedure **n** times, then the quotient and the remainder will be left in register Q and R, respectively. Before the division steps are performed, the dividend and divisor must be loaded into register Q and register P, respectively. Register R must be cleared to zero.

Example 4.12

Write a subroutine implementing the division algorithm that uses the repeated subtraction method for the 32-bit unsigned dividend and divisor. The caller of this subroutine will pass the dividend and divisor in the stack and will allocate space in the stack for this subroutine to return the quotient and remainder. Also, write an instruction to test this subroutine.

Solution: The stack frame of this subroutine is shown in Figure 4.13.

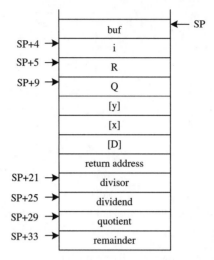

Figure 4.13 ■ Stack frame for Example 4.12

The instruction sequence to test this subroutine and the divide subroutine is as follows:

```
buf           equ    0                ; distance of buf from the top of the stack
i             equ    4                ; distance of i from the top of the stack
R             equ    5                ; distance of R from the top of the stack
Q             equ    9                ; distance of Q from the top of the stack
divisor       equ    21               ; distance of divisor from the top of the stack
dividend      equ    25               ; distance of dividend from the top of the stack
quotient      equ    29               ; distance of quotient from the top of the stack
remainder     equ    33               ; distance of remainder from the top of the stack
local         equ    13               ; number of bytes for local variables
dvdend_hi     equ    $42              ; dividend to be tested
dvdend_lo     equ    $4c15            ;    "
dvsor_hi      equ    $0               ; divisor to be tested
dvsor_lo      equ    $64              ;    "
              org    $800
quo           rmb    4                ; memory locations to hold the quotient
rem           rmb    4                ; memory locations to hold the remainder
              org    $1000            ; starting address of the program
              lds    #$8000           ; initialize stack pointer
              leas   -8,sp            ; make a hole of 8 bytes to hold the result
              ldd    #dvdend_lo
              pshd
              ldd    #dvdend_hi
              pshd
              ldd    #dvsor_lo
              pshd
              ldd    #dvsor_hi
              pshd
              jsr    div32            ; call the divide subroutine
; the following instruction deallocates the stack space used by the divisor and dividend
              leas   8,sp
```

```
                ; the following four instructions get the quotient from the stack
                        puld
                        std         quo
                        puld
                        std         quo+2
                ; the following four instructions get the remainder from the stack
                        puld
                        std         rem
                        puld
                        std         rem+2
                        swi
; *********************************************************************
; The following subroutine divides an unsigned 32-bit integer by another unsigned
; 32-bit integer
; *********************************************************************
div32           pshd
                pshx
                pshy
                ; the following instruction allocates space for local variables
                        leas        -local,sp
                        ldd         #0
                        std         R,sp            ; initialize register R to 0
                        std         R+2,sp
                        ldd         dividend,sp
                        std         Q,sp            ; place dividend in register Q
                        ldd         dividend+2,sp
                        std         Q+2,sp
                        ldaa        #32
                        staa        i,sp            ; initialize loop count
loop                    lsl         Q+3,sp          ; shift register pair Q and R to the right
                        rol         Q+2,sp          ; by 1 bit
                        rol         Q+1,sp          ;      "
                        rol         Q,sp            ;      "
                        rol         R+3,sp          ;      "
                        rol         R+2,sp          ;      "
                        rol         R+1,sp          ;      "
                        rol         R,sp            ;      "
                ; the following 8 instructions subtract the divisor from register R
                        ldd         R+2,sp
                        subd        divisor+2,sp
                        std         buf+2,sp
                        ldaa        R+1,sp
                        sbca        divisor+1,sp
                        staa        buf+1,sp
                        ldaa        R,sp
                        sbca        divisor,sp
                        bcs         smaller
                ; the following 6 instructions store the difference back to R register
                        staa        R,sp
                        ldaa        buf+1,sp
                        staa        R+1,sp
                        ldd         buf+2,sp
                        std         R+2,sp
```

```
              ldaa      Q+3,sp
              oraa      #01                    ; set the least significant bit of Q register to 1
              staa      Q+3,sp
              bra       looptest
smaller       ldaa      Q+3,sp
              anda      #$FE                   ; set the least significant bit of Q register to 0
              staa      Q+3,sp
looptest      dec       i,sp
              lbne      loop
; the following 4 instructions copy the remainder into the hole
              ldd       R,sp
              std       remainder,sp
              ldd       R+2,sp
              std       remainder+2,sp
; the following 4 instructions copy the quotient into the hole
              ldd       Q,sp
              std       quotient,sp
              ldd       Q+2,sp
              std       quotient+2,sp
              leas      local,sp               ; deallocate local variables
              puly
              pulx
              puld
              rts
              end
```

4.9.3 Bubble Sort

Sorting is among the most common ingredients of programming and many sorting methods are used. Sorting makes many efficient search methods possible. The *bubble sort* is a simple, inefficient, but widely known sorting method. Many other more efficient sorting methods require the use of recursive subroutine calls, which fall outside the scope of this book.

The basic idea underlying the bubble sort is to go through the array or file sequentially several times. Each iteration consists of comparing each element in the array or file with its successor (x[i] with x[i+1]) and interchanging the two elements if they are not in proper order (either ascending or descending). Consider the following array:

157 13 35 9 98 810 120 54 10 30

Suppose we want to sort this array in ascending order. The following comparisons are made in the first iteration:

```
x[0] with x[1] (157 and 13) interchange
x[1] with x[2] (157 with 35) interchange
x[2] with x[3] (157 with 9) interchange
x[3] with x[4] (157 with 98) interchange
x[4] with x[5] (157 with 810) no interchange
x[5] with x[6] (810 with 120) interchange
x[6] with x[7] (810 with 54) interchange
x[7] with x[8] (810 with 10) interchange
x[8] with x[9] (810 with 30) interchange
```

Thus, after the first iteration, the array is in the order:

13 35 9 98 157 120 54 10 30 810

Notice that after this first iteration, the largest element (in this case 810) is in its proper position within the array. In general, x[n – i] will be in its proper position after iteration i. The method is called the bubble sort because each number slowly bubbles up to its proper position. After the second iteration the array is in the order:

13 9 35 98 120 54 10 30 157 810

Notice that 157 is now in the second highest position. Since each iteration places a new element into its proper position, an array or a file of n elements requires no more than n – 1 iterations. The complete set of iterations is as follows:

Iteration		
	0 (original array)	157 13 35 9 98 810 120 54 10 30
	1	13 35 9 98 157 120 54 10 30 810
	2	13 9 35 98 120 54 10 30 157 810
	3	9 13 35 98 54 10 30 120 157 810
	4	9 13 35 54 10 30 98 120 157 810
	5	9 13 35 10 30 54 98 120 157 810
	6	9 13 10 30 35 54 98 120 157 810
	7	9 10 13 30 35 54 98 120 157 810
	8	9 10 13 30 35 54 98 120 157 810
	9	9 10 13 30 35 54 98 120 157 810

There are some obvious improvements you can make to the foregoing method:

- First, since all elements in positions greater than or equal to n – i are already in proper position after iteration i, they need not be considered in succeeding iterations. Thus in the first iteration $n - 1$ comparisons are made, on the second iteration $n - 2$ comparisons, and on the (n – 1)th iteration only one comparison is made (between x[0] and x[1]). Therefore the process is sped up as it proceeds through successive iterations.

- Second, although we have shown that n – 1 iterations are sufficient to sort an array or a file of size n, in the preceding sample array of ten elements, the array was sorted after the 7th iteration, making the last two iterations unnecessary. To eliminate unnecessary iterations we must be able to detect the fact that the array is already sorted. An array is sorted if no swaps are made in an iteration. By keeping a record of whether any swaps are made in a given iteration it can be determined whether any further iterations are necessary. The logic flow of the bubble sort algorithm is illustrated in Figure 4.14. The following example implements the bubble sort as a subroutine.

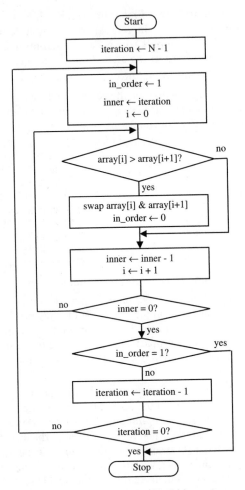

Figure 4.14 ■ Logic flow of bubble sort

Example 4.13

Write a subroutine to implement the bubble sort algorithm and a sequence of instructions along with a set of test data for testing this subroutine. Use an array that consists of n 8-bit unsigned integers for testing purposes.

Solution: The subroutine has several local variables:

- buf: buffer space for swapping adjacent elements
- in_order: flag to indicate whether the array is in order after an iteration
- inner: loop count for each iteration
- iteration: number of iterations remains to be performed

The stack frame of this program is shown in Figure 4.15.

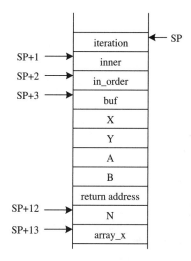

Figure 4.15 ■ Stack frame for bubble sort

```
arr         equ     13              ; distance of the variable arr from stack top
arcnt       equ     12              ; distance of the variable arcnt from stack top
buf         equ     3               ; distance of local variable buf from stack top
in_order    equ     2               ; distance of local variable in_order from stack top
inner       equ     1               ; distance of local variable inner from stack top
iteration   equ     0               ; distance of local variable iteration from stack top
true        equ     1
false       equ     0
n           equ     30              ; array count
local       equ     4               ; number of bytes used by local variables
            org     $800
array_x     db      3,29,10,98,54,9,100,104,200,92,87,48,27,22,71
            db      1,62,67,83,89,101,190,187,167,134,121,20,31,34,54
            org     $1000
            lds     #$8000          ; initialize stack pointer
            ldx     #array_x
            pshx
            ldaa    #n
            psha
            jsr     bubble
            leas    3,sp            ; deallocate space used by outgoing parameters
            swi                     ; break to D-Bug12 monitor
bubble      pshd
            pshy
            pshx
            leas    -local,sp       ; allocate space for local variables
            ldaa    arcnt,sp        ; compute the number of iterations to be performed
            deca                    ;        "
            staa    iteration,sp    ;        "
ploop       ldaa    #true           ; set array in_order flag to true before any iteration
            staa    in_order,sp     ;        "
            ldx     arr,sp              ; use index register X as the array pointer
```

```
                    ldaa      iteration,sp    ; initialize inner loop count for each iteration
                    staa      inner,sp        ;    "
cloop               ldaa      0,x             ; compare two adjacent elements
                    cmpa      1,x             ;    "
                    bls       looptest
; the following five instructions swap the two adjacent elements
                    staa      buf,sp          ; swap two adjacent elements
                    ldaa      1,x             ;    "
                    staa      0,x             ;    "
                    ldaa      buf,sp          ;    "
                    staa      1,x             ;    "
                    ldaa      #false          ; reset the in-order flag
                    staa      in_order,sp     ;    "
looptest            inx
                    dec       inner,sp
                    bne       cloop
                    tst       in_order,sp     ; test array in_order flag after each iteration
                    bne       done
                    dec       iteration,sp
                    bne       ploop
; the following instruction deallocates local variables
done                leas      local,sp        ; deallocate local variables
                    pulx
                    puly
                    puld
                    rts
                    end
```

4.9.4 Finding the Square Root

There are several methods available for finding the square root of a number **q**. One of the methods is based on the following equation:

$$\sum_{i=0}^{n-1} i = \frac{n(n-1)}{2} \qquad (4.1)$$

This equation can be transformed into:

$$n^2 = \sum_{i=0}^{n-1} (2i+1) \qquad (4.2)$$

Equation 4.2 gives us a clue about how to compute the square root of a value. Suppose we want to compute the square root of q, and n is the integer value that is closest to the true square root. One of the following three relationships is satisfied:

$n^2 < q$
$n^2 = q$
$n^2 > q$

The logic flow shown in Figure 4.16 can be used to find a value for n, and this algorithm is translated into the program shown in Example 4.14.

Example 4.14

Write a subroutine to implement the square root algorithm. This subroutine must be able to find the square root of a 32-bit unsigned integer. The parameter q (for which we want to find the square root) is pushed into the stack and the square root is to be returned in the double accumulator D.

Solution: To implement the algorithm illustrated in Figure 4.16, we need to use two local variables:

i: for looping purpose
sum: for accumulating all of the (2i + 1) terms

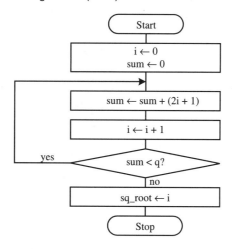

Figure 4.16 ■ Algorithm for finding the square root of integer q.

The stack frame of this subroutine is shown in Figure 4.17.

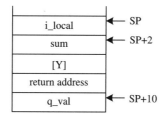

Figure 4.17 ■ Stack frame of Example 4.14

The subroutine and the instruction sequence for testing this subroutine are as follows:

```
q_hi      equ     $000F       ; upper word of q
q_lo      equ     $4240       ; lower word of q
i_local   equ     0           ; distance of local variable i from the top of the stack
sum       equ     2           ; distance of local variable sum from the top of the stack
q_val     equ     10          ; distance of incoming parameter q from the top of stack
local     equ     6           ; number of bytes allocated to local variables
          org     $800
sq_root   rmb     2           ; to hold the square root of q
          org     $1000
```

```
                    ldd         #q_lo
                    pshd
                    ldd         #q_hi
                    pshd
                    jsr         find_sq_root
                    std         sq_root
; the next instruction deallocates the space used by outgoing parameters
                    leas        4,sp
                    swi
find_sq_root
                    pshy                            ; save y in the stack
; the next instruction allocate space to local variables
                    leas        -local,sp
                    ldd         #0                  ; initialize local variable i to 0
                    std         i_local,sp          ;   "
                    std         sum,sp              ; initialize local variable sum to 0
                    std         sum+2,sp            ;   "
loop                ldd         i_local,sp
                    ldy         #2
                    emul                            ; compute 2i
; add 2i to sum
                    addd        sum+2,sp
                    std         sum+2,sp
                    tfr         y,d
                    adcb        sum+1,sp
                    stab        sum+1,sp
                    adca        sum,sp
                    staa        sum,sp
; add one to sum (need to propagate carry to the most significant byte of sum)
                    ldaa        #1
                    adda        sum+3,sp
                    staa        sum+3,sp
                    ldaa        sum+2,sp
                    adca        #0
                    staa        sum+2,sp
                    ldaa        sum+1,sp
                    adca        #0
                    staa        sum+1,sp
                    ldaa        sum,sp
                    adca        #0
                    staa        sum,sp
; increment i by one
                    ldd         i_local,sp
                    addd        #1
                    std         i_local,sp
; compare sum to q_val by performing subtraction (need consider borrow)
                    ldd         sum+2,sp
                    subd        q_val+2,sp
                    ldaa        sum+1,sp
                    sbca        q_val+1,sp
                    ldaa        sum,sp
                    sbca        q_val,sp
                    lblo        loop
```

```
            ldd        i_local,sp          ; place sq_root in D before return
; deallocate space used by local variables
exit        leas       local,sp
            puly
            rts
            end
```

This subroutine will find the exact square root if the given number has one. If the given number does not have an exact integer square root, then the number returned by the subroutine may not be the closest approximation. The algorithm in Figure 4.16 will stop only when $sum > q$. This may have the undesirable effect that the last i value may not be as close to the real square root as is the value $i - 1$. However, this can be fixed easily by comparing the following two expressions:

1. $i^2 - q\_val$
2. $q\_val - (i - 1)^2$

If the first expression is smaller, then i is a better choice. Otherwise, we should choose $i - 1$ as the approximation to the real square root. This will be left as an exercise problem.

4.10 Using the D-Bug12 Functions to Perform I/O Operations

The D-Bug12 monitor provides a few subroutines to support I/O operations. We can utilize these I/O routines to facilitate program developments on a demo board that contains the D-Bug12 monitor. A summary of these routines is depicted in Table 4.2.

Subroutine	Function	Starting address in D-Bug12 version 2.xxx
main ( )	Start of D-Bug12	$F680 ($FE00)
getchar ( )	Get a character from (SCI0) serial I/O port	$F682 ($FE02)
putchar ( )	Send a character out to the serial (SCI0) I/O port	$F684 ($FE04)
printf ( )	Formatted string output	$F686 ($FE06)
GetCmdLine ( )	Get a line of input from the user	$F688 ($FE08)
sscanhex ( )	Convert ASCII hex string to a binary integer	$F68A ($FE0A)
isxdigit ( )	Check if a character (in B) is hex	$F68C ($FE0C)
toupper ( )	Convert lower-case to upper-case character	$F68E ($FE0E)
isalpha ( )	Check if a character is alphabetic	$F690 ($FE10)
strcpy ( )	Copy a NULL-terminated string	$F692 ($FE12)
strlen ( )	Returns the length of a NULL-terminated string	$F694 ($FE14)
out2hex ( )	Output 8-bit number as 2 ASCII hex character	$F696 ($FE16)
out4hex ( )	Output a 16-bit number as 4 ASCII hex characters	$F698 ($FE18)
SetUserVector ( )	Setup a vector to a user's interrupt service routine	$F69A ($FE1A)
WriteEEByte( )	Write a byte to the on-chip EEPROM memory	$F69C ($FE1C)
EraseEE ( )	Bulk erase the EEPROM memory	$F69E ($FE1E)
ReadMem ( )	Read data from the 68HC12 memory map	$F6A0 ($FE20)
WriteMem ( )	Write data to the 68HC12 memory map	$F6A2 ($FE22)

Note. For D-Bug12 version 1.xx the starting addresses are in parenthesis)

Table 4.2 ■ D-Bug12 monitor routines

All of the user-accessible routines are written in C. All except the first parameter are passed to the user-callable functions on the stack. Parameters must be pushed onto the stack in the reverse order they are listed in the function declaration (right-to-left). The first parameter is passed to the function in accumulator D. If a function has only a single parameter, then the parameter is passed in accumulator D. Parameters of type *char* must be converted to an *integer* (16-bit). Parameters of type *char* will occupy the lower order byte (higher address) of a word pushed onto the stack or accumulator B if the parameter is passed in D.

Parameters pushed onto the stack before the function is called remain on the stack when the function returns. It is the responsibility of the *caller* to remove passed parameters from the stack.

All 8- and 16-bit function values are returned in accumulator D. A value of type *char* returned in accumulator D is located in the 8-bit accumulator B. *Boolean* function returns a zero value for false and a non-zero value for true.

None of the CPU register contents, except the stack pointer, are preserved by the called functions. If any of the register values need to be preserved, they should be pushed onto the stack before any of the parameters have been pushed and restored after deallocating the parameters.

4.10.1 Calling D-Bug12 Functions from Assembly Language

Calling the functions from assembly language is a simple matter of pushing the parameters onto the stack in the proper order and loading the first or only function parameter into accumulator D. The function can then be called with a JSR instruction. The code following the JSR instruction should remove any parameters pushed onto the stack. If a single parameter was pushed onto the stack, a simple PULX or PULY instruction is one of the most efficient ways to remove the parameter from the stack. If two or more parameters are pushed onto the stack, the LEAS instruction is the most efficient way to remove the parameters. Any of the CPU registers that were saved on the stack before the function parameters should be restored with the corresponding PULL instruction.

For example, the *WriteEEByte* function has two parameters: the first parameter is the address of the memory location to which the data is to be written; the second is the data itself. An example of an instruction sequence to call this function to write the value #$55 into EEPROM location $D00 is as follows:

```
WriteEEByte    equ      $F69C
               .
               .
               .
               ldab     #$55              ; write $55 to EEPROM
               clra
               pshd
               ldd      #$0D00            ; EEaddress to write data
               jsr      [WriteEEByte,PCR] ; call the routine
               leas     2,sp              ; remove the parameter from stack
               beq      EEWError          ; zero return value means error
               .
               .
```

The addressing mode used in the *jsr* instruction of the above example is a form of indexed indirect addressing that uses the program counter as an index register. The PCR mnemonic used in place of an index register name stands for *Program Counter Relative* addressing. In reality, the 68HC12 does not support PCR. Instead, the PCR mnemonic is used to instruct the assembler to calculate an offset to the address specified by the label *WriteEEByte*. The offset is calculated by subtracting the value of the PC at the address of the first object code byte of the next instruction (in this example, PULX) from the address supplied in the indexed offset field

(*WriteEEByte*). When the JSR instruction is executed, the opposite occurs. The 68HC12 adds the value of the PC at the first object code byte of the next instruction to the offset embedded in the instruction object code. The indirect addressing, indicated by the square brackets, specifies that the address calculated as the sum of the index register (in this case the PC) and the 16-bit offset contains a pointer to the destination of the JSR.

The *MiniIDE* software supports this syntax. However, if you are using an assembler that does not support program-counter-relative indexed addressing, the following two-instruction sequence can be used:

```
ldx    WriteEEByte        ; load the address of WriteEEByte()
jsr    0,x                ; call the subroutine
```

4.10.2 Descriptions of Callable Functions

For each of the callable functions, the prototype declaration, the pointer address (where the starting address of the function is stored), the stack space used, the incoming parameters, and the returned value are listed. Both the pointers for the D-Bug12 version 2.xx and 1.xx will be given (enclosed in parenthesis).

void main (void)

Pointer address:	$F680 ($FE00)
Stack space:	none
Incoming parameter:	none
Returned value:	none

This function simply restarts the D-Bug12 monitor and hence won't be useful for those who are learning the 68HC12 microcontroller. Readers who are really interested in knowing the other uses of this function should refer to the Motorola application node AN1280a/D.

int getchar (void)

Pointer address:	$F682 ($FE02)
Stack space:	2 bytes
Incoming parameter:	none
Returned value:	8-bit character in accumulator B

This function retrieves a single character from the serial communication port SCI0. If an unread character is not available in the receive data register when this function is called, it will wait until one is received. Because the character is returned as an integer, the 8-bit character is placed in accumulator B.

Adding the following instruction sequence to your program will read a character from the SCI0 port:

```
getchar    equ    $F682
           ...
           jsr    [getchar,PCR]
           ...
```

int putchar(int)

Pointer address:	$F684 ($FE04)
Stack space:	4 bytes
Incoming parameter:	character to be output in accumulator B
Returned vale:	the character that was sent (in B)

This function outputs a single character to the serial communication port SCI0. If the SCI0 transmit data register is full when the function is called, *putchar()* will wait until the transmit data register is empty before sending the character. No buffering of characters is provided. *Putchar()* returns the character that was sent. However, it does not detect any error conditions that may occur in the process and therefore will never return *EOF* (end-of-file). Adding the following instruction sequence to your program will output the character *A* to serial port SCI0 (when the program is running on the CME-12BC demo board, the character A will be displayed on the monitor screen):

```
putchar        equ      $F684
               ...
               ldab     #$41
               jsr      [putchar,PCR]
               ...
```

int printf(char *format,...)

Pointer address: $F686 ($FE06)
Stack space: maximum of 64 bytes, does not include parameter stack space
Incoming parameters: zero or more integer data to be output on the stack, D contains the address of the format string. The format string must be terminated with a zero.
Returned value: number of characters printed in D.

This function is used to convert, format, and print its arguments as standard output (the output device could be the monitor screen, printer, LCD, etc) under the control of the format string pointed to by *format*. It returns the number of characters that were sent to standard output (sent through serial port SCI0). All except floating-point data types are supported.

The format string can contain two basic types of objects: ASCII characters which are copied directly from the format string to the display device, and conversion specifications that cause succeeding *printf()* arguments to be converted, formatted, and sent to the display device. Each conversion specification begins with a percent sign (%) and ends with a single conversion character. Optional formatting characters may appear between the percent sign and end with a single conversion character in the following order:

[-] [<FieldWidth>] [.] [<Precision>] [h | l]

These optional formatting characters are explained in Table 4.3.

Character	Description
-(minus sign)	Left justifies the converted argument.
FieldWidth	Integer number that specifies the minimum field width for the converted argument. The argument will be displayed in a field at least this wide. The displayed argument will be padded on the left or right if necessary.
. (period)	Separates the field width from the precision.
Precision	Integer number that specifies the maximum number of characters to display from a string or the minimum number of digits for an integer.
h	To have an integer displayed as a short.
l(letter ell)	To have an integer displayed as a long.

Table 4.3 ■ Optional formatting characters

The *FieldWidth* or *Precision* field may contain an asterisk (*) character instead of a number. The asterisk will cause the value of the argument in the argument list to be used instead.

The formatting characters supported by the *printf ()* function are listed in Table 4.4. If the conversion character(s) following the percent sign are not one of the formatting characters shown in this table or the characters shown in Table 4.3, the behavior of the *printf()* function is undefined.

Character	Argument type; displayed as
d, i	int; signed decimal number
o	int; unsigned octal number (without a leading zero)
x	int; unsigned hex number using abcdef for 10...15
X	int; unsigned hex number using ABCDEF for 10...15
u	int; unsigned decimal number
c	int; single character
s	char *; display from the string until a '\0' (NULL)
p	void *; pointer (implementation-dependent representation)
%	no argument is converted; print a %

Table 4.4 ■ Printf() conversion characters

The *printf()* function can be used to print a message. Here is one example:

```
CR        equ       $0D
LF        equ       $0A
printf    equ       $F686
          ...
          ldd       #prompt
          jsr       [printf,PCR]
          ...
prompt    db        "Flight simulation",CR,LF,0
```

This instruction sequence will cause the message *Flight simulation* to be displayed and the cursor will be moved to the beginning of the next line.

Suppose labels *m*, *n*, and *gcd* represent three memory locations and the memory locations (two bytes) started with the label *gcd* hold the greatest common divisor of two numbers stored at memory locations started with label *m* and *n*. By adding the following instruction sequence to Example 4.11, the computation result can be displayed on the PC monitor screen when the program is executed on the CME-BC12 demo board:

```
CR        equ       $0D
LF        equ       $0A
printf    equ       $F686
          ...
          ldd       gcd
          pshd
          ldd       n
          pshd
          ldd       m
```

```
                    pshd
                    ldd         #prompt
                    jsr         [printf,PCR]
                    leas        6,sp
                    ...
        prompt      db          "The greatest common divisor of %d and %d is %d",CR,LF,0
                    ...
```

This example is probably not easy to follow. In reality, this example is equivalent to the following C statement:

```
printf("The greatest common divisor of %d and %d is %d\n", m, n, gcd);
```

This function call has four parameters: the first parameter is a pointer to the formatting string and should be placed in D (done by the *ldd #prompt* instruction). The other parameters (m, n, and gcd) should be pushed into the stack in order from right to left. There are three formatting characters corresponding to these three variables to be output.

int GetCmdLine(char *CmdLineStr, int CmdLineLen)

Pointer address:	$F688 ($FE08)
Stack space:	11 bytes
Incoming parameters:	a pointer to the buffer where the input string is to be stored and the maximum number of characters that will be accepted by this function
Returned value:	a string from the user (usually from the keyboard)

This function is used to obtain a line of input from the user. *GetCmdLine()* accepts input from the user a single character at a time by calling *getchar()*. As each character is received it is echoed back to the user terminal by calling *putchar()* and placed in the character array pointed to by *CmdLineStr*. A maximum of *CmdLineLen – 1* characters may be entered. Only printable ASCII characters are accepted as input with the exception of the ASCII backspace character ($08) and the ASCII carriage return ($0D). All other non-printable ASCII characters are ignored by the function.

The ASCII backspace character ($08) is used by the *GetCmdLine()* function to delete the previously received character from the command line buffer. When *GetCmdLine()* receives the backspace character, it will echo the backspace to the terminal, print the ASCII space character ($20), and then send a second backspace character to the terminal device. At the same time, the character is deleted from the command line buffer. If a backspace character is received when there are no characters in *CmdLineStr*, the backspace character is ignored.

The reception of an ASCII carriage return character ($0D) terminates the reception of characters from the user. The carriage return is not placed in the command line buffer. Instead, a NULL character ($00) is placed in the next available buffer location.

Before returning, all the entered characters are converted to upper case. *GetCmdLine()* always returns an error code of *noErr*.

We can use this function to request the user to enter a string from the keyboard when running programs on the CME-12BC demo board. Usually the program would output a message so that the user knows when to enter the string. The following instruction sequence will ask the user to enter a string from the keyboard:

```
        printf          equ         $F686
        GetCmdLine      equ         $F688
        cmdlinelen      equ         100
        CR              equ         $0D
        LF              equ         $0A
                        ...
```

```
prompt      db          "Please enter a string: ",CR,LF,0
            ...
inbuf       rmb         100
            ...
            ldd         #prompt             ; output a prompt to remind the user to
            jsr         [printf,PCR]        ; enter a string
            ldd         #cmdlinelen         ; push the CmdLineLen
            pshd                            ;    "
            ldd         #inbuf
            jsr         [GetCmdLine,PCR]    ; read a string from the keyboard
            puld                            ; clean up the stack
```

char *sscanhex(char *HexStr, unsigned int *BinNum)

Pointer address: $F68A ($FE0A)
Stack space: 6 bytes
Incoming parameters: a pointer (*HexStr*) to the string (a hex ASCII string) to be converted and a pointer (*BinNum*) to the memory location to hold the converted binary number
Returned value: The function *sscanhex()* returns either a pointer to the terminating character or a NULL pointer. A NULL pointer indicates that either an invalid hex character was found in the string or that the converted value of the ASCII hex string was greater than $FFFF.

The main purpose of this function is to convert a hex string into a hex number. Suppose that we have entered an ASCII string that represents a number; then the following instruction sequence will convert it into a hex number (or binary number):

```
sscanhex    equ         $F68A
            ...
HexStr      rmb         10                  ; input buffer to hold a hex string
BinNum      rmb         2                   ; to hold the converted number
            ...
            ldd         #BinNum
            pshd
            ldd         #HexStr
            jsr         [sscanhex,PCR]
            leas        2,sp                ; deallocates space used by outgoing parameters
            ...
```

int isxdigit(int c)

Pointer address: $F68C ($FE0C)
Stack space: 4 bytes
Incoming parameter: character to be tested (in B)
Returned value: A nonzero value is returned (in B) when the character c is a hex digit. Otherwise, a zero is returned.

This function would be useful in checking errors of input data. The following instruction sequence illustrates the use of this function:

```
isxdigit    equ         $F68C
            ...
c_buf       rmb         20                  ; buffer that holds data to be validated
            ...
            clra                            ; clear accumulator A
            ldab        c_buf               ; get one character
            jsr         [isxdigit,PCR]
            ...
```

int toupper(int c)

Pointer address:	$F68E ($FE0E)
Stack space:	4 bytes
Incoming parameter:	character to be tested (in B)
Returned value:	If character c is a lower-case letter, [a..z], *toupper()* will return the corresponding upper-case letter. If the character is upper-case, it simply returns c.

The following instruction utilizes this function to convert a character contained in B to upper-case:

```
toupper   equ    $F68E
          ...
c_buf     rmb    20                ; buffer that contains string to be converted to
          ...    ; upper-case
          ldab   c_buf             ; get one character to convert
          clra
          jsr    [toupper,PCR]
          ...
```

int isalpha(int c)

Pointer address:	$F690 ($FE10)
Stack space:	4 bytes
Incoming parameter:	character to be tested (in B)
Returned value:	A nonzero value (in B) is returned if the character c is an alphabet. Otherwise, a zero is returned.

This function would also be useful for validating an input string. The following instruction sequence illustrates the use of this function:

```
isalpha   equ    $F690
          ...
c_buf     rmb    20
          ...
          ldab   c_buf
          clra
          jsr    [isalpha,PCR]     ; check whether the character in B is alphabetic
          ...
```

unsigned int strlen(const char *cs)

Pointer address:	$F692 ($FE12)
Stack space:	4 bytes
Incoming parameter:	a pointer to the character string (in D)
Returned value:	the number of characters in the string pointed by cs (in D)

The function *strlen()* returns the length of the string pointed to by *cs*. The following instruction sequence counts the number of characters contained in the string pointed to by *cs*:

```
strlen    equ    $ F692
          ...
cs        db     "....."
          ...
          ldd    #cs
          jsr    [strlen,PCR]
          ...
```

char *strcpy(char *s1, char *s2)

Pointer address:	$F694 ($FE14)
Stack space:	8 bytes
Incoming parameters:	pointers to the source (s2) and the destination (s1) strings
Returned value:	a pointer to s1

This function copies the string pointed to by *s2* into the string pointed to by *s1* and returns a pointer to *s1*. The following instruction sequence copies the string pointed to by *s1* to the memory location pointed to by *s2*:

```
strcpy      equ     $F694
            ...
s1          db      "......"
s2          rmb     ...

            ...
            ldd     #s2
            pshd
            ldd     #s1
            jsr     [strcpy,PCR]
            leas    2,sp
            ...
```

void out2hex(unsigned int num)

Pointer address:	$F696 ($FE16)
Stack space:	70 bytes
Incoming parameters:	the lower byte (in B) of the integer num
Returned value:	none

This function displays the lower byte of *num* on the terminal screen as two hex characters. The upper byte of *num* is ignored. *Out2hex()* simply calls *printf()* with a format string of "%2.2X".

The following instruction sequence outputs the number in accumulator B as two hex digits to the terminal screen (when running on the CME-12BC demo board):

```
out2hex     equ     $F696
            ...
data   rmb  20

            ...
            ldab    data
            clra
            jsr     [out2hex,PCR]
            ...
```

void out4hex(unsigned int num)

Pointer address:	$F698 ($FE18)
Stack space:	70 bytes
Incoming parameters:	integer num in D
Returned value:	none

This function displays num on the terminal screen as four hex characters. *Out4hex()* simply calls *printf()* with a format string of "%4.4X". The following instruction sequence outputs the 16-bit number stored at memory location *num* as four hex digits:

```
out4hex     equ     $F698
            ...
num         db      ...
            ...
```

```
ldd     num
jsr     [out4hex,PCR]
...
```

int SetUserVector(int VectNum, Address UserAddress)

Pointer address:	$F69A ($FE1A)
Stack space:	8 bytes
Incoming parameters:	the interrupt source vector number and the starting address of the interrupt service routine
Returned value:	A value zero is returned when a correct VectNum is passed. Otherwise, a -1 is returned.

Since interrupts will be discussed in Chapter 6, we will discuss the use of this function in detail there.

Boolean WriteEEByte (Address EEAddress, Byte EEData)

Pointer address:	$F69C ($FE1C)
Stack space:	12 bytes
Incoming parameters:	the address of the EEPROM location to be written and the data to be written into the EEPROM
Returned value:	If the EEPROM data does not match the incoming *EEData* after the write operation, a zero is returned. Otherwise, a one is returned.

The *WriteEEByte()* function provides a mechanism to program individual bytes of the on-chip EEPROM. It does not perform any range checking on the passed *EEAddress*.

IntEraseEE(void)

Pointer address:	$F69E ($FE1E)
Stack space:	4 bytes
Incoming parameters:	none
Returned value:	A zero is returned if every EEPROM byte contains the value $FF after the erasure operation. Otherwise, a nonzero value is returned.

int ReadMem(Address StartAddress, Byte *MemDataP, unsigned int NumBytes)

Pointer address:	$F6A0 ($FE20)
Stack space:	10 bytes
Incoming parameters:	the starting address of the memory block to read, the starting address of the buffer where the read data is to be stored, and the number of bytes to read
Returned value:	A nonzero value is returned whenever there is a problem in reading. Otherwise a zero is returned.

This function is used internally by D-Bug12 for all memory read access. A user program probably won't be benefited by calling this function.

int WriteMem(Address StartAddress, Byte *MemDataP, unsigned int NumBytes)

Pointer address:	$F6A2 ($FE22)
Stack space:	22 bytes
Incoming parameters:	the starting address of the memory block to write, the starting address of the buffer where the data is to be written, and the number of bytes to write
Returned value:	A nonzero value is returned if a problem occurs while writing target memory. Otherwise, a zero is returned.

The *WriteMem()* function is used internally by D-Bug12 for all memory write accesses. If a byte is written to the memory range described by *CustData.EEBase* and *CustData.EESize*, *WriteMem()* calls the *WriteEEByte()* function to program the data into the on-chip EEPROM memory. A nonzero error code is returned if a problem occurs while writing target memory.

4.10.3 Using the D-Bug12 Functions

A useful program usually consists of several functions. That is the focus of this section.

Example 4.15

Write a program that invokes appropriate functions to find the prime number between 100 and 1,000. Output eight prime numbers in one line. To do this, you will need to:

1. Write a subroutine to test if an integer is a prime.
2. Invoke the *printf()* function to output the prime number.
3. Write a loop to test all the integers between 100 and 1,000.

Solution: The logic structure of the program is as follows:

Step 1
Output the message "The prime numbers between 100 and 1,000 are as follows:".

Step 2
For every number between 100 and 1,000, do the following:

1. Call the *test_prime()* function to see if it is a prime.
2. Output the number (call *printf()*) if it is a prime.
3. If there are already eight prime numbers in the current line, then also output a carry return.

The algorithm of the *test_prime()* function is as follows:

Step 1
Let *num*, *i*, and *isprime* represent the number to be tested, the loop index, and the flag to indicate if *num* is prime.

Step 2
isprime ← 0;

Step 3
```
For i = 2 to num/2 do
        if num % i = 0 then
                    return;
isprime ← 1;
return;
```

The stack frame of the function *test_prime()* is shown in Figure 4.18.

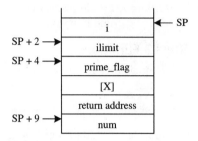

Figure 4.18 ■ Stack frame for the prime test subroutine

The assembly program is as follows:

```
CR          equ     $0D
LF          equ     $0A
i           equ     0                 ; distance of variable i from stack top
ilimit      equ     2                 ; distance of variable ilimit from stack top
prime_flag  equ     4                 ; distance of variable isprime from stack top
num         equ     9                 ; distance of variable num from the stack top
plocal      equ     5                 ; number of bytes allocated to local variables
upper       equ     2000              ; upper limit for testing prime
lower       equ     1000              ; lower limit for testing prime
printf      equ     $F686             ; location where the address of printf() is stored
            org     $800
out_buf     rmb     10
prime_cnt   rmb     1
k           rmb     2
temp        rmb     2
            org     $1000
            ldx     #upper
            stx     temp
            pshx
            ldx     #lower
            stx     k                 ; initialize k to 100 for prime testing
            pshx
            ldd     #format0
            jsr     [printf,PCR]
            leas    4,sp
            clr     prime_cnt
again       ldd     k
            cpd     #upper
            bhi     stop              ; stop when k is greater than upper
            pshd
            jsr     prime_tst         ; test if k is prime
            leas    2,sp              ; deallocate space used by outgoing parameters
            tstb
            beq     next_k            ; test next integer if k is not prime
            inc     prime_cnt         ; increment the prime count
            ldd     k
            pshd
            ldd     #format1
```

```
                    jsr      [printf,PCR]              ; output k
                    leas     2,sp
                    ldaa     prime_cnt
                    cmpa     #8                        ; are there eight prime numbers in the current line?
                    blo      next_k
; output a CR, LF if there are already eight prime numbers in the current line
                    ldd      #format2
                    jsr      [printf,PCR]
                    clr      prime_cnt
                    ldx      k
                    inx
                    stx      k
                    bra      again
next_k              ldx      k
                    inx
                    stx      k
                    jmp      again
stop                swi
prime_tst           pshx
                    leas     -plocal,sp                ; allocate local variable
                    ldaa     #1
                    staa     prime_flag,sp             ; set the flag to true
                    ldd      num,sp
                    cpd      #1
                    beq      not_prime                 ; 1 is not a prime number
                    cpd      #2
                    beq      is_prime                  ; 2 is a prime number
                    lsrd                               ; compute num/2 as the test limit
                    std      ilimit,sp                 ;       "
                    ldd      #2
                    std      i,sp                      ; set first test number to 2
test_loop           ldx      i,sp
                    ldd      num,sp
                    idiv
                    cpd      #0                        ; if num can be divided by i, then
                    beq      not_prime                 ; it is not a prime
                    ldd      i,sp
                    cpd      ilimit,sp
                    beq      is_prime                  ; num is prime at the end of test loop
                    ldx      i,sp
                    inx
                    stx      i,sp
                    bra      test_loop
not_prime           clr      prime_flag,sp
is_prime            ldab     prime_flag,sp             ; put the result in B
                    leas     plocal,sp
                    pulx
                    rts
format0             db       CR,LF,"The prime numbers between %d and %d are as follows:
                    db       ",CR,LF,CR,LF,0
format1             db       " %d ",0
format2             db       " ",CR,LF,0
                    end
```

The program execution output should look like this:

```
>load
**
>g 1000
The prime numbers between 100 and 1000 are as follows:
    101    103    107    109    113    127    131    137
    139    149    151    157    163    167    173    179
    181    191    193    197    199    211    223    227
    229    233    239    241    251    257    263    269
    271    277    281    283    293    307    311    313
    317    331    337    347    349    353    359    367
    373    379    383    389    397    401    409    419
    421    431    433    439    443    449    457    461
    463    467    479    487    491    499    503    509
    521    523    541    547    557    563    569    571
    577    587    593    599    601    607    613    617
    619    631    641    643    647    653    659    661
    673    677    683    691    701    709    719    727
    733    739    743    751    757    761    769    773
    787    797    809    811    821    823    827    829
    839    853    857    859    863    877    881    883
    887    907    911    919    929    937    941    947
    953    967    971    977    983    991    997   User Breakpoint Encountered
    PC    SP     X      Y     D = A:B   CCR = SXHI NZVC
   105F  0A00  03E9  10DD      03:E9         1001 0000
   105F   3F            SWI
```

4.11 Summary

This chapter deals with three commonly used data structures including the array, the stack, and the string. Conceptually, a stack is a list of data items whose elements can be accessed from only one end. A stack has a top and a bottom. The most common operations that can be applied to a stack are pushing and popping (or pulling). The push operation adds a new item to the top of the stack. The pop operation removes the top element of the stack. Physically, a stack is a reserved area in main memory where programs perform only push and pop operations. The 68HC12 has a 16-bit register SP that points to the top byte of the stack. The 68HC12 has instructions for pushing and pulling accumulators A, B, and D, the CCR register, and index registers X and Y.

An array is an indexable data structure. It consists of a sequence of elements of the same type and each element is associated with an index. The index of the array often starts from zero to simplify the address calculation. An array can be one-dimensional or multidimensional. A two-dimensional array is called a *matrix*.

A string is a sequence of characters terminated by a NULL (ASCII code 0) or other special character such as EOT (ASCII code $04). Common operations applied to strings include concatenation, substring insertion and deletion, character and word counting, word or substring matching, and so on.

Good program design is based on the concept of modularity—the partitioning of program code into subroutines. The main module contains the logical structure of the algorithm, while smaller program units execute many of the details.

The principles of program design in high-level languages apply even more to the design of assembly programs. Begin with a simple main program whose steps clearly outline the logical flow of the algorithm, and then assign the execution details to subroutines. Subroutines may themselves call other subroutines.

The 68HC12 provides instructions BSR, JSR, and CALL for making subroutine calls. The instructions BSR and JSR will save the return address in the stack before jumping to the subroutine. The CALL instruction will save the contents of the PPAGE register in the stack in addition to the return address. The CALL instruction is provided to call subroutines located in expanded memory. All subroutines should have RTS (RTC for subroutines in expanded memory) as the last instruction. The RTS instruction will pop the return address onto the PC register from the stack and program control will be returned to the point that called the subroutine. The RTC instruction should be used by subroutines that are in expanded memory. The RTC instruction will restore the PPAGE value pushed onto the stack in addition to the return address.

Issues related to subroutine calls include parameter passing, result returning, and local variables allocation. Several examples are given to illustrate the writing of subroutines, parameter passing, result returning, and local variable allocation.

Parameters can be passed in registers, program memory, the stack, or the global memory. The result computed by the subroutine can be returned in CPU registers, the stack, or the global memory. Local variables must be allocated in the stack so that they are not accessible to the caller and other program units. Local variables come into being only when the subroutine is being executed. The 68HC12 provides instructions to facilitate the access of variables in the stack. The *LEAS* instruction is most effective for local variables allocation and deallocation.

The D-Bug12 monitor provides many functions to support I/O programming on demo boards that include the D-Bug12 monitor. The CME-12BC and many other 68HC12 demo boards from Axiom contain the D-Bug12 monitor. These functions greatly accelerate the program development process.

4.12 Exercises

E4.1 Suppose that we have the following instruction sequence to be executed by the 68HC12. What will the contents of the topmost 4 bytes of the stack be after the execution of these instructions?

```
lds     #$8000
ldaa    #$56
staa    1, -SP
ldab    #22
staa    1, -SP
ldy     #0
sty     2, -SP
```

E4.2 Convert the binary search program into a subroutine so that it can be called by other program units. Let the starting address of the array, array count, and key to be matched be passed via the stack and the result returned in accumulator D.

E4.3 Write a subroutine to create a delay of a multiple of 100 ms. The minimum delay created by this subroutine is 100 ms. The delay to be created is equal to the value in accumulator B multiplied by 100 ms.

E4.4 Convert the program displayed in Example 4.5 into a subroutine. The 16-bit number to be converted and the pointer to the buffer to hold the converted string are passed in the stack.

E4.5 Convert the program shown in Example 4.6 into a subroutine. Let the pointer to the ASCII string to be converted be passed in index register X. This subroutine would return the resultant 16-bit hex number in double accumulator D.

E4.6 The label *array_x* is the starting address of an array of 100 8-bit elements. Trace the following code sequence and describe what the subroutine *sub_x* does.

```
                ldx      #array_x
                ldaa     #100
                jsr      sub_x
                ...
sub_x           psha
                pshx
                deca
                ldab     0,x
                inx
loop            cmpb     0,x
                ble      next
                ldab     0,x
next            inx
                deca
                bne      loop
                pulx
                pula
                rts
```

E4.7 Write a subroutine that can perform a 32-bit by 32-bit multiplication. Both the multiplicand and the multiplier are passed to this subroutine in the stack. The pointer to the buffer to hold the product is passed in index register X.

E4.8 Convert the program used in Example 4.7 into a subroutine. The starting address of the string is passed to this subroutine in index register X, and the character and word count are returned to the caller in registers D and X.

E4.9 Convert the program displayed in Example 4.8 into a subroutine. The caller of this subroutine would push the starting addresses of the string and the word into the stack, and this subroutine would return a one in B if the word is found in the string. Otherwise, a zero is returned in B.

E4.10 Draw the stack frame and enter the value of each stack slot (if it is known) at the end of the following instruction sequence:

```
                leas     -2,sp
                clrb
                ldaa     #20
                psha
                ldaa     #$E0
                psha
                ldx      #$7000
                pshx
                jsr      sub_abc
                ...
sub_abc         pshd
                leas     -12,sp
                ...
```

E4.11 Draw the stack frame and enter the value of each stack slot (if it is known) at the end of the following instruction sequence:

```
          leas    -8,sp
          ldd     #$1020
          psha
          ldx     #$800
          pshx
          bsr     xyz
          ...
xyz       pshd
          pshx
          leas    -10,sp
          ...
```

E4.12 Write a subroutine to convert all the lower-case letters in a string into upper-case. The starting address of the string is passed to this subroutine in index register X.

E4.13 Write a subroutine to convert all the upper-case letters in a string into lower-case. The starting address of the string is passed to this subroutine in index register X.

E4.14 Write a subroutine to generate a random number of 16-bit. The result is returned to the caller in D. Pass any appropriate parameters to this subroutine in the stack.

E4.15 Write a subroutine to compute the *least common multiple* of two 16-bit integers. Incoming parameters are passed in the stack and the result should be returned in a (Y,D) pair, with upper 16-bit in Y and lower 16-bit in D.

E4.16 Write a subroutine to convert the 8-bit signed integer into an ASCII string that represents a decimal number. The 8-bit integer and the pointer to the buffer to hold the ASCII string are passed to this subroutine in accumulator B and index register X, respectively.

E4.17 Write a subroutine that converts a 16-bit unsigned integer into an ASCII string (that represents a decimal number) so that it can be output to the serial communication port. Both the 16-bit integer and the pointer to the buffer to hold the ASCII string are passed to this subroutine in the stack.

E4.18 Give an instruction sequence to call the *out4hex()* function to output the 16-bit integer stored in memory location $800-$801.

E4.19 Give an instruction sequence that outputs a prompt "Please enter a string" and reads the string entered by the user from the keyboard and then echoes this string on the screen again.

E4.20 Write a subroutine that will convert a 32-bit signed integer into a BCD ASCII string so that it can be output to the serial communication port 0 (SCI0) and appear as a BCD string on the screen. The 32-bit integer to be converted and the pointer to the buffer to hold the resultant string are passed to this subroutine via the stack.

4.13 Lab Practices & Assignments

L4.1 Enter, assemble, and download the following program for execution on the demo board using MiniIDE.

```
CR          equ     $0D
LF          equ     $0A
printf      equ     $F686
getcmdline  equ     $F688
cmdlinelen  equ     40
            org     $800
inbuf       rmb     20
```

```
err_flag        rmb     1
sign_flag       rmb     1
                org     $1000
                lds     #$8000
                ldd     #prompt1
                jsr     [printf,PCR]        ; output a prompt to remind the user to enter
                ldd     #cmdlinelen         ; an integer
                pshd
                ldd     #inbuf
                jsr     [getcmdline,PCR]    ; read in a string that represents an integer
                leas    2,sp
                ldd     #prompt3            ; move cursor to the next line
                jsr     [printf,PCR]        ;    "
                ldd     #inbuf
                pshd
                ldd     #prompt2
                jsr     [printf,PCR]        ; output the numbered that you entered
                leas    2,sp
                ldd     #prompt3
                jsr     [printf,PCR]        ; move cursor to the next line
                swi
prompt1         db      "Please enter a number: ",CR,LF,0
prompt2         db      "The entered number is: %s ",0
prompt3         db      " ",CR,LF,0
                end
```

When you see the message *Please enter a number:*, enter an integer following by a carriage return. The screen output should be similar to what appears below:

```
>g 1000
Please enter a number:

The entered number is: 4321
User Breakpoint Encountered
  PC   SP    X    Y     D = A:B   CCR = SXHI NZVC
 1032 8000 106C 106B    00:03           1001 0000
 1032 3F         SWI
```

L4.2 Write a subroutine that will convert the temperature in Fahrenheit to Celsius, accurate to one decimal digit. Write a main program that will:

1. Prompt the user to enter a temperature in Fahrenheit by displaying the following message:

   Please enter a temperature in Fahrenheit:

2. Call the function *GetCmdLine()* to read in the temperature.

3. Call a subroutine to convert the input string (representing a decimal number) into binary number.

4. Call the temperature conversion subroutine to convert it to Celsius.

5. Output the current temperature to the screen in the following format:

   Current temperature:

   xxxx°F   yyyy.y°C

6. Output the next message:

Want to continue? (y/n)

7. Call the *getchar()* function to read in one character. If the character entered by the user is **y**, then repeat the process. Otherwise, return to the D-Bug12 monitor by executing the **swi** instruction.

Note: The ASCII code of the degree character ° is 176 (or $B0).

L4.3 Write a program to display the current time of day. This program will:

1. Output the message: *Please enter current time in the format of hhmmss:*.

2. Read in the time entered by the user by calling the *GetCmdLine()* function.

3. Store the current time in memory and display it the format of:

current time: hh mm ss

4. Call a delay program to delay for one second, update the current time, and also redisplay the time on the screen.

5. When redisplaying the current time, output the appropriate number of backspace characters (ASCII code is $08) so that the screen display can be more stable.

5

C Language Programming

5.1 Objectives

After completing this chapter, you should be able to:

- explain the overall structure of a C language program

- use the appropriate operators to perform desired operations in the C language

- understand the basic data types and expressions of the C language

- write program loops in the C language

- write functions and make function calls in the C language

- use arrays and pointers for data manipulation

- perform basic I/O operations in the C language

- use the ImageCraft ICC12 compiler to compile your C programs

5.2 Introduction to C

This chapter is not intended to provide a complete coverage of the C language. Instead, it provides only a summary of those C language constructs that will be used in this book. You will be able to deal with the basic 68HC12 interfacing programming if you fully understand the contents of this chapter.

The C language is gradually replacing assembly language in many embedded applications because it has several advantages over assembly language. The most important one is that it allows us to work on program logic at a level higher than assembly language, and thus programming productivity is greatly improved.

A C program, whatever its size, consists of functions and variables. A function contains statements that specify the operations to be performed. The types of statements in a function could be a *declaration, assignment, function call, control,* or *null.* A variable stores a value to be used during the computation. The *main( )* function is required in every C program and is the one to which control is passed when the program is executed. A simple C program is as follows:

```
(1)    #include <stdio.h>                              — include information about standard library
(2)    /* this is where program execution begins */
(3)    main (void)                                     — defines a function named main that receives
                                                       — no argument values
(4)    {                                               — statements of main are enclosed in braces
(5)        int  a, b, c;                               — defines three variables of type int
(6)        a = 3;                                      — assigns 3 to variable a
(7)        b = 5;                                      — assigns 5 to variable b
(8)        c = a + b;                                  — adds a and b together and assigns it to c
(9)        printf(" a + b = %d \n", c);                — calls library function printf to print the result
(10)       return 0;                                   — returns 0 to the caller of main
(11)   }                                               — the end of main function
```

The first line of the program:

```
#include <stdio.h>
```

causes the file *stdio.h* to be included in your program. This line appears at the beginning of many C programs. The header file *stdio.h* contains the prototype declarations of all I/O routines that can be called by the user program and constant declarations that can be used by the user program. The C language requires a function prototype be declared before that function can be called if a function is not defined when it is called. The inclusion of the *stdio.h* file allows the function *printf(...)* to be invoked in the program.

The second line is a comment. A comment explains what will be performed and will be ignored by the compiler. A comment in C language starts with /* and ends with */. Everything in between is ignored by the compiler. Comments provide documentation to our code and enhance readability. Comments affect only the size of the text file and do not increase the size of executable code. Many commercial C compilers also allow us to use two slashes (//) for commenting out a single line.

The third line *main()* is where program execution begins. The opening brace on the fourth line marks the start of the *main()* function's code. Every C program must have one, and only one, *main()* function. Program execution is also ended with the main function. The fifth line declares three integer variables *a, b,* and *c.* In C, all variables must be declared before they can be used.

The sixth line assigns 3 to the variable *a.* The seventh line assigns 5 to the variable *b.* The eighth line computes the sum of variables *a* and *b* and assigns it to the variable *c.* You will see that assignment statements are major components in our C programs.

The ninth line calls the library function *printf()* to print the string *a* + *b* = followed by the value of *c* and move the cursor to the beginning of the next line. The tenth line returns a 0 to the caller of *main()*. The closing brace in the eleventh line ends the *main()* function.

5.3 Types, Operators & Expressions

Variables and constants are the basic objects manipulated in a program. Variables must be declared before they can be used. A variable declaration must include the name and type of the variable and may optionally provide its initial value. A variable name may start with a letter (*A* through *Z* or *a* through *z*) or underscore character followed by zero or more letters, digits, or underscore characters. Variable names cannot contain arithmetic signs, dots, apostrophes, C keywords, or special symbols such as @, #, ?, and so on. Adding the underscore character "_" may sometimes improve the readability of long variables. Don't begin variable names with an underscore, however, since library routines often use such names. Upper- and lower-case letters are distinct.

5.3.1 Data Types

There are only a few basic data types in C: *void, char, int, float,* and *double.* A variable of type *void* represents nothing. The type void is used most commonly with functions. A variable of type *char* can hold a single byte of data. A variable of type *int* is an integer, which is normally the natural size for a particular machine. For most commercial 68HC12 C compilers, type integer has the length of 16 bits. The type *float* refers to a 32-bit, single-precision, floating-point number. The type *double* represents a 64-bit double-precision, floating-point number. In addition, there are a number of qualifiers that can be applied to these basic types. *Short* and *long* apply to integers. These two qualifiers will modify the lengths of integers. An integer variable is 16-bit by default for many C compilers, including ICC12. The modifier, short, does not change the length of an integer. The modifier, long, doubles a 16-bit integer to 32-bit. Both the type float and type double are 32 bits in length.

5.3.2 Variable Declarations

All variables must be declared before their use. A declaration specifies a type and contains a list of one or more variables of that type, as in:

```
int       i, j, k;
char      cx, cy;
```

A variable may also be initialized when it is declared, as in:

```
int i = 0;
char echo = 'y';     /* the ASCII code of letter y is assigned to variable echo. */
```

5.3.3 Constants

There are four kinds of constants: *integers, characters, floating-point numbers,* and *strings.* A character constant is an integer, written as one character within single quotes, such as 'x'. A character constant is represented by the ASCII code of the character. A string constant is a sequence of zero or more characters surrounded by double quotes, as in:

```
"68HC912BC32 is a microcontroller made by Motorola"
```

or:

```
""      /* an empty string */
```

Each individual character in the string is represented by its ASCII code.

An integer constant such as 3241 is an *int*. A long constant is written with a terminal l (el) or L, as in 44332211L. The following constant characters are predefined in C language:

\a	alert (bell) character	\\	backslash
\b	backspace	\?	question mark
\f	formfeed	\'	single quote
\n	newline	\"	double quote
\r	carriage return	\ooo	octal number
\t	horizontal tab	\xhh	hexadecimal number
\v	vertical tab		

As in assembly language, a number in C can be specified in different bases. The method to specify the base of a number is to add a prefix to the number. The prefixes for different bases are:

Base	Prefix	Example	
decimal	none	1357	
octal	0	047233	;preceded by a zero
hexadecimal	0x	0x2A	

5.3.4 Arithmetic Operators

There are seven arithmetic operators:

+	add and unary plus
-	subtract and unary minus
*	multiply
/	divide
%	modulus (or remainder)
++	increment
--	decrement

The expression:

a % b

produces the remainder when a is divided by b. The % operator cannot be applied to float or double. The ++ operator adds one to the operand, and the -- operator subtracts one from the operand. The / operator truncates the quotient to integer when both operands are integers.

5.3.5 Bitwise Operators

C provides six operators for bit manipulations; these may only be applied to integral operands, that is, *char, short, int,* and *long,* and whether each is *signed* or *unsigned.*

&	AND
|	OR
^	XOR
~	NOT
>>	right shift
<<	left shift

The & operator is often used to clear one or more bits to zero. For example, the statement:

```
PORTC      = PORTC & 0xBD;        /* PORTC is 8 bits */
```

clears bits 6 and 1 of PORTC to 0.

The | operator is often used to set one or more bits to one. For example, the statement:

```
PORTB      = PORTB | 0x40;        /* PORTB is 8 bits */
```

sets the bit 6 of PORTB to 1.

The XOR operator can be used to toggle a bit. For example, the statement:

```
abc        = abc ^ 0xF0;          /* abc is of type char */
```

toggles the upper four bits of the variable *abc*.

The >> operator shifts the involved operand to the right for the specified number of places. For example:

```
xyz = abc >> 3;
```

shifts the variable *abc* to the right three places and assigns it to the variable *xyz*.

The << operator shifts the involved operand to the left for the specified number of places. For example:

```
xyz = xyz << 4;
```

shifts the variable *xyz* to the left four places.

The assignment operator = is often combined with the operator. For example:

```
PORTD = PORTD & 0xBD;
```

can be rewritten as:

```
PORTD &= 0xBD;
```

The statement:

```
PORTB = PORTB | 0x40;
```

can be rewritten as:

```
PORTB |= 0x40;
```

5.3.6 Relational & Logical Operators

Relational operators are used in expressions to compare the values of two operands. If the result of the comparison is true, then the value of the expression is one. Otherwise, the value of the expression is zero. Here are the relational and logical operators:

==	equal to (two "=" characters)
!=	not equal to
>	greater than
>=	greater than or equal to
<	less than
<=	less than or equal to
&&	and
||	or
!	not (one's complement)

Here are some examples of relational and logical operators:

```
if (!(ADCTL & 0x80))
        statement₁;          /* if bit 7 is 0, then execute statement₁ */
if (i > 0 && i < 10)
        statement₂;          /* if 0 < i < 10 then execute statement₂ */
if (a1 == a2)
        statement₃;          /* if a1 equals a2 then execute statement₃ */
```

5.3.7 Precedence of Operators

Precedence refers to the order in which operators are processed. The C language maintains a precedence for all operators. The precedence for all operators is shown in Table 5.1. Operators at the same level are evaluated from left to right. A few examples that illustrate the precedence of operators are listed in Table 5.2.

Precedence	Operator	Associativity
Highest	( ) [ ] → .	left to right
	! ~ ++ -- - (type) * & sizeof	right to left
	* / %	left to right
	+ -	left to right
	<< >>	left to right
	< <= > >=	left to right
	== !=	left to right
	&	left to right
	^	left to right
	|	left to right
	&&	left to right
	||	left to right
	?:	right to left
	= += -= *= /= %= &= ^= |= <<= >>=	right to left
Lowest	'	left to right

Table 5.1 ■ Table of precedence of operators

Expression	Result	Note
15 - 2 * 7	1	* has higher precedence than +
(13 - 4) * 5	45	
(0x20 | 0x01) != 0x01	1	
0x20 | 0x01 != 0x01	0x20	!= has higher precedence than |
1 << 3 + 1	16	+ has higher precedence than <<
(1 << 3) + 1	9	

Table 5.2 ■ Examples of operator precedence

5.4 Control Flow

The control-flow statements specify the order in which computations are performed. In the C language, the semicolon is a statement terminator. Braces, { and }, are used to group declarations and statements together into a *compound statement*, or *block*, so that they are syntactically equivalent to a single statement.

5.4.1 If Statement

The *if statement* is a conditional statement. The statement associated with the *if statement* is executed based upon the outcome of a condition. If the condition evaluates to nonzero, the statement is executed. Otherwise, it is skipped.

The syntax of the *if statement* is as follows:

```
if (expression)
      statement;
```

Here is an example of an *if statement*:

```
if (a > b)
      sum += 2;
```

The value of the sum will be incremented by two if the variable a is greater than the variable b.

5.4.2 If-Else Statement

The *if-else* statement handles conditions where a program requires one statement to be executed if a condition is nonzero and a different statement if the condition is zero.

The syntax of an *if-else* statement is:

```
if (expression)
      statement₁
else
      statement₂
```

The *expression* is evaluated; if it is true (non-zero), $statement_1$ is executed. If it is false, $statement_2$ is executed. Here is an example of the *if-else* statement:

```
if (a != 0)
      r = b;
else
      r = c;
```

The *if-else* statement can be replaced by the **?:** operator. The statement:

```
r = (a != 0)? b : c;
```

is equivalent to the previous if-statement.

5.4.3 Multiway Conditional Statement

A multiway decision can be expressed as a cascaded series of *if-else* statements. Such a series looks like this:

```
if (expression₁)
      statement₁
else if (expression₂)
      statement₂
else if (expression₃)
      statement₃
```

```
...
else
        statement_n
```

Here is an example of a three-way decision:

```
if (abc > 0) return 5;
else if (abc == 0) return 0;
else return –5;
```

5.4.4 Switch Statement

The *switch* statement is a multiway decision based on the value of a control expression. The syntax of the switch statement is:

```
switch (expression) {
        case const_expr_1:
                statement_1;
                break;
        case const_expr_2:
                statement_2;
                break;
        ...
        default:
                statement_n;
        }
```

As an example, consider the following program fragment:

```
switch (i) {
        case 1: printf("*");
                break;
        case 2: printf("**");
                break;
        case 3: printf("***");
                break;
        case 4: printf("****");
                break;
        case 5: printf("*****");
        default:
                printf("\n");
}
```

The number of * characters printed is equal to the value of i. The break keyword forces the program flow to drop out of the switch statement so that only the statements under the corresponding *case-label* are executed. If any break statement is missing, then all the statements from that case-label until the next break statement within the same switch statement will be executed.

5.4.5 For-Loop Statement

The syntax of a *for-loop* statement is:

```
for (expr1; expr2; expr3)
        statement;
```

where *expr1* and *expr3* are assignments or function calls, and *expr2* is a relational expression. For example, the following *for loop* computes the sum of the squares of integers from one to nine:

```
sum = 0;
for (i = 1; i < 10; i++)
        sum = sum + i * i;
```

The following *for loop* prints out the first 10 odd integers:

```
for (i = 1; i < 20; i++)
        if (i % 2) printf("%d ", i);
```

5.4.6 While Statement

While an expression is nonzero, the *while* loop repeats a statement or block of code. The value of the expression is checked prior to each execution of the statement. The syntax of a *while* statement is:

```
while (expression)
        statement;
```

The *expression* is evaluated. If it is nonzero (true), *statement* is executed and *expression* is re-evaluated. This cycle continues until *expression* becomes zero (false), at which point execution resumes after statement. The statement may be a NULL statement. A NULL statement does nothing and is represented by a semicolon.

Consider the following program fragment:

```
int_cnt = 5;
while (int_cnt);
```

The CPU will do nothing before the variable *int_cnt* is decremented to zero. In microprocessor applications, the decrement of *int_cnt* is often done by external events such as interrupts.

5.4.7 Do-While Statement

The *while* and *for* loops test the termination condition at the beginning. By contrast, the *do-while* statement tests the termination condition at the end of the statement; the body of the statement is executed at least once. The syntax of the statement is:

```
do
        statement
while (expression);
```

The following *do-while* statement displays the integers nine down to one:

```
int  digit = 9;
do
        printf("%d ", digit--);
while  (digit >= 1);
```

5.4.8 GOTO Statement

Execution of a *goto* statement causes control to be transferred directly to the labeled statement. This statement must be located in the same function as the *goto* statement. The use of the *goto* statement interrupts the normal sequential flow of a program and thus makes it harder to follow and decipher. For this reason, the use of *goto's* is not considered good programming style, so it is recommended that you do not use them in your program.

The syntax of the *goto* statement is:

```
goto label
```

An example of the use of the *goto* statement is as follows:

```
if (x > 100)
            goto        fatal_error;
    ...
fatal_error:
        printf("Variable x is out of bound!\n");
```

5.5 Input & Output

Input and output facilities are not part of the C language itself. However, input and output are fairly important in application. The ANSI standard defines a set of library functions that must be included so that they can exist in a compatible form on any system where C exists.

Some of the functions deal with file input and output. Others deal with text input and output. In this section we will look at the following four input and output functions:

1. int *getchar* (). This function returns a character when it is called. The following program fragment returns a character and assigns it to the variable *xch*:

   ```
   char xch;

   xch = getchar ();
   ```

2. int *putchar* (int). This function outputs a character on the standard output device. The following statement outputs the letter *a* from the standard output device:

   ```
   putchar ('a');
   ```

3. int *puts* (const char *s). This function outputs the string pointed to by *s* on the standard output device. The following statement outputs the string *Learning microcontroller is fun!* from the standard output device:

   ```
   puts ("Learning microcontroller is fun! \n");
   ```

4. int *printf* (*formatting string*, arg_1, arg_2, ..., arg_n). This function converts, formats, and prints its arguments on the standard output under control of *formatting string*. arg_1, arg_2, ..., arg_n are arguments that represent the individual output data items. The arguments can be written as constants, single variable or array names, or more complex expressions. The formatting string is composed of individual groups of characters, with one character group associated with each output data item. The character group corresponding to a data item must start with %. In its simplest form, an individual character group will consist of the percent sign followed by a *conversion character* indicating the type of the corresponding data item.

   Multiple character groups can be contiguous, or separated by other characters, including white space characters. These "other" characters are simply transferred directly to the output device where they are displayed. A subset of the more frequently used conversion characters are listed in Table 5.3.

Conversion character	Meaning
c	data item is displayed as a single character
d	data item is displayed as a signed decimal number
e	data item is displayed as a floating-point value with an exponent
f	data item is displayed as a floating-point value without an exponent
g	data item is displayed as a floating-point value using either e-type or f-type conversion, depending on value; trailing zeros, trailing decimal point will not be displayed
i	data item is displayed as a signed decimal integer
o	data item is displayed as an octal integer, without a leading zero
s	data item is displayed as a string
u	data item is displayed as an unsigned decimal integer
x	data item is displayed as a hexadecimal integer, without the leading 0x

Table 5.3 ■ Commonly used conversion characters for data output

Between the % character and the conversion character there may be, in order:

- A minus sign, which specifies left adjustment of the converted argument.
- A number that specifies the minimum field width. The converted argument will be printed in a field at least this wide. If necessary it will be padded on the left (or right, if left adjustment is called for) to make up the field width.
- A period, which separates the field width from precision.
- A number, the precision, that specifies the maximum number of characters to be printed from a string, or the number of digits after the decimal point of a floating-point value, or the minimum number of digits for an integer.
- An *h* if the integer is to be printed as a *short*, or *l* (letter el) if as a *long*.

Several valid printf calls are as follows:

```
printf ("this is a challenging course! \n");      /* outputs only a string */

printf ("%d %d %d", x1, x2, x3);                  /* outputs variables x1, x2, x3 using minimal number
                                                     of digits with one space separating each value */

printf ("Today's temperature is %4.1d \n", temp); /* display the string Today's
                                                     temperature is followed by the value
                                                     of temp. Display one fractional digit
                                                     and use at least four digits for the
                                                     value. */
```

5.6 Functions & Program Structure

Every C program consists of one or more functions. If a program consists of multiple functions, their definitions cannot be embedded within another. The same function can be called from several different places within a program. Generally, a function will process information passed to it from the calling portion of the program and return a single value. Information is passed to the function via special identifiers called *arguments* (also called *parameters*) and returned via the *return* statement. Some functions, however, accept information but do not return anything (for example, the library function *printf*).

The syntax of a function definition is as follows:

```
return_type  function_name (declarations of arguments)
{
       declarations and statements
}
```

The declaration of an argument in the function definition consists of two parts: the *type* and the *name* of the variable. The return type of a function is *void* if it does not return any value to the caller. An example of a function that converts a lower-case letter to an upper-case letter is as follows:

```
char  lower2upper (char cx)
{
       if (cx >= 'a' && cx <= 'z') return (cx – ('a' - 'A'));
       else return cx;
}
```

A character is represented by its ASCII code. A letter is in lower-case if its ASCII code is between 97 (0x61) and 122 (0x7A). To convert a letter from lower-case to upper-case, subtract its ASCII code by the difference of the ASCII codes of letters *a* and *A*.

To call a function, simply put down the name of the function and replace the argument declarations by actual arguments or values and terminate it with a semicolon.

Example 5.1
▼

Write a function to test whether an integer is a prime number.

Solution: The integer one is not a prime number. A number is prime if it is indivisible by any integer between two and its half.

```
/* this function returns a 1 if a is prime. Otherwise, it returns a 0. */
char  test_prime (int a)
{
       int i;
       if (a == 1) return 0;
       for (i = 2; i < a/2; i++)
              if ((a % i) == 0) return 0;
       return 1;
}
```

▲

Example 5.2

▼

Write a program to find out the number of prime numbers between 100 and 1,000.

Solution: We can find the number of prime numbers between 100 and 1,000 by calling the function in Example 5.1 to find out if a number is prime.

```
#include <stdio.h>
char test_prime (int a);        /* prototype declaration for the function test_prime */

main ( )
{
        int  i, prime_count;
        prime_count = 0;
        for (i = 100; i <= 1000; i++) {
            if (test_prime(i))
                    prime_count ++;
        }
        printf("\n The total prime numbers between 100 and 1,000 is %d\n", prime_count);
}
char  test_prime (int a)
{
        int i;
        if (a == 1) return 0;
        for (i = 2; i < a/2; i++)
            if ((a % i) == 0) return 0;
        return 1;
}
```

A function cannot be called before it has been defined. This dilemma is solved by using the function prototype statement. The syntax for a function prototype statement is as follows:

```
return_type function_name (declarations of arguments);
```

The statement:

```
char test_prime (int a);
```

before *main()* is a function prototype statement.

▲

5.7 Pointers, Arrays, Structures & Unions

5.7.1 Pointers & Addresses

A *pointer* is a variable that holds the address of a variable. Pointers are used frequently in C, as they have a number of useful applications. For example, pointers can be used to pass information back and forth between a function and its reference (calling) point. In particular, pointers provide a way to return multiple data items from a function via function arguments. Pointers also permit references to other functions to be specified as arguments to a given function. This has the effect of passing functions as arguments to the given function.

Pointers are also closely associated with arrays and therefore provide an alternate way to access individual array elements.

The syntax for declaring a pointer type is:

type_name *pointer_name;

For example:

int *ax;

declares that the variable *ax* is a pointer to an integer.

char *cp;

declares that the variable *cp* is a pointer to a character.

To access the value pointed to by a pointer, use the *dereferencing* operator *. For example:

```
int     a, *b;       /* b is a pointer to int */
...
a = *b;
```

assigns the value pointed to by *b* to variable *a*.

We can assign the address of a variable to a pointer by using the unary operator *&*. The following example shows how to declare a pointer and how to use & and *:

```
int x, y;
int *ip;            // ip is a pointer to an integer

ip = &x;            // assigns the address of the variable x to ip
y = *ip;            // y gets the value of x
```

Example 5.3

Write the bubble sort function to sort an array of integers.

Solution: The algorithm for the bubble sort is already described in Chapter 4. Here is the C language version:

```
void    swap (int *, int *);
void    bubble (int a[], int n)  /* n is the array count */
{
        int i, j;
        for (i = 0; i < n - 1; i++)
                for (j = 0; j < n - i - 1; j++)
                        if (a[j - 1] > a[j])
                                swap (&a[j - 1], &a[j]);

}
void    swap (int *px, int *py)
{
        int temp;
        temp = *px;
        *px = *py;
        *py = temp;
}
```

5.7.2 Arrays

Many applications require the processing of multiple data items that have common characteristics (e.g., a set of numerical data, represented by $x_1, x_2, ..., x_n$). In such situations it is more convenient to place data items into an *array*, where they will all share the same name. The individual data items can be characters, integers, floating-point numbers, and so on. They must all be of the same type and the same storage class.

Each array element is referred to by specifying the array name followed by one or more *subscripts*, with each subscript enclosed in brackets. Each subscript must be expressed as a non-negative integer. Thus, the elements of an n-element array x are x[0], x[1], ..., x[n − 1]. The number of subscripts determines the dimensionality of the array. For example, x[i] refers to an element of a one-dimensional array. Similarly, y[i][j] refers to an element of a two-dimensional array. Higher dimensional arrays can be formed by adding additional subscripts in the same manner. However, higher dimensional arrays are not used very often in 8- and 16-bit microcontroller applications.

In general, a one-dimensional array can be expressed as:

data-type array_name[expression];

A two-dimensional array is defined as:

data-type array_name[expr1][expr2];

An array can be initialized when it is defined. This is a technique used in table lookup, which can speed up the computation process. For example, a data acquisition system that utilizes an analog-to-digital converter can use the table lookup technique to speed up the conversion (from digital value back to the original physical quantity) process.

5.7.3 Pointers & Arrays

In C, there is a strong relationship between pointers and arrays. Any operation that can be achieved by array subscripting can also be done with pointers. The pointer version will in general be faster but somewhat harder to understand.

For example:

int ax[20];

defines an array *ax* of 20 integral numbers.

The notation ax[i] refers to the *i*-th element of the array. If *ip* is a pointer to an integer, declared as:

int *ip;

then the assignment:

ip = &ax[0];

makes ip contain the address of ax[0]. Now the statement:

x = *ip;

will copy the contents of ax[0] into x.

If *ip* points to ax[0], then ip+1 points to ax[1], and ip+i points to ax[i], etc.

5.7.4 Passing Arrays to a Function

An array name can be used as an argument to a function, thus permitting the entire array to be passed to the function. To pass an array to a function, the array name must appear by itself, without brackets or subscripts, as an actual argument within the function call. When declaring a one-dimensional array as a formal argument, the array name is written with a pair of empty square brackets. The size of the array is not specified within the formal argument declaration. If the array is two-dimensional, then there should be two pairs of empty brackets following the array name.

The following program outline illustrates the passing of an array from the main portion of the program to a function.

```
int    average (int n, int arr[]);
void main ( )
{
      int n, avg;                /* variable declaration */
      int arr[50];               /* array definition */
      ...
      avg = average(n, arr);     /* function call */
      ...
}
int average (int k, int brr[])   /* function definition */
{
      ...
}
```

Within *main* we see a call to the function *average*. This function call contains two actual arguments—the integer variable *n* and the one-dimensional, integer array *arr*. Note that *arr* appears as an ordinary variable within the function call.

In the first line of the function definition, we see two formal arguments: *k* and *brr*. The formal argument declarations establish *k* as an integer variable and *brr* as a one-dimensional integer array. Note that the size of *brr* is not defined in the function definition.

As formal parameters in a function definition:

```
int brr[];
```

and:

```
int *brr;
```

are equivalent.

5.7.5 Initializing Arrays

C allows initialization of arrays. Standard data type arrays may be initialized in a straightforward manner. The syntax for initializing an array is as follows:

```
array_declarator = { value-list }
```

The following statement shows a 5-element integer array initialization:

```
int i[5] = {10, 20, 30, 40, 50};
```

The element i[0] has the value of 10 and the element i[4] has the value of 50.

A string (character array) can be initialized in two ways. One method is to make a list of each individual character:

```
char strgx[5] = {'w', 'x', 'y', 'z', 0};
```

The second method is to use a string constant:

```
char myname [6] = "huang";
```

A null character is automatically appended at the end of "huang". When initializing an entire array, the array size (which is one more than the actual length) must be included:

```
char prompt [24] = "Please enter an integer:";
```

5.7.6 Structures

A structure is a group of related variables that can be accessed through a common name. Each item within a structure has its own data type, which can be different from the other data items. The syntax of a structure declaration is as follows:

```
struct struct_name {                    /* struct_name is optional */
        type1       member1;
        type2       member2;
        ...
};
```

The *struct_name* is optional and if it exists, defines a *structure tag*. A *struct* declaration defines a type. The right brace that terminates the list of members may be followed by a list of variables, just as for any basic type. The following example is for a card catalog in a library:

```
struct catalog_tag {
        char        author [40];
        char        title [40];
        char        pub [40];
        unsigned int date;
        unsigned char rev;
} card;
```

where the variable *card* is of type *catalog_tag*.

A structure definition that is not followed by a list of variables reserves no storage; it merely describes a template or the shape of a structure. If the declaration is tagged (i.e., has a name), however, the tag can be used later in definitions of instances of the structure. For example, suppose we have the following structure declaration:

```
struct point {
        int x;
        int y;
};
```

We can then define a variable *pt* of type *point* as follows:

```
struct point pt;
```

A member of a particular structure is referred to in an expression by a construction of the form:

```
structure-name.member
```

or:

```
structure-pointer→member
```

The structure member operator "." connects the structure name and the member name. As an example, the square of the distance of a point to the origin can be computed as follows:

```
long integer sq_distance;
...
sq_distance = pt.x * pt.x + pt.y * pt.y;
```

Structures can be nested. One representation of a circle consists of the center and radius shown in Figure 5.1.

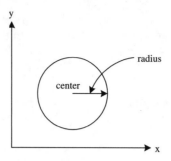

Figure 5.1 ■ A circle

This circle can be defined as:

```
struct circle {
        struct point center;
        unsigned int radius;
};
```

5.7.7 Unions

A *union* is a variable that may hold (at different times) objects of different types and sizes, with the compiler keeping track of size and alignment requirements. Unions provide a way to manipulate different kinds of data in a single area of storage, without embedding any machine-dependent information in the program.

The syntax of the union is as follows:

```
union  union_name {
        type-name1          element1;
        type-name2          element2;
        ...
        type-namen          elementn;
};
```

The field *union_name* is optional. When it exists, it is also called *union-tag*. You can declare a union variable at the same time that you declare a union type. The union variable name should be placed after the right brace "}".

In order to represent the current temperature using both the integer and string, we can use the following declaration:

```
union u_tag {
        int i;
        char c[4];
} temp;
```

Four characters must be allocated to accommodate the larger of the two types. Integer type is good for internal computation, whereas string type is suitable for output. Of course, some conversion may be needed before we make certain kinds of interpretation. Using this method, we can interpret the variable *temp* as an integer or a string, depending on our purpose.

Syntactically, members of a union are accessed as:

```
union-name.member
```

or:

```
union-pointer → member
```

just as for structures.

5.8 Miscellaneous Items

5.8.1 Automatic/External/Static/Volatile

A variable defined inside a function is an *internal variable* of that function. These variables are called *automatic* because they come into existence when the function is entered, and disappear when it is left. Internal variables are equivalent to local variables in assembly language. *External variables* are defined outside of any function, and are thus potentially available to many functions. Because external variables are globally accessible, they provide an alternative to function arguments and return values for communicating data between functions. Any function may access an external variable by referring to it by name, if the name has been declared somehow.

External variables are also useful when two functions must share some data, yet neither calls the other.

The use of *static* with a local variable declaration inside a block or a function causes the variable to maintain its value between entrances to the block or function. Internal static variables are local to a particular function just as automatic variables are, but unlike automatic variables, they remain in existence rather than coming and going each time the function is activated. When a variable is declared static outside of all functions, its scope is limited to the file that contains the definition. A function can also be declared as static. When a function is declared as static, it becomes invisible outside of the file that defines the function.

A *volatile* variable has a value that can be changed by something other than user code. A typical example is an input port or a timer register. These variables must be declared as volatile so the compiler makes no assumptions on their values while performing optimizations. The keyword volatile prevents the compiler from removing apparently redundant references through the pointer.

5.8.2 Scope Rules

The functions and external variables that make up a C program need not all be compiled at the same time; the source text of the program may be kept in several files, and previously compiled routines may be loaded from libraries.

The scope of a name is the part of the program within which the name can be used. For a variable declared at the beginning of a function, the scope is the function in which the name is declared. Local (internal) variables of the same name in different functions are unrelated.

The scope of an external variable or a function lasts from the point at which it is declared to the end of the file being compiled. Consider the following program segment:

```
...
void f1 (...)
{
        ...
}
int a, b, c;
void f2 (...)
{
        ...
}
```

Variables a, b, and c are accessible to function f2 but not to f1.

When a C program is split into several files, it is convenient to put all global variables into one file so that they can be accessed by functions in different files. Functions residing in different files that need to access global variables must declare them as external variables. In addition, we can place the prototypes of certain functions in one file so that they can be called by functions in other files.

The following example is a skeletal outline of a two-file C program that makes use of external variables:

In file1:

```
extern  int  xy;
extern  long arr[];
main ( )
{
        ...
}
void foo (int abc) { ... }
long soo (void) { ... }
```

In file2:

```
int  xy;
long  arr[100];
```

5.9 Using the Imagecraft C Compiler (ICC12)

ICC12 is a C compiler and development environment for the Motorola 68HC12 microcontroller running on Windows platforms. The compiler accepts the ANSI C language with the following exception:

The supplied library is only a subset of what is defined by the ANSI standard.

ICC12 provides a small IDE (Integrated Development Environment) that includes a project builder, a syntax-aware text editor, a terminal program, plus the compiler. The tight integration of tools allows for a fast edit-compile-download development cycle for the 68HC12.

The procedure for using the ICC12 to enter, compile, and download a C program into a 68HC12 demo board for execution is as follows:

Step 1
Invoke the ICC12 by clicking the icon of the ICC12 program. The ICC12 window will appear as shown in Figure 5.2. There are three regions in the command window. The largest region is the *workspace*, which is used to display the files; when the terminal emulator function is active, the workspace is used to display the terminal emulator, etc. The right pane is the *project manager* window. This pane contains the list of C and assembly files in a project. The pane at the bottom of the window is the *status window*. Any compilation status is displayed on this pane. In addition, the bottom status bar displays useful information such as the full file name of the file being edited, the cursor position, and the full file name of the project file.

Step 2
Click the **File** menu and select **New** to create a new file for holding a program. The screen should look like Figure 5.3. This step must be repeated as many times as the number of files required in your program. If you want to edit an existing file rather than create a new one, select **Open** instead of **New**.

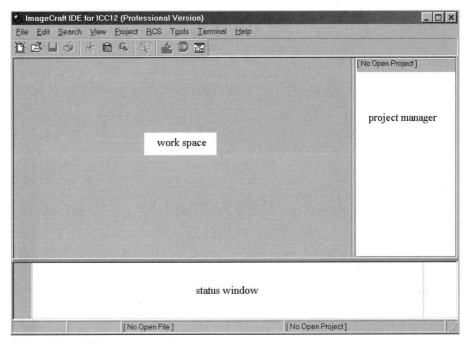

Figure 5.2 ■ ICC12 command window

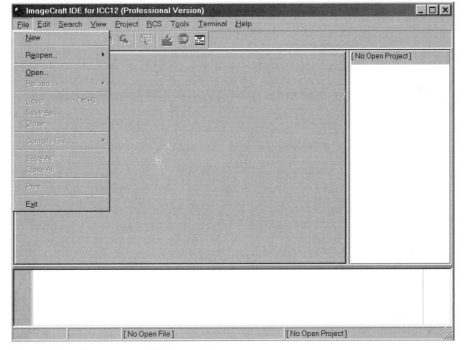

Figure 5.3 ■ Create a new file for entering a program

Step 3

Type in the C program and save it in an appropriate directory (name it gcd.c). An example is shown in Figure 5.4.

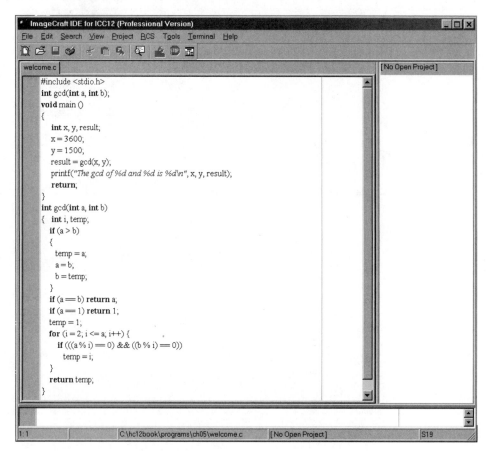

Figure 5.4 ■ Entering a new program in ICC12

Step 4

Create a new project by clicking the **Project** menu and selecting **New**. The screen is shown in Figure 5.5. After this, the screen should change to what appears in Figure 5.6. Type in an appropriate project name (call it *gcd.prj*) that you like and click the **OK** button. The result is shown in Figure 5.7.

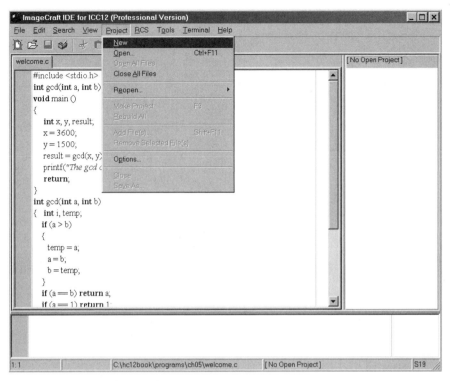

Figure 5.5 ■ Screen for creating a new project

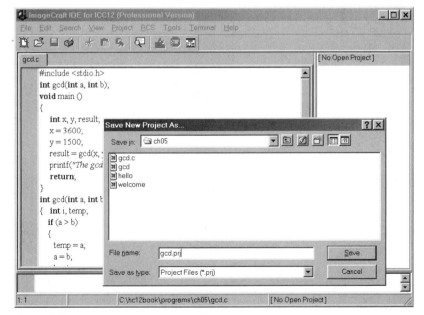

Figure 5.6 ■ Create a new project gcd.prj

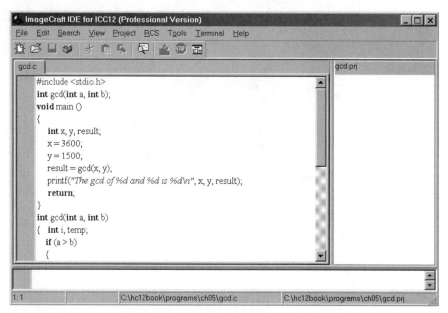

Figure 5.7 ■ Screen after setting the project name

Step 5

Add programs to the newly created project. To add a program to the project, press the **Add** button. A list of files will be displayed, and you can select those files that you want to include in this project. In this example, we will include the file *gcd.c.* The screen should look similar to Figure 5.8. Click on this file and press the **OK** button.

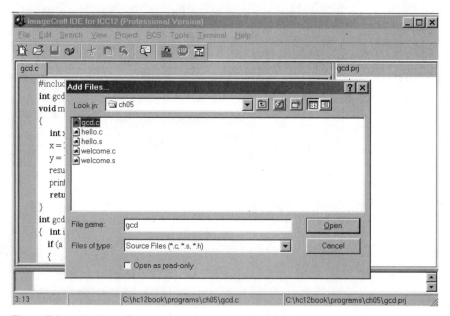

Figure 5.8 ■ Adding a file into the project gcd

Step 6

Setting appropriate options for the project, editor, and terminal. The project options are set by pressing the **Project** menu and selecting **Options**.... The screen should now appear like Figure 5.9. You can set the search path for the included files, compiler options, and target options from here. Click on the **Path** button to set the path. Only the search path for the included files and library need to be set. Usually the default paths are correct and need not be changed. You also need to make sure that the compiler options are set correctly. Here, the default setting shown in Figure 5.10 is acceptable. It is important to set up the target option correctly. The target option screen is shown in Figure 5.11. The device configuration field allows you to select the device. If your target is not on the list, select "**Custom**" and enter the relevant parameters in other fields. Since we are using a demo board in learning the 68HC12, we should choose **Custom**. The starting address of the **Program Memory** of the Axiom CME-12BC demo board can be set to 0x1000. **Data Memory** usually follows immediately after the program. Unless you have a special preference, it can be left blank. The stack pointer also needs to be set. The stack starts from the highest address and grows downward. You can use internal or external SRAM as the stack space. Suppose you use the internal SRAM of the 68HC912BC32 as the stack space, then the stack pointer can be set to 0xA00. After all the options have been set properly, click **OK**. The computer will save the options.

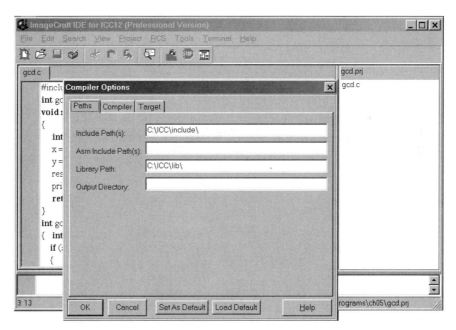

Figure 5.9 ■ Setting the search path

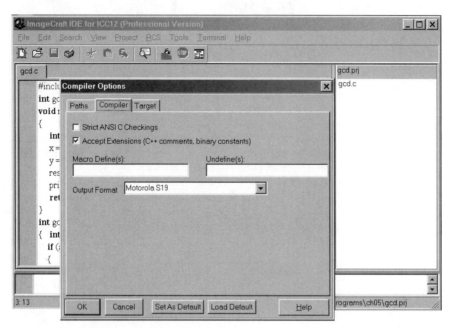

Figure 5.10 ■ Set the compiler option

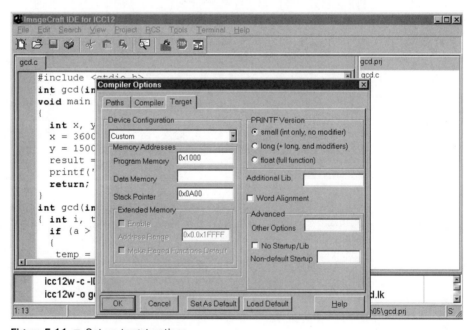

Figure 5.11 ■ Set up target options

Step 7

Set up other options. There are other options to be set up. The most important one is the option for the terminal program. To set the terminal option, press the **Tools** menu and select **Environment Options**. Click on *terminal*. The proper values for communicating with the CME-12BC demo board (or other demo boards made by Axiom) are shown in Figure 5.12.

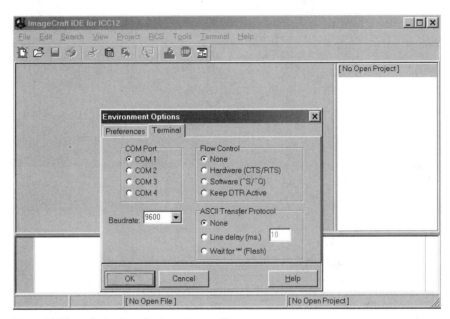

Figure 5.12 ■ Setup terminal program options

Step 8

Build the project. After setting all the options properly, we want to compile the source code into executable code and download it onto the demo board for execution. Press on the **Project** menu and select **Make Project**. The result will look like Figure 5.13 if the project is built correctly.

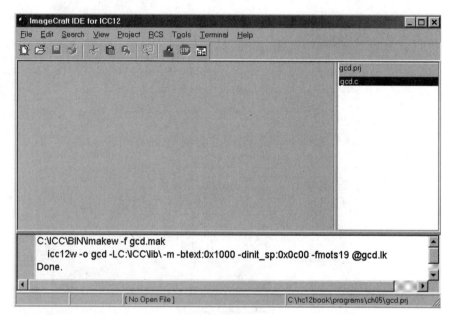

Figure 5.13 ■ A successful make project process

Step 9

Download the program onto the demo board for execution. Press the **Terminal** menu and select **Show Terminal Window** and the screen will change to what you see in Figure 5.14. Click on **Open Com Port** and you will be able to communicate with the D-Bug12 monitor on the CME-12BC demo board. Press the enter key and you will see the D-Bug12 monitor prompt as shown in Figure 5.15. If the monitor prompt did not appear, hit the reset key on the demo board. To download, type **load** and enter key and click on **Browse** to select the file to be downloaded. The screen for browsing the file to be downloaded is shown in Figure 5.16. Click on the name of the file (*gcd.s19*) to be downloaded and click on **Download!**, and the selected file will be sent to the demo board. The screen of a successful download is shown in Figure 5.17.

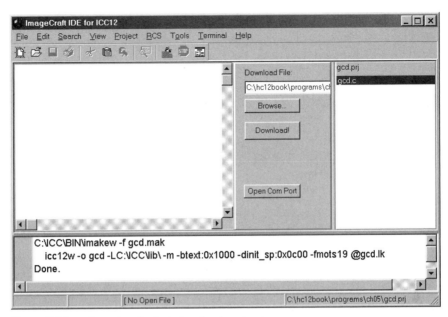

Figure 5.14 ■ Open terminal window for download

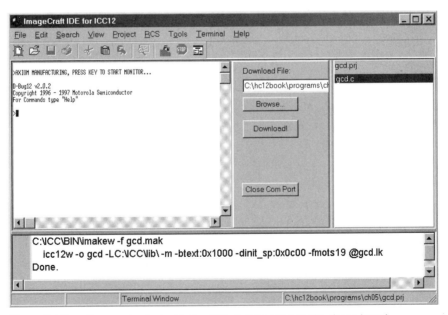

Figure 5.15 ■ Communicating with the D-Bug12 monitor on the demo board

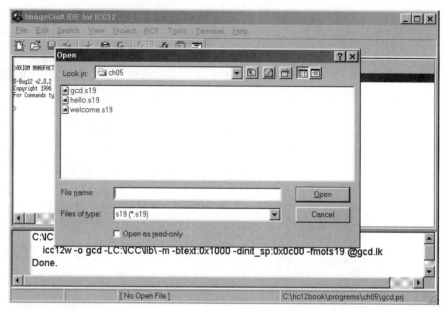

Figure 5.16 ■ Screen for browsing and selecting a file to be downloaded

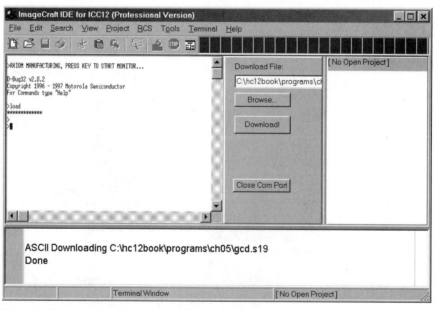

Figure 5.17 ■ A successful download to the demo board

Step 10

Program execution and debugging. The ICC12 terminal program allows you to use all the D-Bug12 commands to debug programs when using the CME-12BC demo board. To run the program that we just downloaded, enter **G 1000**. After running the program *gcd.c*, the screen should look like Figure 5.18. To redisplay the D-Bug12 monitor prompt, press the reset button of the demo board and also hit carriage return.

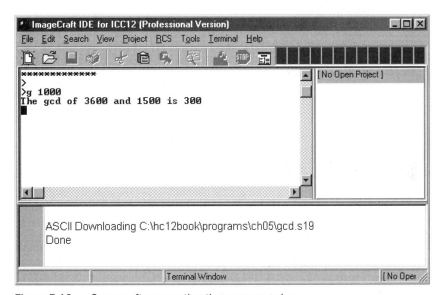

Figure 5.18 ■ Screen after executing thr program gcd.c

The ICC12 saves the options settings for future use so you won't need to set the options every time. There are many details in the menus of different categories. You are encouraged to explore these menus of the ICC12 to make yourself a better user of the ICC12.

The ICC12 has included most of the library functions defined in the ANSI C language standard. Interested readers should refer to Appendix B for further information.

5.10 Inline Assembly Instructions

Inline assembly allows you to write assembly code within your C program. The syntax for inline assembly is:

asm("<string>");

Multiple assembly statements can be separated by the newline character \n. String concatenations can be used to specify multiple statements without using additional *asm* keywords. To access a C variable inside an assembly statement, use the %<*name*> format:

```
unsigned char uc;
asm("ldab %uc\n"
    "stop\n");
```

Any C variable can be referenced this way (excluding the C goto labels).

Inline assembly may be used inside or outside of a C function. The compiler indents each line of the inline assembly for readability. You may get a warning on *asm* statements that are outside of a function. You may ignore these warnings.

5.11 Summary

A C program consists of one or more functions and variables. The *main( )* function is required in every C program. It is the entry point of your program. A function contains statements that specify the operations to be performed. The types of statements in a function could be *declaration, assignment, function call, control,* or *null.*

A *variable* stores a value to be used during the computation. A variable must be declared before it can be used. The declaration of a variable consists of the name and the type of the variable. There are four basic data types in C: int, char, float, and double. Several qualifiers can be added to the variable declarations. They are short, long, signed, and unsigned.

Constants are often needed in forming a statement. There are four types of constants: integers, characters, floating-point numbers, and strings.

There are seven *arithmetic operators*: +, -, *, /, %, ++, --. There are six *bitwise operators*: &, |, ^, ~, >>, and <<. Bitwise operators can only be applied to integers. *Relational operators* are used in control statements. They are ==, !=, >, >=, <, <=, &&, ||, and !.

The *control-flow statements* specify the order in which computations are performed. Control-flow statements include the if-else statement, multi-way conditional statement, switch statement, for-loop statement, while-statement, and do-while statement.

Every C program consists of one or more functions. If a program consists of multiple functions, their definitions cannot be embedded within another. The same function can be called from several different places within a program. Generally, a function will process information passed to it from the calling portion of the program and return a single value. Information is passed to a function via special identifiers called *arguments* (also called *parameters*) and returned via the *return* statement. Some functions, however, accept information but do not return anything (for example, the library function *printf*).

A *pointer* holds the address of a variable. Pointers can be used to pass information back and forth between a function and its reference (calling) point. In particular, pointers provide a way to return multiple data items from a function via function arguments. Pointers also permit references to other functions to be specified as arguments to a given function. Two operators are related with pointers: * and &. The * operator returns the value of the variable pointed to by the pointer. The & operator returns the address of a variable.

Data items that have common characteristics are placed in an *array*. An array may be one-dimensional or multi-dimensional. The dimension of an array is specified by the number of square bracket pairs ([]) following the array name. An array name can be used as an argument to a function, thus permitting the entire array to be passed to the function. To pass an array to a function, the array name must appear by itself, without brackets or subscripts. An alternate method to pass arrays to a function is to use pointers.

A variable defined inside a function is an *internal variable* of that function. *External variables* are defined outside of any function and are thus potentially available to many functions. The *scope* of a name is the part of the program within which the name can be used. The scope of an external variable or a function lasts from the point at which it is declared to the end of the file being compiled.

A tutorial for using the ImageCraft C compiler (ICC12) is provided at the end of this chapter. The ICC12 is window-driven and is straightforward to use.

5.12 Exercises

E5.1 Write the C function *power(int x, char n)* that computes the value of x^n.

E5.2 Write a C program to find the median and mode of an array of integers. When the array has an even number of elements, the median is defined as the average of the middle two elements. Otherwise, it is defined as the middle element of the array. The mode is the element that occurs most frequently. You need to sort the array in order to find the median.

E5.3 Write a function that tests if a given number is a multiple of eight. A one is returned if the given number is a multiple of eight. Otherwise, a zero is returned. The number to be tested is an integer and is an argument to this function.

E5.4 Write a function that computes the least common multiple (LCM) of two integers m and n.

E5.5 What is a function prototype? What is the difference between a function prototype and function declaration?

E5.6 Write a *switch* statement that will examine the value of an integer variable *xx* and store one of the following messages in an array of seven characters, depending on the value assigned to *xx* (terminate the message with a NULL character):

    a. Cold, if xx == 1
    b. Chilly, if xx == 2
    c. Warm, if xx == 3
    d. Hot, if xx == 4

E5.7 Write a C program to clear the screen and then move the cursor to the middle line and output the message "Microcontroller is fun to use!". Outputting the character "\f" can clear the screen.

E5.8 Write a function that will convert an uppercase letter to lowercase.

E5.9 Write a C program that swaps the first column of a matrix with the last column, swaps the second column of the matrix with the second to last column, and so on.

E5.10 Write a loop to compute the sum of the squares of the first 100 odd integers.

E5.11 An *Armstrong number* is a number of n digits that is equal to the sum of each digit raised to the nth power. For example, 153 (which has three digits) equals $1^3 + 5^3 + 3^3$. Write a function to store all three-digit Armstrong numbers in an array.

E5.12 Write a C function to perform a binary search on a sorted array. The binary search algorithm is given in Example 4.3. The starting address, key, and array count are parameters to this function. Both the key and array count are integers.

E5.13 Write a program to find the first five numbers that, when divided by two, three, four, five, or six, leave a remainder of one, and when divided by seven have no remainder.

E5.14 Take a four-digit number. Add the first two digits to the last two digits. Now, square the sum. Surprise, you've got the original number again. Of course, not all four-digit numbers have this property. Write a C program to find the three numbers that have this special property.

5.13 Lab Exercises & Assignments

L5.1 Enter, compile, and download the following C program onto the CME-12BC demo board for execution using the procedure described in section 5.9:

```c
#include <stdio.h>
char test_prime (int a);          /* prototype declaration for the function test_prime */
void main ( )
{
        int i, prime_count;
        prime_count = 0;
        printf("The following are prime numbers between 100 and 1000.\n\n");
        for (i = 100; i <= 1000; i++) {
                if (test_prime(i)) {
                        prime_count ++;
                        printf("%d  ",i);
                        if ((prime_count % 8) == 0)
                        putchar('\n');
                }
        }
        printf("\n\nThere are %d prime numbers between 100 to 1000.\n",prime_count);
        return;
}
char  test_prime (int a)
{
        int i;
        if (a == 1) return 0;
        for (i = 2; i < a/2; i++)
                if ((a % i) == 0) return 0;
        return 1;
}
```

L5.2 Write a C program that will generate every third integer, beginning with i = 2 and continuing for all integers that are less than 300. Calculate the sum of those integers that are divisible by five. Store those integers and their sum in an array and an integer variable, respectively.

L5.3 Write a C function that calculates the least common multiple (LCM) of two integers *m* and *n*. Integers m and n are parameters to this function. Also, write a main program to test this function with several pairs of integers. Use the *printf* function to output the results.

L5.4 Write a C function to compute the root-mean-square (RMS) value of an unknown signal. The RMS value will be approximated by *n* (at least 32) samples. The method for computing the square root value is given in Example 4.14. The input parameters to this function include the starting address of the array and the array count. Also, write a main program to test this function. Compile and download the program onto the CME-12BC demo board for execution.

6

Interrupts, Resets & Operation Modes

6.1 Objectives

After completing this chapter, you should be able to:

- explain the difference between interrupts and resets

- describe the handling procedures for interrupts and resets

- raise one of the 68HC12 maskable interrupts to the highest priority

- enable and disable maskable interrupts

- use one of the low-power modes to reduce power consumption

- use the COP watchdog timer reset to detect software failure

- set up the interrupt vector jump table for demo boards that have the D-Bug12 monitor

- distinguish the 68HC12 operation modes

6.2 Fundamental Concepts of Interrupts

Interrupts and resets are among the most useful mechanisms that a computer system provides. With interrupts and resets, I/O operations are performed more efficiently, errors are handled more smoothly, and CPU utilization is improved. This chapter will begin with a general discussion of interrupts and resets and then focus on the specific features of the 68HC12 interrupts and resets.

6.2.1 What is an Interrupt?

Without an interrupt, our application programs will follow a certain order to execute. An *interrupt* is an event that requires the CPU to stop normal program execution and perform some service to this event. An interrupt can be generated internally or externally. An external interrupt is generated when external circuitry asserts an interrupt signal to the CPU. An internal interrupt can be generated by the hardware circuitry in the CPU or caused by software errors. Most microcontrollers have timers, I/O interface functions, and the CPU incorporated on the same chip. These subsystems can generate hardware interrupts to the CPU. Abnormal situations that occur during program execution, such as illegal opcodes, overflow, division by zero, and underflow, are called *software interrupts.* The terms *traps* and *exceptions* are both used to refer to software interrupts. Since there is no universally agreed-upon name for these special events, we will use the word *exception* to refer to interrupts and *resets* when there is no need to differentiate them in this text.

6.2.2 Why are Interrupts Used?

Interrupts are useful in many applications, such as:

- *Coordinating I/O activities and preventing the CPU from being tied up during the data transfer process.* As we will explain in more detail in later chapters that are dealing with I/O functions, the interrupt mechanism enables the CPU to perform other functions during an I/O activity (when the I/O device is busy). CPU time can thus be utilized more efficiently because of the interrupt mechanism.

- *Providing a graceful way to exit from an application when a software error occurs.* The service routine for a software interrupt may also output useful information about the error so that it can be corrected.

- *Reminding the CPU to perform routine tasks.* Keeping track of time-of-day is a common function of a computer. Without interrupts, the CPU would need to use program loops to create the required time delay in order to update the current time or check the current time from a dedicated time-of-day chip. With timer interrupts, the CPU can perform its regular operations without examining the current time until the timer interrupt occurs. The CPU will update the current time-of-day when servicing the timer interrupt. Another well-known application of interrupts is the implementation of the modern *multitasking operating system.* In modern computer-operating systems, multiple application programs are resident in the main memory, and the CPU time is divided into slots of about 10 to 20 ms. The operating system assigns a program to be executed for one time slot. At the end of a time slot or when a program is waiting for the completion of I/O, the operating system takes over and assigns another program for execution. This technique is called *multitasking.* Because input/output operations are quite slow, CPU utilization is improved dramatically by multitasking because the CPU can execute other programs while one program is waiting for the completion of input/output. Multitasking is made possible by the timer interrupt. Multitasking computer systems incorporate timers that periodically interrupt the CPU. On a timer interrupt, the operating system updates the system resource utilization status and switches the CPU from one program to another.

6.2.3 Interrupt Maskability

Some interrupts can be ignored by the CPU while others cannot. Interrupts that can be ignored by the CPU are called *maskable interrupts.* Interrupts that the CPU cannot ignore are called *nonmaskable interrupts.* A program can request the CPU to service or ignore a maskable interrupt by setting or clearing an *enable bit.* When an interrupt is *enabled,* it will be serviced by the CPU. When an interrupt is *disabled,* it will be ignored by the CPU. An interrupt request is said to be *pending* when it is active but not yet serviced by the CPU. A pending interrupt may or may not be serviced by the CPU, depending on whether or not it is enabled.

To make the interrupt system more flexible, a computer system normally provides a global and local interrupt masking capability. When none of the interrupts are desirable, the processor can disable all of the interrupts. This is achieved by clearing the global interrupt enable bit (or setting the global interrupt disable bit for some other processor). In other situations, the processor can selectively enable certain interrupts while at the same time disable other undesirable interrupts. This is achieved by providing each interrupt source an enable bit in addition to the global interrupt mask. Whenever any interrupt is undesirable, it can be disabled while at the same time allowing other interrupt sources to be serviced (attended) by the processor. Today, almost all commercial processors are designed to provide this two-level (or even three-level) interrupt enabling capability.

6.2.4 Interrupt Priority

If there are multiple interrupts pending at the same time, the CPU needs to decide which interrupt to service first. The solution to this problem is to prioritize all interrupt sources. When occurring roughly at the same time, an interrupt with higher priority always receives service before interrupts at lower priorities. Interrupt priorities are not programmable for most processors.

6.2.5 Interrupt Service

The CPU provides service to an interrupt by executing a program called the *interrupt service routine.* An interrupt is a special event for the CPU. After providing service to an interrupt, the CPU must resume normal program execution. How does the CPU stop execution of a program and resume it later? It does this by saving the program counter value and the CPU status information before executing the interrupt service routine and then restoring the program counter and the CPU status before exiting the interrupt service routine. The complete interrupt service cycle is as follows:

1. When an interrupt occurs, save the program counter value in the stack.

2. Save the CPU status (including the CPU status register and some other registers) in the stack.

3. Identify the cause of the interrupt.

4. Resolve the starting address of the corresponding interrupt service routine.

5. Execute the interrupt service routine.

6. Restore the CPU status and the program counter from the stack before exiting the interrupt service routine.

7. Restart the interrupted program.

Another issue is related to the time when the CPU begins to service an interrupt. For all hardware-maskable interrupts, the microprocessor starts to provide service when it completes execution of the instruction being executed when an interrupt occurs. For some nonmaskable interrupts, the CPU may start the service without completing the current instruction. Many software interrupts are caused by an error in an instruction execution that prevents the instruction from being completed. The service to this type of interrupt is simply to output an error message and abort the program.

For some other types of interrupts, the CPU resolves the problem in the service routine and then re-executes the instruction that caused the interrupt. One example of this type of interrupt is the *page-fault* interrupt in a *virtual-memory operating system*. In a virtual memory operating system, all programs are divided into pages of equal size. The size of a page can range from 512 bytes to 32 KB. When a program is first started, the operating system loads only a few pages (possibly only one page) from secondary storage (generally hard disk) into main memory and then starts to execute the program. Before long, the CPU may need to execute an instruction that hasn't been loaded into the main memory yet. This causes an interrupt called a *page-fault interrupt*. The CPU then executes the service routine for the page-fault interrupt, which brings in the page that caused the page fault into main memory. The program that caused the page fault is then restarted. The virtual memory operating system gets its name from the fact that it allows execution of a program without loading the entire program into main memory. With a virtual memory operating system, a program that is much larger than the physical main memory can be executed.

6.2.6 Interrupt Vector

The term *interrupt vector* refers to the starting address of the interrupt service routine. In general, interrupt vectors are stored in a table called an *interrupt-vector table*. The starting address of each entry (holds one interrupt vector) in the interrupt vector table is called a *vector address*. The interrupt-vector table is fixed for some microprocessors and may be relocated for other microprocessors (for example, the AMD29000 family of microprocessors).

The CPU needs to determine the interrupt vector before it can provide service. One of the following methods can be used by a microprocessor/microcontroller to determine the interrupt vector:

1. *Predefined*. In this method, the starting address of the service routine is predefined when the microcontroller is designed. The processor uses a table to store all the interrupt service routines. The Microchip PIC17/PIC18 microcontrollers use this approach. Each interrupt is allocated the same number of bytes to hold its service routine. The PIC17/PIC18 allocates eight words to each interrupt service routine. When the service routine requires more than eight words, the solution is to place a jump instruction in the pre-defined location to jump to the actual service routine.

2. *Fetch the vector from a predefined memory location*. In this approach, the interrupt vector of each interrupt source is stored at a predefined location in the interrupt vector table, where the microprocessor can get it directly. The Motorola 68HC12 and most other Motorola microcontrollers use this approach.

3. *Execute an interrupt acknowledge cycle to fetch a vector number in order to locate the interrupt vector*. During the interrupt acknowledge cycle, the microprocessor performs a read bus cycle, and the external I/O device that requested the interrupt places a number on the data bus to identify itself. This number is called an *interrupt vector number*. The CPU can figure out the starting address of the interrupt service

routine by using this number. The CPU needs to perform a read cycle in order to obtain it. The Motorola 68000 and Intel x86 family microprocessors support this method. The Motorola 68000 family of microprocessors also uses the second method. This method is not used by microcontrollers because of its incurred latency.

6.2.7 Interrupt Programming

Interrupt programming deals with how to provide service to the interrupt. There are three steps in interrupt programming:

Step 1

Initialize the interrupt vector table. (This step is not needed for microprocessors that have predefined interrupt vectors). This can be done by using the assembler directive ORG (or its equivalent):

```
ORG     $xxxx       ; xxxx is the vector table address
FDB     service_1   ; store the starting address of interrupt source 1
FDB     service_2   ;
    .
    .
    .
FDB     service_n
```

where *service_i* is the starting address of the service routine for interrupt source i. The assembler syntax and the number of bytes needed to store an interrupt vector on your particular microcontroller may be different.

Step 2

Write the service routine. An interrupt service routine should be as short as possible. For some interrupts, the service routine may only output a message to indicate that something unusual has occurred. A service routine is similar to a subroutine—the only difference is the last instruction. An interrupt service routine uses the *return from interrupt (or return from exception)* instruction instead of the *return from subroutine* instruction to return to the interrupted program. The following is an example of an interrupt service routine (in 68HC12 instructions):

```
irq_isr  ldx    #msg
         jsr    putst        ; putst outputs a string pointed by X
         rti                 ; return from interrupt
msg      fcc    "This is an error"
```

The service routine may or may not return to the interrupted program, depending on the cause of the interrupt. It makes no sense to return to the interrupted program if the interrupt is caused by a software error such as division-by-zero or overflow, because the program is unlikely to generate correct results under these circumstances. In such situations the service routine would return to the monitor program or the operating system instead. Returning to a program other than the interrupted program can be achieved by changing the saved program counter (in the stack) to the desired value. Execution of the *return from interrupt* instruction will then return CPU control to the new address.

Step 3

Enable the interrupts to be serviced. An interrupt can be enabled by clearing the global interrupt mask and setting the local interrupt enable bit in the I/O control register. It is a common mistake to forget enabling interrupts when writing interrupt-driven application programs.

6.2.8 Overhead of Interrupts

Although the interrupt mechanism provides many advantages, it also involves some overhead. The overhead of the 68HC12 interrupt includes:

1. Saving the CPU registers, including accumulators (A:B), index registers X and Y, and the condition code register (CCR), and fetching the interrupt vector. This takes at least 13 E cycles.

2. The execution time of RTI instruction. This instruction restores all the CPU registers that have been stored in the stack by the CPU during the interrupt and takes from 8 to 12 E clock cycles to complete for the 68HC12.

3. Execution time of instructions of the interrupt service routine. This depends on the type and the number of instructions in the service routine.

The total overhead is thus at least 25 E clock cycles, which amounts to 3.125 µs for an 8 MHz E clock. We should be aware of the overhead involved in interrupt processing when deciding whether to use the interrupt mechanism.

6.3 Resets

The initial values of some CPU registers, flip-flops, and the control registers in I/O interface chips must be established before the computer can operate properly. Computers provide a reset mechanism to establish initial conditions.

There are at least two types of resets in each microprocessor: the *power-on reset* and the *manual reset*. A power-on reset allows the microprocessor to establish the initial values of registers and flip-flops and to initialize all I/O interface chips when power to the microprocessor is turned on. A manual reset without power-down allows the computer to get out of most error conditions (if hardware hasn't failed) and reestablish the initial conditions. The computer will *reboot* itself after a reset.

The reset service routine has a fixed starting address and is stored in the read-only memory of all microprocessors. At the end of the service routine, control should be transferred to either the monitor program or the operating system.

Like nonmaskable interrupts, resets are also nonmaskable. However, resets are different from the nonmaskable interrupts in that no registers are saved by resets because resets establish the values of registers.

6.4 68HC12 Exceptions

The 68HC12 exceptions can be classified into the following categories:

- *Maskable interrupts*. These include the $\overline{IRQ}$ pin interrupt and all peripheral function interrupts. Since different 68HC12 members implement a different number of peripheral functions, they have a different number of maskable interrupts.

- *Nonmaskable interrupts*. These include the $\overline{XIRQ}$ pin interrupt, the SWI instruction interrupt, and the unimplemented opcode trap.

- *Resets*. These include the power-on reset, the reset pin manual reset, the COP (computer operate properly) reset, and the clock monitor reset.

6.4.1 Maskable Interrupts

Since different 68HC12 members implement a different number of peripheral functions, they have a different number of maskable interrupts. The I flag in the CCR register is the global mask of all maskable interrupts. Whenever the I flag is set, all maskable interrupts are disabled. All maskable interrupts have a local enable bit that allow them to be selectively enabled. They are disabled (I flag is set to 1) when the 68HC12 gets out of reset state.

Like any other microcontrollers, all 68HC12 exceptions are prioritized. The priorities of resets and non-maskable interrupts are not programmable. However, one can raise one of the maskable interrupts to the highest priority within the group of maskable interrupts so that it can get quicker attention from the CPU. The relative priorities of the other sources remain the same. The bits 5 to 1 of the HPRIO register select the maskable interrupt at the highest priority within the group of maskable interrupts. The contents of the HPRIO register are shown in Figure 6.1.

Figure 6.1 ■ Highest priority I interrupt register (HPRIO)

The priorities and vector addresses of all 68HC12 exceptions are listed in Table 6.1. To raise a maskable interrupt source to the highest priority, simply write the low byte of the vector address of this interrupt to the HPRIO register. For example, to raise the Timer channel 0 interrupt to the highest priority, write the value of $EE to the HPRIO register.

Non Maskable

Vector address	Interrupt source	CCR mask	Local Enable	HPRIO value to elevate to highest I bit
$FFFE	Reset	none	none	-
$FFFC	Clock monitor reset	none	COPCTL(CME,FCME)	-
$FFFA	COP failure reset	none	COP rate selected	-
$FFF8	Unimplemented instruction trap	none	none	-
$FFF6	SWI	none	none	-
$FFF4	XIRQ	X bit	none	-
$FFF2	IRQ	I bit	INTCR(IRQEN)	$F2
$FFF0	Real time interrupt	I bit	RTICTL(RTIE)	$F0
$FFEE	Timer channel 0	I bit	TMSK1(C0I)	$EE
$FFEC	Timer channel 1	I bit	TMSK1(C1I)	$EC
$FFEA	Timer channel 2	I bit	TMSK1(C2I)	$EA
$FFE8	Timer channel 3	I bit	TMSK1(C3I)	$E8
$FFE6	Timer channel 4	I bit	TMSK1(C4I)	$E6
$FFE4	Timer channel 5	I bit	TMSK1(C5I)	$E4
$FFE2	Timer channel 6	I bit	TMSK1(C6I)	$E2
$FFE0	Timer channel 7	I bit	TMSK1(C7I)	$E0
$FFDE	Timer overflow	I bit	TMSK2(TOI)	$DE
$FFDC	Pulse accumulator overflow	I bit	PACTL(PAOVI)	$DC
$FFDA	Pulse accumulator input edge	I bit	PACTL(PAI)	$DA
$FFD8	SPI serial transfer complete	I bit	SP0CR1(SPIE)	$D8
$FFD6	SCI0	I bit	SC0CR2(TIE,TCIE,RIE,ILIE)	$D6
$FFD4	SCI1	I bit	SC1CR2(TIE,TCIE,RIE,ILIE)	$D4 (1,3,4)
$FFD2	ATD0 or ATD1	I bit	ATDxCTL2(ASCIE)	$D2
$FFD0	MSCAN 0 wakeup	I bit	C0RIER(WUPIE)	$D0 (1*,2,2*)
$FFCE	Key wakeup J or H	I bit	KWIEJ[7:0] and KWIEH[7:0]	$CE (1,3,4)
$FFCC	Modulus down counter underflow	I bit	MCCTL(MCZI)	$CC
$FFCA	Pulse accumulator B overflow	I bit	PBCTL(PBOVI)	$CA
$FFC8	MSCAN 0 errors	I bit	C0RIER(RWRNIE,TWRNIE, RERRIE,TERRIE,BOFFIE,OVRIE)	$C8 (2*,3,4)
$FFC6	MSCAN 0 receive	I bit	C0RIER(RXFIE)	$C6 (2*,3,4)
$FFC4	MSCAN 0 transmit	I bit	C0TCR(TXEIE[2:0])	$C4 (2*,3,4)
$FFC2	CGK lock and limp home	I bit	PLLCR(LOCKIE, LHIE)	$C2 (3,4)
$FFC0	IIC Bus	I bit	IBCR(IBIE)	$C0 (4)
$FFBE	MSCAN 1 wakeup	I bit	C1RIER(WUPIE)	$BE (4)
$FFBC	MSCAN 1 errors	I bit	C1RIER(RWRNIE,TWRNIE, RERRIE,TERRIE,BOFFIE,OVRIE)	$BC (4)
$FFBA	MSCAN 1 receive	I bit	C1RIER(RXFIE)	$BA (4)
$FFB8	MSCAN 1 transmit	I bit	C1TCR(TXEIE[2:0])	$B8 (4)
$FFB6	Reserved	I bit		$B6
$FF80-$FFB5	Reserved	I bit		$80-$B4

Note. 1. Available in 812 A4   1*. Used as wake up key J in 812A4
2. Used as BDLC interrupt vector for 912B32, 912BE32.   2*. Available in 912BC32
3. Available in D60
4. Available in DG128 (DT128)

Table 6.1 ■ Interrupt vector map

In Table 6.1, exceptions that have higher vector addresses are at higher priorities. Not all the exceptions are available in all HC12 members. For example, the I2C interrupt is available only in 912DG128. The vector address $FFD0 is the vector address for BDLC in 912B32 and 912BE32, but is the vector address for MSCAN 0 wake-up in other members that implement the CAN function.

$\overline{\text{IRQ}}$ PIN INTERRUPT

The $\overline{\text{IRQ}}$ pin is the only external maskable interrupt signal. The $\overline{\text{IRQ}}$ interrupt can be edge-triggered or level-triggered. The triggering method is selected by programming the IRQE bit of the INTCR register. The $\overline{\text{IRQ}}$ interrupt has a local enable bit called the IRQEN bit, which is the bit 6 the Interrupt Control Register (INTCR). The contents of the INTCR register are shown in Figure 6.2. The $\overline{\text{IRQ}}$ interrupt may be delayed for 4096 E cycles after exiting the stop mode by setting the DLY bit of the INTCR register.

address: $1E	7	6	5	4	3	2	1	0
read: write:	IRQE	IRQEN	DLY	0	0	0	0	0
value after reset:	0	1	1	0	0	0	0	0

IRQE -- $\overline{\text{IRQ}}$ edge sensitive only bit
      IRQE can be written once in normal mode. In special modes,
      it can be written any time, but the first write is ignored.
      1 = $\overline{\text{IRQ}}$ pin responds only to falling edge
      0 = $\overline{\text{IRQ}}$ pin responds to low level.

IRQEN -- $\overline{\text{IRQ}}$ enable bit
      IRQEN bit can be written any time in all modes. The $\overline{\text{IRQ}}$
      pin has an internal pullup.
      1 = $\overline{\text{IRQ}}$ pin interrupt enabled
      0 = $\overline{\text{IRQ}}$ pin interrupt disabled

DLY -- Oscillator startup delay on exit from stop mode
      DLYcan be written once in normal modes. In special
      modes, DLY can be written anytime.
      1 = Stabilization delay on exit from stop mode
      0 = No stabilization delay on exit from stop mode

Figure 6.2 ■ Interrupt control register (INTCR)

The advantage of making the $\overline{\text{IRQ}}$ interrupt *level-sensitive* (active low) is that it allows multiple external interrupt sources to be tied to this pin. Whenever one of the interrupt sources (that are tied to the $\overline{\text{IRQ}}$ pin) is low, an interrupt request will be detected by the 68HC12. The user of this method must make sure that the $\overline{\text{IRQ}}$ signal is de-asserted (goes high) before the 68HC12 exits the interrupt service routine if there are no other pending interrupts connected to the $\overline{\text{IRQ}}$ pin.

The major advantage of making the $\overline{\text{IRQ}}$ interrupt *edge-sensitive* (falling edge) is that the user does not need to worry about the duration of the assertion time of the $\overline{\text{IRQ}}$ signal. However, this approach is not appropriate for a noisy environment. In a noisy environment, any noise spike could generate an undesirable interrupt request on the $\overline{\text{IRQ}}$ pin.

INTERRUPT RECOGNITION

Once enabled, an interrupt request can be recognized at any time after the I mask bit is cleared. When an interrupt service request is recognized, the CPU responds at the completion of the instruction being executed. Interrupt latency varies according to the number of cycles required to complete the current instruction. The 68HC12 has implemented a few instructions to support fuzzy logic rule evaluation. These instructions take a much longer time to complete. These instructions include fuzzy logic rule evaluation (REV), fuzzy logic rule evaluation weighted (REVW), and weighted average (WAV) instructions. The 68HC12 does not wait until the completion of these instructions to service the interrupt request. These instructions will resume at the point that they were interrupted.

Before the 68HC12 starts to service an interrupt, it will set the I mask to disable other maskable interrupts. When the CPU begins to service an interrupt, the instruction queue is refilled, a return address is calculated, and then the return address and the contents of all CPU registers (except SP) are saved in the stack in the order shown in Figure 6.3.

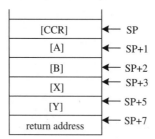

Figure 6.3 ■ Stack order on entry to interrupts

THE RTI INSTRUCTION

RTI is used to terminate interrupt service routines. RTI is an 8-cycle instruction when no other interrupt is pending and a 10-cycle instruction when another interrupt is pending. In either case, the first five cycles are used to restore the CCR, B:A, X, Y, and the return address from the stack. The 68HC12 then clears the I mask to enable further maskable interrupts.

If no other interrupt is pending at this point, three program words are fetched to refill the instruction queue from the area of the return address and processing proceeds from there.

If another interrupt is pending after registers are restored, a new vector is fetched, and the stack pointer is adjusted to point at the CCR value that was just recovered (SP = SP – 9). This makes it appear that the registers have been stacked again. After the SP is adjusted, three programs words are fetched to refill the instruction queue, starting at the address the vector points to. Processing then continues with the execution of the instruction that is now at the head of the queue.

6.4.2 Nonmaskable Interrupts

There are three interrupts in this category: the $\overline{\text{XIRQ}}$ pin, the unimplemented opcode trap, and the SWI instruction.

$\overline{\text{XIRQ}}$ PIN INTERRUPT

The $\overline{\text{XIRQ}}$ pin interrupt is disabled during a system reset and upon entering the interrupt service routine for an $\overline{\text{XIRQ}}$ interrupt.

During reset, both the I and X bit in the CCR register are set. This disables maskable interrupts and interrupt requests made by asserting the $\overline{\text{XIRQ}}$ pin (pulled to low). After minimum system initialization, software can clear the X bit using an instruction such as *andcc #$BF*. Software cannot reset the X bit from 0 to 1 once it has been cleared, and hence the interrupt requests made via the $\overline{\text{XIRQ}}$ pin become nonmaskable.

When a nonmaskable interrupt is recognized, both the X and I bits are set after the CPU registers are saved. The X bit is not affected by maskable interrupts. The execution of an RTI instruction at the end of the $\overline{\text{XIRQ}}$ service routine will restore the X and I bits to the pre-interrupt request state.

UNIMPLEMENTED OPCODE TRAP

The 68HC12 uses up to 16 bits (2 pages) to encode the opcode. All 256 combinations in the page 1 opcode map have been used. However, only 54 of the 256 positions on page 2 of the opcode map are used. If 68HC12 attempts to execute one of the 202 unused opcodes on page 2, an unimplemented opcode trap occurs. The 202 unimplemented opcodes are essentially interrupts that share a common interrupt vector address, $FFF8:$FFF9.

The 68HC12 uses the next address after an unimplemented page 2 opcode as a return address.

SOFTWARE INTERRUPT INSTRUCTION (SWI)

Execution of the SWI instruction causes an interrupt without an interrupt request signal. SWI is not inhibited by the global mask bits in the CCR. So far, we have been using this instruction to jump back to the D-Bug12 monitor.

The SWI instruction is commonly used in the debug monitor to implement *breakpoints* and to transfer control from a user program to the debug monitor. A breakpoint in a user program is a memory location where we want program execution to be stopped and information about instruction execution (in the form of register contents) to be displayed. To implement breakpoints, the debug monitor sets up a breakpoint table. Each entry of the table holds the address of the breakpoint and the opcode byte at the breakpoint. The monitor also replaces the opcode byte at the breakpoint with the opcode of the SWI instruction. When the instruction at the breakpoint is executed, it causes an SWI interrupt. The service routine of the SWI interrupt will look up the breakpoint table and take different actions depending on whether the saved PC value is in the breakpoint table:

Case 1
The saved PC value is not in the breakpoint table. In this case, the service routine will simply replace the saved PC value (in the stack) with the address of the monitor program and return from the interrupt.

Case 2
The saved PC is in the breakpoint table. In this case, the service routine will:

1. Replace the SWI opcode with the opcode in the breakpoint table
2. Replace the saved PC value (in the stack) with the address of the monitor program
3. Display the contents of the CPU registers
4. Return from the interrupt

6.4.3 Resets

There are four possible sources of resets:

- Power-on reset (POR)
- External reset
- COP reset
- Clock monitor reset

Power-on reset and external reset share the same reset vector. The COP reset and clock monitor reset each have a separate vector.

POWER-ON RESET

The 68HC12 has circuitry to detect a positive transition in the V_{DD} supply and initialize the microcontroller by asserting the reset signal internally. The reset signal is released after a delay that allows the device clock generator to stabilize.

EXTERNAL RESET

The 68HC12 distinguishes between internal and external resets by sensing how quickly the signal on the RESET pin rises to logic high after it has been asserted. When the 68HC12 senses any of the four reset conditions, internal circuitry drives the $\overline{RESET}$ pin low for 16 clock cycles, and then releases. Eight clock cycles later, the CPU samples the state of the signal applied to the RESET pin. If the signal is still low, an external reset has occurred. If the signal is high, the reset has been initiated internally by either the COP system or the clock monitor.

The power supply to an embedded system may drop below the required level. If the microcontroller keeps working under this situation, the contents of the EEPROM might be corrupted. The common measure is to pull the reset signal to low so that the microcontroller cannot execute instructions. A low-voltage-inhibit (LVI) circuit such as the Motorola MC34064 can be used to protect against the EEPROM corruption.

Figure 6.4 shows an example of a reset circuit with a manual reset and LVI circuit.

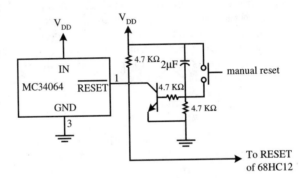

Figure 6.4 ■ A typical external reset circuit

COP RESET

The 68HC12 includes a computer operating properly (COP) system to help protect against software failures. In general, if our software was written correctly, it should follow a certain sequence of execution which means the execution time can also be predicted. When the COP is enabled, software must write $55 and $AA (in this order) to the COPRST register to keep a watchdog timer from timing out. Other instructions may be executed between these writes. If our software was not written properly, then it may not write $55 and $AA to the COPRST (located at $17) before the COP times out and the CPU will be reset. The software problems can therefore be detected.

The operation of the COP timer circuit is configured by the COPCTL register. The contents of the COPCTL register are shown in Figure 6.5. The COP system is driven by a constant frequency of $E/2^{13}$. The bits 2, 1, and 0 specify an additional division factor to arrive at the COP timeout rate.

Address: $0016

	7	6	5	4	3	2	1	0
read: write:	CME	FCME	FCM	FCOP	DISR	CR2	CR1	CR0
normal reset:	0	0	0	0	0	0	0	1
special reset:	0	0	0	0	1	0	0	1

Read: anytime

Write: varies on a bit by bit basis

CME: Clock monitor enable bit

Write anytime

if FCME is set, this bit has no meaning nor effect

0 = Clock monitor is disabled; slow clock and STOP instruction may be used.

1 = Slow or stopped clocks (including the STOP instruction) cause a clock reset sequence

FCME: Force clock monitor enable bit

Write once in normal mode, anytime in special modes.

In normal modes, when this bit is set, the clock monitor function cannot be disabled until a reset occurs.

0 = Clock monitor follows the state of the CME bit

1 = Slow or stopped clocks cause a clock reset sequence.

To use both STOP and clock monitor, the CME bit should be cleared prior to executing a STOP instruction and set after recovery from STOP. Always keep FCME = 0, if STOP will be used.

FCM: Force clock monitor reset bit

Writes are not allowed in normal modes, anytime in special modes.

If DISR is set, this bit has no effect.

0 = normal operation

1 = force a clock monitor reset, if clock monitor is enabled.

FCOP: Foce COP watchdog reset bit.

Writes are not allowed in normal modes; can be written anytime in special modes. If DISR is set, this bit has no effect.

0 = normal operation

1 = force a COP reset, if COP is enabled.

DISR: Disable resets from COP watchdog and clock monitor do not generate a system reset.

Writes are not allowed in normal modes, anytime in special modes.

0 = normal operation

1 = regardless of other control bit states, COP and clock monitor do not generate a system reset.

CR2, CR1, and CR0: COP watchdog timer rate select bit. The rates are shown in Table 6.2. These three bits can be written once in normal modes but can be written many times in special modes.

Figure 6.5 ■ COP control register

CR2	CR1	CR0	Divide E by	At E = 4.0 MHz timeout 0 to 2.048 ms	At E = 8.0 MHz timeout 0 to 1.024 ms
0	0	0	off	off	off
0	0	1	2^{13}	2.048 ms	1.024 ms
0	1	0	2^{15}	8.192 ms	4.096 ms
0	1	1	2^{17}	32.768 ms	16.384 ms
1	0	0	2^{19}	131.072 ms	65.536 ms
1	0	1	2^{21}	524.288 ms	262.144 ms
1	1	0	2^{22}	1.048 s	524.288 ms
1	1	1	2^{23}	2.097 s	1.048576 s

Table 6.2 ■ COP watchdog rates (RTBYP=0)

CLOCK MONITOR RESET

The clock monitor circuit uses an internal RC circuit to determine whether clock frequency is above a predetermined limit. If no EXTALi clock edges are detected within this time delay, the clock monitor can optionally generate a system reset.

The clock monitor function is enabled/disabled by the CME control bit in the COPCTL register. Clock monitor time-outs are shown in Table 6.3. The corresponding EXTALi clock period with an ideal 50% duty cycle is twice this time-out value.

Supply	Range
5V +/- 10%	2 - 20 µs
3V +/- 10%	2 - 50 µs

Table 6.3 ■ Clock monitor time-outs

The three forced bits (FCME, FCM, and FCOP) are provided for fabrication testing and debugging purposes and should not be used in normal modes.

6.4.4 Low Power Modes

When a microcontroller is performing normal operations, power consumption is unavoidable. However, the microcontroller in an embedded system may not always be performing useful operations. Under this situation, it would be ideal if the power consumption could be reduced to the minimum. This issue is especially important for those embedded products powered by battery.

The 68HC12 has two low-power modes that can reduce power consumption drastically: the *wait* and *stop* modes.

THE WAIT INSTRUCTION

The *wait* instruction pushes all CPU registers (except the stack pointer) and the return address into the stack and enters a wait state. During the wait state, CPU clocks are stopped (clock signals that drive the ALU and register file), but other clocks in the microcontroller (clock signals that drive peripheral functions) continue to run.

The CPU leaves the wait state when it senses one or more of the following events:

- maskable interrupts that are not masked
- nonmaskable interrupts
- resets

Upon leaving the wait state, the 68HC12 CPU sets the appropriate interrupt mask bit(s), fetches the vector corresponding to the exception sensed, and continues instruction execution at the location the vector points to.

THE STOP INSTRUCTION

When the S bit in the CCR register is cleared and a *stop* instruction is executed, the 68HC12 saves all CPU registers (except the stack pointer) in the stack, stops all system clocks, and puts the microcontroller in standby mode.

Standby operation minimizes system power consumption. The contents of registers and the states of I/O pins remain unchanged.

Asserting $\overline{RESET}$, $\overline{XIRQ}$, or $\overline{IRQ}$ signals ends the standby mode. If it is the $\overline{XIRQ}$ signal that ends the stop mode and the X mask bit is 0, instruction execution resumes with a vector fetch for the $\overline{XIRQ}$ interrupt. If the X mask bit is 1 ($\overline{XIRQ}$ disabled), a 2-cycle recovery sequence is used to adjust the instruction queue, and execution continues with the next instruction after the stop instruction.

6.5 Exception Programming for the 68HC12

The major steps of interrupt programming have been discussed briefly in section 6.2.7. In this section, we will apply these procedures to the 68HC12 microcontroller.

6.5.1 Initializing the Interrupt Vector Table

The interrupt vectors (starting addresses of all interrupt service routines) should be stored in the interrupt vector table. The vector address for each interrupt source is listed in Table 6.1.

The interrupt vector table can be set up by using the following assembler directives:

```
org     $FF80
...
fdb     pbov_isr    ; pulse accumulator B overflow interrupt service
                    ; routine
...
fdb     atd_isr     ; A/D conversion complete interrupt service routine
...
fdb     pai_isr     ; PAI interrupt service routine
...
fdb     c7_isr      ; output compare channel 7 interrupt service routine
fdb     c6_isr      ; input capture channel 6 interrupt service routine
...
```

Each entry of the table holds the starting address of the corresponding exception service routine.

When designing an embedded product, we probably will use only a subset of the available peripheral functions of the 68HC12 and may need to use a subset of the available interrupt sources. For those undesired interrupts, we can simply put in the starting address of a common dummy interrupt service routine that simply returns from interrupt:

```
dum_isr          rti         ; a dummy interrupt service routine
```

Using this method, we can avoid the embedded software being crashed by the undesired interrupts that occur during the normal use of the product. Suppose the output compare channel 7 and input capture channel 6 interrupts are not needed, then the interrupt vector table will look like:

```
org        $FF80
...
fdb        pbov_isr        ; pulse accumulator B overflow interrupt service
                           ; routine

...
fdb        atd_isr         ; A/D conversion complete interrupt service routine

...
fdb        pai_isr         ; PAI interrupt service routine

...
fdb        dum_isr         ; output compare channel 7 interrupt service routine
fdb        dum_isr         ; input capture channel 6 interrupt service routine

...
```

In C language, the interrupt vector table of a standalone system (based on 912BC32) can be set up by including the following file in your *main()* function:

```
#define dum_vec          (void(*)(void))0xFFFF
#pragma abs_address       0xFFC4
extern   void can0_tr_isr();
extern   void can0_rcv_isr();

...
extern   void atod_isr();
extern   void sci1_isr();
extern   void sci0_isr();

...
extern   void _start(void);

void (*interrupt_vectors[])(void) =
         can0_tr_isr,     /* CAN0 transmit interrupt vector */
         can0_rcv_isr,    /* CAN0 receive interrupt vector */

         ....
         dum_vec,         /* dummy vector */

         ...
         atod_isr,        /* AtoD0 or AtoD1 interrupt vector */
         sci1_isr,        /* SCI1 interrupt vector */
         sci0_isr,        /* SCI0 interrupt vector */

         ...
         _start           /* reset vector */
};
#pragma end_abs_address
```

In the previous vector table, the dummy vector *dum_vec* is used for those unused interrupt sources. This dummy vector is simply the reset vector. Depending on the application, the dummy vector can be something else.

Can we use this method to set up the interrupt vector table in the CME-12BC or other demo boards that contain a debug monitor such as D-Bug12? No. Because the memory space for the interrupt vector table is in the ROM.

The D-Bug12 monitor provides an interrupt vector jump table in the on-chip SRAM and the *SetUserVector()* function for the user to set up an interrupt vector in the jump table. The *SetUserVector()* function prototype is declared as follows:

    int SetUserVector(int VectNum, Address UserAddress);

The first parameter to the function is a vector number used to identify the source of the interrupt, and the second parameter is the starting address of the corresponding interrupt service routine. The vector numbers defined by the D-Bug12 monitor are shown in Table 6.4.

Vector number	D-Bug12 Version 1.xxx	D-Bug12 Version 2.xxx
7	Port H key Wakeup	Port H key Wakeup
8	Port J key Wakeup	Port J key Wakeup
9	Analog-to-Digital converter	Analog-to-Digital converter
10	Serial Communication Interface 1	Serial Communication Interface 1
11	Serial Communication Interface 0	Serial Communication Interface 0
12	Serial Peripheral Interface 0	Serial Peripheral Interface 0
13	Timer Channel 0	Pulse Accumulator Edge
14	Timer Channel 1	Pulse Accumulator Overflow
15	Timer Channel 2	Timer Overflow
16	Timer Channel 3	Timer Channel 7
17	Timer Channel 4	Timer Channel 6
18	Timer Channel 5	Timer Channel 5
19	Timer Channel 6	Timer Channel 4
20	Timer Channel 7	Timer Channel 3
21	Pulse Accumulator Overflow	Timer Channel 2
22	Pulse Accumulator Edge	Timer Channel 1
23	Timer Overflow	Timer Channel 0
24	Real Time Interrupt	Real Time Interrupt
25	IRQ interrupt	IRQ interrupt
26	XIRQ interrupt	XIRQ interrupt
27	SWI instruction	SWI instruction
28	Unimplemented Instruction Trap	Unimplemented Instruction Trap
-1	Return to the starting address of the RAM vector table	Return to the starting address of the RAM vector table

Table 6.4 ■ D-Bug12 monitor exception vector numbers

Suppose that the pulse accumulator PAI pin interrupt service routine starts with the label *pai_isr*. Then the following instruction sequence will set up the interrupt vector table entry for the PAI interrupt:

```
pa_vec_no      equ     13
setuservector  equ     $F69A

               ldd     #pai_isr
               pshd                    ; pass the PAI interrupt vector via stack
               ldab    #pa_vec_no      ; interrupt vector number of PAI pin
```

```
           clra
           ldx       setuservector
           jsr       0,x                    ; call D-Bug12 function to setup the vector
           leas      2,sp                   ; deallocate stack space
```

In C language, we can use embedded assembly instructions to set up the interrupt vectors for a demo board that contains a monitor. The C compiler usually adds a underscore character as the prefix to the function name during the compiling process. For example, the function *pai_isr* will be translated to *_pai_isr*. The following embedded assembly instructions will set up the interrupt vector for the PAI interrupt:

```
#define    pa_vec_no 13
#define    setuservector 0xF69A

           asm ("ldd #_pai_isr");
           asm ("pshd");
           asm ("ldab #pa_vec_no");
           asm ("clra");
           asm ("ldx setservector");
           asm ("jsr 0,x");
           asm ("leas 2,sp");
```

6.5.2 Writing Exception Service Routine

An interrupt service routine is no different from an ordinary subroutine except that the last instruction must be the RTI (return from interrupt) instruction.

The rule for writing an interrupt service routine is to minimize the number of operations performed by the routine. For example, the service routine for the output compare 7 (OC7) interrupt might be as follows:

```
oc7_isr    bclr      tflg1,$7F      ; clear the C7F flag
           dec       oc7_cnt        ; decrement a counter
           rti
```

In C, we need to tell the compiler that a function is an interrupt handler so that the compiler will add an RTI instruction as the last instruction of the function. The following statement will inform the C compiler that *oc7_isr()* is an interrupt handler:

```
#pragma interrupt_handler oc7_isr
```

Multiple interrupt handlers can be declared in one statement. So, the following statement declares that *oc7_isr()*, *pai_isr()*, and *atod_isr()* are interrupt handlers:

```
#pragma interrupt_handler oc7_isr pai_isr atod_isr
```

The OC7 interrupt handler in the C language is as follows:

```
#pragma  interrupt_handler oc7_isr
void oc7_isr (void)
{
      TFLG1 = 0x80;        /* clear PAIF flag */
      toc7_cnt--;
}
```

Of course, you also need to provide a prototype declaration of the interrupt service routine:

```
void  oc7_isr(void);
...
main ( )
...
```

6.5.3 Enabling the Interrupt

All maskable interrupts have local interrupt masks. To enable them, we need to set the local interrupt mask and clear the I mask of the CCR register. For example, we can use the following two instructions to enable the OC7 interrupt:

```
bset     tmsk1,$80    ; enable OC7 interrupt
cli
```

In C, we can use the following two statements to clear the global interrupt mask and set the local interrupt mask:

```
TMSK1 | = 0x80;    /* enable OC7 interrupt */
asm ("cli");
```

The ICC12 compiler defines a macro *INTR_ON()* as the replacement for the embedded assembly instruction *asm ("cli")*. You can use either method to clear a global interrupt mask.

When interrupts are not desirable, we can use the *sei* instruction to disable them. ICC12 also defines a macro replacement *INTR_OFF()* for the *sei* instruction.

Example 6.1

Write a program that uses the output compare function of timer channel 7 (TC7) to create time delays. The program should operate like this:

1. Configure the timer channel 7 so that it uses the E clock ÷ 8 as the clock input of the timer counter. With this configuration, the timer counter will increment at the rate of 1 MHz when the E clock is 8 MHz. Also, configure the timer counter so that it resets itself to zero when its count value is equal to that of the TC7 output compare register.

2. Perform 40 output compare operations on the TC7 with each operation creating 25 ms delay. Forty such operations will create a delay of one second.

3. Enable the TC7 interrupt so that whenever the contents of the timer counter equal that of the output compare register 7, an interrupt is generated.

4. The interrupt service routine simply clears the TC7 interrupt flag, decrements a counter value, and returns from interrupt.

5. Repeat step 2 to 4 10 times. Print out the message *xx seconds have passed* each time a one-second delay is created.

Solution: After performing the initialization of timer channel 7, the program stays in a two-layer loop. The inner loop repeats the output compare operation (on TC7) 40 times with each operation creating a 25 ms delay so that a one-second delay is created after 40 iterations. The outer loop repeats 10 times so that a 10-second delay is created. Since we haven't discussed the operation of the timer function yet, we will not discuss the details of this program. This example simply serves as a demonstration of the timer interrupt. Enter, assemble, and download the following program to the CME-12BC demo board for execution using the MiniIDE:

```
CR       equ     $0D          ; ASCII code of carriage return
LF       equ     $0A          ; ASCII code of line feed
tios     equ     $80          ; timer IC/OC select register
tcnt     equ     $84          ; timer count register
tscr     equ     $86          ; timer system control register
tctl1    equ     $88          ; timer control register 1
tmsk1    equ     $8C          ; timer mask register 1
```

```
tmsk2          equ     $8D               ; timer mask register 2
tflg1          equ     $8E               ; timer flag register 1
tc7            equ     $9E               ; timer output compare register 7
pactl          equ     $A0               ; pulse accumulator control register

printf         equ     $F686             ; memory location to hold the starting address of printf()
setuservector  equ     $F69A             ; memory location to hold the starting address of SetUserVector()

               org     $800              ; starting address of internal SRAM
oc7_cnt        rmb     1                 ; number of output compare operations remained to be performed
seconds        rmb     1                 ; number of seconds passed since the beginning
total          rmb     1                 ; number of seconds remained

               org     $1000             ; starting address of the program
               lds     #$8000            ; initialize the stack pointer
               clr     seconds           ; start from 0 seconds
               jsr     init_oc7
               ldaa    #10
               staa    total
; the following 6 instructions set up oc7 interrupt vector
               ldd     #oc7_isr
               pshd
               ldab    #16
               clra
               ldx     setuservector
               jsr     0,x
               puld                      ; clean up the stack

; the following two instructions enable oc7 interrupt
               bset    tmsk1,$80         ; enable oc7 interrupt
               cli                       ; enable interrupt globally
; the outer loop make sure 10 second delays are created
outer_loop     tst     total
               beq     done
               movb    #40,oc7_cnt       ; set up the number of oc7 compares to be performed
               ldd     #25000
               std     tc7               ; initialize tc7 to 25000 so that every match creates
                                         ; 25 ms delay
;****************************************************************************
; the inner loop performs 40 oc7 output compare operations to create a 1 second delay
; with each operation creates 25 milliseconds delay
;****************************************************************************
inner_loop     tst     oc7_cnt           ; wait for 40 oc7 operations to be performed
               bne     inner_loop        ; and oc7 interrupt will occur here
               inc     seconds           ; one second delay has been created
               clra
               ldab    seconds
               pshd
               ldd     #time_msg
               ldx     printf
               jsr     0,x               ; print out the message
               leas    2,sp
               dec     total
```

```
                jmp       outer_loop
        done    swi


; *******************************************************************
; the following subroutine initializes the oc7 parameters
; *******************************************************************
        init_oc7   movb    #$80,tscr    ; enable timer, timer runs during wait state, and while
                                        ; in background mode, also clear flags normally
                   clr     tmsk1        ; disable all output compare interrupts
                   movb    #$80,tios    ; select channel 7 to act as output compare
                   movb    #$2B,tmsk2   ; set the timer counter prescale factor to 8 and reset
                                        ; tcnt on a successful output compare 7 event
                   movb    #$00,pactl   ; choose timer prescaler clock as timer counter clock
                   rts


; *******************************************************************
; oc7 interrupt service routine
; *******************************************************************
        oc7_isr    bclr    tflg1,$7F    ; clear c7F flag
                   dec     oc7_cnt
                   rti


        time_msg   db      " %d seconds have passed",CR,LF,0
                   end
```

After the execution of this program, the terminal window of the MiniIDE should display the following messages (each line of the message appears to be delayed from the previous one by one second):

```
>load
*
>g 1000
 1 seconds have passed
 2 seconds have passed
 3 seconds have passed
 4 seconds have passed
 5 seconds have passed
 6 seconds have passed
 7 seconds have passed
 8 seconds have passed
 9 seconds have passed
10 seconds have passed
User Breakpoint Encountered

PC    SP    X     Y     D = A:B   CCR = SXHI NZVC
104C  8000  1086  1085    00:19         1000 0100
104C  3F          SWI
```

The C language version of the previous program is as follows:

```c
#include <stdio.h>
#include <hc12.h>
unsigned int oc7_cnt, seconds;
void init_oc7();            /* prototype declaration of init_oc7() */
void tc7_isr(void);         /* prototype declaration of tc7_isr() */
```

```c
void main (void)
{
 int total;
 seconds = 0;
/* The following embedded instructions set up the interrupt vector for TC7 */
 asm ("ldd #_tc7_isr");
 asm ("pshd");
 asm ("ldab #16");            /* pass the vector number of TC7 */
 asm ("clra");
 asm ("ldx $F69A");           /* call the SetUserVector() function */
 asm ("jsr 0,x");             /*          "          */
 asm ("leas 2,sp");
 init_oc7();
 TFLG1 = 0x80;               /* clear the tc7f flag /
 TMSK1 |= 0x80;              /* enable TC7 interrupt */
 INTR_ON();                  /*          "          */
 for (total = 0; total < 10; total ++) { /* this loop creates a 10-second delay */
    TC7 = 25000;
    oc7_cnt = 40;
    while (oc7_cnt);          /* wait for one second */
    seconds++;
    printf(" %d seconds have passed\n", seconds);
 }
 INTR_OFF();
 asm ("swi");
}
void init_oc7 (void)
{
 TSCR = 0x80;
 TMSK1 = 0;
 TIOS = 0x80;               /* select channel 7 to act as output compare */
 TMSK2 = 0x2B;              /* set timer counter prescale factor to 8 and reset */
                            /* timer counter TCNT on a successful output compare */
                            /* 7 event */
 PACTL = 0x00;             /* choose timer prescale clock as timer counter clock */
}

#pragma interrupt_handler tc7_isr
void tc7_isr (void)
{
 TFLG1 = 0x80;             /* clear tc7F flag */
 oc7_cnt--;                /* decrement oc7_cnt */
}
```

▲

Example 6.2

Generate external interrupts to the 68HC12 by:

1. Using the 555 timer chip to generate a digital waveform with frequency equal to approximately 1 Hz. The circuit connection of the 555 timer is illustrated in Figure 6.6. Connect the 555 timer output (pin 3) to the PT7 pin (pulse accumulator input) of the 68HC912BC32.

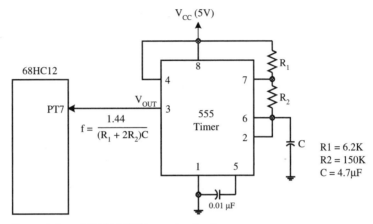

Figure 6.6 ■ 68HC12 PAI interrupt circuit

2. Configure the pulse accumulator so that it operates in event-counting mode and interrupts the CPU whenever rising edges arrive at the PAI (PT7) pin.

3. Write a service routine for the PAI interrupt that increments the interrupt count and outputs the message *interrupt x*, where *x* can be from 1 to 10.

4. Write a loop to check the interrupt count. Exit the loop when the interrupt count reaches 10.

Solution: The program consists of a main program and an interrupt service routine. The main program performs the following operations:

1. Initializes the pulse accumulator function (this part won't be clear until you have learned the pulse accumulator function).

2. Sets up the interrupt vector entry for the PAI interrupt.

3. Enables the PAI interrupt.

4. Stays in a wait loop until the interrupt count (variable *total*) equals 10 and exits.

The PAI interrupt service routine performs two operations:

1. Clears the PAIF flag.

2. Outputs the message "interrupt x".

The assembler program is as follows:

```
CR        equ     $0D        ; ASCII code of carriage return
LF        equ     $0A        ; ASCII code of line feed
pactl     equ     $A0        ; pulse accumulator control register
paflg     equ     $A1        ; pulse accumulator flag register
```

```
        printf          equ     $F686          ; memory location to hold the starting address of printf()
        setuservector   equ     $F69A          ; memory location to hold the starting address of setuservector()
        pa_vect_no      equ     13             ; vector number of PAI edge interrupt

                        org     $800           ; starting address of internal SRAM
        total           rmb     1              ; number of seconds passed

                        org     $1000          ; starting address of the program
                        lds     #$8000         ; initialize the stack pointer
                        sei                    ; disable interrupt
; the next instruction enable PA function, select event counting mode,
; and select rising edge as the active edge
                        movb    #$50,pactl     ; initialize pa function
; the following 6 instructions set up PAI interrupt vector
                        ldd     #pai_isr
                        pshd
                        ldab    #pa_vect_no    ; interrupt vector number of PAI pin
                        clra
                        ldx     setuservector
                        jsr     0,x
                        puld                   ; clean up the stack
                        ldaa    #01
                        staa    paflg          ; clear the PAIF flag
                        ldaa    #0             ; initialize the second count to 0
                        staa    total          ;       "
; the following two instructions enable PAI interrupt
                        bset    pactl,$01      ; enable PAI interrupt
                        cli                    ; enable interrupt globally
; the outer loop makes sure 10 second delays are created
        wait_loop       ldaa    total
                        cmpa    #10
                        beq     done
                        jmp     wait_loop
        done            swi
;************************************************************************
; the PAI interrupt service routine cleared the PAIF flag, incremented the interrupt
; output count, and the message "interrupt x", where x can be 1, 2, 3, ...
;************************************************************************
        pai_isr         ldaa    #01            ; clear PAIF flag
                        staa    paflg          ;       "
                        inc     total          ; one second delay has been created
                        clra                   ; print out the message
                        ldab    total          ;       "
                        pshd                   ;       "
                        ldd     #time_msg      ;       "
                        ldx     printf         ;       "
                        jsr     0,x            ;       "
                        leas    2,sp           ; clean up the stack
                        rti
        time_msg        db      " interrupt %d ",CR,LF,0
                        end
```

After the execution of this program, the terminal window of the MiniIDE should display the following messages (each line of message appears to be delayed from the previous one by one second):

```
>g 1000
 interrupt 1
 interrupt 2
 interrupt 3
 interrupt 4
 interrupt 5
 interrupt 6
 interrupt 7
 interrupt 8
 interrupt 9
 interrupt 10
User Breakpoint Encountered

PC    SP    X     Y     D = A:B   CCR = SXHI NZVC
102F  8000  8000  0000    0A:30         1000 0100
102F  3F          SWI
```

The C language version of the previous program is as follows:

```c
#include <stdio.h>
#include <hc12.h>

void pai_isr(void);
int total;
void main (void)
{
  total = 0;
/* the following embedded assembly instructions set up the interrupt vector for
   PAI input interrupt */
  asm ("ldd #_pai_isr");
  asm ("pshd");
  asm ("ldab #13");              /* pass vector number of PAI */
  asm ("clra");
  asm ("ldx $F69A");            /* call SetUserVector () */
  asm ("jsr 0,x");             /*        "              */
  asm ("leas 2,sp");
  PACTL = 0x50; /*configure PA function with event-counting mode,
                    rising edge active, disable PAI edge interrupt */
  PAFLG = 0x01; /* clear the PAIF flag */
  PACTL |= 0x01; /* enable PAI edge interrupt /
  INTR_ON();
  while (total < 10); /* wait for interrupt */
  INTR_OFF();
  asm ("swi");
}

#pragma interrupt_handler pai_isr
void pai_isr (void)
{
  PAFLG = 0x01; /* clear PAIF flag /
  total++;          /* increment interrupt count */
  printf(" interrupt %d \n", total);
}
```

6.6 Operation Modes

As listed in Table 6.5, the 68HC12 can operate in eight different modes. Each mode has a different default memory map and external bus configuration. After reset, most system resources can be mapped to other addresses by writing to the appropriate control registers.

BKGD	MODB	MODA	Mode	Port A	Port B
0	0	0	Special single chip	general-purpose I/O	general-purpose I/O
0	0	1	Special expanded narrow	ADDR[15:8]DATA[7:0]	ADDR[7:0]
0	1	0	Special peripheral	ADDR/DATA	ADDR/DATA
0	1	1	Special expanded wide	ADDR/DATA	ADDR/DATA
1	0	0	Normal single chip	General-purpose I/O	General-purpose I/O
1	0	1	Normal expanded narrow	ADDR[15:8]DATA[7:0]	ADDR[7:0]
1	1	0	Reserved (forced to peripheral)	--	--
1	1	1	Normal expanded wide	ADDR/DATA	ADDR/DATA

Table 6.5 ■ 68HC12 mode selection

The states of the BKGD, MODB, and MODA pins when the reset signal is low determine the operating mode after the CPU leaves the reset state. The SMODN, MODB, and MODA bits in the MODE register show the current operation mode and provide limited mode switching during the operation. The states of the BKGD, MODB, and MODA pins are latched into these bits on the rising edge of the reset signal. During reset an active pull-up (on-chip transistor) is connected to the BKGD pin (as input) and active pull-downs (on-chip transistors) are connected to the MODB and MODA pins. If an open occurs on any of these pins, the device will operate in normal single-chip mode.

The two basic types of operation modes are:

1. *Normal modes*—Some registers and bits are protected against accidental changes.
2. *Special modes*—Greater access for special purposes such as testing and emulation to protected control registers and bits are allowed.

The background debug mode (BDM) is a system development and debug feature and is available in all modes. In special single-chip mode, BDM is active immediately after reset.

6.6.1 Normal Operation Modes

These modes provide three operating configurations. Background debugging is available in all three modes, but must first be enabled for some operations by means of a BDM command. BDM can then be made active by another command.

NORMAL EXPANDED WIDE MODE

In this mode, ports A and B are used as 16-bit address and data buses. ADDR[15..8] and DATA[15..8] are multiplexed on port A. ADDR[7..0] and DATA[7..0] are multiplexed on port B.

NORMAL EXPANDED NARROW MODE

The 16-bit external address bus uses port A for the high byte and port B for the low byte. The 8-bit external data bus uses port A. ADDR[15..8] and DATA[7..0] are multiplexed on port A.

NORMAL SINGLE-CHIP MODE

Normal single-chip mode has no external buses. Ports A, B, and E are configured for general-purpose input/output. Port E bits 1 and 0 are input only with internal pull-ups and the other 22 pins are bidirectional I/O pins that are initially configured as high-impedance inputs. Port E pull-ups are enabled on reset. Port A and B pull-ups are disabled on reset.

6.6.2 Special Operation Modes

Special operation modes are commonly used in factory testing and system development.

SPECIAL EXPANDED WIDE MODE

This mode is for emulation of normal expanded wide mode and normal single-chip mode with a 16-bit bus. The bus-control pins of port E are all configured for their bus-control output functions rather than general-purpose I/O.

SPECIAL EXPANDED NARROW MODE

This mode is for emulation of normal expanded narrow mode. External 16-bit data is handled as two back-to-back bus cycles, one for the high byte followed by one for the low byte. Internal operations continue to use full 16-bit data paths.

SPECIAL SINGLE CHIP MODE

This mode can be used to force the microcontroller to active BDM mode to allow a system debug through the BKGD pin. The 68HC12 CPU does not fetch the reset vector nor execute application code as it would in other modes. Instead, the active background mode is in control of CPU execution and BDM firmware waits for additional serial commands through the BKGD pin. There are no external address and data buses in this mode. The microcontroller operates as a standalone device and all program and data space are on-chip. External port pins can be used for general purpose I/O.

SPECIAL PERIPHERAL MODE

The 68HC12 CPU is not active in this mode. An external master can control on-chip peripherals for testing purposes. It is not possible to change to or from this mode without going through reset. Background debugging should not be used while the microcontroller is in special peripheral mode as internal bus conflicts between the BDM and the external master can cause improper operation of both modes.

6.6.3 Background Debug Mode

Background debug mode (BDM) is an auxiliary operating mode that is used for system development. This mode will be discussed in detail in Chapter 11.

6.7 Summary

Interrupt is a special event that requires the CPU to stop normal program execution and provide certain services to the event. The interrupt mechanism has many applications including coordinating I/O activities, exiting from software errors, and reminding the CPU to perform routine works, and so on.

Some interrupts are *maskable* and can be ignored by the CPU. Other interrupts are *nonmaskable* and cannot be ignored by the CPU. Nonmaskable interrupts are often used to handle emergent events such as power failure.

Multiple interrupts may be pending at the same time. The CPU needs to decide which one to service first. The solution to this issue is to *prioritize* all of the interrupt sources. The pending interrupt with the highest priority will receive service before other pending interrupts.

The CPU provides service to an interrupt request by executing an *interrupt service routine*. The current program counter value will be saved in the stack before the CPU executes the service routine so that CPU control can be returned to the interrupted program when the interrupt service routine is completed.

In order to provide service to the interrupt, the CPU must have some way to find out the starting address of the interrupt service routine. There are three methods to determine the starting address (called the *interrupt vector*) of the interrupt service routine:

1. Each interrupt vector is predefined when the microcontroller was designed. In this method, the CPU simply jumps to the predefined location to execute the service routine.

2. Each interrupt vector is stored in a predefined memory location. When an interrupt occurs, the CPU fetches the interrupt vector from that predefined memory location. The 68HC12 uses this approach.

3. The interrupt source provides an interrupt vector number to the CPU so that the CPU can figure out the memory location where the interrupt vector is stored. The CPU needs to perform a read bus cycle to obtain the interrupt vector number.

There are three steps in the interrupt programming:

Step 1
Initialize the interrupt vector table that holds all the interrupt vectors. This step is not needed for those microcontrollers that use the first method to resolve the interrupt vector.

Step 2
Write the interrupt service routine.

Step 3
Enable the interrupt to be serviced.

Reset is a mechanism for:

1. Setting up the operation mode for the microcontroller.

2. Setting up initial values for the control registers.

3. Exiting from software errors and some hardware errors.

All 68HC12 microcontrollers have the same number of resets and nonmaskable interrupt sources despite the fact that they may not have the same number of maskable interrupts. The 68HC12 has two low-power modes that are triggered by the execution of WAIT and STOP instructions. Power consumption will be reduced dramatically in either low-power mode. The 68HC12 has a COP timer reset mechanism to detect the software error. A software program that behaves properly will reset the COP timer before it times out and prevent it from resetting the CPU.

The 68HC12 has a clock monitor reset mechanism that can detect the slowing down or loss of clock signals. Whenever the clock frequency gets too low, the clock monitor will detect it and reset the CPU.

Two examples are given to illustrate the procedure of interrupt programming.

The 68HC12 has seven different operation modes divided into two basic categories: normal modes and special modes. Normal modes are used for embedded applications, whereas special modes are used in fabrication testing and development debugging activities.

6.8 Exercises

E6.1 What is the name given to a routine that is executed in response to an interrupt?

E6.2 What are the advantages of using interrupts to handle data inputs and output?

E6.3 What are the requirements of interrupt processing?

E6.4 How do you enable other interrupts when the 68HC12 is executing an interrupt service routine?

E6.5 Why would there be a need to promote one of the maskable interrupts to highest priority among all maskable interrupts?

E6.6 Write the assembler directives to initialize the $\overline{IRQ}$ interrupt vector located at $2000.

E6.7 What is the last instruction in most interrupt service routines? What does this instruction do?

E6.8 Suppose the 68HC12 is executing the following instruction segment and the $\overline{IRQ}$ interrupt occurs when the TSY instruction is being executed. What will the contents of the top ten bytes in the stack be?

```
ORG      $1000
LDS      #$8000
CLRA
LDX      #$0
BSET     10,X $48
LDAB     #$40
INCA
TAP
PSHB
TSY
ADDA     #10
```

E6.9 Suppose the E clock frequency is 6 MHz. Compute the COP watchdog timer timeout period for all the possible combinations of the CR2, CR1, and CR0 bits in the COPCTL register.

E6.10 Suppose the starting address of the service routine of the timer overflow interrupt is at $3000. Write the assembler directives to initialize its vector table entry on the CME-12BC demo board.

E6.11 Write an instruction sequence to clear the X and I bits in the CCR. Write an instruction sequence to set the S, X, and I bits in the CCR.

E6.12 Why does the 68HC12 need to be reset when the power supply is too low?

E6.13 Write the instruction sequence to prevent the COP timer from timing out and resetting the microcomputer.

6.9 Lab Exercises & Assignments

L6.1 *PAI input interrupt experiment.* Use the 555 timer to generate a digital waveform with frequency equal to about 1 Hz as shown in Example 6.2. Enter the program to a file, assemble/compile, and download the S-record file onto the CME-12BC demo board for execution. Follow the procedure illustrated in Example 6.2.

L6.2 *Simple interrupts.* Connect the $\overline{IRQ}$ pin of the demo board to a debounced switch that can generate a negative-going pulse. Write a main program and an $\overline{IRQ}$ service routine. The main program initializes the variable *irq_cnt* to 10, stays in a loop, and keeps checking the value of *irq_cnt*. When *irq_cnt* is decremented to zero, the main program jumps back to monitor. The $\overline{IRQ}$ service routine simply decrements *irq_cnt* by one and returns.

The lab procedure is as follows:

Step 1

Connect the $\overline{\text{IRQ}}$ pin of the demo board to a debounced switch that can generate a clean negative-going pulse.

Step 2

Enter the main program and $\overline{\text{IRQ}}$ service routine, assemble them, and then download them to the single-board computer. Remember to enable the $\overline{\text{IRQ}}$ interrupt in your program.

Step 3

Pulse the switch 10 times.

If everything works properly, you should see the D-Bug12 monitor prompt after 10 pulses applied to the $\overline{\text{IRQ}}$ pin.

L6.3 *Time-of-Day update using interrupt.* Use the timer channel 7 (TC7) interrupt to create a one-second time delay as demonstrated in Example 6.1. Write a program that performs the following operations:

1. Prompt the user to enter the current time in the format of hhmmss using the D-Bug12 built-in functions, where hh, mm, and ss represent the current hours, minutes, seconds.

2. Clear the screen by outputting the form-feed character (ASCII code is 0x0C) and output appropriate number of CR/LF pairs to move the cursor to the center of the screen.

3. Enable the TC7 interrupt and stay in an infinite loop waiting for interrupt to update the current time.

4. When updating the current time, output the backspace character several times to make the display as stable as possible.

7

Parallel Ports

7.1 Objectives

After completing this chapter, your should be able to:

- define I/O addressing methods
- explain the data transfer synchronization methods between the CPU and I/O interface chip
- explain the data transfer synchronization methods between the I/O interface chip and the I/O device
- explain input and output handshake protocols
- input data from simple switches
- input data from keypads and keyboards
- output data to LED and LCD displays
- interface with a D/A converter to generate waveforms
- explain the operation and programming of a stepper motor
- explain the operation, application, and programming of key wakeup ports

7.2 Basic Concepts of I/O

I/O devices are also called *peripheral* devices. They are mainly used to exchange data with a computer. Examples of I/O devices include switches, light-emitting diodes (LED), cathode-ray tube (CRT) screens, printers, modems, keyboards, disk drives, and so on. The speed and electrical characteristics of I/O devices are very different from the CPU and hence it is not feasible to connect them directly to the CPU. Instead, interface chips are used to resolve the differences between the microprocessor and I/O devices.

The major function of the interface chip is to synchronize data transfer between the CPU and I/O devices. An interface chip consists of control registers, data registers, status registers, data direction registers, and control circuitry. Control registers allow us to set up parameters for the desired I/O operations. The status registers report the progress and status of the I/O operations. The data direction registers allow us to select the data transfer direction for each I/O pin. A data register may hold the data to be sent to the output device or the new data placed by the input device.

In an input operation, the input device places data in the data register, which holds data until they are read by the CPU. In an output operation, the CPU places data in the data register, which holds the data until they are fetched by the output device.

An interface chip has data pins that are connected to the microprocessor data bus and I/O port pins that are connected to the I/O device, as illustrated in Figure 7.1.

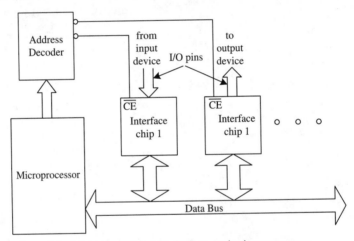

Figure 7.1 ■ Interface chip, I/O devices and microprocessor

In a large embedded system, many I/O devices are attached to the data bus via interface chips. Only one device is allowed to drive data to the data bus at a time. Otherwise, *data bus contention* can result and the system can be damaged. The address decoder in Figure 7.1 makes sure that one and only one device is allowed to drive data to the data bus or accept data from the data bus. An interface chip is allowed to respond to the data transfer request from the microprocessor when its chip enable (CE) input is low. Otherwise, the interface chip is electrically isolated from the data bus.

Data transfer between the I/O device and the interface chip can be proceeded bit-by-bit (serial) or in multiple bits (parallel). Data are transferred serially in low-speed devices such as modems and low-speed printers. Parallel data transfer is mainly used by high-speed I/O devices. Only parallel I/O transfer will be discussed in this chapter.

7.3 I/O Addressing

An interface chip normally has several registers. An address must be assigned to each of these registers so that they can be accessed. Two issues are related to the access of I/O registers:

1. *Address space.* I/O devices can share the same memory space with memory devices (such as SRAM, ROM, and so on) or they can have their own memory space. In the earlier days of microprocessors, some manufacturers (including Intel and Zilog) used a separate memory space for I/O devices. The current trend shows that most microprocessor and microcontroller manufacturers use the same memory space for memory and I/O devices.

2. *Instruction set and addressing modes.* I/O devices could have their own instruction set and addressing modes or use the same instruction set and addressing modes with memory devices. In the first approach, the microprocessor may use the following instructions for input and output:

   out 3: sends data in accumulator to I/O device at address 3

   in 5: inputs a byte from input device 5 to the accumulator

   In the second approach, the microprocessor uses the same instructions and addressing modes to perform input and output operations. Again, the current trend is to use the second approach.

Traditionally, Motorola microprocessors and microcontrollers use the same addressing modes and instruction sets to access I/O and memory devices. Memory and I/O devices share a single memory space.

7.4 I/O Synchronization

The role of an interface chip is shown in Figure 7.2. The microprocessor deals with the interface chip rather than the I/O devices. The electronics in the I/O devices converts electrical signals into mechanical actions or vice versa. The functions of an interface chip and the CPU are incorporated onto the same chip in a microcontroller.

When inputting data, the microprocessor reads data from the interface chip, so there must be a mechanism to make sure that data are valid when the microprocessor reads it. When outputting data, the microprocessor writes data into the interface chip. Again, there must be some mechanism to make sure that the output device is ready to accept data when the microprocessor outputs. There are two aspects in the I/O synchronization: the synchronization between the microprocessor and the interface chip and the synchronization between the interface chip and the I/O device.

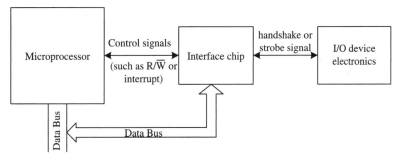

Figure 7.2 ■ The role of an interface chip

7.4.1 Synchronizing the Microprocessor & the Interface Chip

To input valid data from an input device, the microprocessor must make sure that the interface chip has correctly latched the data from the input device. There are two ways to do that:

1. *The polling method.* The interface chip uses a status flag to indicate whether it has valid data (stored in a data register) for the microprocessor. The microprocessor knows that the interface chip has valid data when the status flag is set to one. The microprocessor keeps checking (reading) the status flag until the status flag is set to one, and then it reads the data from the data register. When this method is used, the microprocessor is tied up and cannot do anything else when it is waiting for the data. However, this method is very simple to implement and is often used when the microprocessor has nothing else to do when waiting for the completion of the input operation.

2. *Interrupt-driven method.* In this method, the interface chip asserts an interrupt signal to the microprocessor when it has valid data in the data register. The microprocessor then executes the service routine associated with the interrupt to read the data.

To output data successfully, the microprocessor must make sure that the output device is not busy. There are also two ways to do this:

1. *The polling method.* The interface chip has a data register that holds data to be output to an output device. New data should be sent to the interface chip only when the data register of the interface chip is empty. In this method, the interface chip uses a status bit to indicate whether the output data register is empty. The microprocessor keeps checking the status flag until it indicates that the output data register is empty and then writes data into it.

2. *The interrupt-driven method.* The interface chip asserts an interrupt signal to the microprocessor when the output data register is empty and can accept new data. The microprocessor then executes the service routine associated with the interrupt and outputs the data.

The M6821 from Motorola and the i8255 from Intel support both methods. The port C of the 68HC11 also supports both methods. However, the 68HC12 supports neither method (there is no data status bit associated with any I/O port). Serial interface functions of the 68HC12 still implement polling and interrupt methods as mechanisms for synchronization between the processor and the interface function.

7.4.2 Synchronizing the Interface Chip & I/O Device

The interface chip is responsible for making sure that data are properly transferred to and from I/O devices. The following methods have been used to synchronize data transfer between the interface chip and I/O devices:

1. *The brute-force method.* Nothing special is done in this method. For input, the interface chip returns the voltage levels on the input pins to the microprocessor. For output, the interface chip makes the data written by the microprocessor directly available on output pins. This method is useful in situations in which the timing of data is unimportant. It can be used to test the voltage level of a signal, set the voltage of an output pin to high or low, or drive LEDs. All I/O ports of the 68HC12 can perform brute force I/O.

2. *The strobe method.* This method uses strobe signals to indicate that data are stable on input or output port pins. During input, the input device asserts a strobe signal when the data are stable on the input port pins. The interface chip latches data into

the data register using the strobe signal. For output, the interface chip first places data on the output port pins. When the data become stable, the interface chip asserts a strobe signal to inform the output device to latch data on the output port pins. This method can be used if the interface chip and I/O device can keep up with each other. None of the 68HC12 parallel ports support this method.

3. *The handshake method.* The previous two methods cannot guarantee correct data transfer between an interface chip and an I/O device when the timing of data is critical. For example, it takes a much longer time to print a character than it does to send a character to the printer electronics, so data shouldn't be sent to the printer if it is still printing. The solution is to use a handshake protocol between the interface chip and the printer electronics. There are two handshake methods: the *interlocked hand-shake* and the *pulse-mode handshake.* Whichever handshake protocol is used, two handshake signals are needed—one (call it H1) is asserted by the interface chip and the other (call it H2) is asserted by the I/O device. The handshake signal transactions for input and output are described in the following subsections. Note that the hand-shake operations of some interface chips may differ slightly from what we describe here. None of the 68HC12 parallel ports support the handshake I/O method.

INPUT HANDSHAKE PROTOCOL

The signal transaction of the input handshake protocol is illustrated in Figure 7.3.

Step 1
The interface chip asserts (or pulses) H1 to indicate its intention to input new data.

Step 2
The input device puts valid data on the data port and also asserts (or pulses) the handshake signal H2.

Step 3
The interface chip latches the data and de-asserts H1. After some delay, the input device also de-asserts H2.

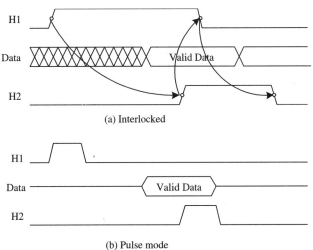

(a) Interlocked

(b) Pulse mode

Figure 7.3 ■ Input handshakes

The whole process will be repeated if the interface chip wants more data.

OUTPUT HANDSHAKE PROTOCOL

The signal transaction of the output handshake protocol is shown in Figure 7.4. It also takes place in three steps:

Step 1

The interface chip places data on the data port and asserts (or pulses) H1 to indicate that it has data to be output.

Step 2

The output device latches the data and asserts (or pulses) H2 to acknowledge the receipt of data.

Step 3

The interface chip de-asserts H1 following the assertion of H2. The output device then de-asserts H2.

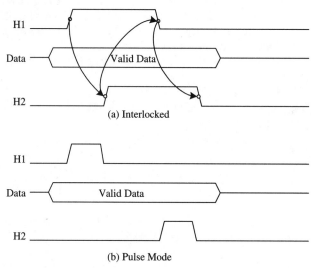

Figure 7.4 ■ Output handshaking

The whole process will be repeated if the microprocessor has more data to be output.

The handshake is a very common synchronization mechanism for older interface chips and I/O devices. However, the microcontrollers (including the 68HC12) designed in the last 10 years rarely provide this mechanism. This is probably due to the shift of applications of the parallel I/O ports.

7.5 Overview of 68HC12 Parallel Port

The 68HC12 members have from 80 to 112 pins arranged in either a quad flat pack (QFP) or a thin quad flat pack (TQFP). A summary of the I/O ports and their characteristics are in Table 7.1. There are several signal pins that are not part of any ports. They will be covered in the appropriate chapters.

Port name	Functions	Available in
Port A	General-purpose I/O in single-chip mode. External address ADDR[15:8] or multiplexed address/data bus ADDR[15:8]/DATA[15:8] in expanded mode.(1)	all members
Port B	General-purpose I/O in single-chip mode. External address ADDR[7:0] or multiplexed address/data bus ADDR[7:0]/DATA[7:0] in expanded mode.(2)	all members
Port C	General-purpose I/O in single-chip modes. External data bus DATA[15:8] in expanded wide modes; external data bus DATA[15:8]/DATA[7:0] in expanded narrow mode.	812A4
Port D	General-purpose I/O in single-chip modes and expanded narrow modes. External data bus DATA[7:0] in expanded wide mode. As key wakeup can cause an interrupt on a high to low transition.	812A4
Port E	Mode selection, bus control signals and interrupt service request signals; or general-purpose I/O.	all members
Port F	Chip select and general-purpose I/O	812A4
Port G	Memory expansion and general-purpose I/O	812A4
Port H	Key wakeup and general-purpose I/O, can cause interrupt when an input transitions from high to low	812A4, 912D60, 912DG128, 912DT128
Port J	Key wakeup and general-purpose I/O, can cause interrupt when an input transitions from high to low or from low to high	812A4, 912DG128, 912DT128
Port K	General-purpose I/O or page index emulation in expanded or peripheral mode	912DG128, 912DT128
Port P	General-purpose I/O or PWM outputs	the B series, 912D60, 912DG128, 912DT128
Port S	Serial communications interface and serial peripheral interface subsystems and general-purpose I/O	All members
Port T	Timer system and general-purpose I/O	All members
Port AD	Analog-to-digital converter and general-purpose input	812A4 and B series
Port AD0	Analog-to-digital converter and general-purpose input	912D60, 912DG128/DT128
Port AD1	Analog-to-digital converter and general-purpose input	912D60, 912DG128/DT128
Port DLC	Byte data link communication (BDLC) subsystem and general-purpose I/O	912B32, 912BE32
Port CAN	Control area network pins and general-purpose I/O	912BC32, 912D60
Port CAN0	Control area network pins	912DG128, 912DT128
Port CAN1	Control area network pins	912DG128, 912DT128
Port CAN2	Control area netwrok pins	912DT128
Port IIC	Inter-Integrated Circuit	912DG128, 912DT128

Note (1) 812A4 uses this port as address bus ADDR[15:8] in expanded modes. Other members use this
         port as multiplexed address/data bus ADDR[15:8]/DATA[15:8].
      (2) 812A4 uses this port as address bus ADDR[7:0] in expanded modes. Other members use this
         port as multiplexed address/data bus ADDR[7:0]/DATA[7:0].

Table 7.1 ■ Summary of the 68HC12 ports

The names and addresses of all 68HC12 registers are listed in Appendix C. In general, a HC12 I/O port has in addition to port pins:

- a data register to hold the data (for output) to be applied to the port pins or the current signal levels (for input) of the port pins
- a data direction register to select the I/O direction (input or output). Set a bit in the data direction register to 1 (0) will force the corresponding I/O pin to output (input)

It is considered a good design practice to tie unused input pins to either a high or a low logic level. The 68HC12 has a *Pull-Up Control Register (PUCR)* to enable pull-up devices on any of the ports that are configured as inputs. On reset, all pull-up devices are enabled. You may disable any of them if you wish to improve speed performance and reduce power consumption.

The contents of the PUCR register are shown in Figure 7.5. The pull-up control has no effect when a port is being used as an address or data bus. When a pull-up control bit is set to one, the pull-up devices of the corresponding port are enabled. Otherwise, they are disabled.

	7	6	5	4	3	2	1	0	$000C
value after	PUPH	PUPG	PUPF	PUPE	PUPD	PUPC	PUPB	PUPA	PUCR for 812A4
reset:	1	1	1	1	1	1	1	1	
value after	0	0	0	PUPE	0	0	PUPB	PUPA	PUCR for B series
reset:	0	0	0	1	0	0	1	1	
value after	PUPH	PUPG	0	PUPE	0	0	PUPB	PUPA	PUCR for 912D60
reset:	1	1	0	1	0	0	1	1	
value after	PUPK	PUPJ	PUPH	PUPE	0	0	PUPB	PUPA	PUCR for 912DG128/ 912DT128
reset:	1	1	1	1	0	0	1	1	

Figure 7.5 ■ PUCR register

The amount of current available at an output pin is referred to as the *drive* of that output pin. High drive current is an advantage when an output pin must drive a large capacitive load because it results in higher switching speed. Unfortunately, it also means higher power consumption and possible radio frequency interference (RFI). The 68HC12 provides the *Reduced Drive of I/O Lines (RDRIV)* feature that allows you to reduce the drive level to reduce power consumption and RFI emission. The contents of the RDRIV register are shown in Figure 7.6. When a bit is set to one, the corresponding port is set to reduced drive mode.

	7	6	5	4	3	2	1	0	$000D
value after	RDPJ	RDPH	RDPG	RDPF	RDPE	RDPD	RDPC	RDPB	RDRIV for 812A4
reset:	0	0	0	0	0	0	0	0	
value after	0	0	0	0	RDPE	0	RDPB	RDPA	RDRIV for B series
reset:	0	0	0	0	0	0	0	0	
value after	0	RDPH	RDPG	0	RDPE	0	RDPB	RDPA	RDRIV for 912D60
reset:	0	0	0	0	0	0	0	0	
value after	RDPK	RDPJ	RDPH	RDPE	0	0	RDPB	RDPA	RDRIV for 912DG128 912DT128
reset:	0	0	0	0	0	0	0	0	

Figure 7.6 ■ RDRIV register

7.5.1 Port A & Port B

In expanded modes, these two ports are used for the address bus for the 812A4 but are used for multiplexed address/data buses for other members. In single chip modes, these two ports are used for general I/O ports. Data direction registers DDRA and DDRB set the data direction of each port A and port B pin. When a bit in DDRA or DDRB is set to one, the corresponding port A or port B pin becomes an output pin. Otherwise it is an input pin.

Example 7.1

▼

Write an instruction sequence to output the contents of accumulator A to port A pins (assume that the 68HC12 is operating in single-chip mode).

Solution: We need to write the value 0xFF into the DDRA register to configure port A as an output port. The appropriate instructions are as follows:

```
PORTA     equ     $00
DDRA      equ     $02
          ...
          movb    #$FF,DDRA       ; configure PORT A as an output port
          staa    PORTA           ; output the contents of A to PORTA
```

▲

7.5.2 Port C & Port D

These two ports are available only in the 68HC812A4. In expanded wide mode, Port C and Port D are used for the upper and lower bytes of the 16-bit data bus. In expanded narrow mode, Port C is used to carry both the upper and lower bytes of data whereas port D is used as a general-purpose I/O port.

When in single-chip mode, these two ports can be used as general I/O ports. The direction of each Port C or Port D pin is set by the corresponding bit in the DDRC or the DDRD register. Port D can also be used as a key wakeup port.

Example 7.2

▼

Write an instruction sequence to read the current signal levels of Port C and Port D into accumulator A and B registers.

Solution: We need to write the value zero into the DDRC and DDRD registers before we can read the PORTC and PORTD pins. Because PORTC and PORTD are adjacent, we can read them using a 16-bit load instruction. The following instruction sequence performs the desired operation:

```
PORTC     equ     $04
PORTD     equ     $05
DDRC      equ     $06
DDRD      equ     $07
          ...
          movw    #$00,DDRC       ; configure Port C and Port D for input
          ldd     PORTC           ; read in 16 bits in one operation
```

▲

7.5.3 Port E

Port E pins are used for bus control and interrupt service request signals. When a pin is not used for one of these specific functions, it can be used as a general-purpose I/O. However, two of the pins (PE[1:0]) can only be used for input, and the states of these pins can be read in the port data register even when they are used for $\overline{\text{IRQ}}$ and $\overline{\text{XIRQ}}$.

The contents of the PORTE data register are shown in Figure 7.7. For 812A4, bit seven is used for auxiliary reset (ARST) input. For other members, this pin is used for a data bus enable signal during an external read in expanded modes. Pin seven can also be used for calibration reference (CAL) when the *slow mode programmable clock* is enabled. Pins corresponding to bit six (IPIPE1/MODB) and five (IPIPE0/MODA) are used to set up the operation mode of the 68HC12 when the CPU exits the reset state. During normal operation, these two pins can be used to show the instruction queue status.

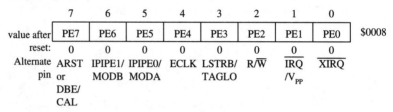

Figure 7.7 ■ PORTE register

The PEAR register determines pin functions, and register DDRE determines whether a pin is an input or output when it is used for general-purpose I/O. PEAR settings override DDRE settings. Because PE[1:0] are input pins, only DDRE[7:2] have effects.

The contents of the register PEAR are shown in Figure 7.8. The PEAR register is not accessible for reads or writes in peripheral mode.

	7	6	5	4	3	2	1	0	
value after reset:	NDBE	0	PIPOE	NECLK	LSTRE	RDWE	0	0	$000A
normal expanded	0	-	0	0	0	0	-	-	
special expanded	0	-	1	0	1	1	-	-	
peripheral	1	-	0	1	0	0	-	-	
normal single-chip	1	-	0	1	0	0	-	-	
special single-chip	0	-	1	0	1	1	-	-	

NDBE: *No data bus enable.* Can be read/written any time.

When set to 0, PE7 is used for external control of data enables on memories. When set to 1, PE7 is used for general-purpose I/O

PIPOE: *Pipe signal output enable.*

In normal mode: write once. Special mode: write anytime except the first time. This bit has no effect in single chip modes.

0 = PE[6:5] are general-purpose I/O

1 = PE[6:5] are outputs and indicate the state of the instruction queue.

NECLK: *No external E clock.* Can be read anytime.

In expanded mode, writes to this bit has no effect. E clock is required for de-multiplexing the external address. NECLK can be written once in normal single chip mode and can be written anytime in special single chip mode.

0 = PE4 is the external E-clock. 1 = PE4 is a general-purpose I/O pin.

LSTRE: *Low strobe ($\overline{LSTRB}$) enable.* Can be read anytime.

In normal modes: write once; special modes: write anytime except the first time. This bit has no effect in single-chip modes or normal expanded narrow mode.

0 = PE3 is a general-purpose I/O pin.

1 = PE3 is configured as the $\overline{LSTRB}$ bus-control output, provided the 68HC12 is not in single chip or normal expanded narrow modes.

RDWE: *Read/write enable.* Can be read anytime.

In normal modes: write once; special modes: write anytime except the first time. This bit has no effect in single-chip modes.

0 = PE2 is a general-purpose I/O pin

1 = PE2 is configured as the R/$\overline{W}$ pin. In single-chip mode, RDWE has no effect and PE2 is a general-purpose I/O pin.

R/W is used for external writes. After reset in normal expanded mode, it is disabled. If needed it should be enabled before any external writes.

Figure 7.8a ■ PEAR register (bits common to all members)

	7	6	5	4	3	2	1	0	
value after reset:	--	CGMTE	--	--	--	--	CALE	DBENE	$000A
normal expanded	-	0	-	-	-	-	0	0	
special expanded	-	0	-	-	-	-	0	0	
peripheral	-	1	-	-	-	-	0	0	
normal single-chip	-	0	-	-	-	-	0	0	
special single-chip	-	0	-	-	-	-	0	0	

CGMTE: *Clock Generator Module Testing Enable*. Can only be written special mode.

0 = PE6 is a general-purpose I/O or pipe output

1 = PE6 is a test signal output from the CGM module (no effect in single chip or normal expanded modes). PIPOE = 1 overrides this function and forces PE6 to be a pipe status output signal.

CALE: *Calibration Reference Enable*.

Read and write anytime.

0 = Calibration reference is disabled and PE7 is general-purpose I/O in single-chip or peripheral modes or if the NDBE bit is set.

1 = Calibration reference is enabled on PE7 in single chip and peripheral modes or if the NDBE bit is set.

DBENE: $\overline{DBE}$ *or Inverted E clock on Port E[7]*. Normal modes: write once. Special modes: write anytime EXCEPT the first; read anytime.

0 = PE7 pin used for $\overline{DBE}$ external control of data enable on memories in expanded modes when NDBE = 0.

1 = PE7 pin used for inverted ECLK output in expanded modes when NDBE = 0.

Figure 7.8b ■ PEAR register (bits available in 912D60, 912DG128 & 912DT128 only)

7.5.4 Port F & Port G

These two ports are available in the 812A4 and are used for chip select signals and memory expansion address bits in expanded modes. In single chip mode, these two ports are used for general-purpose I/O ports. Each of them has an associated data direction register (DDRF and DDRG) and port data register (PORTF and PORTG). Port G is also available in the 912D60. The Port G key wakeup interrupt feature is available in the 912D60 but not in the 812A4.

7.5.5 Port H & J

These two ports are available in the 812A4, the 912DG128, and the 912DT128. Port H is also available in the 912D60. Port H and J pins are used for key wake-ups that can be used with the pins configured as inputs or outputs. The key wake-ups are triggered with either a rising or falling edge signal (specified in registers KWPH and KWPJ). An interrupt is generated if the corresponding bit is enabled (specified in registers KWIEH and KWIEJ). If any of the interrupts is not enabled, the corresponding pin can be used as a general-purpose I/O pin.

Setting a bit in the register KWPH (or KWPJ) makes the corresponding key wake-up input pin trigger at rising edges and loads a pull-down in the corresponding port H (or J) input pin. Clearing a bit in KWPH (or KWPJ) makes the corresponding key wake-up input pin trigger at falling edges and loads a pull-up in the corresponding port H (or J) pin. Like other ports, the direction of any pin of Port H or J is determined by the setting of its associated port data direction register (DDRH and DDRJ).

7.5.6 Port K

Port K pins can be used for general-purpose I/O in single-chip mode or when the EMK bit of the MODE register is not set in other modes. In expanded or peripheral modes, these pins are used for page index emulation.

7.5.7 Port P

The four pulse-width modulation channel outputs share general-purpose port P pins. The PWM function is enabled with the PWEN register. Enabling PWM pins takes precedence over the general-purpose port. When pulse-width modulation is not in use, the port pins may be used for general-purpose I/O. The PORTP register and DDRP are the Port P data register and data direction register, respectively.

7.5.8 Port S

Port S is the 8-bit interface to the standard serial interface consisting of the serial communications interfaces (only one SCI channel in the B family, two channels SCI0 and SCI1 in the 812A4, the 912D60, the 912DG128, and the 912DT128) and the serial peripheral interface (SPI) subsystems. Port S pins are available for general-purpose I/O when standard serial functions are not enabled. The PORTS register and DDRS are the Port S data register and data direction register, respectively.

7.5.9 Port T

This port provides eight general-purpose I/O pins when not enabled for input-capture and output-compare functions in the timer and pulse accumulator subsystem. The TEN bit in the TSCR register enables the timer function. The pulse accumulator subsystem is enabled with the PAEN bit in the PACTL register. The PORTT register and DDRT are the Port T data register and data direction register, respectively.

7.5.10 Port AD

Pins of this port are the input to the analog-to-digital subsystem and general-purpose input. When analog-to-digital functions are not enabled, the port has eight general-purpose input pins, PAD[7:0]. The ADPU bit in the ADCTL2 register enables the A/D system.

Port AD pins are inputs; no data direction register is associated with this port. The port has no resistive input loads and no reduced drive controls. This port is available in the 812A4 and all members in the B family. PORTAD is the data register of this port.

7.5.11 Port AD0 & AD1

These two ports are analog input interfaces to the analog-to-digital subsystem. When analog-to-digital functions are not enabled, these two ports are available for general-purpose I/O. The ADPU bit in ATD0CTL2 and ATD1CTL2 registers enables the AD0 and AD1 functions.

Port AD0 and AD1 pins are inputs; no data direction register is associated with this port. The port has no resistive input loads and no reduced drive controls. PORTAD0 and PORTAD1 are data registers of the Port AD0 and AD1, respectively.

7.5.12 Port CAN

CAN stands for *controller area network*. It is a serial network initially proposed to be used for data communication in automotive applications. The port CAN has five general-purpose I/O pins, PCAN[6:2:]. The MSCAN12 receive pin, RxCAN, and transmit pin, TxCAN, cannot be configured as a general-purpose I/O on port CAN.

Register DDRCAN determines whether each port CAN pin PCAN[6:2] is an input or output. The 912BC32 and 912D60 have one CAN port.

7.5.13 Port CAN0, CAN1 & CAN2

The 912DG128 has two control area network ports CAN0 and CAN1. The 912DT128 has three CAN modules: CAN0, CAN1, and CAN2. CAN0 uses two external pins, one input (RxCAN0), and one output (TxCAN0). CAN1 uses pins RxCAN1 and TxCAN1. CAN2 uses pins RxCAN2 and TxCAN2. The CAN ports in the 912DG128 and 912DT128 cannot be used as general-purpose I/O ports.

7.5.14 Port IB

The 912DG128 and 912DT128 also implement the Inter-Integrated Circuit (I2C) protocol. This is a low speed serial communication protocol proposed by Philips to facilitate data communications between microcontrollers and peripheral chips. The I2C bus interface uses a serial data line (SDL) and a serial clock line (SCL) for data transfer. Both pins are pulled up to a positive supply via a resistor. The pull-ups can be enabled and disabled. The Port IB of the 912DG128 has four pins whereas that of the 912DT128 has two pins. The remaining two pins of the Port IB in the 912DG128 are used for general-purpose input and output. When the SDL and SCL pins are not used for the I2C function, they can be used for general-purpose I/O.

7.6 Interfacing with Simple Input & Output Devices

Many embedded systems only require simple input and output devices such as switches, light-emitting devices (LEDs), keypads, and seven-segment displays.

7.6.1 Interfacing with LEDs

The light emitting diode (LED) is one of the most often used output devices in an embedded system. In many embedded systems, LEDs are simply used to indicate that the system is operating properly.

An LED can illuminate if it is forward-biased and has sufficient current flowing through it. The current required to light an LED may range from a few to more than 10 mA. The voltage drop across the LED when it is forward-biased can range from about 1.6V to more than 2.2V.

LED indicators are easy to interface with microcontrollers. All you need is enough current to drive the LED and a series resistor to absorb the voltage drop. The circuit in Figure 7.9 is often used to interface with an LED. The series resistor is used to limit the current to some acceptable value. Using a 5-volt supply and assuming that the LED has a 2.0-volt drop across it, a 300-Ω resistor will limit the current to 10 mA. An I/O port pin of a microcontroller generally does not have enough drive to supply the current. So an inverter is often used as a switch to turn the LED on and off. When the inverter's output is low (close to 0V), the diode has a 2.0V voltage drop, and by ohm's law:

$$5V = 2.0V + I_{Rx} \times Rx$$

When setting I_{Rx} to 10 mA, the resistor Rx is solved to be 300 Ω.

V_{CC}

74HC04

Figure 7.9 ■ An LED connected to a CMOS inverter
through a current limiting resistor

Example 7.3

Use the 68HC12 Port P to drive green, yellow, red, and blue LEDs. Light each of them for half of a second in turn and repeat. The 68HC12 uses a 16-MHz crystal oscillator to generate internal clock signals.

Solution: We can use Port P to drive LEDs only when Port P is not used for other purposes. Port P must be configured for output in this application. The circuit connection is shown in Figure 7.10. The 74HC04 can sink more than 20 mA of current when its output is low.

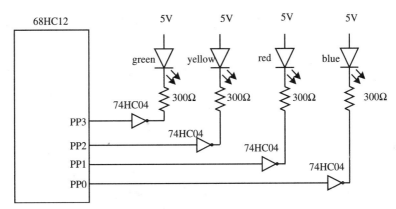

Figure 7.10 ■ Circuit connection for Example 7.3

To turn an LED on or off, we need to output appropriate values to Port P. A two-layer loop is used to create a half-second delay. The inner loop creates a 100-ms delay whereas the outer loop repeats the inner loop five times to create a time delay of a half second. The program is as follows:

```
PORTP      equ     $56
DDRP       equ     $57
           org     $1000
           ldaa    #$FF         ; configure PORTP for output
           staa    DDRP         ;      "
forever    ldaa    #$08         ; turn on green LED and turn off other LEDs
           staa    PORTP        ;      "
           jsr     delay_hs     ; wait for a half second
           ldaa    #$04         ; turn on yellow LED and turn off other LEDs
           staa    PORTP        ;      "
           jsr     delay_hs     ; wait for a half second
           ldaa    #$02         ; turn on red LED and turn off other LEDs
           staa    PORTP        ;      "
           jsr     delay_hs     ; wait for a half second
           ldaa    #$01         ; turn on blue LED and turn off other LEDs
           staa    PORTP        ;      "
           jsr     delay_hs     ; wait for a half second
           jmp     forever      ; repeat
           swi
;****************************************************************
; The following subroutine creates a delay of a half second
;****************************************************************
delay_hs   ldab    #5           ; 1 E cycle
```

```
out_loop    ldx     #20000        ; 2 E cycles
inner_loop  psha                  ; 2 E cycles
            pula                  ; 3 E cycles
            psha
            pula
            psha
            pula
            psha
            pula
            psha
            pula
            psha
            pula
            psha
            pula
            nop                   ; 1 E cycle
            nop
            dbne    x,inner_loop  ; 3 E cycles
            dbne    b,out_loop
            rts                   ; 5 E cycles
            end
```

7.6.2 Interfacing with Seven-Segment Displays

A seven-segment display consists of seven LED segments (a, b, c, d, e, f, and g) along with an optional decimal point (segment h). There are two types of seven-segment displays. In a *common-cathode seven-segment display*, the cathodes of all seven LEDs are tied together (must be connected to low voltage), and a segment will be lighted whenever a high voltage is applied at the corresponding segment input. In a *common-anode seven-segment display*, the anodes of all seven LEDs are tied together (must be connected to high voltage), and a segment will be lighted whenever a low voltage is applied at the corresponding segment input. Diagrams of a common-anode and common-cathode seven-segment display are shown in Figure 7.11. The current required to light a segment is similar to that required to light an LED.

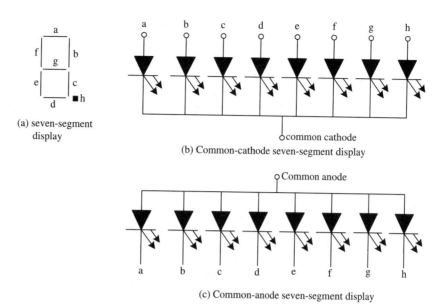

(a) seven-segment display

(b) Common-cathode seven-segment display

(c) Common-anode seven-segment display

Figure 7.11 ■ Seven-segment display

Seven-segment displays are mainly used to display BCD digits and a subset of letters. Any output port of the 68HC12 can be used to drive a seven-segment display as long as it is not used for other functions. One port will be adequate if only one display needs to be driven. The total supply current of the 68HC12 (shown in Table 7.2) is not enough to drive the seven-segment display directly, but this problem can be resolved by adding a buffer chip such as the 74HC244. In Figure 7.12, a common-cathode seven-segment display is driven by PORTP.

Characteristic[1]	Symbol	2 MHz (E)	4 MHz (E)	8 MHz (E)	Unit
Maximum total supply current					
RUN	I_{DD}				
single-chip mode		15	25	45	mA
expanded mode		25	45	70	mA
WAIT	W_{IDD}				
single-chip mode		1.5	3	5	mA
expanded mode		4	7	10	mA
STOP	S_{IDD}				
single-chip, no clock					
-40 to + 85		10	10	10	μA
+85 to 105		25	25	25	μA
+105 to +125		50	50	50	μA
Maximum power consumption (2)	P_D				mW
single-chip mode		75	125	225	
expanded mode		125	225	350	

(1) V_{DD} = 5.0V±10%, V_{SS} = 0V
(2) Includes I_{DD} and I_{DDA}

Table 7.2 ■ Supply current of the 68HC12

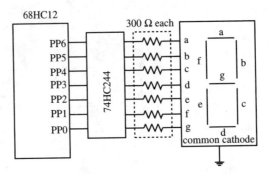

Figure 7.12 ■ Driving a single seven-segment display

The output high voltage (V_{OH}) of the 74HC244 is 5V. Since the voltage drop of one LED segment is about 2.0V, a 300-Ω current-limiting resistor could limit the current to about 10 mA, which would be sufficient to light an LED segment. The light patterns corresponding to the ten BCD digits are shown in Table 7.3. The numbers in Table 7.3 require that segments a, ..., g be connected from the most significant pin to the least significant pin of the output port.

BCD Digit	Segments							Corresponding Hex Number
	a	b	c	d	e	f	g	
0	1	1	1	1	1	1	0	$7E
1	0	1	1	0	0	0	0	$30
2	1	1	0	1	1	0	1	$6D
3	1	1	1	1	0	0	1	$79
4	0	1	1	0	0	1	1	$33
5	1	0	1	1	0	1	1	$5B
6	1	0	1	1	1	1	1	$5F
7	1	1	1	0	0	0	0	$70
8	1	1	1	1	1	1	1	$7F
9	1	1	1	1	0	1	1	$7B

Table 7.3 ■ BCD to seven-segment decoder

There is often a need to display multiple BCD digits, and this can be achieved by using the time-multiplexing technique. An example of a circuit that displays five BCD digits is shown in Figure 7.13. In Figure 7.13, the common cathode of a seven-segment display is connected to the collector of an NPN transistor. When a PORT CAN pin voltage is high, the connected NPN transistor will be driven into the saturation region. The common cathode of the display will then be pulled down to low (about 0.1V), allowing the display to be lighted. By turning the five NPN transistors on and off in turn many times in one second, multiple digits can be displayed. A 2N2222 transistor can sink from 100 mA to 300 mA of current. The maximum current that flows into the common cathode is about 70 mA (7×10 mA) and hence can be sunk by a 2N2222 transistor. The resistor R should be selected so that the transistor 2N2222 can be driven into the saturation region and at the same time won't draw too much current from the 68HC12. A value of several hundred to 1K ohms will work.

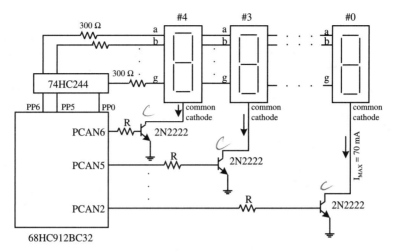

Figure 7.13 ■ Port P and Port CAN together drive five seven-segments displays (HC912BC32)

Example 7.4

▼

Write a sequence of instructions to display 4 on the seven-segment display #4 in Figure 7.13.

Solution: To display the digit 4 on the display #4, we need to:

1. Output the hex value $33 to port P
2. Set the PCAN6 pin to 1
3. Clear pins PCAN5..PCAN2 to 0

The instruction sequence is as follows:

```
PORTP     equ     $56
DDRP      equ     $57
PORTCAN   equ     $13E
DDRCAN    equ     $13F
four      equ     $33            ; seven-segment pattern of digit 4
          ...
          movb    #$7C,DDRCAN    ; configure PORT CAN for output
          movb    #$FF,DDRP      ; configure PORT P for output
          ldaa    PORTCAN        ; set PCAN6 to 1 and clear PCAN5..2 to 0
          ora     #$40           ;          "
          anda    #$C3           ;          "
          staa    PORTCAN        ;          "
          movb    #four,PORTP    ; output the seven-segment pattern to PORTP
          ...
```

In C, this can be achieved by the following statements:

```
DDRCAN = 0x7C;              /* configure PORT CAN for output *
DDRP = 0xFF;               /* configure PORTP for output */
PORTCAN |= 0x40;           /* turn on display number #4 */
PORTCAN &= 0xC3;           /* turn off other displays */
PORTP = 0x33;              /* output the segment pattern of digit 4 */
```

By now, you are probably tired of entering those *equ* statements that allow you to use symbolic names for registers in your program. The MiniIDE freeware allows you to use the *include* statement to include the definitions of peripheral registers in your program. Add the following statement in your assembly program so that you don't need to enter those equate directives:

#include "hc12.inc"

The file *hc12.inc* contains *equ* statements for all peripheral registers. The file *hc12.inc* that comes with the MiniIDE freeware contains only those peripheral registers contained in the 812A4 and 912B32. You can add those missing definitions so that it includes all the peripheral registers contained in the 68HC12 member that you want to use. The CD enclosed with this text provides two files to facilitate your assembly programming:

- hc12.inc: contains definitions for peripheral registers of the 812A4 and all of the B series members
- hc912dg128.inc: contains definitions for peripheral registers of the 912D60 and the 912DG128

The C compiler ICC12 has an include file, *hc12.h* that contains the register definitions of the 812A4 and the 912B32. The CD enclosed with this text includes a file (also called *hc12.h*) that contains the register definitions for the 812A4 and all the B-series members. The register definitions for 912D60 and 912DG128 are contained in the file *hc912dg128.h*. You can copy these two files into the appropriate directory (for example, under ..\icc\include) to make them accessible to your programs.

The circuit in Figure 7.13 can display five digits simultaneously by using the time-multiplexing technique, in which each seven-segment display is lighted in turn for a short period of time and then turned off. When one display is lighted, all the other displays are turned off. Within one second, each seven-segment display is lighted and then turned off many times. Because of the *persistence of vision*, the five displays will appear to be lighted simultaneously.

Example 7.5

Write a program to display 12345 on the five seven-segment displays shown in Figure 7.13.

Solution: We will display the digits 1, 2, 3, 4, and 5 on display #4, #3, ..., and #0, respectively. To selectively light these digits, we can clear pins PCAN6..PCAN2 and then set one of them to 1 (by using a mask for the OR operation). This can be implemented by using the table-lookup technique. The display patterns are shown in Table 7.4.

Seven-segment Display	Displayed BCD digit	Port P	Port CAN
#4	1	$30	$40
#3	2	$6D	$20
#2	3	$79	$10
#1	4	$33	$08
#0	5	$5B	$04

Table 7.4 ■ Table of display patterns for Example 7.5

This table can be created by the following assembler directives:

```
display     db          $30,$40
            db          $6D,$20
            db          $79,$10
            db          $33,$08
            db          $5B,$04
```

The program logic of this example is shown in Figure 7.14.

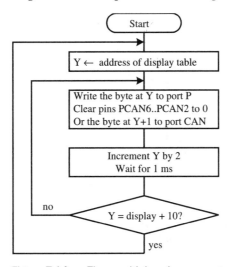

Figure 7.14 ■ Time-multiplexed seven-segment display algorithm

The assembly program that implements this algorithm is as follows:

```
#include     "d:\miniide\hc12.inc"
             org         $1000
             ldaa        #$FF
             staa        DDRP             ; configure PORT P for output
             ldaa        #$7C
             staa        DDRCAN           ; configure PORT CAN for output
forever      ldy         #display         ; use Y as a pointer to the display table
next         ldaa        0,y              ; get the digit pattern
             staa        PORTP            ; output the digit pattern
             ldaa        PORTCAN
             anda        #$83
             staa        PORTCAN          ; clear PORTCAN6..PORTCAN2 pins
             ldaa        1,y
             oraa        PORTCAN
             staa        PORTCAN          ; turn on the appropriate seven-segment display
; the following instructions create a delay of 1 ms for 16MHz oscillator
             ldx         #200
again        psha                         ; 3 E cycles
             pula                         ; 2 E cycles
             psha
             pula
```

```
                   psha
                   pula
                   psha
                   pula
                   psha
                   pula
                   psha
                   pula
                   psha
                   pula
                   nop
                   nop
                   dbne      x,again           ; 3 E cycles
                   cpy       #display+10       ; reach the end of the display table?
                   lbeq      forever
                   jmp       next
display            db        $30,$40
                   db        $6D,$20
                   db        $79,$10
                   db        $33,$08
                   db        $5B,$04
                   end
```

The C language version of this program is as follows:

```c
#include        <hc12.h>
char display[5][2] = {{0x30,0x40}, {0x6D,0x20}, {0x79,0x10}, {0x33,0x08},
                {0x5B,0x04}};
void delay_1ms();
void main ()
{
            int i;
            DDRP = 0xFF;   /* configure PORTP for output */
            DDRCAN = 0x7C; /* configure PORTCAN6..PORTCAN2 for output*/
            while (1) {
              for (i = 0; i < 5; i++) {
                  PORTP = display[i][0];
                  PORTCAN &= 0x83;           /* clear pins PORTCAN6..PORTCAN2 */
                  PORTCAN |= display[i][1]; /* enable the appropriate display to light */
                  delay_1ms();
                  }
              }
}
/* the following subroutine uses output compare function to create 1 ms delay */
/* the output compare function will be explained in the next chapter */
void delay_1ms()
{
            TSCR = 0x80;/* enable timer to function normally */
            TFLG1 = 0xFF;                 /* clear all timer flags */
            TC6 = TCNT + 8000; /* start an output compare operation with 1 ms delay */
            while (!(TFLG1 & 0x40));
}
```

7.6.3 Liquid Crystal Displays (LCDs)

Liquid crystal displays (LCDs) are among the most familiar output devices. The main advantages of the LCD include high contrast, low power consumption, and a small footprint (compare the LCD display and CRT display for a PC). LCDs can be used to display characters and graphics.

The basic construction of an LCD display is illustrated in Figure 7.15. The most common type of LCDs allows light to pass through it when activated. A segment is activated when a low frequency bipolar signal in the range of 30Hz to 1000Hz is applied to it. The polarity of the voltage must alternate or else the LCD will not be able to change very quickly.

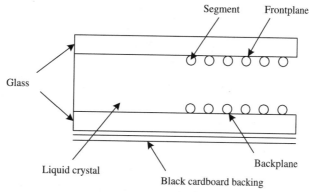

Figure 7.15 ■ A liquid crystal display (LCD)

The LCD functions in the following manner. When voltage is placed across the segment, it sets up an electrostatic field that aligns the crystals in the liquid. This alignment allows the light to pass through the segment. If no voltage is applied across a segment, the crystals appear to be opaque because they are randomly aligned. Random alignment is assured by the AC excitation voltage applied to each segment. In a digital watch, the segments appear to darken when they are activated because light passes through the segment to a back cardboard backing that absorbs all light. The area surrounding the activated segment appears brighter in color because the randomly aligned crystals reflect much of the light. In a backlit computer display, the segment appears to grow brighter because of a light placed behind the display; the light is allowed to pass through the segment when it is activated.

Many manufacturers have developed color LCD displays for use with computers and small color television sets. These displays use three segments (dots) for each picture element (pixel), and the three dots are filtered so that they pass red, blue, and green light. By varying the number of dots and the amount of time each dot is active, just about any color and intensity can be displayed. White light consists of 59% green, 30% red, and 11% blue light. Secondary colors are magenta (red and blue), cyan (blue and green), and yellow (red and green). Any other colors, not just secondary colors, can be obtained by mixing red, green, and blue light.

LCDs are often sold in a module with LCDs and the controller unit built in. The Hitachi HD44780 is one of the most popular LCD display controllers in use today. In the following section we will describe an LCD module that uses this controller.

7.6.4 The Optrex DMC-20434 LCD Kit

The DMC-20434 manufactured by Optrex is a 4×20 LCD kit that uses the HD44780 as its display controller. This LCD kit can plug into the LCD port of the CME-12BC demo board and many other demo boards made by Axiom. The block diagram and the photograph of the DMC-20434 are shown in Figures 7.16 and 7.17.

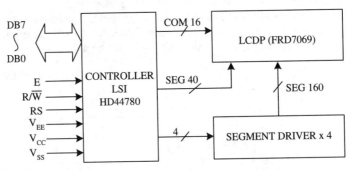

Figure 7.16 ■ Block diagram of the DMC-20434 LCD kit

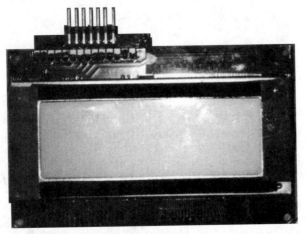

Figure 7.17 ■ The DMC-20434 LCD kit

The DB7~DB0 pins are used to exchange data with the microcontroller. The E pin is an enable signal to the kit. The $R/\overline{W}$ pin determines the direction of the data transfer. The RS signal selects the register to be accessed. The value one selects the data register whereas the value zero selects the control register. The V_{EE} pin is used to control the brightness of the display and is often connected to a variable resistor. The V_{EE} input should not be set to maximum value (= V_{CC}) for an extended period of time to avoid burning the LCD.

The DMC-20434 can be used as a memory-mapped device and be enabled by an address decoder. The signal E should be connected to one of the address decoder outputs in this configuration. The DMC-20434 can also be interfaced directly with an I/O port. In this configuration, we must use I/O pins to control the signals E, $R/\overline{W}$, and RS.

The LCD_PORT interfaced on the CME-12BC32 is connected to the data bus and memory-mapped to 0x270-0x27F. For a standard display such as the DMC-20434, address 0x270 is the command register; address 0x271 is the data register.

The interface supports all OPTREX DMC series displays in 8-bit bus mode with up to 80 characters and provides the most common pin-out for a dual-row rear-mounted display connector.

THE SETUP OF THE DMC-20434 LCD KIT

All LCD kits need to be set up before they can be used. To set up the DMC-20434, we need to write commands into the command register. Commands that are used to configure the DMC-20434 are listed in Table 7.5. An LCD command takes much longer to execute than a normal 68HC12 instruction does. No new command should be sent to the LCD before the current command completes. The most significant bit of the command register indicates whether the LCD is still busy (1 = busy, 0 = not busy).

The initialization of the LCD includes four operations:

1. Set cursor to home and/or clear display

2. Set up entry mode

3. Configure display control

4. Program cursor/display shift

An example subroutine that performs the LCD initialization is as follows:

```
initlcd  ldaa    #$3C        ; configure display format to 2x40
         jsr     lcdcmd      ;      "
         ldaa    #$0f        ; turn on display and cursor
         jsr     lcdcmd      ;      "
         ldaa    #$14        ; shift cursor right
         jsr     lcdcmd      ;      "
         ldaa    #$01        ; clear display and return cursor to home
         jsr     lcdcmd      ;      "
         rts

; Send a command in A to the LCD command register
lcdcmd   staa    cmd_reg     ; write command
         jsr     delay40     ; wait
         rts
```

Instruction	Execution Command code	Description	Time
Clear display	00000001	Clears display and returns cursor to the home position	1.64 ms
Cursor home	0000001×	Returns cursor to home position (address 0). Also return display being shifted to the original position. DDRAM contents remains unchanged.	1.64 ms
Entry mode set	000001 I/D S	Set cursor move direction (I/D̄), specifies to shift the display (S). These operations are performed during data read/write.	40 μs
Display of/off control	00001 D C B	Sets on/off of all display (D), cursor on/off (C) and blink of cursor position character (B).	40 μs
Cursor/display shift	0001 S/C R/L * *	Sets cursor-move or display-shift (S/C̄), shift direction (R/L̄), DDRAM contents remains unchanged.	40 μs
Function set	001 DL N F * *	Sets interface data length (DL), number of display lines (N) and character font (F).	40 μs
Set CGRAM address	01 CGRAM addr	Sets the CGRAM address. CGRAM data is sent and received after this setting.	40 μs
Set DDRAM address	1 DDRAM addr	Sets the DDRAM address. DDRAM data is sent and received after this setting.	40 μs
Read busy-flag and address counter	BF CGRAM/ DDRAM addr	Reads busy flag (BF) indicating internal operation is being performed and reads CGRAM or DDRAM address counter contents (depending previous instruction)	0 μs
Write to CGRAM or DDRAM	write data	Wrttes data to CGRAM or DDRAM	40 μs
Read from CGRAM or DDRAM	read data	Reads data from CGRAM or DDRAM.	40 μs

Note.
DDRAM: display data RAM.   CGRAM: Character generator RAM.
DDRAM address corresponds to cursor position.
I/D̄: increments (1) or decrement (0) cursor position.
S: display shift. 0 = no shift, 1 = shift.
D: display on/off. 0 = off, 1 = on
C: cursor on/off. 0 = off, 1 = 0n
B: cursor blink on/off. 0 = off, 1 = on.
S/C̄: 0 = move cursor, 1 = shift display
R/L̄: shift direction. 0 = shift left, 1 = shift right
DL: data length. 0 = 4-bit interface, 1 = 8-bit interface
N: number of lines. 0 = 1 line, 1 = 2 lines.
F: font size. 0 = 5×7 dots, 1 = 5×10 dots
BF: busy flag. 0 = can accept instruction. 1 = internal operation in progress.

Table 7.5 ■ HD44780 LCD controller instruction set

The address $270 and $271 are not I/O register addresses of the 68HC12. We will add the following statements to the *hc12.h* file of the ICC12 compiler so that we can use the symbols *cmd_reg* and *dat_reg* to access the registers of the LCD kit:

```
#define        cmd_reg  *(unsigned char volatile *)(0x0270)
#define        dat_reg  *(unsigned char volatile *)(0x271)
```

The C language version of the LCD initialization routine is as follows:

```
void lcd_init()
{
        while (cmd_reg & 0x80);        /* wait until LCD is ready */
        cmd_reg = 0x3C;
        while (cmd_reg & 0x80);        /* wait until LCD is ready */
        cmd_reg = 0x0f;
        while (cmd_reg & 0x80);        /* wait until LCD is ready */
        cmd_reg = 0x14;
        while (cmd_reg & 0x80);        /* wait until LCD is ready */
        cmd_reg = 0x01;
        while (cmd_reg & 0x80);
}
```

OUTPUT DATA ON THE DMC-20434 LCD KIT

The DMC20434 can display four rows of characters. Each row can hold up to 20 characters. The position of the character to be displayed is controlled by the DDRAM address. The addresses of the four rows of the LCD are not arranged in sequential order:

- Row 0 DDRAM addresses: from 0x00 to 0x13

- Row 1 DDRAM addresses: from 0x40 to 0x53

- Row 2 DDRAM addresses: from 0x14 to 0x27

- Row 3 DDRAM addresses: from 0x54 to 0x67

After the initialization, the LCD DDRAM address is reset to zero. If we keep outputting characters to the LCD, it will display the message in the order of row 0, row 2, row 1, and then row 3. In order to force the LCD to display information in sequential order, we need to:

1. Set the DDRAM address to 0x40 when the DDRAM address is 0x13 after outputting a character

2. Set the DDRAM address to 0x14 when the DDRAM address is 0x53 after outputting a character

3. Set the DDRAM address to 0x54 when the DDRAM address is 0x27 after outputting a character

In addition, the LCD does not display the 20th, 40th, and 60th characters. To avoid losing these characters, we will output these characters twice. The following subroutine will output one character to the LCD and take care of resetting the DDRAM address so that the message is output in sequential rows:

```
lcdout      staa      dat_reg      ; output data to lcd
            jsr       delay100us   ; wait (if this is too long, use a shorter delay)
            ldab      cmd_reg
            cmpb      #$13
            beq       lcd1
            cmpb      #$53
            beq       lcd2
            cmpb      #$27
            beq       lcd3
            rts
lcd1        ldab      #$40
            orab      #$80
            stab      cmd_reg
```

```
                    staa        dat_reg
                    jsr         delay100us
                    rts
lcd2                ldab        #$14
                    orab        #$80
                    stab        cmd_reg
                    staa        dat_reg
                    jsr         delay100us
                    rts
lcd3                ldab        #$54
                    orab        #$80
                    stab        cmd_reg
                    staa        dat_reg
                    jsr         delay100us
                    rts
; Send a command in A to the LCD command register
lcdcmd              staa        cmd_reg         ; write command
                    jsr         delay40         ; wait
                    rts


; wait 100 microseconds (approx.)
delay100us          psha
                    ldaa        #200
loop100             deca
                    bne         loop100
                    pula
                    rts


; wait appropriate time for LCD command process
delay40             pshx
                    ldx         #0000
loop40              dex
                    bne         loop40
                    pulx
                    rts
                    end
```

A NULL-terminated string can be output to the LCD using the following subroutine:

```
lcd_strg            ldaa        0,x
                    beq         done
                    jsr         lcdout
                    inx
                    bra         lcd_strg
done                rts
```

The C language versions of these two subroutines are as follows:

```c
/* The following subroutine outputs a character to the DMC-20434 LCD kit */
void putc2lcd (char ch)
{
        dat_reg = ch;
        while (cmd_reg & 0x80);

        if (cmd_reg == 0x13) {
                cmd_reg = 0x40 | 0x80;          /* correct line 1 wrap from line 3 to line 2 */
                while (cmd_reg & 0x80);
```

```
                                dat_reg            = ch;          /* re-output the same character */
                                while (cmd_reg & 0x80);
        }
        if (cmd_reg == 0x53) {                    /* correct line 2 wrap from line 4 to line 3 */
                        cmd_reg = 0x14 | 0x80;
                        while (cmd_reg & 0x80);
                        dat_reg            = ch;          /* re-output the same character */
                        while (cmd_reg & 0x80);
        }
        if (cmd_reg == 0x27) {          /* correct line 3 wrap from line 2 to line 4 */
                        cmd_reg = 0x54 | 0x80;
                        while (cmd_reg & 0x80);
                        dat_reg            = ch;          /* re-output the same character */
                        while (cmd_reg & 0x80);
        }
}
/* The following subroutine outputs a string to the DMC-20434 LCD kit */
void puts2lcd (char *ptr)
{
        while (*ptr) {
                        putc2lcd (*ptr);
                        ptr++;
        }
}
```

Example 7.6

Write a C program to test the two subroutines discussed above.

Solution: Add the following statements and main program:

```
#include <hc12.h>
#include <stdio.h>
char msg = "this is a very very very long string in the desert storm";
void putc2lcd(char ch);
void puts2lcd(char ptr);
void lcd_init();

void main(void)
{
        lcd_init();
        puts2lcd(msg);
        asm("swi");                     /* break to D-Bug12 monitor */
}
void lcd_init()
{
        while(cmd_reg & 0x80);          /* wait until LCD is ready */
        cmd_reg = 0x3C;                 /* set 4 x 20 display */
        while(cmd_reg & 0x80);
        cmd_reg = 0x0f;                 /* shift cursor right */
        while(cmd_reg & 0x80);
        cmd_reg = 0x14;                 /* turn on display and cursor */
```

```
        while(cmd_reg & 0x80);
        cmd_reg = 0x01;                /* clear screen and return cursor to home */
        while(cmd_reg & 0x80);
}
```

7.6.5 Interfacing with DIP Switches

A switch is probably the simplest input device we can find. To make input more efficient, we often use a set of eight switches organized as a dual inline package (DIP). A DIP package can be connected to any input port with eight pins such as PORTP, PORTT, and PORTAD, as shown in Figure 7.18. When a switch is closed, the associated port P input is zero. Otherwise, the associated port P pin has a value of one. Each port P pin is pulled up to high via a 10 KΩ resistor when the associated switch is open.

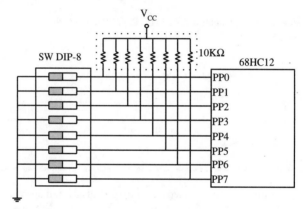

Figure 7.18 ■ Connecting a set of eight DIP switches to port P of the 68HC12

Example 7.7

Write a sequence of instructions to read the value from an eight-switch DIP connected to PORTP of the 68HC12 into accumulator A.

Solution:

```
#include "d:\miniide\hc12.inc"

        ldaa       PORTP
        ...
```

In C language, this operation can be achieved as follows:

```
#include <hc12.h>
void main ()
{
        char       xx;

        ...
        xx = PORTP;          /* read a byte from the DIP switch */
        ...
}
```

7.7 Interfacing Parallel Ports to a Keyboard

A keyboard is arranged as an array of switches, which can be mechanical, membrane, capacitors, or Hall-effect in construction. In mechanical switches, two metal contacts are brought together to complete an electrical circuit. In membrane switches, a plastic or rubber membrane presses one conductor onto another; this type of switch can be made to be very thin. Capacitive switches internally comprise two plates of a parallel plate capacitor; pressing the key cap effectively increases the capacitance between the two plates. Special circuitry is needed to detect this change in capacitance. In a Hall-effect key switch, the motion of the magnetic flux lines of a permanent magnet perpendicular to a crystal is detected as voltage appearing between the two faces of the crystal—it is this voltage that registers a switch closure.

Because of their construction, mechanical switches have a problem called *contact bounce*. Instead of producing a single, clean pulse output, closing a mechanical switch generates a series of pulses because the switch contacts do not come to rest immediately. This phenomenon is illustrated in Figure 7.19, where it can be seen that a single physical push of the button results in multiple electrical signals being generated and sent to the computer. When the key is not pressed, the voltage output to the computer is 5V. The response time of the switch is several orders of magnitude slower than that of a computer, so the computer could read a single switch closure many times during the time the switch is operated, interpreting each low signal as a new input when in fact only one input is being sent.

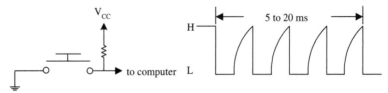

Figure 7.19 ■ Contact bounce

Because of the contact bounce in the keyboard, a *debouncing* process is needed. A keyboard input program can be divided into three stages, which will be discussed in the following subsections:

1. Keyboard scanning to find out which key was pressed

2. Key debouncing to make sure a key was indeed pressed

3. Table lookup to find the ASCII code of the key that was pressed

7.7.1 Keyboard Scanning Techniques

A keyboard can be arranged as a linear array of switches if it has only a few keys. A keyboard with more than a few keys is often arranged as a matrix of switches that uses two decoding and selecting devices to determine which key was pressed by coincident recognition of the row and column of the key.

As shown in Figure 7.20, a CMOS analog multiplexor MC14501 and a 3-to-8 decoder are used to scan the keyboard to determine which key is pressed. A scanning program is invoked to search for the pressed key, and then the debouncing routine is executed to verify that the key is closed. For a 64-key keyboard, 6 bits are needed to select the rows and columns (3 bits for each). The MC14501 has three select inputs (A, B, and C), eight data inputs (X1-X7), one inhibit input, and one common output (X). The common output is active high. When a data input is selected by the select signals, its voltage will be passed to the common output.

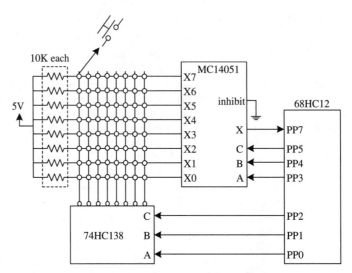

Figure 7.20 ■ Keyboard structure

In Figure 7.20, the MC14501 is used to select the row and the 74HC138 is used to select the column. The data inputs (X7-X0) of the MC14501 are pulled up to high. When a key is pressed, the row and the column where the keyboard is pressed are shorted together by the switch. Since the 3-to-8 decoder has low output, the common output of the MC14501 will be driven to low and can be detected by the microcontroller.

Port P is used to handle the keyboard scanning in the following program. Since port P is a bi-directional port, it must be configured properly. In our example, we will use the PP7 pin to detect whether a key has been pressed. Therefore, the pin PP7 should be configured for input and the lower six pins (PP5-PP0) should be configured for output. The pin PP6 is not used. The following instruction will configure port P as desired:

```
#include "d:\miniide\hc12.inc"
    ...
    movb    #$7F,DDRP
```

If two adjacent keys in the same matrix row are pressed at the same time, the output pins of the 74HC138 will be shorted together and may damage the decoder. Practical keyboard circuits include diodes to prevent such damage to the interface circuitry.

The algorithm for keyboard scanning is shown in Figure 7.21. Since the lowest 6 bits of the PORTP register are used to specify the row number and column number, these two indices can be incremented by a single INC instruction. If the keyboard is not pressed and the lowest six bits of the PORTP registers are all ones, the row number and column number must be reset to zero. The scanning program is as follows:

```
#include "d:\miniide\hc12.inc"
keyboard     equ      PORTP
             ...
resetp       clr      keyboard
scan         brclr    keyboard,$80,debnce    ; is the key pressed?
             brset    keyboard,$3F,resetp    ; if reach the last row & column, then reset
             inc      keyboard
             bra      scan
             ...
```

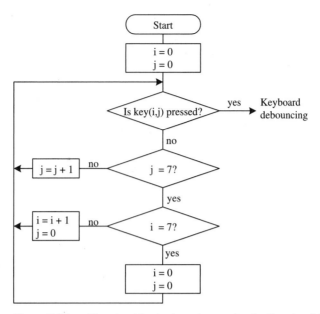

Figure 7.21 ■ Flowchart for keyboard scanning for the circuit in Figure 7.20

Keyboard decoding can be done strictly within software without using any hardware decoders or multiplexors.

7.7.2 Keyboard Debouncing

Contact bounce is due to the dynamics of a closing contact. The signal falls and rises a few times within a period of about 5 ms as a contact bounces. Since a human being cannot press and release a switch in less than 20 ms, a debouncer will recognize that the switch is closed after the voltage is low for about 10 ms and will recognize that the switch is open after the voltage is high for about 10 ms.

Both hardware and software solutions to the key bounce problem are available. Using a good switch can reduce keyboard bouncing. A mercury switch is much faster, and optical and Hall-effect switches are free of bounce. Hardware solutions to contact bounce include an analog circuit that uses a resistor and capacitor to smooth the voltage and two digital solutions that use set-reset flip-flops or CMOS buffers and double-throw switches.

In practice, hardware and software debouncing techniques are both used (but not at the same time). Dedicated hardware scanner chips are also used frequently—typical ones are the National Semiconductor 74C922 and 94C923. These two chips perform keyboard scanning and debouncing.

HARDWARE DEBOUNCING TECHNIQUES

- *Set-reset flip-flops.* Before being pressed, the key is touching the set input. When it is pressed, the key moves toward the reset position. Before it has settled down, the key bounces. When the key touches the reset position, the voltage at Q goes low. When the key is bouncing, it touches neither the set terminal nor the reset terminal. In this situation, both the set and reset inputs are pulled down to low by the pull-down resistor. Since both set and reset inputs are grounded, the output Q will not change. Therefore, the key will be recognized as closed. This solution is shown in Figure 7.22a.

- *Noninverting CMOS gates with high impedance.* When the switch is pressed, the input of the buffer 4050 is grounded and hence V_{out} is forced to the ground level. When the switch is bouncing, the feedback resistor R keeps the input at the same voltage level as the output and hence the output voltage stays at low. This is due to the high input impedance of the 4050, which causes a negligible voltage drop on the feedback resistor. Thus the output is debounced. This solution is shown in Figure 7.22b.

- *Integrating debouncers.* The RC constant of the integrator (smoothing filter) determines the rate at which the capacitor charges up toward the supply voltage once the ground connection via the switch has been removed. As long as the capacitor voltage does not exceed the logic zero threshold value, the V_{OUT} signal will continue to be recognized as a logic zero. This solution is shown in Figure 7.22c.

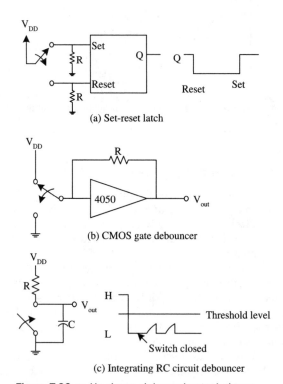

(a) Set-reset latch

(b) CMOS gate debouncer

(c) Integrating RC circuit debouncer

Figure 7.22 ■ Hardware debouncing techniques

SOFTWARE DEBOUNCING TECHNIQUES

An easy software solution to the key bounce problem is the wait-and-see technique. When the input drops, indicating that the switch might be closed, the program waits 10 ms and looks at the input again. If it is high, the program decides that the input signal was noise or that the input is bouncing—if it is bouncing, it will certainly be cleared later when the key was actually pressed. In either case, the program returns to wait for the input to drop. The following sample program uses this technique to do the debouncing:

```
#include "d:\miniide\hc12.inc"
keyboard    equ     PORTP
debnce      ldx     #2000
wait        psha                            ; wait for 10 ms at 16 MHz of crystal oscillator
            pula
            psha
            pula
            psha
            pula
            psha
            pula
            psha
            pula
            psha
            pula
            psha
            pula
            nop
            nop
            dbne    x,wait                  ; is index register X decremented to zero yet?
            ldaa    keyboard
            lbmi    scan                    ; is the key not pressed?
            jmp     getcode                 ; yes, go and lookup ASCII code
            ...
```

7.7.3 ASCII Code Table Lookup

After the key has been debounced, the keyboard should look up the ASCII table and send the corresponding ASCII code to the CPU. The instruction sequence that performs ASCII code table lookup is as follows:

```
keytab      db      "0123456789"
            db      ...
            ...
getcode     ldab    keyboard
            andb    #$7F            ; compute the address of the ASCII code
            ldx     #keytab         ;          "
            ldaa    b,x             ;          "
            ...
```

Example 7.8

Write an assembly routine that reads a character from the keyboard. This subroutine will perform keyboard scanning, debouncing, and ASCII code lookup, and return the ASCII code to the caller.

Solution: The subroutine is simply the combination of the code segments described in the previous section.

```
          #include "d:\miniide\hc12.inc"
keyboard  equ     PORTP
get_char  clr     keyboard            ; start from row 0, column 0
scan      brclr   keyboard,$80,debnce ; detect a pressed key?
scan1     brset   keyboard,$37,get_char ; reach the last row, last column?
          inc     keyboard
          bra     scan
; the following instruction sequence wait for 10 ms
debnce    ldx     #2000
wait      psha
          pula
          psha
          pula
          psha
          pula
          psha
          pula
          psha
          pula
          psha
          pula
          psha
          pula
          psha
          pula
          nop
          nop
          dbne    x,wait
; recheck the detected key
          ldaa    keyboard
          lbmi    scan1               ; go and check the next key if the key is not pressed
          ldab    keyboard            ; compute the address of the ASCII code
          andb    #$7F                ; of the pressed key
          ldx     #keytab             ; "
          ldaa    b,x                 ; get the ASCII code of the pressed key
          rts
keytab    db      "0123456789"
          db      ...
```

7.8 Interfacing the 68HC12 to a Keypad

A 64-key keyboard is probably too much for many embedded applications. A 12- to 24-key keypad comes in handy for many applications that need only limited input options. A membrane keypad with no more than 16 keys can easily be interfaced with an 8-bit bi-directional I/O port such as PORTP. An example is shown in Figure 7.23. In this example, the rows and columns of the keypad are simply conductors. Port P pins PP0-PP3 are pulled up to high by the pull-up resistors. Whenever a key is pressed, the corresponding row and column will be shorted together just like a keyboard. To scan a row, we will set that row to low. The row selection of the 16-key keypad is shown in Table 7.6.

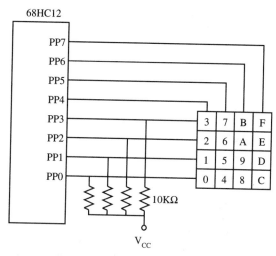

Figure 7.23 ■ Sixteen-key keypad connected to 68HC12

PP7	PP6	PP5	PP4	Selected keys			
1	1	1	0	0,	1,	2,	and 3
1	1	0	1	4,	5,	6,	and 7
1	0	1	1	8,	9,	A,	and B
0	1	1	1	C,	D,	E,	and F

Table 7.6 ■ Sixteen-key keypad row selections

For the circuit shown in Figure 7.23, we need to use a software-only technique to perform keypad scanning and debouncing. The keypad is scanned from the row controlled by PP4 toward the row controlled by PP7 and repeat.

Example 7.9

Write an assembly subroutine that reads a character from the keypad shown in Figure 7.23. This subroutine will perform keypad scanning, debouncing, and ASCII code generation. The ASCII code will be returned in accumulator A.

Solution: This subroutine is straightforward but lengthy.

```
#include "d:\miniide\hc12.inc"
keyboard     equ     PORTP

get_char     ldaa    #$F0
             staa    DDRP                ; configure pins PP7..PP4 for output
scan_r0      bset    keyboard,$E0        ; select the row containing keys 0123
             bclr    keyboard,$10        ;          "
scan_k0      brclr   keyboard,$01,key0   ; is key 0 pressed?
```

```
scan_k1    brclr    keyboard,$02,key1        ; is key 1 pressed?
scan_k2    brclr    keyboard,$04,key2        ; is key 2 pressed?
scan_k3    brclr    keyboard,$08,key3        ; is key 3 pressed?
           bra      scan_r1
key0       jmp      db_key0
key1       jmp      db_key1
key2       jmp      db_key2
key3       jmp      db_key3
scan_r1    bset     keyboard,$D0             ; select the row containing keys 4567
           bclr     keyboard,$20             ;              "
scan_k4    brclr    keyboard,$01,key4        ; is key 4 pressed?
scan_k5    brclr    keyboard,$02,key5        ; is key 5 pressed?
scan_k6    brclr    keyboard,$04,key6        ; is key 6 pressed?
scan_k7    brclr    keyboard,$08,key7        ; is key 7 pressed?
           bra      scan_r2
key4       jmp      db_key4
key5       jmp      db_key5
key6       jmp      db_key6
key7       jmp      db_key7
scan_r2    bset     keyboard,$B0             ; select the row containing keys 89AB
           bclr     keyboard,$40             ;              "
scan_k8    brclr    keyboard,$01,key8        ; is key 8 pressed?
scan_k9    brclr    keyboard,$02,key9        ; is key 9 pressed?
scan_kA    brclr    keyboard,$04,keyA        ; is key A pressed?
scan_kB    brclr    keyboard,$08,keyB        ; is key B pressed?
           bra      scan_r3
key8       jmp      db_key8
key9       jmp      db_key9
keyA       jmp      db_keyA
keyB       jmp      db_keyB
scan_r3    bset     keyboard,$70             ; select the row containing keys 89AB
           bclr     keyboard,$80             ;              "
scan_kC    brclr    keyboard,$01,keyC        ; is key C pressed?
scan_kD    brclr    keyboard,$02,keyD        ; is key D pressed?
scan_kE    brclr    keyboard,$04,keyE        ; is key E pressed?
scan_kF    brclr    keyboard,$08,keyF        ; is key F pressed?
           jmp      scan_r0
keyC       jmp      db_keyC
keyD       jmp      db_keyD
keyE       jmp      db_keyE
keyF       jmp      db_keyF
; debounce key 0
db_key0    jsr      wait_10ms
           brclr    keyboard,$01,getc0
           jmp      scan_k1
getc0      ldx      #hextab
           ldab     #0
           ldaa     b,x                      ; get the ASCII code of 0
           rts
; debounce key 1
db_key1    jsr      wait_10ms
           brclr    keyboard,$02,getc1
           jmp      scan_k2
getc1      ldx      #hextab
```

```
                    ldab       #1
                    ldaa       b,x                          ; get the ASCII code of 1
                    rts
db_key2             jsr        wait_10ms
                    brclr      keyboard,$04,getc2
                    jmp        scan_k3
getc2               ldx        #hextab
                    ldab       #2
                    ldaa       b,x                          ; get the ASCII code of 2
                    rts
db_key3             jsr        wait_10ms
                    brclr      keyboard,$08,getc3
                    jmp        scan_r1
getc3               ldx        #hextab
                    ldab       #3
                    ldaa       b,x                          ; get the ASCII code of 3
                    rts
db_key4             jsr        wait_10ms
                    brclr      keyboard,$01,getc4
                    jmp        scan_k5
getc4               ldx        #hextab
                    ldab       #4
                    ldaa       b,x                          ; get the ASCII code of 4
                    rts
db_key5             jsr        wait_10ms
                    brclr      keyboard,$02,getc5
                    jmp        scan_k6
getc5               ldx        #hextab
                    ldab       #5
                    ldaa       b,x                          ; get the ASCII code of 5
                    rts
db_key6             jsr        wait_10ms
                    brclr      keyboard,$04,getc6
                    jmp        scan_k7
getc6               ldx        #hextab
                    ldab       #6
                    ldaa       b,x                          ; get the ASCII code of 6
                    rts
db_key7             jsr        wait_10ms
                    brclr      keyboard,$08,getc7
                    jmp        scan_r2
getc7               ldx        #hextab
                    ldab       #7
                    ldaa       b,x                          ; get the ASCII code of 7
                    rts
db_key8             jsr        wait_10ms
                    brclr      keyboard,$01,getc8
                    jmp        scan_k9
getc8               ldx        #hextab
                    ldab       #8
                    ldaa       b,x                          ; get the ASCII code of 8
                    rts
db_key9             jsr        wait_10ms
                    brclr      keyboard,$02,getc9
```

```
                  jmp         scan_kA
getc9             ldx         #hextab
                  ldab        #9
                  ldaa        b,x                          ; get the ASCII code of 9
                  rts
db_keyA           jsr         wait_10ms
                  brclr       keyboard,$04,getcA
                  jmp         scan_kB
getcA             ldx         #hextab
                  ldab        #10
                  ldaa        b,x                          ; get the ASCII code of A
                  rts
db_keyB           jsr         wait_10ms
                  brclr       keyboard,$08,getcB
                  jmp         scan_r3
getcB             ldx         #hextab
                  ldab        #11
                  ldaa        b,x                          ; get the ASCII code of B
                  rts
db_keyC           jsr         wait_10ms
                  brclr       keyboard,$01,getcC
                  jmp         scan_kD
getcC             ldx         #hextab
                  ldab        #12
                  ldaa        b,x                          ; get the ASCII code of C
                  rts
db_keyD           jsr         wait_10ms
                  brclr       keyboard,$02,getcD
                  jmp         scan_kE
getcD             ldx         #hextab
                  ldab        #13
                  ldaa        b,x                          ; get the ASCII code of D
                  rts
db_keyE           jsr         wait_10ms
                  brclr       keyboard,$04,getcE
                  jmp         scan_kF
getcE             ldx         #hextab
                  ldab        #14
                  ldaa        b,x                          ; get the ASCII code of E
                  rts
db_keyF           jsr         wait_10ms
                  brclr       keyboard,$08,getcF
                  jmp         scan_r0
getcF             ldx         #hextab
                  ldab        #15
                  ldaa        b,x                          ; get the ASCII code of F
                  rts
; the following subroutine creates a delay of 10 ms
wait_10ms ldx#2000
again             psha
                  pula
                  psha
                  pula
```

```
                    psha
                    pula
                    psha
                    pula
                    psha
                    pula
                    psha
                    pula
                    psha
                    pula
                    nop
                    nop
                    dbne        x,again
                    rts
hextab              db          "0123456789ABCDEF"
```

The C language version of this subroutine will be left as an exercise problem.

7.9 Interfacing with a D/A Converter

A *digital-to-analog (D/A) converter* converts a digital code into an analog signal (can be voltage or current). A D/A converter has many applications. Examples include digital gain and offset adjustment, programmable voltage and current sources, programmable attenuators, digital audio, closed-loop positioning, robotics, and so on.

Although there are a few microcontrollers that incorporate the D/A converter on the chip, most microcontrollers still need to use an off-chip D/A converter to perform the D/A conversion function. The 68HC12 is no exception.

A D/A converter may use a serial or parallel interface to obtain digital code from the microprocessor or microcontroller.

7.9.1 The AD7302 D/A Converter

The AD7302 is a dual-channel 8-bit D/A converter chip from Analog Devices that has a parallel interface with the microcontroller. The AD7302 converts an 8-bit digital value into an analog voltage.

The block diagram of the AD7302 is shown in Figure 7.24. The AD7302 is designed to be a memory-mapped device. In order to send data to the AD7302, the CS signal must be pulled to low. On the rising edge of the WR signal, the values on D7-D0 will be latched into the Input Register. When the signal LDAC is low, the data in Input Register will be transfer to the DAC Register and a new D/A conversion is started. The AD7302 needs a reference voltage to perform the D/A conversion. The reference voltage can come from either the external REFIN input or the internal V_{DD}. The A/B signal selects the channel (A or B) to perform the D/A conversion. The PD pin puts the AD7302 in power-down mode and reduces the power consumption to 1 μW.

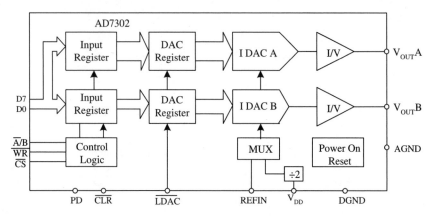

Figure 7.24 ■ Functional block diagram of the AD7302

The AD7302 operates from a single +2.7 to 5.5V supply and typically consumes 15 mW at 5V, making it suitable for battery-powered applications. Each digital sample takes about 2 μs to convert. The output voltage ($V_{OUT}A$ or $V_{OUT}B$) from either DAC is given by:

$$V_{OUT} A/B = 2 \times V_{REF} \times (N/256)$$

where:

V_{REF} is the voltage applied to the external REFIN pin or $V_{DD}/2$ when the internal reference is selected. If the voltage applied to the REFIN pin is within 1V of the V_{DD}, V_{DD} is used as the reference voltage automatically. Otherwise, the voltage applied at the REFIN pin is used as the reference voltage.

N is the decimal equivalent of the code loaded to the DAC register and ranges from 0 to 255.

7.9.2 Interfacing the AD7302 with the 68HC12

Interfacing the AD7302 with the 68HC12 can be very simple. Both the CS and LDAC can be tied to ground permanently. The value to be converted must be sent to the AD7302 via a parallel port (connect to pins D7..D0). An output pin can be used (as the WR signal) to control the transferring of data to the Input Register. A typical connection between the 68HC12 and the AD7302 is shown in Figure 7.25.

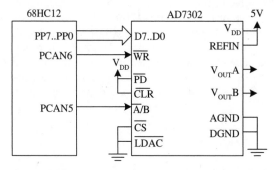

Figure 7.25 ■ Circuit connection between the AD7302 and the 68HC12

Example 7.10

Write a program to generate a sawtooth waveform from the $V_{OUT}A$ pin.

Solution: The procedure for generating a sawtooth waveform is as follows:

Step 1
Configure PP7..PP0, PCAN6, and PCAN5 for output.

Step 2
Output the digital value from 0 to 255 and repeat. For each value, pull the PCAN6 to low and then to high so that the value on pins PP7..PP0 can be transferred to the AD7302. Pull the signal PCAN5 to low during the process.
The assembly program is as follows:

```
#include "D:\miniide\hc12.inc"
        org     $1000
        ldaa    #$FF
        staa    DDRP             ; configure PORTP for output
        ldaa    #$6F
        staa    DDRCAN           ; configure PCAN6..PCAN5 for output
        bclr    PORTCAN,$20      ; select VOUTA output
loop    inc     PORTP            ; increase the output by one step
        bclr    PORTCAN,$40      ; generate a rising edge on PORTCAN6 pin
        bset    PORTCAN,$40      ;        "
        nop                      ; add NOP so that each step has 2 microsecond
        bra     loop             ; to complete the D/A conversion
        end
```

The C language version of the program is as follows:

```
#include <hc12.h>
void main()
{
        DDRP = 0xFF;                    /* configure PORTP for output */
        DDRCAN = 0x6F;                  /* configure PORTCAN6..PORTCAN5 for output */
        PORTCAN &= 0xDF;                /* pull the signal A/B to low */
        while (1) {
                PORTP = PORTP + 1;
                PORTCAN &= 0xBF;        /* generate a rising edge */
                PORTCAN |= 0x40;        /*    "        */
        }
}
```

The AD7302 can be used to generate many interesting waveforms. Two examples are given as exercise problems.

7.10 Stepper Motor Control

Stepper motors are digital motors. They are convenient for applications where a high degree of positional control is required. Printers, tape drives, disk drives, and robot joints, for example, are typical applications of stepper motors. In its simplest form, a stepper motor has a permanent magnet rotor and a stator consisting of two coils. The rotor aligns with the stator coil that is energized. By changing which coil is energized, as illustrated in the following figures, the rotor is turned.

In Figure 7.26a to 7.26d, the permanent magnet rotor lines up with the coil pair that is energized. The direction of the current determines the polarity of the magnetic field, and thus the angular position of the rotor.

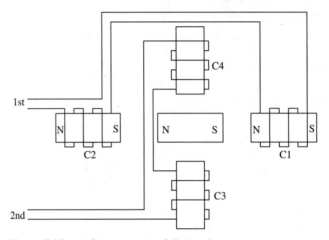

Figure 7.26a ■ Stepper motor full step 1

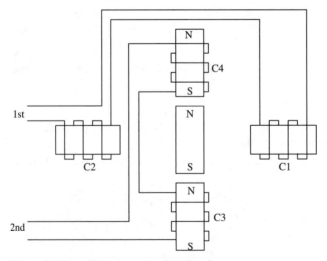

Figure 7.26b ■ Stepper motor full step 2

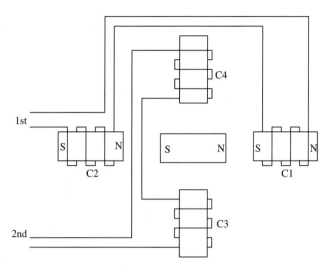

Figure 7.26c ■ Stepper motor full step 3

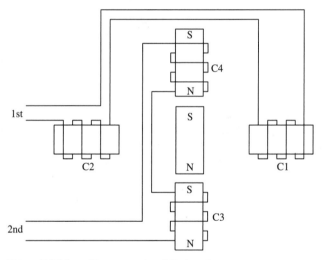

Figure 7.26d ■ Stepper motor full step 4

Energizing coil pair C3-C4 causes the rotor to rotate 90 degrees. Again, the direction of the current determines the magnetic polarity and thus the angular position of the rotor. In this example, the direction of the current causes the rotor to rotate in a clockwise direction, as shown in Figure 7.26b.

Next, coils C1-C2 are energized again, but with a current opposite to that in Step 1. The rotor moves 90 degrees in a clockwise direction as shown in Figure 7.26c. The last full step moves the rotor another 90 degrees in a clockwise direction. Note that again the coil pair C3-C4 is energized, but with a current opposite to that in Step 2.

We can also rotate the stepper motor in the counterclockwise direction. This can be done by reversing the polarities of coils C3 and C4 in Figure 7.26b and 7.26d. Figure 7.27 shows the counterclockwise sequence.

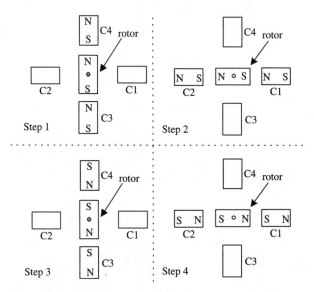

Figure 7.27 ■ Full-step counterclockwise operation of step motor

The stepper motor may also be operated with half steps. A *half step* occurs when the rotor (in a four-pole step) is moved to eight discrete positions (45°).

To operate the stepper motor in half steps, sometimes both coils may have to be on at the same time. When two coils in close proximity are energized there is a resultant magnetic field whose center will depend on the relative strengths of the two magnetic fields. Figure 7.28 illustrates the half-stepping sequence.

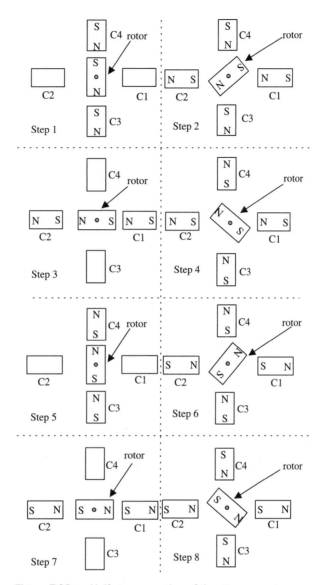

Figure 7.28 ■ Half-step operation of the stepper motor

The step sizes of stepper motors vary from approximately 0.72° to 90°. However, the most common step sizes are 1.8°, 7.5°, and 15°. The steps of 90° or 45° are too crude for many applications.

The actual stator (the stationary electromagnets) of a real motor has more segments on it than previously indicated. One example is shown in Figure 7.29. The rotor is also a little bit different and is also shown in Figure 7.29.

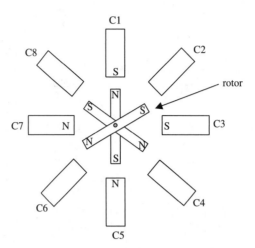

Figure 7.29 ■ Actual internal construction of step motor

In Figure 7.29, the stator has eight individual sections (coils) on it and hence the angle between two adjacent sections is 45°. The rotor has six sections on it and hence there is 60° between two adjacent sections. Using the principle of a Vernier mechanism, the actual movement of the rotor for each step would be 60° – 45° or 15°. Interested readers should try to figure out how these sections are energized to rotate the motor in the clockwise and counterclockwise directions.

7.10.1 Stepper Motor Drivers

Driving a step motor involves applying a series of voltages to the coils of the motor. A subset of coils is energized at a time to cause the motor to rotate one step. The pattern of coils energized must be followed exactly for the motor to work correctly. The pattern will vary depending on the mode used on the motor. A microcontroller can easily time the duration that the coil is energized, and hence control the speed of the stepper motor in a precise manner.

The circuit in Figure 7.30 shows how the transistors are used to switch the current to each of the four coils of the stepper motor. The diodes in Figure 7.30 are called *fly-back diodes* and are used to protect the transistors from reverse bias. The transistor loads are the windings in the stepper motor. The windings are inductors, storing energy as a magnetic field. When the current is cut off, the inductor dispenses its stored energy in the form of an electric current. This current attempts to flow through the transistor, reversely biasing its collector-emitter pair. The diodes are placed to prevent this current from going through the transistors.

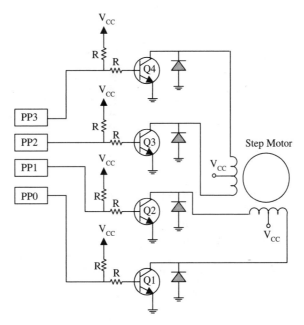

Figure 7.30 ■ Driving a step motor

For higher-torque applications the normal full-step sequence is used. This sequence is shown in Table 7.7. For lower-torque applications the half-step mode is used and its sequence is shown in Table 7.8. To control the motor, the microcontroller must output the values in the table in the shown sequence. Tables 7.7 and 7.8 are circular in that after the last step, the next output must be the first step. The values may be output in the order shown to rotate the motor in one direction or in reverse order to rotate in the reverse direction.

Step	Q1 PP0	Q2 PP1	Q3 PP2	Q4 PP3	Value
1	on	off	on	off	1010
2	on	off	off	on	1001
3	off	on	off	on	0101
4	off	on	on	off	0110
1	on	off	on	off	1010

Table 7.7 ■ Full-step sequence for clockwise rotation

Step	Q1 PP0	Q2 PP1	Q3 PP2	Q4 PP3	Value
1	on	off	on	off	1010
2	on	off	off	off	1000
3	on	off	off	on	1001
4	off	off	off	on	0001
5	off	on	off	on	0101
6	off	on	off	off	0100
7	off	on	on	off	0110
8	off	off	on	off	0010
1	on	off	on	off	1010

Table 7.8 ■ Half-step sequence for clockwise rotation

It is essential that the order be preserved even if the motor is stopped for a while. The next step to restart the motor must be the next sequential step following the last step used. The mechanical inertia of the motor will require a short delay (usually 5 to 15 ms) between two steps to prevent the motor from missing steps.

Example 7.11

Suppose the pins PP3..PP0 are used to drive the four transistors shown in Figure 7.30. Write a subroutine to rotate the stepper motor using the half-step sequence clockwise one cycle.

Solution: The assembly language subroutine is as follows:

```
#include "D:\miniide\hc12.inc"
step1       equ     $FA
step2       equ     $F8
step3       equ     $F9
step4       equ     $F1
step5       equ     $F5
step6       equ     $F4
step7       equ     $F6
step8       equ     $F2
half_step   movb    #$FF,DDRP        ; configure PORTP for output
            movb    #step1,PORTP
            bsr     delay
            movb    #step2,PORTP
            bsr     delay
            movb    #step3,PORTP
            bsr     delay
            movb    #step4,PORTP
            bsr     delay
            movb    #step5,PORTP
            bsr     delay
            movb    #step6,PORTP
            bsr     delay
            movb    #step7,PORTP
            bsr     delay
            movb    #step8,PORTP
            bsr     delay
```

```
                    movb        #step1,PORTP
                    bsr         delay
                    rts
; the following subroutine waits for 10 ms
delay       ldx         #2000
again       psha
            pula
            psha
            pula
            psha
            pula
            psha
            pula
            psha
            pula
            psha
            pula
            psha
            pula
            psha
            pula
            nop
            nop
            dbne        x,again
            rts
```

The C language version of the routine is straightforward and is left for you as an exercise.

7.11 Key Wakeups

Most embedded products are powered by batteries. In order for batteries to last longer, most microcontrollers have incorporated power saving modes such as the wait mode or the stop mode in the 68HC12. Whenever there is no activity from the end user over a period of time, the application software can put the embedded product in one of the power-saving modes and reduce its power consumption. Whenever the end user wants to use the product, a keystroke would wake up the microcontroller and put the embedded product back to normal operation mode. As we have learned in Chapter 6, both the reset and unmasked interrupts can put the microcontroller back to normal operation mode. However, reset is not recommended because it will restart the microcontroller, which would delay the response to the user request. An unmasked interrupt does not have this drawback. Some of the 68HC12 members have incorporated the key wake-up feature that will issue an interrupt to wake up the CPU when it is in the STOP or WAIT mode.

At the time of this writing, all except the B family members of the 68HC12 have key wakeup capability. The 812A4 has three key wakeup ports: Port D, Port H, and Port J. The 912D60 has two key wakeup ports: Port G and Port H. The 912DG128 and 912DT128 also have two key wakeup ports: Port H and Port J. A wakeup port is designed to issue interrupts to wake up the CPU when it has gone to sleep following a STOP or WAIT instruction.

After the appropriate configuration, an active edge on these pins will generate an interrupt to the CPU and force the CPU to exit from the low-power mode.

7.11.1 Key Wakeup Registers

Each wakeup port has a port data register, port data direction register, key wakeup interrupt enable register, key wakeup flag register, and key wakeup polarity register (port H and J only). A summary of all wakeup registers is shown in Table 7.9.

Port	Register	Register name	Interrupt vector	Assertion level
D	PORTD	Port D data register	$FFF2:$FFF3	falling edge
	DDRD	Port D data direction register		
	KWIED	Port D key wakeup interrupt enable register		
	KWIFD	Port D key wakeup interrupt flag register		
	INTCR	Interrup control register		
G	PORTG	Port G data register	$FFCE:$FFCF	falling edge
	DDRG	Port G data direction register		
	KWIEG	Port G key wakeup interrupt enable register		
	KWIFG	Port G key wakeup interrupt flag register		
H	PORTH	Port H data register	$FFCE:$FFCF	falling edge[1], and rising edge
	DDRH	Port H data direction register		
	KWIEH	Port H key wakeup interrupt enable register		
	KWIFH	Port H key wakeup interrupt flag register		
	KWPH	Port H key wakeup polarity register		
J	PORTJ	Port J data register	$FFCE:$FFCF	falling or rising edge
	DDRJ	Port J data direction register		
	KWIEJ	Port J key wakeup interrupt enable register		
	KWIFJ	Port J key wakeup interrupt flag register		
	KPOLJ	Port J key wakeup polarity register		
	PUPSJ	Port J pullup/pulldown select register		
	PULEJ	Port J pullup/pulldown enable register		

Note 1. Falling edge for 912D60. Falling or rising edge for 912DG128

Table 7.9 ■ 68HC12 key wakeup registers

PORT DATA REGISTERS

PORTD, PORTG, PORTH, and PORTJ are port data registers of these key wakeup ports.

Since the port register mapping for 812A4, 912D60, 912DG128, and 912DT128 are different, you need to use different header files for 812A4, 912D60, 912DG128, and 912DT128 when writing the program.

PORT DATA DIRECTION REGISTERS

DDRD, DDRG, DDRH, and DDRJ are data direction registers for these key wakeup ports. To configure a port pin for output (input), set the corresponding bit of the associated data direction register to one (zero). Again the address mapping of DDRH and DDRJ for the 812A4 is different from those of the 912D60, the 912DG128, and the 912DT128.

KEY WAKEUP INTERRUPT ENABLE REGISTERS

KWIED, KWIEG, KWIEH, and KWIEJ are key wakeup interrupt enable registers. When a bit of these registers is set to one (zero), it enables (disables) the interrupt request from the associated port pin. An interrupt is requested by the active (rising or falling) edge of the associated pin. Only the port J of the 812A4 and the port H and J of the 912DG128/DT128 can be programmed to select the falling or the rising edge for waking up the CPU. Other port pins can only use the falling edge to wake up the CPU.

KEY WAKEUP INTERRUPT FLAG REGISTERS

KWIFD, KWIFG, KWIFH, and KWIFJ are key wakeup interrupt flag registers. Each bit of these registers indicates whether the associated interrupt request signal has occurred (when it equals one) or not (when it equals zero).

A flag bit must be cleared when its associated interrupt request is being serviced. To clear a flag bit, write a one to it. There are two methods:

1. Use the store instruction. For example, the following two instructions will clear the bit 7 of the KWIFH register:

```
movb        #$80,KWIFH
```

2. Use the bclr instruction with a mask that has a zero at the bit positions to be cleared. For example, the following instruction will clear the bit 7 of the KWIFH register:

```
bclr        KWIFH,$7F
```

PORT H & J RISING & FALLING EDGE SELECTION FOR THE 912DG128/DT128

The active edge of each wakeup signal in port J can be programmed to be either rising or falling. It uses the KPOLJ register to specify the active edge of each port J pin. To use the wakeup function of port H or port J, each must be enabled for pull-up or pull-down by setting the corresponding bits (PUPH and PUPJ) of the PUCR register. The 912DG128 and the 912DT128 can program the wakeup signal to be either rising or falling edge active on both the port H and port J. The 912DG128 and the 912DT128 use registers KWPH and KWPJ to specify the active edge of each port H and J pin. Setting a bit to zero selects the falling edge. Otherwise, the rising edge is selected.

PORT J PULL-UP & PULL-DOWN SELECTION FOR 812A4

For the 812A4, each port J pin can be programmed to be pull-up or pull-down by setting the corresponding bit in the PUPSJ register to one or zero.

For the 812A4, each port J pin wakeup pull-up/pull-down must be enabled. The enabling of pull-up/pull-down is done via the PULEJ register. When a bit in the PULEJ register is set to one (zero), its corresponding port J pin has a (no) pull-up or pull-down device.

PORT D & PORT H WAKEUP OF 812A4

After being enabled, each pin of these two ports will wake up the CPU on the falling signal edge.

7.11.2 Key Wakeup Initialization

In order to use the 68HC12 key wakeup feature, you must initialize the wakeup port properly. The procedure for using the key wakeup feature is as follows:

Step 1
For the 812A4 port D key wakeup, set the IRQ enable bit in the INTCR register to one to enable the interrupt. (This step is not required for other key wakeup port.)

Step 2
Set the direction of the key wakeup bits to input by writing zeros in the data direction register.

Step 3
For the 812A4 port J key wakeup, select the bits for which pull-up or pull-down registers are needed by writing ones or zeros to the pull-up/pull-down select register (PUPSJ). Enable the bits that have pull-up/pull-down resistors by writing ones to the pull-up/pull-down enable register.

For the 912DG128 port H and J key wakeup, set the appropriate bit in the PUCR register to enable pull-up/pull-down. Choose the falling (rising) edge to wake up the CPU by clearing (setting) the associated bit in the KWPH or the KWPJ register.

Step 4
Initialize the key wakeup interrupt vector.

Step 5
Clear any flags that have been set in the key wakeup flag register.

Step 6
Enable the key wakeup bit by setting the appropriate bits in the wakeup interrupt enable register.

Step 7
Clear the global interrupt mask.

7.11.3 Considerations of the Key Wakeup Application

The main application of the key wakeup feature is to support the power saving modes of the 68HC12. Application software puts the microcontroller in low power mode by executing a STOP or a WAIT instruction when the inactivity of the end user has exceeded the preset time.

Many applications are designed to be a wait loop that waits for the end user to enter a request for service. When a request is entered, the application calls an appropriate routine to provide the service. After the service is done, the routine returns to the wait loop.

After the completion of a service to a user request, the application software starts a timer. If the user enters another command before the timer times out, the application software resets the timer and responds to the user request. If the timer times out before the user makes another service request, the application software puts the microcontroller in low power mode to save power. As long as there is no user request for service, the microcontroller will stay in the low power mode. The timer time-out interval could be a few minutes or longer depending on how much power the user wants to save. The timer *output compare* function is often used to implement the time-out interval.

When the user presses the key, the microcontroller will exit the low power mode and continue to execute the instruction following the STOP (or WAIT) instruction and another cycle of the normal application loop is started.

In order to be used in the key wakeup application, port pins must be configured for input. Port J of the 812A4 and ports H and J of the 912DG128 and 912DT128 can be configured to use the rising or falling edge to wake up the microcontroller. The choice of signal edge will dictate the choice of pull-up or pull-down resistive device. As shown in Figure 7.31, a rising edge is resulted when a high voltage is applied to a pull-down resistor; a falling edge is resulted when a low voltage is connected to a pull-up resistor. Therefore, the user should enable the pull-down resistor when a rising edge is selected and enable a pull-up resistor when a falling edge is chosen to wakeup the microcontroller.

Since the purpose of the key wakeup feature is to enable the microcontroller to resume its normal operation mode, the interrupt service routine needs only to perform minimal operations. The minimal operations to be performed would be to clear the interrupt flag set by the wakeup interrupt. The 812A4 Port D key wakeup interrupt shares the interrupt vector with the $\overline{IRQ}$ pin. Therefore, the interrupt service routine for Port D wakeup interrupts must check the interrupt flags to identify the source of the interrupt.

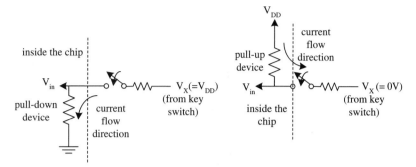

Figure 7.31 ■ (a) Pull-down resistor creates rising edge (b) Pull-up resistor creates falling edge

The logic flow of an embedded application that incorporates a key wakeup interrupt is illustrated in Figure 7.32.

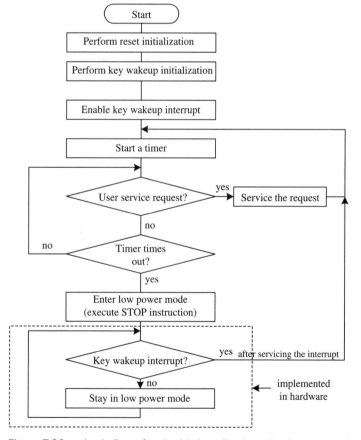

Figure 7.32 ■ Logic flow of embedded applications that incorporate key wakeup feature

Example 7.12

Write an instruction sequence to configure the 912D60 Port G for the key wakeup feature. Program the port G so that pins KWG3..KWG0 generate interrupt whenever there is a falling edge applied to any one of these four pins.

Solution:

```
#include     "d:\miniide\hc912dg128.inc"
...
            movb     #$00,DDRG     ; configure Port G for input
            movb     #$0F,KWIEG    ; enable key wakeup to Port G
            movb     #$0F,KWIFG    ; clear the Port G key wakeup flags
            cli                    ; enable key wakeup interrupt globally
```

7.12 Summary

Peripheral devices are pieces of equipment that exchange data with a computer. Examples of peripheral devices are switches, keypads, keyboards, LEDs, seven-segment displays, LCDs, printers, and so on. The speed and electrical characteristics of peripheral devices are very different from those of a microprocessor. Therefore they are not connected to the processor directly. Instead, interface chips are used to resolve the difference between the microprocessor and peripheral devices.

The major function of an interface chip is to synchronize the data transfer between the processor and an I/O device. An interface chip consists of control registers, status registers, data registers, data direction registers, and control circuitry. There are two aspects in the I/O transfer synchronization: the synchronization between the CPU and the interface chip and the synchronization between the interface chip and the I/O device. There are two methods to synchronize the CPU and the interface chip: the *polling* method and the *interrupt* method. There are three methods for synchronizing the interface chip and the I/O device: the *brute force*, *strobe*, and *handshake* methods.

Different processors deal with I/O differently. Some processors have dedicated instructions and addressing modes for performing I/O operations. In these processors, I/O devices occupy a separate memory space from the main memory. Other processors use the same instructions and addressing modes to access I/O devices. I/O devices and main memory share the same memory space.

The 68HC12 members have either 80 or 112 signal pins. Most of these pins serve multiple purposes and can be used as general I/O pins. The 812A4 and the 912DG128/DT128 have 112 pins. All of the B family members have 80 pins. The 912D60 and the MC9S12DP256 have either 80 or 112 pins. These I/O pins are divided up to 12 ports. Each I/O port has a port data register and data direction register. Except for Port AD, all I/O port pins can be configured for either input or output.

Input devices including DIP switches, keypads, and keyboards are discussed in detail. Both the assembly and C languages are used to illustrate I/O programming. The key bouncing issue is examined in detail and solutions to the problem are also explored. The characteristics and interfacing of *LEDs* and *LCDs* are explained and their programming is demonstrated.

D/A converters are often used to convert digital values into voltages. Sawtooth, triangular, and sinusoidal waveforms can be generated from a D/A converter easily.

A stepper motor is a digital motor in the sense that each step of the rotation rotates a fixed number of degrees of angle. It is most suitable for applications that require a high degree of positional control such as plotters, disks, magnetic tapes, robot joints, and so on. The resolution of one step of a stepper motor can be as small as 0.72° and as large as 90°. The simplest stepper motor has two pairs of coils.

Driving a stepper motor involves applying a series of voltages to the coils of the motor. A subset of coils is energized at one time to cause the motor to rotate one step. The pattern of coils energized must be followed exactly for the motor to work correctly. The pattern will vary depending on the mode used on the motor. A microcontroller can easily time the duration the coil is energized, and hence control the speed of the stepper motor in a precise manner.

Saving power is a major concern in most embedded applications. When the user is not using an embedded system, the microcontroller should be switched to a low-power mode. The 68HC12 has two low-power modes: the *wait mode* and *stop mode*. The 68HC12 consumes the least power in the stop mode. The stop (wait) mode can be entered by executing the STOP (WAIT) instruction. To facilitate exiting the STOP or WAIT mode, the 68HC12 members 812A4, 912D60, 912DG128, and 912DT128 implement the key wakeup feature. Whenever the 68HC12 is in the WAIT or the STOP mode and a selected signal edge arrives at one of the key wakeup port pins, an interrupt request will be generated and the 68HC12 will be woken up. The service for the key wakeup interrupt is simply to clear the interrupt flag and resume the execution of the instruction following the WAIT or STOP instruction.

The signal edge to wake up the microcontroller could be rising or falling. When the rising edge is selected, the pull-down resistor should be enabled. When the falling edge is chosen, the pull-up device should be enabled.

7.13 Exercises

E7.1 What is a handshake?

E7.2 Write an instruction sequence to output the value $35 to Port P.

E7.3 Write an instruction sequence to input the current signal levels on Port H pins and place the value in accumulator B.

E7.4 Suppose the Port T pins T4..T0 are connected to green, yellow, red, blue, and purple LEDs. The circuit connection is similar to that in Figure 7.10. Write a program to light the green, yellow, red, blue, and purple LEDs in turn for 0.1, 0.2, 0.3, 0.4, and 0.5 seconds forever. Use both the assembly and C languages.

E7.5 *Traffic light controller simulation.* Use the port P pins PP5..PP0 of the CME-12BC and green, yellow, and red LEDs to simulate a traffic-light controller. The traffic-light patterns and durations for traffic heading east–west and north–south are given in Table 7E.1. Write an assembly and C program to control the light patterns, and connect the circuit to demonstrate the changes in the lights.

East-west			North-south			Duration
Green	Yellow	Red	Green	Yellow	Red	
1	0	0	0	1	0	25
0	0	1	0	1	0	5
0	1	0	1	0	0	20
0	1	0	0	0	1	4

Table 7E.1 ■ Traffic light pattern and duration

E7.6 Suppose that Port P pins drive the seven-segment pattern and Port T pins drive six 2N2222 transistors with a connection similar to that in Figure 7.13. Write an assembly and C program to display digits 1, 2, 3, 4, 5, and 6, in turn, each for one second, repeatedly.

E7.7 Write an assembly and C program to display the following information in four rows in the LCD connected to the CME-12BC demo board:

Date: 08 02 1952
Time: 10:20:10
Temperate: 30 F
Humidity: 40%

E7.8 Calculate the period of the sawtooth waveform generated in Example 7.10. What are the voltages corresponding to the digital value 20, 30, 50, 127, and 192?

E7.9 Write a program using the D/A converter to generate a sine wave from $V_{OUT}A$ in Figure 7.25. (*Hint:* Use the table lookup method.)

E7.10 Write a C program to read a character from the keyboard shown in Figure 7.20. This program will perform keyboard scanning, debouncing, and ASCII code lookup.

E7.11 Write a C program to read a character from the keypad shown in Figure 7.23. This program will perform keypad scanning, key debouncing, and ASCII code lookup.

E7.12 Write a program to generate the waveform shown in Figure 7E.12 from the pin $V_{OUT}A$ in Figure 7.25.

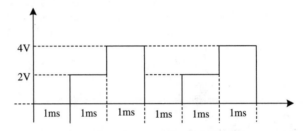

Figure 7E.12 ■ A wave form to be generated

E7.13 Write an assembly and a C program to generate a triangular waveform from the $V_{OUT}A$ pin in Figure 7.25. What is the period of the waveform generated by your program?

E7.14 Write an instruction sequence to rotate the stepper motor shown in Figure 7.30 clockwise for one cycle using the full-step sequence.

E7.15 Write a C program to rotate the stepper motor one cycle in the counterclockwise direction using the half-step sequence. The circuit connection of the stepper motor is shown in Figure 7.30.

E7.16 Write a C program to rotate the stepper motor clockwise one cycle using the full-step sequence. The circuit connection of the stepper is shown in Figure 7.30.

E7.17 Write a C subroutine to initialize the Port H key wakeup function for the 912D60 that interrupts the microcontroller on the falling edge (automatic) of any Port H pin. Also write an interrupt service routine for this interrupt that simply clears the key wakeup interrupt flags and return.

E7.18 According to Motorola, we must write a one to the key wakeup interrupt flag in order to clear it. Why shouldn't we use the *bset* instruction with a mask that has ones at the positions corresponding to the flags that we want to clear?

7.14 Lab Exercises & Assignments

L7.1 Use the Port P to drive two 74HC48s. The 74HC48 is a common cathode seven-segment display driver. It accepts a 4-bit binary number and converts it into the corresponding seven-segment pattern. Write a program to display the value from 00 to 99 on these two seven-segment displays three times and stop. Display each value for .25 seconds.

L7.2 *Keypad input and LCD output exercises.* Use Port P to drive a 16-key keypad and connect a DMC-20434 LCD kit to the CME-12BC demo board. Write a C program to perform keypad scanning, debouncing, and ASCII code lookup. Output the character read from the keypad to the LCD kit. Stop keypad input whenever an *F* character is entered.

8

Timer Functions

8.1 Objectives

After completing this chapter you should be able to:

- explain the overall structure of the 68HC12 timer system

- use the input-capture function to measure the duration of a pulse or the period of a square wave

- use the input-capture function to measure the duty cycle of a waveform or the phase difference of two waveforms having the same frequency

- use the input-capture function to measure the frequency of a signal

- use the output-compare function to create a time delay, generate a pulse, or periodic waveform

- use the forced output-compare function

- use the pulse accumulator to measure the frequency of an unknown signal, count the number of events occurred in an interval

- use the real-time interrupt function to generate periodic interrupts

- use the pulse-width-modulation (PWM) function to generate waveforms with certain frequency and duty cycle

- use the features provided by the enhanced capture timer available in the 912BE32, the 912D60, the 912DG128, and the 912DT128

8.2 Why are Timer Functions Important?

There are many applications that require a dedicated timer system, including:

- time-delay creation and measurement
- period and pulse width measurement
- frequency measurement
- event counting
- arrival time comparison
- time-of-day tracking
- waveform generation
- periodic interrupt generation

These applications will be very difficult to implement without a dedicated timer system. The 68HC12 implements a very complicated timer system to support the implementation of these applications.

At the heart of the 68HC12 timer system is the 16-bit *timer counter* (TCNT). This timer can be started or stopped, as you like. One of the timer functions is called *input capture*. The input-capture function latches the contents of the 16-bit timer into a latch when the predefined *event* arrives. An event is represented by a signal edge. The signal edge could be a rising or a falling edge. By capturing the timer value, many measurements can be made. Some of them include:

- pulse width measurement
- period measurement
- duty cycle measurement
- event arrival time recording
- timing reference

Another timer function is called the *output compare*. The output-compare circuit compares the 16-bit timer value with that of the output-compare register in each clock cycle and performs the following operations when they are equal:

- (optionally) triggers an action on a pin (set to high, set to low, or toggle its signal level)
- sets a flag in a register
- (optionally) generates an interrupt request

The output-compare function is often used to generate a time delay, trigger an action at some future time, and generate a digital waveform. The key to using the output-compare function is to make a copy of the 16-bit timer, add a delay to it, and store the sum in an *output-compare register*. The 68HC12 has eight output-compare channels, which share the signal pins and registers with input-capture channels.

The third timer function is the *real-time interrupt* (RTI) function. The RTI function can generate a periodic interrupt to remind the CPU to perform routine tasks. The CPU will respond to the RTI interrupt immediately unless one of the following exceptions is pending:

- IRQ interrupt
- Nonmaskable interrupts
- Resets

Because of the nature of an RTI interrupt, the required tasks will be handled within a specified time limit (hence the name *real-time*).

The fourth timer function is the *computer operating properly* (COP) subsystem. As we have discussed in Chapter 6, if the software did not take care of the COP timer before it times out, the COP system will reset the 68HC12 and expose the software bug to the user. We will not discuss COP in this chapter.

The fifth timer function is the *pulse accumulator*. This circuit is often used to count the events that have arrived in certain intervals or measure the frequency of an unknown signal.

Although the output-compare function can be used to generate digital waveforms, it takes up a lot of CPU overhead. For many applications, the frequency and duty cycle of the generated waveform need not be changed so often. The required overhead in using an output-compare function in waveform generation is not justified. The designers of the 68HC12 added the *Pulse-width-modulation* (PWM) circuit to the system. This circuit allows the user to specify the duty cycle and period of the waveform to be generated. Once the waveform parameters have been set up, the CPU can work on other chores unless there is a need to change the duty cycle and/or the period of the waveform. The 812A4 is the only 68HC12 member that does not implement the PWM function.

8.3 Standard Timer Module

The earliest three members of the 68HC12 family, the 812A4, the 912B32, and the 912BC32, have implemented the *standard timer module* (TIM). The standard timer module consists of a 16-bit timer counter (TCNT) driven by a prescaler. It contains eight complete 16-bit input-capture/output-compare channels and one 16-bit pulse accumulator.

Later members, including the 912BE32, the 912D60, the 912DG128, and the 912DT128, implement an enhanced capture timer module.

In the following three sections, we will discuss the function and applications of the standard timer module. After that, we will explore the enhancements made by the enhanced capture timer module.

8.4 Timer Counter Register

The *timer counter register* (TCNT), a 16-bit register, is required for the functioning of all input-capture and output-compare functions. The user must access this register in one access, rather than two separate accesses, to its high byte and low byte. Because the TCNT counter does not stop during the access operation, the value accessed in one 16-bit read won't be the same as two separate accesses to its high byte and low byte.

There are three other registers related to the operation of the TCNT counter. They are:

1. Timer system control register (TSCR)
2. Timer interrupt mask 2 register (TMSK2)
3. Timer interrupt flag 2 register (TFLG2)

8.4.1 Timer System Control Register (TSCR)

The contents of the TSCR register are shown in Figure 8.1. The timer counter must be enabled before it can count. Setting the bit 7 enables the timer counter to count (up). The timer counter can be optionally stopped (by setting bit 6 of the TSCR) during the wait mode to save more power. If the timer function is not needed, it can also be stopped during the background debug mode.

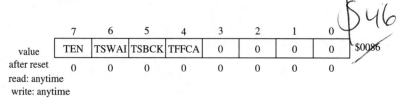

	7	6	5	4	3	2	1	0	
value	TEN	TSWAI	TSBCK	TFFCA	0	0	0	0	$0086
after reset	0	0	0	0	0	0	0	0	

read: anytime
write: anytime

TEN -- timer enable bit

    0 = disable timer; this can be used to save power consumption

    1 = allows timer to function normally

TSWAI -- timer stops while in wait mode bit

    0 = allows timer to continue running during wait mode

    1 = disables timer when MCU is in wait mode

TSBCK -- timer stops while in background mode bit

    0 = allows timer to continue running while in background mode

    1 = disables timer when MCU is in background debug mode

TFFCA -- timer fast flag clear all bit

    0 = allows timer flag clearing to function normally

    1 = For TFLG1 ($8E), a read from an input capture or a write to
the output compare channel ($90-$9F) causes the
corresponding channel flag, CnF, to be cleared.
For TFLG2 ($8F), any access to the TCNT register clears
the TOF flag. Any access to the PACNT register clears the
PAOVF and PAIF flags in the PAFLG register.

Figure 8.1 ■ Timer system control register (TSCR)

All timer interrupt flags can be cleared by writing a one to them. However, there is a faster way to clear timer flags. When the bit 4 of the TSCR register is set to one, a read from an input-capture register or a write to an output-compare register will clear the corresponding flag in the TFLG1 register; any access to the TCNT register will clear the TOF flag and any access to the PACNT register will clear the PAOVF and PAIF flags in the PAFLG register. This feature reduces the software overhead in a separate clear sequence.

8.4.2 Timer Interrupt Mask Register 2 (TMSK2)

Timer interrupt mask register 2 is another register that controls the operation of the timer counter. Its contents are shown in Figure 8.2.

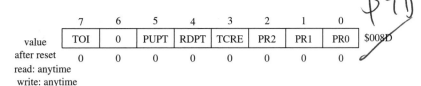

	7	6	5	4	3	2	1	0	
value	TOI	0	PUPT	RDPT	TCRE	PR2	PR1	PR0	$008D

after reset   0   0   0   0   0   0   0   0

read: anytime
write: anytime

TOI -- timer overflow interrupt enable bit

     0 = interrupt inhibited
     1 = interrupt requested when TOF flag is set

PUPT -- timer pullup resistor (when configured for input) enable bit

     0 = disable pullup resistor function
     1 = enable pullup resistor function

RDPT -- timer drive reduction bit

     0 = normal output drive capability
     1 = enable output drive reduction function

TCRE -- timer counter reset enable bit

     0 = counter reset inhibited and counter free runs
     1 = counter reset by a successful output compare 7
         If TC7 = $0000 and TCRE = 1, TCNT stays at $0000
         continuously. If TC7 = $FFFF and TCRE = 1, TOF never
         gets set even though TCNT counts from $0000 to $FFFF.

Figure 8.2 ■ Timer interrupt mask 2 register (TMSK2)

An interrupt will be requested when the TCNT overflows (when TCNT rolls over from $FFFF to $0000) and the bit 7 of TMSK2 is set to one.

The bit 5 of TMSK2 allows the user to control pull-up resistors on the timer port pins when the pins are configured as inputs. Setting this bit enables the pull-up resistor function.

It might be desirable to reduce the output driver size of the timer port output (when the microcontroller needs to drive a lot of peripheral devices) to reduce the power supply current. The bit 4 of the TMSK2 allows you to do just that. Setting this bit to one enables the output drive reduction function.

The timer counter needs a clock signal to operate. The clock input to the timer counter could be the M clock prescaled by a factor or the PAI pin input prescaled by a factor. The user has the option to choose the prescale factor. When the bits 3 and 2 of the PACTL register are 00, the clock to the TCNT is the M clock prescaled by a factor. The lowest three bits of the TMSK2 register specify the prescaled factor for the E clock as shown in Table 8.1.

PR2	PR1	PR0	Prescale Factor
0	0	0	1
0	0	1	2
0	1	0	4
0	1	1	8
1	0	0	16
1	0	1	32
1	1	0	reserved
1	1	1	reserved

Table 8.1 ■ Timer counter prescale factor

8.4.3 Timer Interrupt Flag 2 Register (TFLG2)

Only the bit 7 (TOF) of this register is implemented. When the TCNT counter rolls over from $FFFF to $0000, the bit 7 of this register is set to one. This flag can be cleared by writing a one to it.

8.5 Input Capture Function

Some applications need to know the arrival time of events. In a computer, *physical time* is represented by the count in a counter, while the occurrence of an event is represented by a signal edge (either the rising or falling edge). The time when an event occurs can be recorded by latching the count value when a signal edge arrives, as illustrated in Figure 8.3.

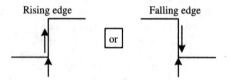

Figure 8.3 ■ Events represented by signal edges

The 68HC12 timer system has eight input-capture channels that implement this operation. Each input-capture channel includes a 16-bit input capture register, an input pin, input edge-detection logic, and an interrupt generation circuit. In the 68HC12, physical time is represented by the count in the timer counter (TCNT).

8.5.1 Input-Capture/Output-Compare Selection

Since input-capture functions and output-compare functions share signal pins and registers, they cannot be enabled simultaneously. When one is enabled, the other is disabled. The selection is done by the Timer Input-Capture/Output-Compare Select Register (TIOS), as shown in Figure 8.4.

	7	6	5	4	3	2	1	0	
value	IOS7	IOS6	IOS5	IOS4	IOS3	IOS2	IOS1	IOS0	$0080
after reset	0	0	0	0	0	0	0	0	

read: anytime
write: anytime

IOS[7:0] -- Input capture or output compare channel cinfiguration bits

0 = The corresponding channel acts as an input capture
1 = The corresponding channel acts as an output compare

Figure 8.4 ■ Timer input capture/output compare select register (TIOS)

Example 8.1

▼

Write an instruction sequence to enable the output-compare channels 7..4 and input-capture channels 3..0.

Solution: The following instruction sequence will achieve the desired configuration:

```
#include "d:\miniide\hc12.inc"
    ...
    movb    #$F0,TIOS
```

▲

8.5.2 Pins for Input Capture

Port T has eight signal pins (PT7..PT0) that can be used as input-capture/output-compare or general I/O pins. PT7 can also be used as the *pulse accumulator input* (PAI). The user must make sure that the PT7 pin is enabled for one and only one of these three functions (OC7, IC7, and PA). When these pins are not used for timer functions, they can also be used as general-purpose I/O pins. When being used as general I/O pins, the user must use the DDRT register to configure their direction (input or output). When a port T pin is used as a general-purpose I/O pin, its value is reflected in the corresponding bit in the PORTT register.

8.5.3 Registers Associated with Input Capture

The user needs to specify what signal edge to capture. The edge-selection is done via the timer control registers 3 and 4, as shown in Figure 8.5.

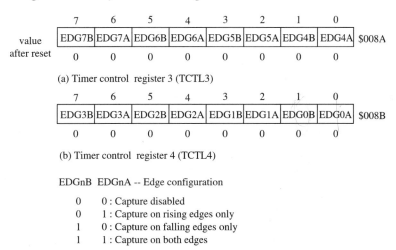

	7	6	5	4	3	2	1	0	
value	EDG7B	EDG7A	EDG6B	EDG6A	EDG5B	EDG5A	EDG4B	EDG4A	$008A
after reset	0	0	0	0	0	0	0	0	

(a) Timer control register 3 (TCTL3)

	7	6	5	4	3	2	1	0	
	EDG3B	EDG3A	EDG2B	EDG2A	EDG1B	EDG1A	EDG0B	EDG0A	$008B
	0	0	0	0	0	0	0	0	

(b) Timer control register 4 (TCTL4)

EDGnB EDGnA -- Edge configuration

0 0	: Capture disabled
0 1	: Capture on rising edges only
1 0	: Capture on falling edges only
1 1	: Capture on both edges

Figure 8.5 ■ Timer control register 3 and 4 (TCTL3 & TCTL4)

When an input-capture channel is selected but capture is disabled, the associated pin can be used as a general-purpose I/O pin.

An input-capture channel can optionally generate an interrupt request on the arrival of a selected edge if it is enabled. The enabling of an interrupt is controlled by the timer interrupt mask register 1 (TMSK1). The enabling of input capture 7 through 0 is controlled by bits 7 through 0 of TMSK1. When a selected edge arrives at the input-capture pin, the corresponding flag in the timer flag register 1 (TFLG1) will be set. The contents of TMSK1 and TFLG1 are shown in Figures 8.6 and 8.7.

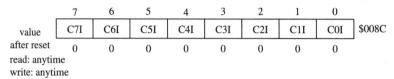

	7	6	5	4	3	2	1	0	
value	C7I	C6I	C5I	C4I	C3I	C2I	C1I	C0I	$008C
after reset	0	0	0	0	0	0	0	0	

read: anytime
write: anytime

C7I-C0I: input capture/output compare interrupt enable bits
    0 = interrupt disabled
    1 = interrupt enabled

Figure 8.6 ■ Timer interrupt mask 1 register (TMSK1)

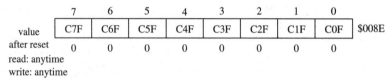

	7	6	5	4	3	2	1	0	
value	C7F	C6F	C5F	C4F	C3F	C2F	C1F	C0F	$008E
after reset	0	0	0	0	0	0	0	0	

read: anytime
write: anytime

C7F-C0F: input capture/output compare interrupt flag bits
    0 = interrupt condition has not occurred
    1 = interrupt condition has occurred

Figure 8.7 ■ Timer interrupt flag 1 register (TFLG1)

To clear a flag in the TFLG1 register, write a one to it. However, there is a better way to clear the flag that incurs less overhead. Setting the bit 4 (TFFCA) of the TSCR allows us to clear a flag by reading the corresponding input-capture register or writing a new value into the output-compare register. This operation is normally needed for the normal operation of the input-capture/output-compare function.

Each input-capture channel has a 16-bit register (TCx, x = 0 to 7) to hold the count value when the selected signal edge arrives at the pin. This register is also used as the output-compare register when the output-compare function is selected instead.

8.5.4 Input-Capture Applications

There are many applications for the input-capture function. Examples include:

- *Event arrival time recording.* Some applications, for example, a swimming competition, need to compare the arrival times of several different swimmers. The input-capture function is very suitable for this application. The number of events that can be compared is limited by the number of input-capture channels.

- *Period measurement.* To measure the period of an unknown signal, the input-capture function should be configured to capture the timer values corresponding to two consecutive rising or falling edges, as illustrated in Figure 8.8.

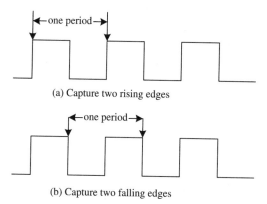

(a) Capture two rising edges

(b) Capture two falling edges

Figure 8.8 ■ Period measurement by capturing two consecutive edges

■ *Pulse-width measurement.* To measure the width of a pulse, two adjacent rising and falling edges are captured, as shown in Figure 8.9.

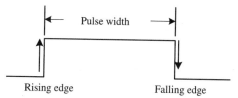

Rising edge   Falling edge

Figure 8.9 ■ Pulse-width measurement using input capture

■ *Interrupt generation.* All input-capture inputs can serve as edge-sensitive interrupt sources. Once enabled, interrupts will be generated on the selected edge(s).

■ *Event counting.* An event can be represented by a signal edge. An input-capture channel can be used in conjunction with an output-compare function to count the number of events that occur during an interval. An event counter can be set up and incremented by the input-capture interrupt service routine. This application is illustrated in Figure 8.10.

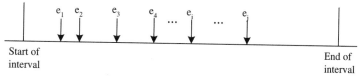

Start of interval   End of interval

Figure 8.10 ■ Using an input-capture function for event counting

■ *Time reference.* In this application, an input-capture function is used in conjunction with an output-compare function. For example, if the user wants to activate an output signal a certain number of clock cycles after detecting an input event, the input-capture function would be used to record the time at which the edge is detected. A number corresponding to the desired delay would be added to this captured value and stored to an output-compare register. This application is illustrated in Figure 8.11.

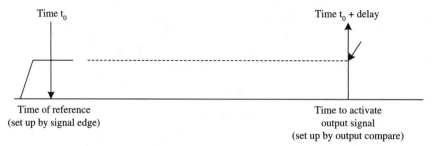

Figure 8.11 ■ A time reference application

- *Duty-cycle measurement.* The duty cycle is the percent of time that the signal is high within a period in a periodic digital signal. The measurement of the duty cycle is illustrated in Figure 8.12.

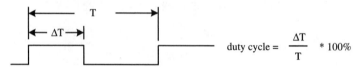

Figure 8.12 ■ Definition of duty cycle

- *Phase difference measurement.* Phase difference is defined as the difference of arrival times (in percentage of a period) of two signals that have the same frequency but do not coincide in their rising and falling edges. The definition of phase difference is illustrated in Figure 8.13.

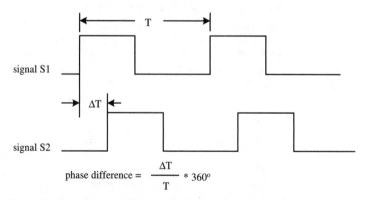

Figure 8.13 ■ Phase difference definition for two signals

The unit used in most of the measurements is the number of clock cycles. When it is desirable, the unit should be converted into an appropriate unit such as seconds.

When using a timer pin for the input-capture function, we need to configure it for input by clearing the corresponding bit of the DDRT register.

Example 8.2

Period measurement. Use the input-capture channel 0 to measure the period of an unknown signal. The period is known to be shorter than 128 ms. Assume the E clock frequency is 8 MHz. Use the number of clock cycles as the unit of the period.

Solution: Since the input-capture register is 16-bit, the longest period of the signal that can be measured with the prescale factor to the timer counter equals set to one is:

$2^{16} \div 8\text{MHz} = 8.192$ ms.

To measure a period that is equal to 128 ms, we have two options:

1. Set the prescale factor to one and keep track of the number of times that the timer counter overflows.

2. Set the prescale factor to 16 and do not keep track of the number of times that the timer counter overflows.

In this example, we will adopt the second method to make the programming easier. The result of this measurement will be in number of clock cycles and the period of each clock cycle is 2 μs. The circuit connection for the period measurement is shown in Figure 8.14, and the logic flow for the period measurement is shown in Figure 8.15.

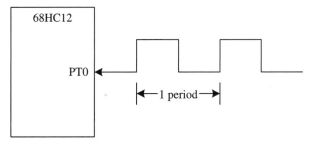

Figure 8.14 ■ Period measurement signal connection

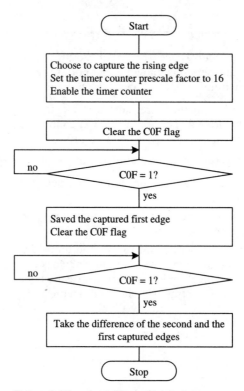

Figure 8.15 ■ Logic flow of period measurement program

The assembly program that measures the period using this method is as follows:

```
#include "d:\miniide\hc12.inc"
                org       $800
edge_1st        rmb       2                  ; memory locations to hold the first edge
period          rmb       2                  ; memory locations to store the period
                org       $1000
                movb      #$90,TSCR          ; enable timer counter and fast timer flags clear
                bclr      TIOS,$01           ; select input capture 0
                movb      #$24,TMSK2         ; disable TCNT overflow interrupt, set prescale
                                             ; factor to 16, enable pullup
                movb      #$01,TCTL4         ; choose to capture the rising edge of PT0 pin
                bclr      TFLG1,$FE          ; clear C0F flag
                brclr     TFLG1,$01,*        ; wait for the arrival of the first rising edge
                ldd       TC0                ; save the first edge and clear C0F flag
                std       edge_1st
                brclr     TFLG1,$01,*        ; wait for the arrival of the second edge
                ldd       TC0
                subd      edge_1st
                std       period
                swi
                end
```

The C language version of the program is as follows:

```
#include <hc12.h>
#include <stdio.h>

void main( )
{
  unsigned int edge1, period;

    TSCR = 0x90;                    /* enable timer counter, enable fast flag clear*/
    TIOS &= 0xFE;                   /* select input capture 0 /
    DDRT &= 0xFE;                   /* configure pin PT0 for input */
    TMSK2 = 0x24;                   /* disable TCNT overflow interrupt, set prescale factor to 16 */
    TCTL1 = 0x01;                   /* capture the rising edge of PT0 pin */
    TFLG1 = 0x01;                   /* clear the COF flag */
    while (!(TFLG1 & 0x01));        /* wait for the arrival of the first rising edge */
    edge1 = TC0;                    /* save the first captured edge and clear COF flag */
    while (!(TFLG1 & 0x01));        /* wait for the arrival of the second rising edge */
    period = TC0 - edge1;
    asm ("swi");
}
```

Example 8.3

Write a program to measure the pulse width of a signal connected to the PT0 pin. Assume the E clock frequency is 16 MHz.

Solution: We will set the prescale factor to eight and use the clock cycle as the unit of measurement. The period of one clock cycle is 1 μs. Since the pulse width could be much longer than 2^{16} clock cycles, we will need to keep track of the number of times that the TCNT counter overflows. Each TCNT overflow adds 2^{16} clock cycles to the pulse width.

Let:

ovcnt	= TCNT counter overflow count
diff	= the difference between two consecutive edges
edge1	= the captured time of the first edge
edge2	= the captured time of the second edge

The pulse width can be calculated by the following equations:

Case 1

edge2 ≥ edge1

pulse width = ovcnt × 2^{16} + diff

Case 2

edge2 < edge 1

pulse width = (ovcnt – 1) × 2^{16} + diff

In Case 2, the timer overflows at least once even if the pulse width is shorter than $2^{16} - 1$ clock cycles. Therefore, we need to subtract one from the timer overflow count in order to get the correct result. The pulse width is obtained by appending the difference of the two captured edges to the TCNT overflow count.

The logic flow of the program is shown in Figure 8.16.

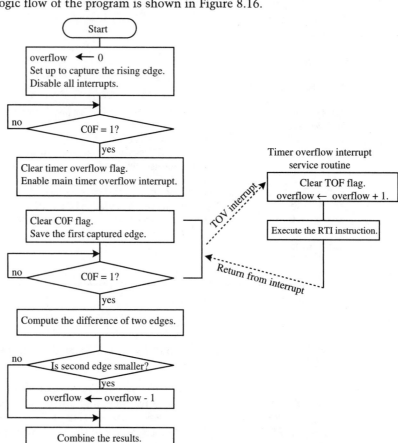

Figure 8.16 ■ Logic flow for measuring pulse width of slow signals

The assembly program that implements this algorithm is as follows:

```
#include "d:\miniide\hc12.inc"
setuservector equ    $F69A
tov_vec_no equ       23


              org       $800
edge1         rmb       2
overflow      rmb       2
pulse_width   rmb       2
              org       $1000
; the following 7 instructions set up TOV interrupt jump vector
              ldd       #tov_isr
              pshd
              ldab      #tov_vec_no
              clra
```

```
                        ldx         setuservector
                        jsr         0,x
                        leas        2,sp
                        ldd         #0
                        std         overflow
                        movb        #$90,TSCR
; disable TCNT overflow interrupt, enable pullup, set prescale factor to 16
                        movb        #$24,TMSK2
; select input capture 0
                        bclr        TIOS,$01
                        bclr        DDRT,$01
; capture the rising edge on PTO pin
                        movb        #$01,TCTL4
                        bclr        TFLG1,$FE          ; clear COF flag
                        brclr       TFLG1,$01,*        ; wait for the arrival of the first rising edge
                        ldd         TC0                ; save the captured first edge & clear COF flag
                        std         edge1
                        bclr        TFLG2,$7F          ; clear TOF flag
; enable TCNT overflow interrupt
                        bset        TMSK2,$80
                        cli
; capture the falling edge on PTO pin
                        movb        #$02,TCTL4
                        brclr       TFLG1,$01,*        ; wait for the arrival of the falling edge
                        ldd         TC0
                        subd        edge1
                        std         pulse_width
                        bcc         next               ; is the second edge smaller?
; second edge is smaller, so decrement overflow count by one
                        ldx         overflow
                        dex
                        stx         overflow
next                    swi
tov_isr                 bclr        TFLG2,$7F          ; clear TOF flag
                        ldx         overflow
                        inx
                        stx         overflow
                        rti
                        end
```

The C language version of this program is as follows:

```
#include <hc12.h>
unsigned int diff, edge1, overflow;
unsigned long int pulse_width;
void tov_isr(void);
void main()
{
        asm ("ldd  #_tov_isr");
        asm ("pshd");
        asm ("ldab #23");
        asm ("clra");
        asm ("ldx 0xF69A");
        asm ("jsr 0,x");
        asm ("leas 2,sp");
```

```
                overflow = 0;
                TSCR = 0x90;              /* enable timer and fast flag clear */
                TMSK2 = 0x24;             /* prescale factor 16, no timer overflow interrupt */
                TIOS &= 0xFE;             /* select input capture 0 */
                DDRT &= 0xFE;
                TCTL4 = 0x01;             /* prepare to capture the rising edge */
                TFLG1 = 0x01;             /* clear C0F flag */
                while(!(TFLG1 & 0x01));   /* wait for the arrival of the rising edge */
                TFLG1 = 0x01;
                TFLG2 = 0x80;             /* clear TOF flag */
                TMSK2 |= 0x80;            /* enable TCNT overflow interrupt */
                INTR_ON();
                edge1 = TC0;              /* save the first edge */
                TCTL4 = 0x02;             /* prepare to capture the falling edge */
                while (!(TFLG1 & 0x01));  /* wait for the arrival of the falling edge */
                diff = TC0 - edge1;
                if (TC0 < edge1)
                    overflow = overflow - 1;
                pulse_width = overflow * 65536u + diff;
                asm ("swi");
        }
        #pragma interrupt_handler tov_isr
        void tov_isr(void)
        {
                TFLG2 = 0x80;             /* clear TOF flag */
                overflow = overflow + 1;
        }
```

▲

8.6 Output-Compare Function

The 68HC12 has eight output-compare channels. Each output-compare channel consists of:
- a 16-bit comparator
- a 16-bit compare register TCx (also used as an input-capture register)
- an output action pin (PTx—can be pulled up to high, pulled down to low, or toggled)
- an interrupt request circuit
- a forced-compared function (CFORCx)
- control logic

8.6.1 Operation of the Output-Compare Function

One of the major applications of an output-compare function is performing an action at a specific time in the future (when the 16-bit timer counter reaches a specific value). The action might be to toggle a signal, turn on a switch, or turn off a valve. To use an output-compare function, the user:

1. makes a copy of the current contents of the TCNT register
2. adds to this copy a value equal to the desired delay
3. stores the sum into an output-compare register (TCx, x = 0..7)

The user has the option of specifying the action to be activated on the selected output-compare pin by programming the TCTL1 and TCTL2 registers. The comparator compares the value of the TCNT and that of the specified output-compare register (TCx) in every clock cycle (the clock input to the TCNT). If they are equal, the specified action on the output-compare pin is activated and the associated status bit in TFLG1 will be set to one. An interrupt request will be generated if it is enabled. The 16-bit output-compare register can be read and written any time.

8.6.2 Registers Related to the Output-Compare Function

The actions that can be activated on an output-compare pin are:

- pull-up to high
- pull-down to low
- toggle

The action of an OC pin can be selected by programming the TCTL1 and TCTL2 registers as shown in Figure 8.17. When either OMn or OLn is 1, the pin associated with OCn becomes an output tied to OCn regardless of the state of the associated DDRT bit.

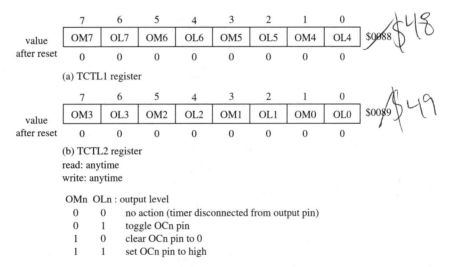

	7	6	5	4	3	2	1	0	
value	OM7	OL7	OM6	OL6	OM5	OL5	OM4	OL4	$0088
after reset	0	0	0	0	0	0	0	0	

(a) TCTL1 register

	7	6	5	4	3	2	1	0	
value	OM3	OL3	OM2	OL2	OM1	OL1	OM0	OL0	$0089
after reset	0	0	0	0	0	0	0	0	

(b) TCTL2 register
read: anytime
write: anytime

OMn	OLn	: output level
0	0	no action (timer disconnected from output pin)
0	1	toggle OCn pin
1	0	clear OCn pin to 0
1	1	set OCn pin to high

Figure 8.17 ■ Timer control register 1 and 2 (TCTL1 & TCTL2)

A successful compare will set the corresponding flag bit in the 8-bit TFLG1 register. An interrupt will be generated if it is enabled. An output-compare interrupt is enabled by setting the corresponding bit in the TMSK1 register. The same bit will enable either the input-capture or output-compare interrupt depending on which one is being selected.

8.6.3 Applications of the Output-Compare Function

An output-compare function can be programmed to perform a variety of functions. Generation of a single pulse, a square wave, and a specific delay are among the most popular applications.

Example 8.4

Generate an active high 1 KHz digital waveform with a 30 percent duty cycle from the PT0 pin. Use the polling method to check the success of the output-compare operation. The frequency of the E clock is 8 MHz.

Solution: An active high 1 KHz waveform with a 30 percent duty cycle is shown in Figure 8.18.

Figure 8.18 ■ 1 KHz 30 percent duty cycle waveform

The logic flow of this problem is illustrated in Figure 8.19. Suppose we set the prescale factor to eight so that the period of the clock input to TCNT is set to 1 μs. Then the intervals of PT0 signal to be high and low during one period is equal to 300 and 700 clock cycles, respectively.

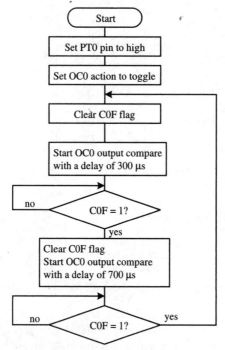

Figure 8.19 ■ The program logic flow for digital waveform generation

The following assembly program implements this algorithm using the OC0 function:

```
#include "d:\miniide\hc12.inc"
high_time    equ        300
low_time     equ        700
             org        $1000
             bset       TIOS,$01        ; select OC0 function
             movb       #$01,TCTL2      ; select toggle as the output compare action
```

```
                    bset        PORTT,$01           ; set PT0 pin to high
                    ldd         TCNT
                    addd        #high_time
                    std         TC0                 ; start an OC0 operation with 300 cycles as the delay
high                brclr       TFLG1,$01,high      ; wait until COF flag is set
                    ldd         TC0
                    addd        #low_time
                    std         TC0
low                 brclr       TFLG1,$01,low       ; wait until COF flag is set
                    ldd         TC0
                    addd        #high_time
                    std         TC0
                    bra         high
                    end
```

The C language version of the program is as follows:

```c
#include <hc12.h>
#define high_time 300
#define low_time 700
void tov_isr (void);
void main ()
{
        TIOS |= 0x01;               /* enable OC0 function */
        TSCR = 0x90;                /* enable TCNT, fast timer flags clear */
        TMSK2 = 0x03;               /* set timer prescale factor to eight */
        PORTT |= 0x01;              /* pull the PT0 pin to high */
        TCTL2 = 0x01;               /* select toggle as the action */
        TC0 = TCNT + high_time;
        while (1) {
            while(!(TFLG1 & 0x01)); /* wait for PT0 signal to go low */
            TC0 = TC0 + low_time;
            while(!(TFLG1 & 0x01)); /* wait for PT0 signal to go high */
            TC0 = TC0 + high_time;
        }
}
```

Example 8.5

Write a function to generate a time delay of 1 second. Assume that the E clock frequency is 8 MHz. Also write an instruction sequence to test this function.

Solution: There are many ways to create a 1-second time delay using the output-compare function. One method is:

- set the prescale factor to TCNT to 8 so that each clock period becomes 1 μs
- perform 20 output-compare operations with each operation creating 50 ms time delay

The corresponding assembly function is as follows:

```
delay_1s    pshx
            movb        #$90,TSCR           ; enable TCNT & fast flags clear
            movb        #$03,TMSK2          ; configure prescale factor to 8
            movb        #$01,TIOS           ; enable OC0
```

```
            ldx        #20                        ; prepare to perform 20 OC0 actions
            ldd        TCNT
again       addd       #50000                     ; start an output compare operation
            std        TC0                        ; with 50 ms time delay
            brclr      TFLG1,$01,*
            ldd        TC0
            dbne       x,again
            pulx
            rts
```

The following instruction sequence calls the previous subroutine to create a delay of 10 seconds and outputs a message at the beginning and end of 10 seconds:

```
printf      equ        $F686
CR          equ        $0D
LF          equ        $0A
            org        $1000
            ldy        #10
            ldd        #msg1
            pshy
            jsr        [printf,PCR]               ; output "Start of 10 seconds !"
            puly
loop        jsr        delay_1s
            dbne       y,loop
            pshy
            ldd        #msg2
            jsr        [printf,PCR]               ; output "End of 10 seconds!"
            puly
            swi
            ...
msg1        db         CR,LF,"Start of 10 seconds!",CR,LF,0
msg2        db         CR,LF,"End of 10 seconds! ",CR,LF,0
            end
```

The C version of the function and the test program is as follows:

```c
#include <hc12.h>
#include <stdio.h>
void delay_1s();

void main ()
{
        int i;
        for (i = 1; i < 11; i++) {
          delay_1s();
          printf("\n %d seconds passed!",i);
        }
        asm ("swi");
}
void delay_1s (void)
{
        int i;
        TSCR = 0x90; /* enable TCNT & fast timer flag clear */
        TIOS = 0x01; /* select OC0 function */
        TMSK2 = 0x03; /* set the clock prescale factor of TCNT to eight */
        TC0 = TCNT + 50000u;
```

```
                    i = 20;
                    while (i) {
                                while (!(TFLG1&0x01));
                                TC0 = TC0 + 50000u;
                                i--;
                    }
                    }
```

Example 8.6

Combining the use of input-capture and output-compare functions. Suppose a signal with unknown frequency is connected to the PT0 pin. Write a program to measure the frequency of this signal.

Solution: One method for measuring the frequency is using one of the output-compare functions to create a 1-second time interval and keep track of the total number of rising (or falling) edges arrived at the PT0 pin. The number of rising edges arrived in 1 second gives the frequency of the signal. By enabling the PT0 interrupt and writing an interrupt-service routine for the PT0 interrupt that increases the edge count, the frequency can be measured.

The assembly program that implements this algorithm is as follows:

```
#include "d:\miniide\hc12.inc"
CR              equ         $0D
LF              equ         $0A
printf          equ         $F686

                org         $800
oc_cnt          rmb         1
frequency       rmb         2
                org         $1000
; install the interrupt vector to the RAM of CME-12BC demo board
                ldd         #tc0_isr
                pshd
                ldab        #23                 ; timer overflow vector number
                clra
                ldx         $F69A
                jsr         0,x
                leas        2,sp

                movb        #$90,TSCR           ; enable TCNT and fast timer flags clear
                movb        #$00,TMSK2          ; set prescale factor to 1
                movb        #$02,TIOS           ; select OC1 and IC0
                ldaa        #200
                staa        oc_cnt              ; prepare to perform 200 OC1 operation, each
                                                ; creates 5 ms delay and total 1 second
                ldd         #0
                std         frequency           ; initialize frequency count to 0
                movb        #$01,TCTL4          ; prepare to capture PT0 rising edges
                bclr        TFLG1,$FE           ; clear the C0F flag
                bset        TMSK1,$01           ; enable IC0 interrupt
                cli                             ;            "
```

```
                    ldd         TCNT
        continue    addd        #40000
                    std         TC1                    ; start the OC1 operation with 5 ms delay
                    brclr       TFLG1,$02,*            ; wait for 5 ms
                    ldd         TC1
                    dec         oc_cnt
                    bne         continue
                    ldd         frequency
                    pshd
                    ldd         #msg
                    jsr         [printf,PCR]
                    leas        2,sp
                    swi
        msg         db          CR,LF,"The frequency is %d",CR,LF,CR,LF,0
        tc0_isr     ldd         TC0             ; clear COF flag
                    ldx         frequency       ; increment frequency count by one
                    inx                         ;         "
                    stx         frequency       ;         "
                    rti
                    end
```

The C language version of the program is as follows:

```
#include <hc12.h>
#include <stdio.h>
unsigned int frequency;
void tc0_isr(void);
void main()
{
        int i, oc_cnt;
        unsigned frequency;
/* install interrupt pseudo vector for IC0 interrupt */
        asm ("ldd #_tc0_isr");
        asm ("pshd");
        asm ("ldab #23");
        asm ("clra");
        asm ("ldx $F69A");
        asm ("jsr 0,x");
        asm ("leas 2,sp");
        TSCR = 0x90;                 /* enable TCNT and fast flag clear */
        TMSK2 = 0x00;                /* set prescale factor to 1 */
        TIOS = 0x02;                 /* select OC1 and IC0 */
            DDRT &= 0xFE;            /* configure the pin PT0 for input */
        oc_cnt = 200;                /* prepare to perform 200 OC2 operations */
        frequency = 0;
        TCTL4 = 0x01;                /* prepare to capture PT0 rising edge */
        TFLG1 = 0x01;                /* clear COF flag */
        TMSK1 |= 0x01;               /* enable IC0 interrupt */
        INTR_ON();
        TC1 = TCNT + 40000;
        while (oc_cnt) {
            while(!(TFLG1 & 0x02));
            TC1 = TC1 + 40000;
            oc_cnt = oc_cnt - 1;
```

```
        }
                printf("\nThe frequency of the signal is %d\n",frequency);
                INTR_OFF();
                asm("swi");
        }
        #pragma interrupt_handler tc0_isr
        void tc0_isr(void)
        {
                TFLG1 = 0x01; /* clear COF flag */
                frequency ++;
        }
```

8.6.4 Using OC7 to Control Multiple Output-Compare Functions

The output-compare function OC7 is special because it can control up to eight output-compare functions at the same time. The register OC7M specifies the output-compare functions to be controlled by OC7. The value that any PTx (x = 0,...,6) pin assumes when the value of TC7 equals that of TCNT is specified by the OC7D register. To control an output-compare pin using OC7, the user sets the corresponding bit in OC7M. When a successful OC7 compare is made, each affected pin assumes the value of the corresponding bit of OC7D. The contents of the OC7M and OC7D registers are shown in Figures 8.20 and 8.21.

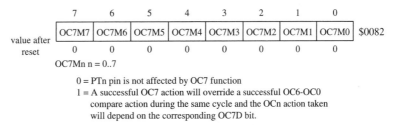

OC7Mn n = 0..7

    0 = PTn pin is not affected by OC7 function
    1 = A successful OC7 action will override a successful OC6-OC0
        compare action during the same cycle and the OCn action taken
        will depend on the corresponding OC7D bit.

Figure 8.20 ■ Output-compare 7 mask register (OC7M)

Figure 8.21 ■ Output-compare 7 data register (OC7D)

For the OC7 pin, the 68HC12 document does not specify what will happen when both the TCTL1 register and the OC7M:OC7D register pair specify the OC7 pin action on a successful OC7 compare. Avoid using this combination.

For OC0-OC6, when the OC7Mn (n = 0..6) bit is set, a successful OC7 action will override a successful OC6-OC0 compare action during the same cycle; therefore, the OCn action taken will depend on the corresponding OC7D bit. This feature allows an OC pin to be controlled by two output-compare functions simultaneously (OC7 and OCn (n = 0..6)).

Example 8.7

What value should be written into OC7M and OC7D if we want pins PT2, PT3, and PT4 to assume the values of one, zero, and one, respectively, when an OC7 compare succeeds?

Solution: Bits 4, 3, and 2 of OC7M must be set to one, and bits 4, 3, 2, of OC7D should be set to one, zero, and one, respectively. The following instruction sequence will set up these values:

```
movb  #$1C,OC7M
movb  #$14,OC7D
```

The following C statements will achieve the same goal:

```
OC7M = 0x1C;
OC7D = 0x14;
```

Example 8.8

Suppose PORTT pins are connected to LEDs as shown in Figure 8.22. Use an OC7 action to flash these LEDs as follows:

1. Light all LEDs for ¼ second and off for ¼ second—repeat this pattern four times.
2. Light one LED at a time for 1 second—from the LED controlled by the PT7 to the LED controlled by PT0.
3. Reverse the order of display at step 2.
4. Turn off all of the LEDs.

Write a C program to implement the requirement.

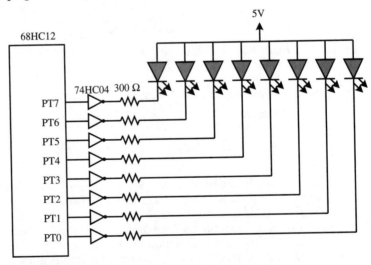

Figure 8.22 ■ An LED-flashing circuit driven by port T

Solution: This display pattern can be handled by the OC7 function beautifully. We will use the timer counter reset feature, which allows the timer counter to be reset by a successful output-compare 7 event. The C program is as follows:

```
#include <hc12.h>
unsigned char flash_tab[18] = {0x80,0x40,0x20,0x10,0x08,0x04,0x02,0x01,0x00,
                  0x01,0x02,0x04,0x08,0x10,0x20,0x40,0x80,0x00};

void main ()
{
    int i, j;

    TCTL1 = 0x00;
    TCTL2 = 0x00;
    TIOS = 0xFF; /* select OC7..OC0 */
/* set prescale factor to 32, enable timer reset by a successful
   OC7 event, disable TCNT overflow interrupt  */
            TMSK2 = 0x2D;
            OC7M = 0xFF;
            OC7D = 0xFF;
            TSCR = 0x90;         /* enable TCNT & fast flag clear */
            TC7 = TCNT + 4;
            while (!(TFLG1 & 0x80)); /* turn on all LEDs in 16 microseconds */
/* turn on all LEDs for 1/4 seconds and off for 1/4 seconds, repeat four times */
        for (i = 0; i < 4; i++) {
            TC7 = 62500u;       /* start the next OC7 action with 250 microseconds delay */
            OC7D = 0x00;         /* prepare to turn off LEDs */
            printf("\nTurn on LEDs 1/4 seconds! \n");
            while (!(TFLG1 & 0x80));
            TC7 = 62500u;
            OC7D = 0xFF;         /* turn on LEDs for 1/4 seconds */
            printf("\nTurn off LEDs 1/4 seconds!\n");
            while (!(TFLG1 & 0x80));
        }
        OC7D = 0x00;
        TC7 = 0x04;
/* turn off all LEDs for 1 second */
        printf("\n Turn off all LEDs for 1 second\n");
        for (i = 0; i < 5; i++) {
            while(!(TFLG1 & 0x80));
            TC7 = 50000u;        /* start the next OC7 operation with 200 ms delay */
        }
/* display LEDs according to the pattern lookup from the table */
        for (i = 0; i < 18; i++) {
            OC7D = flash_tab[i];
            switch (flash_tab[i]) {
                    case 0x80: printf("LED 7 is turned on, others off, 1 second. \n");
                            break;
                    case 0x40: printf("LED 6 is turned on, others off, 1 second. \n");
                            break;
                    case 0x20: printf("LED 5 is turned on, others off, 1 second. \n");
                            break;
                    case 0x10: printf("LED 4 is turned on, others off, 1 second. \n");
                            break;
                    case 0x08: printf("LED 3 is turned on, others off, 1 second. \n");
                            break;
```

```
                    case 0x04: printf("LED 2 is turned on, others off, 1 second. \n");
                                break;
                    case 0x02: printf("LED 1 is turned on, others off, 1 second. \n");
                                break;
                    case 0x01: printf("LED 0 is turned on, others off, 1 second. \n");
                                break;
                    default:    printf("All LEDs are turned off, 1 second. \n");
                                break;
            }
            for (j = 0; j < 5; j++) {
                    while (!(TFLG1 & 0x80));
                    TC7 = 50000u;
            }
        }
        asm("swi");
}
```

Two output-compare functions can control the same pin simultaneously. Thus the OC7 can be used in conjunction with one or more other output-compare functions to achieve even more timer flexibility. We can generate a digital waveform with a given duty cycle by using the OC7 and any other output-compare functions.

▲

Example 8.9

▼

Use OC7 and OC0 together to generate a 2 KHz digital waveform with a 40 percent duty cycle on the PT0 pin. Assume the E clock frequency is 8 MHz.

Solution: We will set the prescale factor to four so that the frequency of the clock input to the TCNT counter is 2 MHz. For this clock frequency, the period of the waveform (2 KHz) to be generated corresponds to 1,000 cycles of this clock input. The high interval in one period of the waveform is 400 clock cycles whereas the low interval is 600 clock cycles.

The idea behind using two OC functions in generating the waveform is as follows:

- Use OC0 (or OC7) to pull the PT0 pin to high every 1,000 clock cycles.
- Use OC7 (or OC0) to pull the PT0 to low after it is high for 400 clock cycles.

We will use an interrupt-driven approach to implement this algorithm. After starting the first action on OC7 and OC0, the program will stay in a wait loop to wait for interrupts to be requested from OC7 and OC0. The interrupt service routine will simply start the next output-compare operation with 1,000 as the delay. The assembly program that implements this idea is as follows:

```
#include "d:\miniide\hc12.inc"

high_cnt        equ     400
setuservector   equ     $F69A
                org     $1000
                lds     #$8000
; set up OC7 interrupt pseudo vector under D-Bug12 monitor
                ldd     #oc7_isr
                pshd
                ldab    #16
                clra
```

```
                    ldx      $F69A
                    jsr      0,x
                    leas     2,sp
; set up OC0 interrupt pseudo vector under D-Bug12 monitor
                    ldd      #oc0_isr
                    pshd
                    ldab     #23
                    clra
                    ldx      setuservector
                    jsr      0,x
                    leas     2,sp

                    movb     #$90,TSCR       ; enable TCNT and fast flags clear
                    movb     #$81,TIOS       ; select OC7 & OC0
                    movb     #$01,OC7M       ; allow OC7 to control OC0 pin
                    movb     #$01,OC7D       ; OC7 action on PT0 pin is to pull high
                    movb     #$22,TMSK2      ; enable pullup resistor, set prescale factor to 4
                    movb     #$02,TCTL2      ; select pull low as the OC0 action
                    movb     #$81,TMSK1      ; enable OC7 and OC0 to interrupt
                    ldd      TCNT
                    addd     #1000
                    std      TC7             ; start an OC7 operation
                    addd     #high_cnt
                    std      TC0             ; start an OC0 operation
                    cli                      ; enable interrupt
loop                bra      loop            ; infinite loop to wait for interrupt
                    swi

oc7_isr             ldd      TC7             ; start the next OC7 action with 1000
                    addd     #1000           ; clock cycles delay, also clear the C7F
                    std      TC7             ; flag
                    rti
oc0_isr             ldd      TC0             ; start the next OC0 action with 1000
                    addd     #1000           ; clock cycles delay, also clear the C0F
                    std      TC0             ; flag
                    rti
                    end
```

The C language version of this program is as follows:

```c
#include <hc12.h>
#define high_cnt 400

void oc7_isr (void);
void oc0_isr (void);

void main ()
{
/* set up OC7 interrupt pseudo vector for D-Bug12 monitor */
        asm ("ldd #_oc7_isr");
        asm ("pshd");
        asm ("ldab #16");
        asm ("clra");
        asm ("ldx $F69A");
```

```
                    asm ("jsr 0,x");
                    asm ("leas 2,sp");
      /* set up OC0 interrupt pseudo vector for D-Bug12 monitor */
                    asm ("ldd #_oc0_isr");
                    asm ("pshd");
                    asm ("ldab #23");
                    asm ("clra");
                    asm ("ldx $F69A");
                    asm ("jsr 0,x");
                    asm ("leas 2,sp");

                    TSCR = 0x90;            /* enable TCNT and fast flag clear */
                    TIOS = 0x81;           /* select OC7 and OC0 */
                    OC7M = 0x01;           /* allow OC7 to control PT0 pin */
                    OC7D = 0x01;           /* on a success OC7 action, the PT0 will be pulled high */
                    TMSK2 = 0x22;          /* enable pullup resistor, set prescale factor to 4 */
                    TCTL2 = 0x02;          /* set OC0 action to pull low */
                    TMSK1 = 0x81;          /* enable OC7 and OC0 interrupt */
                    TC7 = TCNT + 1000;     /* start an OC7 action with delay equal to 1000 */
                    TC0 = TC7 + high_cnt;
                    INTR_ON();
                    while (1);             /* infinite loop, wait for interrupt */
                    asm ("swi");
      }

      #pragma interrupt_handler oc7_isr oc0_isr
      void oc7_isr (void)
      {
                    TC7 = TC7 + 1000;
      }
      void oc0_isr (void)
      {
                    TC0 = TC0 + 1000;
      }
```

8.6.5 Forced Output-Compare

There may be applications in which the user requires an output-compare to occur immediately instead of waiting for a match between the TCNT and the proper output-compare register. This situation arises in the spark-timing control in some automotive engine control applications. To use the forced output-compare mechanism, the user would write to the CFORC register with ones in the bit positions corresponding to the output-compare channels to be forced. At the next timer count after the write to CFORC, the forced channels will trigger their programmed pin actions to occur.

The forced actions are synchronized to the timer counter clock input. The forced output-compare signal causes pin action but does not affect the CnF flag or generate interrupt. Normally, the force mechanism would not be used in conjunction with the automatic pin action that toggles the corresponding output-compare pin. The contents of CFORC are shown in Figure 8.23. CFORC always reads as all zeros.

	7	6	5	4	3	2	1	0	
	FOC7	FOC6	FOC5	FOC4	FOC3	FOC2	FOC1	FOC0	$0081
value after reset	0	0	0	0	0	0	0	0	

Figure 8.23 ■ Contents of the CFORC register

Example 8.10

▼

Suppose the contents of the TCTL1 and TCTL2 registers are $D6 and $6E, respectively. The contents of the TFLG1 are $00. What would occur on pins PT7 to PT0 on the next clock cycle if the value $7F is written into the CFORC register?

Solution: The TCTL1 and TCTL2 configure the output-compare actions as shown in Table 8.2. Since the contents of the TFLG1 are $0, none of the started output-compare operations have succeeded yet.

Register	Bit positions	Value	Action to be triggered
TCTL1	7 6	1 1	set the PT7 pin to high
	5 4	0 1	toggle the PT6 pin
	3 2	0 1	toggle the PT5 pin
	1 0	1 0	pull the PT4 pin to low
TCTL2	7 6	0 1	toggle the PT3 pin
	5 4	1 0	pull the PT2 pin to low
	3 2	1 1	set the PT1 pin to high
	1 0	1 0	pull the PT0 pin to low

Table 8.2 ■ Pin actions on PT7-PT0 pins

Because the CFORC register specifies that the output-compare channels 6 to 0 are to be forced immediately, the actions specified in the fourth column in Table 8.2 will occur immediately.

▲

8.7 Real-Time Interrupt (RTI)

When enabled, the real-time interrupt (RTI) function of the 68HC12 can be used to generate periodic interrupts to the CPU. The RTI function has been used to remind the CPU to perform routine tasks in many applications. These tasks cannot be delayed and must be performed regularly. Otherwise, the embedded system may not function properly.

The operation of the RTI function is configured by the *real-time interrupt control register* (RTICTL). The contents of this register are shown in Figure 8.24.

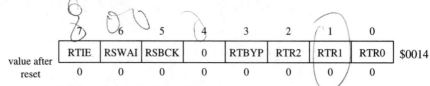

	7	6	5	4	3	2	1	0	
	RTIE	RSWAI	RSBCK	0	RTBYP	RTR2	RTR1	RTR0	$0014
value after reset	0	0	0	0	0	0	0	0	

RTIE: real-time interrupt enable bit
    0 = disable RTI interrupt
    1 = enable RTI interrupt
RSWAI: RTI and COP stop while in wait bit
    0 = allows the RTI and COP to continue running in wait.
    1 = disable both the RTI and COP when the part goes into wait.
RSBCK: RTI and COP stop while in background debug mode bit
    0 = allows the RTI and COP to continue running while in
        background mode.
    1 = disables RTI and COP when the device is in background
        debug mode (useful for emulation).
RTBYP: real-time interrupt divider chain bypass bit
    0 = divider chain functions normally
    1 = divider chain is bypassed, allows faster testing. The divider
        chain is normally P divided by 213, when bypass becomes P
        divided by 4.
RTR2..RTR0: real-time interrupt rate select bits. The rate selection is
        shown in Table 8.3.

Figure 8.24 ■ Real-time interrupt control register (RTICTL)

RTR2	RTR1	RTR0	Divide E by	Timeout period E = 4.0 MHz	Timeout period E = 8.0 MHz
0	0	0	off	off	off
0	0	1	213	2.048 ms	1.024 ms
0	1	0	214	4.096 ms	2.048 ms
0	1	1	215	8.192 ms	4.096 ms
1	0	0	216	16.384 ms	8.192 ms
1	0	1	217	32.768 ms	16.384 ms
1	1	0	218	65.536 ms	32.768 ms
1	1	1	219	131.72 ms	65.536 ms

Table 8.3 ■ Real-time interrupt rates

8.8 Pulse Accumulator

The 68HC12 standard timer system has a 16-bit pulse accumulator (PACNT) that has two operation modes:

1. *Event counting mode.* In this mode, the PACNT counter increments on the active edge of the PT7 pin. Many microcontroller applications are involved in counting things. These things are called *events*, but in real applications they might be anything: pieces on an assembly line, cycles of an incoming signal, or units of time. To be counted by the accumulator, these things must be translated into rising or falling edges on the PAI pin (PT7 pin). A trivial example of an event might be to count pieces on the assembly line: a light emitter/detector pair could be placed across a conveyor so that as each piece passes the sensor, the light beam is interrupted and a logic-level signal that can be connected to the PAI pin is produced.

2. *Gated time-accumulation mode.* The 16-bit PACNT counter is clocked by a free-running E÷64 clock signal, subject to the PAI signal being active. One common use of this mode is to measure the duration of a single pulse. The counter is set to zero before the pulse starts, and the resultant pulse time is read directly when the pulse is finished.

The PAI pin must be configured for input and the PA function must be enabled before the pulse accumulator can function. Interrupts are often used in pulse accumulator applications. There are two interrupt sources:

1. PAI-edge interrupt
2. PACNT overflow

8.8.1 Registers Related to the Pulse Accumulator

The following three registers are related to the operation of the pulse accumulator:

- Pulse accumulator control register (PACTL)
- Pulse accumulator flag register (PAFLG)
- Pulse accumulator count register (PACNT)

The PACTL register is used to set up all of the pulse accumulator parameters. Its contents are shown in Figure 8.25.

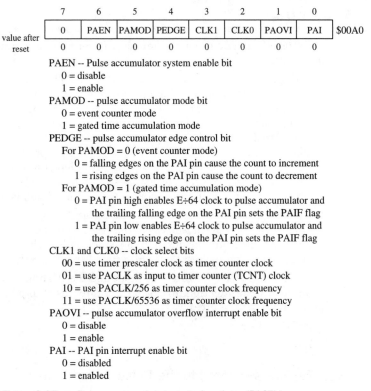

7	6	5	4	3	2	1	0	
0	PAEN	PAMOD	PEDGE	CLK1	CLK0	PAOVI	PAI	$00A0
0	0	0	0	0	0	0	0	

value after reset

PAEN -- Pulse accumulator system enable bit
   0 = disable
   1 = enable
PAMOD -- pulse accumulator mode bit
   0 = event counter mode
   1 = gated time accumulation mode
PEDGE -- pulse accumulator edge control bit
   For PAMOD = 0 (event counter mode)
      0 = falling edges on the PAI pin cause the count to increment
      1 = rising edges on the PAI pin cause the count to decrement
   For PAMOD = 1 (gated time accumulation mode)
      0 = PAI pin high enables E÷64 clock to pulse accumulator and
         the trailing falling edge on the PAI pin sets the PAIF flag
      1 = PAI pin low enables E÷64 clock to pulse accumulator and
         the trailing rising edge on the PAI pin sets the PAIF flag
CLK1 and CLK0 -- clock select bits
   00 = use timer prescaler clock as timer counter clock
   01 = use PACLK as input to timer counter (TCNT) clock
   10 = use PACLK/256 as timer counter clock frequency
   11 = use PACLK/65536 as timer counter clock frequency
PAOVI -- pulse accumulator overflow interrupt enable bit
   0 = disable
   1 = enable
PAI -- PAI pin interrupt enable bit
   0 = disabled
   1 = enabled

Figure 8.25 ■ Pulse accumulator control register (PACTL)

The PACTL register also controls the clock source for the timer counter (TCNT). When CLK1 and CLK0 bits are not 00, the PACLK signal (from the PAI pin) is prescaled by 1, 256, or 65,536, and used as the clock input to the timer counter.

The bit 1 and 0 of the PAFLG register keep track of the status of the operation of the pulse accumulator system, as shown in Figure 8.26.

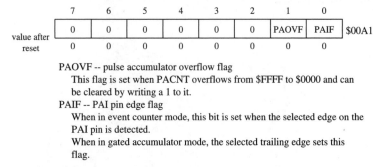

7	6	5	4	3	2	1	0	
0	0	0	0	0	0	PAOVF	PAIF	$00A1
0	0	0	0	0	0	0	0	

value after reset

PAOVF -- pulse accumulator overflow flag
   This flag is set when PACNT overflows from $FFFF to $0000 and can
   be cleared by writing a 1 to it.
PAIF -- PAI pin edge flag
   When in event counter mode, this bit is set when the selected edge on the
   PAI pin is detected.
   When in gated accumulator mode, the selected trailing edge sets this
   flag.

Figure 8.26 ■ Pulse accumulator flag register (PAFLG)

8.8.2 Pulse Accumulator Applications

The pulse accumulator has a few interesting applications, such as interrupting after N events, frequency measurement, and pulse duration measurement.

Example 8.11

▼

Suppose external events are converted into pulses and connected to the PAI pin. Write a program so that the pulse accumulator generates an interrupt to the 68HC12 when N events have occurred. Assume N is less than 65,536.

Solution: By writing the two's complement of N into PACNT, it will overflow after N events and generate an interrupt. The assembly program is as follows:

```
#include  "d:\miniide\hc12.inc"
N                equ        1350
setuservector    equ        $F69A
                 org        $1000
; set up PACNT overflow interrupt pseudo vector for D-Bug12 monitor
                 ldd        #paov_isr
                 pshd
                 ldab       #14
                 clra
                 ldx        setuservector
                 jsr        0,x
                 leas       2,sp
; the next five instructions place the two's complement in PACNT
                 ldd        #N
                 coma
                 comb
                 addd       #1
                 std        PACNT
; configure PA function: enable PA, select event counting mode, rising edge
; of PAI signal increments the PACNT counter
                 movb       #$52,PACTL
                 cli                         ; enable PAOV interrupt
                 ...
                 swi
paov_isr         movb       #$02,PAFLG       ; clear the PAOVF flag
                 end
```

The C language version of the program is as follows:

```
#include <hc12.h>
#define N  1252
void paov_isr (void);
void main ()
{
// setup PAOV interrupt pseudo vector
        asm ("ldd  #_paov_isr");
        asm ("pshd");
        asm ("ldab #14");
        asm ("clra");
        asm ("ldx $F69A");
        asm ("jsr 0,x");
        asm ("leas 2,sp");
        PACNT = ~N + 1;
        PACTL = 0x52;
        INTR_ON();
        ....
}
```

```
#pragma interrupt_handler paov_isr
void paov_isr (void)
{
        PAFLG = 0x02;
}
```

We use the input-capture function to measure the frequency of a signal in Example 8.6. The drawback in that approach is that the interrupt overhead of every input signal edge sets the upper limit of the frequency that can be measured. Using the pulse accumulator system to measure frequency dramatically reduces the incurred interrupt overhead.

The procedure for measuring the frequency of a signal using the pulse accumulator system is as follows:

Step 1
Set up the pulse accumulator system to operate in event counting mode.

Step 2
Connect the unknown signal to the PAI (PT7) pin.

Step 3
Use one of the output-compare functions to create a 1-second time interval.

Step 4
Use a memory location to keep track of the number of pulse accumulator counter overflow interrupts.

Step 5
Enable the PAOV interrupt.

Step 6
Disable the PAOV interrupt at the end of 1 second.

▲

Example 8.12

▼

Write a program to measure the frequency of a signal connected to the PAI pin.

Solution: We will use the OC0 function to create a 1-second delay. We will perform 200 OC0 operations with each OC0 operation creating 5 ms delay. The service routine for the PACNT overflow interrupt will increate the overflow count by one. Let *paov_cnt* represent the PACNT overflow count. At the end of 1 second, the frequency is equal to the following expression:

$$\text{frequency} = \text{paov\_cnt} \times 2^{16} + \text{PACNT}$$

The assembly program that implements this procedure is as follows:

```
#include  "d:\miniide\hc12.inc"
setuservector equ        $F69A
              org        $800
oc_cnt     rmb      1
paov_cnt   rmb      2                     ; use to keep track PACNT overflow count
frequency  rmb      4                     ; hold the signal frequency


              org        $1000
; set up PACNT overflow interrupt pseudo vector for D-Bug12 monitor
              ldd        #paov_isr
              pshd
```

```
                ldab        #14
                clra
                ldx         setuservector
                jsr         0,x
                leas        2,sp

                ldaa        #200
                staa        oc_cnt          ; prepare to perform OC0 actions
                ldd         #0
                std         PACNT           ; let PACNT count up from 0
                std         paov_cnt        ; initialize PACNT overflow count to 0
                std         frequency       ; initialize frequency to 0
                std         frequency+2     ;       "
                movb        #$90,TSCR       ; enable TCNT and fast flag clear
                bset        TIOS,$01        ; select OC0 function
                movb        #$20,TMSK2      ; enable pullup resistor & set prescale factor to 1
                movb        #$00,DDRT       ; configure all timer pins for input
; configure PA function: enable PA, select event counting mode, rising edge
; of PAI signal increments the PACNT counter, enable PAOV interrupt
                movb        #$52,PACTL
                cli                         ; enable PAOV interrupt
                ldd         TCNT
sec_loop        addd        #50000
                std         TC0
                brclr       TFLG1,$01,*
                ldd         TC0
                dec         oc_cnt
                bne         sec_loop
                movb        #0,PACTL        ; disable PA function
                sei                         ; disable interrupt
                ldd         PACNT
                std         frequency+2
                ldd         paov_cnt
                std         frequency
                swi
paov_isr        movb        #$02,PAFLG      ; clear the PAOVF flag
                ldx         paov_cnt        ; increment PACNT overflow
                inx                         ; count by 1
                stx         paov_cnt        ;       "
                end
```

The C language version of this program is as follows:

```
#include <hc12.h>
unsigned long int frequency;
unsigned int paov_cnt;
void paov_isr (void);
void main ()
{
        int oc_cnt;
// setup PAOV interrupt pseudo vector
        asm ("ldd #_paov_isr");
        asm ("pshd");
        asm ("ldab #14");
```

```
                    asm ("clra");
                    asm ("ldx $F69A");
                    asm ("jsr 0,x");
                    asm ("leas 2,sp");

                    PACNT = 0;
                    frequency = 0;
                    paov_cnt = 0;
                    TSCR = 0x90;                    /* enable TCNT and fast flag clear */
                    TIOS = 0x01;                    /* select OC0 function */
                    TMSK2 = 0x20;                   /* set prescale factor to one */
                    PACTL = 0x52;                   /* enable PA function, enable PAOV interrupt */
                    DDRT = 0x00;                    /* configure all port T pins for input */
                    INTR_ON();
                    oc_cnt = 200;
                    TC0 = TCNT + 50000u;
                    while (oc_cnt) {
                            while(!(TFLG1 & 0x01));
                            TC0 = TC0 + 50000u;
                            oc_cnt --;
                    }
                    PACTL = 0x00;                   /* disable PA function */
                    INTR_OFF();
                    frequency = paov_cnt * 65536u + PACNT;
                    asm("swi");
            }
#pragma interrupt_handler paov_isr
void paov_isr (void)
{
                    PAFLG = 0x02;                   /* clear PAOVF flag */
                    paov_cnt = paov_cnt + 1;
}
```

The pulse accumulator system can be set up to measure the duration of a pulse using the gated time-accumulation mode. When the active level is applied to the PAI pin, the PACNT can count and will stop counting on the trailing edge of the PAI signal. The clock input to the pulse accumulator is E ÷ 64. The procedure for measuring the duration of a pulse is as follows:

Step 1
Set up the pulse accumulator system to operate in the gated time-accumulation mode, and initialize PACNT to zero.

Step 2
Select the falling edge as the active edge (for measuring positive pulse). In this setting, the pulse accumulator counter will increment when the signal connected to the PAI pin is high and generate an interrupt to the 68HC12 on the falling edge.

Step 3
Enable the PAI active edge interrupt and wait for the arrival of the active edge of PAI.

Step 4
Stop the pulse accumulator counter when the interrupt arrives.

Without keeping track of the PACNT overflows, the longest pulse width (E = 8 MHz) that can be measured is:

$$\text{pulse\_width} = 2^{16} \times 64 \, T_E = 2^{16} \times 64 \times 1/8 \, \mu s = 524.288 \, ms$$

To measure longer pulse width, we will need to keep track of the number of times that the PACNT counter overflows in the duration of the pulse. Let *paov_cnt* be the overflow count of the PACNT counter, then:

pulse_width = $[(2^{16} \times$ paov_cnt$) +$ PACNT$] \times 64T_E$

▲

Example 8.13

▼

Write a program to measure the duration of an unknown signal connected to the PAI pin.

Solution: The assembly program that implements the previous algorithm is as follows:

```
#include   "d:\miniide\hc12.inc"
setuservector   equ       $F69A
                org       $800
paov_cnt        rmb       1                   ; use to keep track of the PACNT overflow count
pulse_width     rmb       3                   ; hold the signal frequency

                org       $1000
; set up PACNT overflow interrupt pseudo vector for D-Bug12 monitor
                ldd       #paov_isr
                pshd
                ldab      #14
                clra
                ldx       setuservector
                jsr       0,x
                leas      2,sp

                ldd       #0
                std       PACNT               ; let PACNT count up from zero
                clr       paov_cnt            ; initialize PACNT overflow count to zero
                movb      #$20,TMSK2          ; enable timer port pullup resistor
; configure PA function: enable PA, select gated time accumulator mode, high level
; of PAI signal enables PACNT counter, enable PAOV interrupt
                movb      #$72,PACTL
                bclr      DDRT,$80            ; configure PAI pin for input
                cli                           ; enable PAOV interrupt
                brclr     PAFLG,$01,*         ; wait for the arrival of the falling edge of PAI
                movb      #0,PACTL            ; disable PA function
                sei                           ; disable interrupt
                ldd       PACNT
                std       pulse_width+1
                ldaa      paov_cnt
                staa      pulse_width
                swi
paov_isr        movb      #$02,PAFLG          ; clear the PAOVF flag
                inc       paov_cnt            ; increment PACNT overflow count by one
                end
```

The C language version of the program is as follows:

```
#include       <hc12.h>
unsigned int paov_cnt;
```

```
long unsigned int pulse_width;
void paov_isr (void);

void main ( )
{
              asm ("ldd #_paov_isr");
              asm ("pshd");
              asm ("ldab #14");
              asm ("clra");
              asm ("ldx $F69A");
              asm ("jsr 0,x");
              asm ("leas 2,sp");

              PACNT = 0;
              paov_cnt = 0;
              pulse_width = 0;
              TMSK2 = 0x20;              /* enable timer port pull-up resistor */
              DDRT = 0x00;               /* configure all timer port pins for input */
/* configure PA function: enable PA, select gated time accumulator mode,  high-level
    of PAI enables PACNT to count, enable PAOV interrupt */
              PACTL = 0x72;
              INTR_ON();
              while(!(PAFLG & 0x01));    /* wait for the arrival of the PAI falling edge */
              PACTL = 0x00;             /* disable PA system */
              INTR_OFF();
              pulse_width = paov_cnt * 65536u + PACNT;
              asm ("swi");
}
#pragma interrupt_handler paov_isr
void paov_isr (void)
{
              PAFLG = 0x20;             /* clear PAOVF flag */
              paov_cnt ++;
}
```

8.9 Pulse-width-modulation (PWM) Function

One of the major applications of the output-compare function is to generate digital wave-forms. As illustrated in Section 8.6.3, the CPU needs to attend to the output-compare channel frequently. When the polling method is used as shown in examples in Section 8.6.3, the CPU is tied up from performing other tasks. Although this problem can be resolved by using the inter-rupt-driven approach, the overhead incurred in servicing interrupts cannot be ignored. The addition of a *PWM circuit* allows the user to generate the waveform while at the same time avoid the overhead of using the output-compare function.

The 68HC12 pulse-width modulation subsystem provides four independent 8-bit PWM wave-forms or two 16-bit PWM waveforms or a combination of one 16-bit and two 8-bit PWM wave-forms. Each channel has a programmable period and a programmable duty cycle as well as a dedicated counter. The PWM outputs can be programmed as left-aligned or center-aligned as illus-trated in Figures 8.27 and 8.28. The duty cycle can range from 0 to 100 percent. The clock source of the counters used in PWM is the E clock prescaled (divided) by a factor as shown in Figure 8.29.

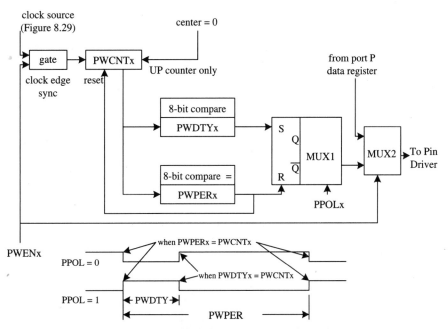

Figure 8.27 ■ Block diagram of PWM left-aligned output channel

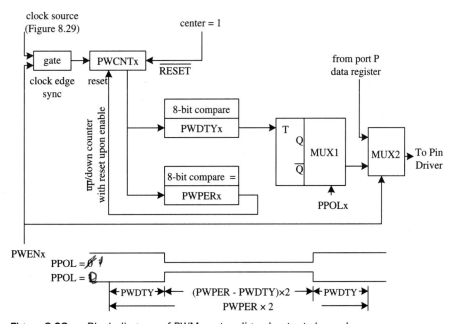

Figure 8.28 ■ Block diagram of PWM center-aligned output channel

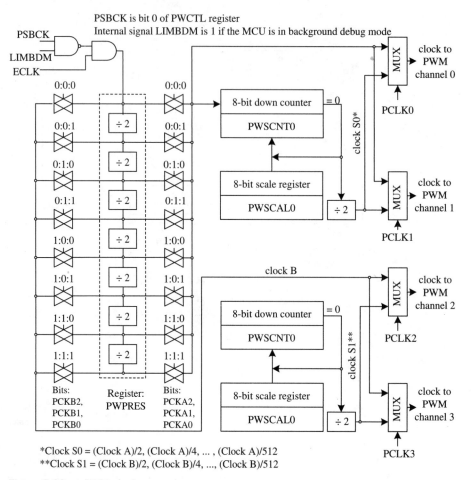

Figure 8.29 ■ PWM clock sources

The period and duty registers are double buffered so that if they change while the channel is enabled, the change does not take effect until the counter rolls over or the channel is disabled. If the channel is not enabled, then writes to the period and/or duty register go directly to the latches as well as the buffer, thus ensuring that the PWM output is always either the old waveform or the new waveform, not some variation in between.

A change in duty cycle or period can be forced into immediate effect by writing the new value to the duty and/or period registers and then writing to the counter. This causes the counter to reset and the new duty cycle and/or period values to be latched. In addition, since the counter is readable it is possible to know where the count is with respect to the duty value and software can be used to make adjustment by turning the enable bit off and on.

The four PWM channel outputs share general-purpose port P pins. Enabling PWM pins takes precedence over general-purpose ports. When PWM outputs are not in use, the port P pins may be used for discrete input/output.

8.9.1 Operation of the PWM

In terms of the combinations of polarity and alignment, there are four types of PWM outputs. We will discuss two of them. The other two modes are similar and hence will not be discussed.

PPOL = 1 and Left Alignment:

As shown in Figure 8.27, the multiplexor MUX1 outputs the value of !Q when PPOLx = 1 and MUX2 outputs the value from MUX1 when PWENx = 1.

The block labeled PWCNTx is an 8-bit counter that counts up. The block labeled PWDTYx is the register that holds the duty value, whereas the block labeled PWPERx is the register that holds the period value.

When the contents of the PWCNTx counter and PWPERx are equal, the R input to the SR latch is one. As a result, both MUX1 and MUX2 output a one. Therefore, the waveform output is high. The PWCNTx counter is reset to zero.

Another comparator compares the contents of PWCNTx and PWDTYx. When they are equal, the S signal is set to one and the value of Q is set to one. At this point, both the MUX1 and MUX2 output low.

Therefore, from the moment that PWCNTx is reset (when PWCNTx equals PWPERx) until the moment that PWCNTx equals PWDTYx is the interval that the waveform is high. The waveform output will be low for the rest of the period.

PPOL = 1 and Center Alignment

As shown in Figure 8.28, the multiplexor MUX1 outputs the value of Q when PPOLx = 1 and MUX2 outputs the value from MUX1 when PWENx = 1. The signal Q is low out of reset. After the user has written values into the PWDTYx and PWPERx and enabled the PWM function, the PWCNTx counter counts up. When the count value equals the duty value in the PWDTYx register, the output of the duty comparator goes high and asserts the signal T, which toggles the value Q. Therefore the PWM output goes high. The PWCNTx continues to count up until it equals the value of the PWPERx register. At that point PWCNTx starts to count down (the up/down input to the PWCNTx counter goes high). When the value of the PWCNTx counter equals that of PWDTYx again, it toggles the T flip-flop and hence the PWM output goes low. The same process repeats.

The duration of PWM output high is equal to twice the difference of the value in PWPERx and that in PWDTYx.

8.9.2 Registers Related to PWM

Clock signals play the center role in the function of PWM. The clock signals to the PWM channels are derived from the E clock, as shown in Figure 8.29.

In Figure 8.29, the dotted rectangle is the PWM prescale counter PWPRES. This counter is a free-running counter that divides down the E clock by the power of two.

Each 8-bit PWM channel has an 8-bit down counter PWSCNTn (n = 0,..,3) and an 8-bit scale register PWSCALn. The scale-down counter reloads the value in PWSCALn when it counts down to $00. A write to the PWSCALn register also causes the scaler counter PWSCNTn to be reloaded from PWSCALn. When PWSCALx = $FF, clock A (B) is divided by 256, and then divided by 2 to generate clock S0 (S1).

The PWM channels 0 and 1 and channels 2 and 3 can be concatenated into a single 16-bit PWM channel. The user can specify whether the concatenation is desirable by programming the PWCLK register. This register also specifies the prescale factor to the PWM channels. The contents of the PWCLK register are shown in Figure 8.30.

	7	6	5	4	3	2	1	0	
value	CON23	CON01	PCKA2	PCKA1	PCKA0	PCKB2	PCKB1	PCKB0	$0040
after reset	0	0	0	0	0	0	0	0	

CON23 -- Concatenate PWM channels 2 and 3 bit
  When concatenated, channel 2 becomes the high-order byte and
  channel 3 becomes the low-order byte. Channel 2 output pin is used as
  the output for this 16-bit PWM. Channel 3 clock-select control bits
  determine the clock source.
  0 = Channels 2 and 3 are separate 8-bit PWMs.
  1 = Channels 2 and 3 are concatenated into a 16-bit PWM channel.
CON01 -- Concatenate PWM channels 0 and 1 bit
  When concatenated, channel 0 becomes the high-order byte and
  channel 1 becomes the low-order byte. Channel 0 output pin is used as
  the output for this 16-bit PWM. Channel 1 clock-select control bits
  determine the clock source.
  0 = Channels 0 and 1 are separate 8-bit PWMs.
  1 = Channels 0 and 1 are concatenated into a 16-bit PWM channel.
PCKA2--PCKA0 -- Prescaler for clock A
  Clock A is one of two clock sources which may be used for channels 0
  and 1. These 3 bits determine the rate of clock A as shown in Table 8.4.
PCKB2--PCKB0 -- Prescaler for clock B
  Clock B is one of two clock sources which may be used for channels 2
  and 3. These 3 bits determine the rate of clock B as shown in Table 8.4.

Figure 8.30 ■ PWM clocks and concatenate register (PWCLK)

PCKA2 (PCKB2)	PCKA1 (PCKB1)	PCKA0 (PCKB0)	Value of Clock A (B)
0	0	0	E
0	0	1	E ÷ 2
0	1	0	E ÷ 4
0	1	1	E ÷ 8
1	0	0	E ÷ 16
1	0	1	E ÷ 32
1	1	0	E ÷ 64
1	1	1	E ÷ 128

Table 8.4 ■ Clock A and Clock B prescaler

Each PWM channel can choose from one of two clock sources as its clock signal. As shown in Figure 8.31, the PWPOL register allows the user to choose the clock source and the polarity of the PWM output.

Depending on the polarity bit, the duty registers may contain the count of either the high time or the low time. If the polarity bit is zero and left alignment is selected, the duty cycle registers contain a count of the low time. If the polarity bit is one, the duty registers contain a count of the high time.

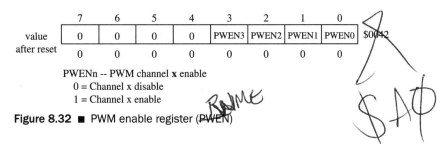

	7	6	5	4	3	2	1	0
value	PCLK3	PCLK2	PCLK1	PCLK0	PPOL3	PPOL2	PPOL1	PPOL0
after reset	0	0	0	0	0	0	0	0

PCLK3 -- PWM channel 3 clock select bit
  0 = Clock B is the clock source for channel 3
  1 = Clock S1 is the clock source for channel 3
PCLK2 -- PWM channel 2 clock select bit
  0 = Clock B is the clock source for channel 2
  1 = Clock S1 is the clock source for channel 2
PCLK1 -- PWM channel 1 clock select bit
  0 = Clock A is the clock source for channel 1
  1 = Clock S0 is the clock source for channel 1
PCLK0 -- PWM channel 0 clock select bit
  0 = Clock A is the clock source for channel 0
  1 = Clock S0 is the clock source for channel 0
If a clock select is changed while a PWM signal is being generated, a
truncated or stretched pulse may occur during the transition.
PPOL3..PPOL0 -- PWM channel 3..0 polarity
  0 = channel 3..0 output is low at the beginning of the period, high
    when the duty count is reached.
  1 = channel 3..0 output is high at the beginning of the period, low
    when the duty count is reached.

Figure 8.31 ■ PWM Clock select and polarity register (PWPOL)

A PWM channel must be enabled before it can be used to generate a digital waveform. As shown in Figures 8.27 and 8.28, the PWEN register enables or disables all of the PWM channels. Setting any of the PWENx bits causes the associated port P line to become an output, regardless of the state of the associated data direction register (DDRP) bit. On the front end of the PWM channel, the scaler clock is enabled to the PWM circuit by the PWENx enable bit being high.

	7	6	5	4	3	2	1	0	
value	0	0	0	0	PWEN3	PWEN2	PWEN1	PWEN0	$0042
after reset	0	0	0	0	0	0	0	0	

PWENn -- PWM channel **x** enable
  0 = Channel x disable
  1 = Channel x enable

Figure 8.32 ■ PWM enable register (PWEN)

There is an edge-synchronizing gate circuit to guarantee that the clock is only enabled or disabled at an edge. The pulse-modulated signal will be available at the port pin when its clock source begins its next cycle.

Some miscellaneous parameters are controlled by the *PWM control register* (PWCTL). The contents of PWCTL are shown in Figure 8.33.

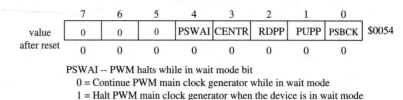

	7	6	5	4	3	2	1	0	
value	0	0	0	PSWAI	CENTR	RDPP	PUPP	PSBCK	$0054
after reset	0	0	0	0	0	0	0	0	

PSWAI -- PWM halts while in wait mode bit
  0 = Continue PWM main clock generator while in wait mode
  1 = Halt PWM main clock generator when the device is in wait mode
CENTR -- Center-aligned output mode bit
  0 = PWM channels operate in left-aligned output mode
  1 = PWM channels operate in center-aligned output mode
RDPP -- Reduced drive of port P bit
  0 = full drive for all port P output pins
  1 = reduced drive for all port P output pins
PUPP -- Pullup port P enable bit
  0 = disable port P pullups
  1 = enable port P pullups for all port P input pins
PSBCK -- PWM stops while in background debug mode bit
  0 = allows PWM to continue while in background mode
  1 = disable PWM input clock while in background mode

Figure 8.33 ■ PWM control register (PWCTL)

PWTST stands for *pulse-width test*. This register is used only in special mode for testing purposes. The contents of this register are shown in Figure 8.34.

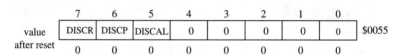

	7	6	5	4	3	2	1	0	
value	DISCR	DISCP	DISCAL	0	0	0	0	0	$0055
after reset	0	0	0	0	0	0	0	0	

DISCR -- Disable channel counter reset bit
  This bit disables the normal operation of resetting the channel
  counter when the channel counter is written.
  0 = normal operation
  1 = write to PWM channel counter does not reset channel counter
DISCP -- disable compare count period bit
  0 = normal operation
  1 = In left-aligned output mode, match of the period does not reset
     associated PWM counter register
DISCAL -- disable scale counter loading bit
  This bit disables the normal operation of loading scale counters on
  a write to the associated scale register.
  0 = normal operation
  1 = Write to PWSCAL0 and PWSCAL1 does not load scale counters

Figure 8.34 ■ PWM special mode register (PWTST)

8.9.3 PWM Duty Cycle & Period

The value in the period register determines the period of the associated PWM channel. If it is written while the channel is enabled, the new value takes effect when the existing period terminates, forcing the counter to reset. The new period is then latched and is used until a new period value is written. To start a new period immediately, write the new period value and then write the counter, forcing a new period to start with the new period value.

The period of the PWM output in the left-aligned mode is:

Period = channel-clock-period $\times$ (PWPER + 1) (8.1)

where PWPER is the value stored in the associated PWM period register.

The period of the PWM output in the center-aligned mode is:

Period = channel-clock-period $\times$ PWPER $\times$ 2 (8.2)

The value in each duty register and its corresponding period register determine the duty cycle of the associated PWM channel. When the duty value is equal to the counter value, the output changes state. If the register is written while the channel is enabled, the new value is held in a buffer until the counter rolls over or the channel is disabled.

If the value in the duty register is greater than or equal to the value in the period register, there is no duty change in state. If the duty register is set to $FF, the output is always in the state which would normally be the state opposite the PPOLx value (in PPOL register).

The duty cycle of the PWM output in the left-aligned mode is:

Duty cycle = [(PWDTYx + 1) ÷ (PWPERx + 1)] $\times$ 100%   (PPOLx = 1)   (8.3)

or:

Duty cycle = [(PWPERx – PWDTYx) ÷ (PWPERx + 1)] $\times$ 100%   (PPOLx = 0) (8.4)

The duty cycle of the PWM output in the center-aligned mode is:

Duty cycle = [(PWPERx – PWDTYx) ÷ PWPERx] $\times$ 100%   (PPOLx = 0)   (8.5)

or:

Duty cycle = (PWDTYx ÷ PWPERx) $\times$ 100%   (PPOLx = 1)   (8.6)

The boundary conditions for the PWM channel duty registers and the PWM channel period registers cause the results shown in Table 8.5.

PWDTYx	PWPERx	PPOLx	Output
$FF	>$00	1	low
$FF	>$00	0	high
≥PWPERx	--	1	high
≥PWPERx	--	0	low
--	$00	1	high
--	$00	0	low

Table 8.5 ■ PWM boundary conditions

Example 8.14

Write an instruction sequence to program the PWM channel 0 to output a waveform with a 50% duty cycle and 100 KHz frequency. Assume the E clock is 8 MHz.

Solution: To achieve the 50% duty cycle, 100 KHz PWM output, we will use the following parameters:

- clock source prescale factor: set to 1
- select clock A as the clock input to PWM channel 0
- left aligned mode
- write the value 79 into the PWPER0 register (frequency is $8MHz \div (79+1) = 100KHz$)
- write the value 39 into the PWCNT0 register

The following instruction sequence will perform the configuration:

```
#include "d:\miniide\hc12.inc"
    ...
        movb    #$0,PWCLK       ; choose 8-bit PWM, set clock A prescale factor =1
        movb    #$01,PWPOL      ; choose clock A as clock source, channel 0 output
                                ; high at the beginning of the period
        movb    #$10,PWCTL      ; select left-aligned, halt PWM clock in wait mode,
                                ; full drive for all port P pins, disable pullup for
                                ; port P input pins, allows PWM to operate in BDM
                                ; mode
        movb    #79,PWPER0      ; set period value
        movb    #39,PWDTY0      ; set duty value
        bset    PWEN,$01        ; enable PWM channel 0
```

Example 8.15

Assume the E clock frequency is 8 MHz. Write an instruction sequence to generate a square wave with a period of 20 μs and a 60% duty cycle using PWM channel 0. Use center-aligned mode.

Solution: Using equation 8.2 and choosing clock A with a prescale factor of 0 as the clock source to PWM channel, we get:

$20\ \mu s = 125\ ns \times PWPER0 \times 2 \Rightarrow PWPER0 = 80$

Using equation 8.6, we obtain:

$60\% = (PWDTY0 \div PWPER0) \times 100\% \Rightarrow PWDTY0 = 48$

We will need to choose the following parameters for this waveform:

- clock source prescale factor: set to 1
- select clock A as the clock input to PWM channel 0
- center aligned mode with PPOL0 equals 1
- write the value 80 into the PWPER0 register
- write the value 48 into the PWDTY0 register

The following instruction sequence will configure the PWM0 properly:

```
movb    #$0,PWCLK       ; choose 8-bit PWM, set clock A prescale factor =1
movb    #$01,PWPOL      ; choose clock A as clock source, channel 0 output
                        ; high at the beginning of the period
```

```
movb    #$18,PWCTL          ; select center-aligned, halt PWM clock in wait
                            ; mode, full drive for all port P pins, disable pullup
                            ; for port P input pins, allows PWM to operate in BDM mode
movb    #80,PWPER0          ; set period value
movb    #48,PWDTY0          ; set duty value
bset    PWEN,$01            ; enable PWM channel 0
```

The following C language statements will configure PWM0 properly:

```
PWCLK = 0x0;
PWPOL = 0x01;
PWCTL = 0x18;
PWPER0 = 80;
PWDTY0 = 48;
PWEN |= 0x01;   /* enable PWM0 channel */
```

The PWM function can be used in many applications that require the average value of output voltages. Some of the examples are the dimming of lamps, LEDs, and DC motor speed.

Example 8.16

Using PWM in dimming the light. Suppose we are using the PWM0 of the 68HC12 to control the brightness of a light bulb. The circuit connection is shown in Figure 8.35. Write a program so that the light is turned down to 10% brightness gradually over a 5 second timespan.

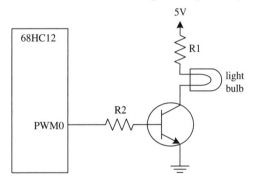

Figure 8.35 ■ Using PWM function to dim the light

Solution: We will dim the light in the following manner:

Use the PWM channel 0 to control the brightness of the light bulb. Set the duty cycle to 100% from the beginning and then dim the brightness by 10% in the first second, and then 20% per second in the following 4 seconds. Use 99 as the initial duty and period value.

Since the brightness of the light is proportional to the average current that flows through the bulb, we can dim the light bulb by reducing the duty cycle of the PWM output from 100% down to 10% in 5 seconds. We will reduce the duty cycle in steps. Within 1 second, we will reduce the duty value 10 times. Therefore, we will reduce the duty value by 1 every 100 ms in the first second and reduce the duty value by 2 every 100 ms in the following 4 seconds.

The assembly program that implements this idea is as follows:

```
#include "d:\miniide\hc12.inc"
            org         $800
dim_cnt     rmb         1
```

```
                org         $1000
                movb        #$90,TSCR           ; enable TCNT, fast flag clear
                movb        #$02,TIOS           ; select OC1 function
                movb        #$24,TMSK2          ; enable pullup, set prescale factor of E to 16

                movb        #$18,PWCLK            ; choose 8-bit PWM, set clock A prescale factor = 8
; choose clock A as clock source, channel 0 output high at the beginning of the period
; set the frequency of clock A to 1 MHz
                movb        #$01,PWPOL
; ****************************************************************************
; The next instruction selects left-aligned, halt PWM clock in wait mode, full drive for all
; port P pins, disable pullup for port P input pins, allows PWM to operate in BDM mode
; ****************************************************************************
                movb        #$10,PWCTL
                movb        #99,PWPER0          ; set period value (period equals 0.1 ms)
                movb        #99,PWDTY0          ; set duty value (equals 100% initially)
                bset        PWEN,$01            ; enable PWM channel 0
; The following instruction segment reduces duty count by 1 per 100 ms
                movb        #10,dim_cnt
                ldd         TCNT
loop1           addd        #50000
                std         TC1
                brclr       TFLG1,$02,*         ; wait for 100 ms
                ldd         TC1
                dec         PWDTY0              ; decrement duty cycle by 1%
                dec         dim_cnt
                bne         loop1
; The following instruction segment reduces duty count by 2 per 100 ms in four seconds
                movb        #40,dim_cnt
loop2           ldd         TC1
                addd        #50000
                std         TC1
                brclr       TFLG1,$02,*         ; wait for 100 ms
                dec         PWDTY0              ; decrement duty cycle by 2%
                dec         PWDTY0              ; per 100 ms
                dec         dim_cnt
                bne         loop2
                swi
                end
```

The C language version of this program is as follows:

```
#include <hc12.h>
void main ()
{
        int      dim_cnt;

        TSCR = 0x90;      /* enable TCNT and fast flag clear */
        TIOS = 0x02;      /* select OC1 function */
        TMSK2 = 0x24;     /* set prescale factor of E to 16 and disable TOV interrupt */

        PWCLK = 0x18;     /* choose 8-bit PWM, and set clock A prescaler to 8 */
/* choose clock A as clock source, PWM channel 0 output high at the beginning of the period */
        PWPOL = 0x01;
/* Select left-aligned, halt PWM clock in wait mode, full drive for all port P pins, disable pullup for port P input pins,
allows PWM to operate in BDM mode */
```

```
            PWCTL = 0x10;

            PWPER0 = 99;      /* set period of PWM0 to 0.1 ms */
            PWDTY0 = 99;      /* set duty cycle to 100% */
            PWEN | = 0x01;    /* enable PWM0 channel */
            TC1 = TCNT + 50000u;
    /* reduce duty cycle 1 % per 100 ms in the first second */
            for (dim_cnt = 0; dim_cnt < 10 ; dim_cnt ++) {
                while (!(TFLG1 & 0x02));
                TC1 = TC1 + 50000u;
                PWDTY0 = PWDTY0 - 1;
            }
    /* reduce duty cycle 2% per 100 ms in the next 4 seconds */
            for (dim_cnt = 0; dim_cnt < 40; dim_cnt ++) {
                TC1 = TC1 + 50000u;
                while (!(TFLG1 & 0x02));
                PWDTY0 = PWDTY0 - 2;
            }
            asm ("swi");
    }
```

▲

8.9.4 DC Motor Control

DC motors are used extensively in control systems as positional devices because their speeds and their torques can be precisely controlled over a wide range. The DC motor has a permanent magnetic field and its armature is a coil. When a voltage and a subsequent current flow are applied to the armature, the motor begins to spin. The voltage level applied across the armature determines the speed of rotation.

The microcontroller can digitally control the angular velocity of a DC motor by monitoring the feedback lines and driving the output lines. Almost every application that uses a DC motor requires it to reverse its direction of rotation or vary its speed. Reversing the direction is simply done by changing the polarity of the voltage applied to the motor. Changing the speed requires varying the voltage level of the input to the motor, and that means changing the input level to the motor driver. In a digitally controlled system, the analog signal to the driver must come from some form of D/A converter. However, adding a D/A converter to the circuit increases the chip count, which means increasing the system cost and power consumption. The other alternative is to vary the pulse width of a digital signal input to the motor. By varying the pulse width the average voltage delivered to the motor changes and so does the speed of the motor. The 68HC12 PWM subsystem can be used to control the DC motor.

The 68HC12 can interface with a DC motor through a driver, as shown in Figure 8.36. This circuit takes up only three I/O pins. The pin that controls the direction can be an ordinary I/O pin but the pin that controls the speed must be a PWM pin. The pin that receives the feedback must be an input-capture pin.

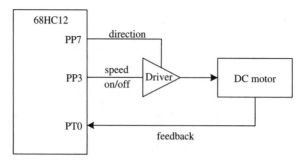

Figure 8.36 ■ Simplified circuit for DC motor control

Although some DC motors can operate at 5V or less, the 68HC12 cannot supply the necessary current to drive a motor directly. The minimum current required by any practical DC motor is much higher than any microcontroller can supply. Depending on the size and rating of the motor, a suitable driver must be selected to take control signals from the 68HC12 and deliver the necessary voltage and current to the motor.

DRIVERS

Standard motor drivers are available in many current and voltage ratings. Examples are the L292 and L293 made by SGS Thompson Inc. The L293 has four channels and can output up to 1 A of current per channel with a supply of 36 V. It has a separate logic supply and takes a logic input (0 or 1) to enable or disable each channel. The L293D also includes clamping diodes needed to protect the driver from the back electromagnetic frequency (EMF) generated during the motor reversal. The pin assignment and block diagram of the L293 are shown in Figure 8.37. There are two supply voltages: V_{ss} and V_s. V_{ss} is the logic supply voltage, which can be from 4.5 to 36 V (normally 5.0 V). V_s is the analog supply voltage and can be as high as 36 V.

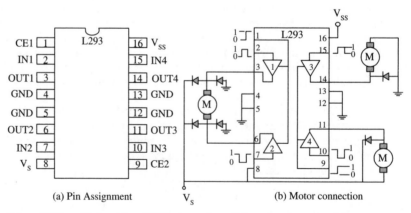

Figure 8.37 ■ Motor driver L293 pin assignment and motor connection

FEEDBACK

The DC motor controller needs information to adjust the voltage output to the motor driver circuit. The most important information is the speed of the motor which must be fed back from the motor by a sensing device. The sensing device may be an optical encoder, infrared detector, Hall-effect sensor, and so on. Whatever the means of sensing, the result is a signal, which is fed back to the microcontroller. The microcontroller can use the feedback to determine the speed and position of the motor. Then it can make adjustments to increase or decrease the speed, reverse the direction, or stop the motor.

Assume a Hall-effect transistor is mounted on the shaft (rotor) of a DC motor and two magnets are mounted on the armature (stator). As shown in Figure 8.38, every time the Hall-effect transistor passes through the magnetic field, it generates a pulse. The input-capture function of the 68HC12 can capture the passing time of the pulse. The time between two captures is half of a revolution. Thus the motor speed can be calculated. By storing the value of the capture registers each time, and comparing it with its previous value, the controller can constantly measure and adjust the speed of the motor. Using this method a motor can be run at a precise speed or synchronized with another event.

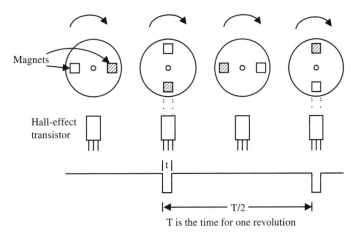

Figure 8.38 ■ The output waveform of the Hall effect transistor

The schematic of a motor-control system is illustrated in Figure 8.39. The PWM output from the PP3 pin is connected to one end of the motor, whereas the PP7 pin is connected to the other end of the motor. The circuit is connected so that the motor will rotate clockwise when the voltage of the PP7 pin is zero while the PWM output is nonzero (positive). The direction of motor rotation is illustrated in Figure 8.40. By applying appropriate voltages on PP7 and PP3 (PWM3), the motor can rotate clockwise, counterclockwise, or even stop. Input-capture channel 0 is used to capture the feedback from the Hall-effect transistor.

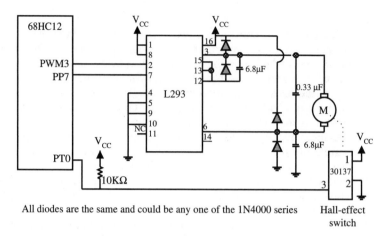

All diodes are the same and could be any one of the 1N4000 series · Hall-effect switch

Figure 8.39 ■ Schematic of a 68HC12-based motor-control system

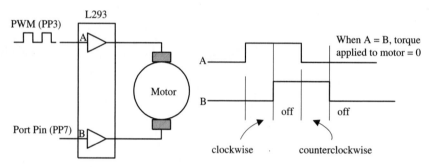

Figure 8.40 ■ The L293 motor driver

When a motor is first turned on, it cannot reach its final steady speed immediately. Some amount of startup time (say 20 seconds or more) should be allowed for the motor to get into speed. In the normal use of a DC motor, we desire the motor speed to be constant. However, when a load is applied to the motor, it will slow down. To keep the speed constant, we need to increase the duty cycle of the voltage that is applied to the motor. When the load gets lighter, the motor will accelerate and run faster than we like. To slow down the motor, the duty cycle of the voltage should be reduced.

To be responsive, the change of the duty cycle should be large. However, a large variation in the duty cycle tends to cause it to overreact and makes the motor unstable. We should increase and/or decrease the duty cycle by a small amount at a time (say 5%, this number should be experimented) to avoid oscillation.

The DC motor does not respond to the change in duty cycle instantaneously due to its inertia. Take a motor that rotates at 7,200 rpm as an example; a revolution will take slightly over 8 ms. After increasing the duty cycle, enough time should be allowed (that allows the motor to make a few revolutions, say 50 ms) for the motor to respond before we measure its new speed and decide whether we should increase or decrease the duty cycle of the PWM output.

Example 8.17

▼

Write a subroutine in the C language to measure the motor speed (in rpm).

Solution: To measure the motor speed, we need to capture two consecutive rising edges.

Let the difference of two consecutive edges be *diff* and the period of the timer is set to 1 μs, then the motor speed (rpm) is:

Speed = $60 \times 10^6 \div (2 \times diff)$ (8.7)

The C function that measures the motor speed is as follows:

```c
#include <hc12.h>

unsigned int motor_speed (void)
{
        unsigned int edge1, diff, rpm;
        long unsigned int temp;

        TSCR = 0x90;                    /* enable TCNT and fast flag clear */
        TIOS &= 0xFE;                   /* select IC0 function */
        TMSK2 = 0x23;                   /* enable input pullup and set TCNT prescale factor to 8 */
        TCTL4 = 0x01;                   /* select to capture the rising edge of PT0 */
        TFLG1 = 0x01;                   /* cleared C0F flag */
        while (!(TFLG1 & 0x01));        /* wait for the first edge */
        edge1 = TC0;
        while (!(TFLG1 & 0x01));        /* wait for the second edge */
        diff = TC0 - edge1;
        temp = 1000000ul / (2 * diff);
        rpm = temp * 60;
        return rpm;
}
```

▲

ELECTRICAL BRAKING

Once a DC motor is running, it picks up speed. Turning off the voltage to the motor does not make it stop immediately because the momentum will keep it turning. After the voltage is shut off, the momentum will gradually wear off due to friction. If the application does not require an abrupt stop, then the motor can be brought to a gradual stop by removing the driving voltage.

An abrupt stop may be required by certain applications in which the motor must run a few turns and stop quickly at a predetermined point. This could be achieved by electrical braking.

Electrical braking is done by reversing the voltage applied to the motor. The length of time that the reversing voltage is applied must be precisely calculated to ensure a quick stop while not starting it in the reverse direction. There is no simple formula to calculate when to start and how long to maintain braking. It varies from motor to motor and application to application. But it can be perfected through trial and error.

In a closed-loop system, the feedback can be used to determine where or when to start and stop braking and when to discontinue. Again, this is application-dependent.

In Figure 8.40, the motor can be braked by (1) reducing the PWM duty-count to 0, and (2) setting port pin PP7 output to high for an appropriate amount of time.

8.10 Enhanced Capture Timer (ECT) Module

The later members of the 68HC12 family, including 912BE32, 912D60, 912DG128, and 912DT128, implement an *enhanced capture timer* (ECT) module that has the features of the *standard timer* (TIM) module enhanced by additional features in order to enlarge the field of applications. These additional features are:

- 16-bit buffer register for four input capture (IC) channels
- Four 8-bit pulse accumulators. Each of these 8-bit pulse accumulators has an associated 8-bit buffer. Two of these 8-bit pulse accumulators can be concatenated into a single 16-bit pulse accumulator.
- 16-bit modulus down-counter with 4-bit prescaler
- Four user-selectable delay counters for increasing input noise immunity
- Main timer prescaler extended to 7 bits (the original timer prescaler is 5 bits)

8.10.1 Enhanced Capture Timer Modes of Operation

The enhanced capture timer has eight input-capture, output-compare (IC/OC) channels, the same as on the 68HC12 standard timer module.

Four IC channels are the same as the standard timer, with one capture register that memorizes the timer value captured by an action on the associated input pin. Four other IC channels, in addition to the capture register, also have one buffer called the *holding register*. This permits the register to memorize two different timer values without generation of any interrupt. This capability can reduce the software overhead in the measurement of period, pulse width, and duty cycle.

Four 8-bit pulse accumulators are associated with the four IC buffered channels. Each pulse accumulator has a holding register to memorize its value by an action on its external input. Each pair of pulse accumulators can be used as a 16-bit pulse accumulator.

The 16-bit modulus down-counter can control the transfer of the IC register's contents and the pulse accumulators to the respective holding register for a given period, every time the count reaches zero. The modulus down-counter can also be used as a standalone timebase with periodic interrupt capability.

As shown in Table 8.1, the last two combinations of PR2-PR1 (110 & 111) are not used. Therefore, the TCNT timer prescaler ranges from 1 to 32. In the enhanced capture timer, the combinations 110 and 111 specify the prescale factors 64 and 128, respectively.

8.10.2 Why Enhanced Capture Timer Module?

There are applications that require the capture of two consecutive edges (both could be rising or falling, or one rising and the other falling) at very high frequencies. In Example 8.2, the following instructions are executed before we have time to wait for the arrival of the second edge (after we detect the firsts edge):

```
ldd     TCx
std     ....
```

It takes five E clock cycles to execute these two instructions, which sets the upper limit on the signal frequency that can be dealt with. By providing the capability of setting the interrupt flag (or interrupting the CPU) after two signal edges have been captured, the upper limit of the signal frequency that can be handled can be significantly improved.

The input-capture function of the original standard timer module allows the newly captured value to overwrite the old one even if the CPU has not read the old value yet. This can cause problems when the event frequency is very high. The enhanced input-capture function allows the user to prevent the overwriting of captured values by enabling the non-overwrite feature.

The 68HC12 is often used in noisy environments such as automotive applications. Due to the noise, false signal edges often occur in input-capture pins. To distinguish between true edges and false edges (usually very short), a delay counter is added to the enhanced input-capture function. Any detected edge with duration shorter than the preprogrammed value is ignored.

8.10.3 Registers Associated with Enhanced Input Capture Function

The enhanced input-capture function has all the registers in the input-capture function of the standard timer module. New registers are added to implement the additional features. An input-capture register is called *empty* if its value has been read or latched into its associated holding register yet. A holding register is *empty* if its value has been read. An enhanced input-capture channel can be configured to operate in either *latch* mode or *queue* mode. The block diagrams of latch mode and queue mode are shown in Figures 8.41 and 8.42.

Figures 8.41 and 8.42 illustrate the registers related to the operation of latch mode and queue mode, respectively. In both diagrams, channels IC0 to IC3 are identical and IC4 to IC7 are identical. Only one channel in each group is shown in the figure.

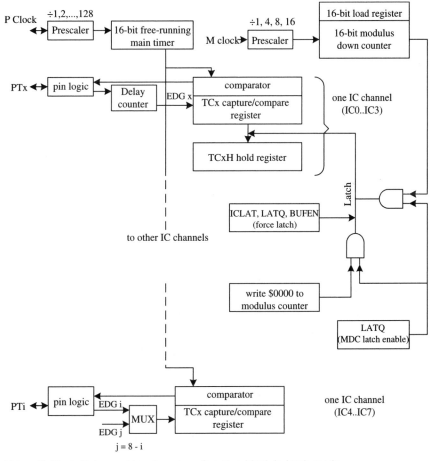

Figure 8.41 ■ Enhanced input capture function block in latch mode

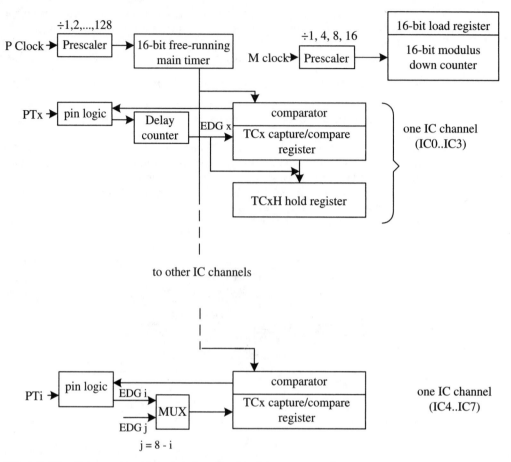

Figure 8.42 • Enhanced input capture function block in queue mode
(channels IC0..IC3 block diagram)

INPUT CONTROL OVERWRITE REGISTER (ICOVW)

This register allows the user to choose between automatic overwrite and non-automatic overwrite to the input-capture register. The contents of the ICOVW register are shown in Figure 8.43.

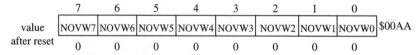

	7	6	5	4	3	2	1	0	
value	NOVW7	NOVW6	NOVW5	NOVW4	NOVW3	NOVW2	NOVW1	NOVW0	$00AA
after reset	0	0	0	0	0	0	0	0	

NOVWx -- No input capture overwrite bits
    0 = The contents of the related capture register or holding register
        can be overwritten when a new input capture or latch occurs.
    1 = The related capture register or holding register cannot be written
        by an event unless they are empty.

Figure 8.43 ■ Input control overwrite register (ICOVW)

INPUT CONTROL SYSTEM CONTROL REGISTER (ICSYS)

This register provides very important control on the input-capture function, as shown in Figure 8.44.

	7	6	5	4	3	2	1	0	
value	SH37	SH26	SH15	SH04	TFMOD	PACMX	BUFEN	LATQ	$00AB
after reset	0	0	0	0	0	0	0	0	

NSHxy -- share input action of input capture channel x and y bits
    0 = normal operation
    1 = The channel input x causes the same action on the channel y.
TFMOD -- timer flag-setting mode bit
    0 = The timer flags C3F-C0F in TFLG1 are set when a valid input
        capture transition on the corresponding port pin occurs.
    1 = If in queue mode (BUFEN = 1 and LATQ = 0), the timer flags
        C3F-C0F in TFLG1 are set only when a latch on the
        corresponding holding register occurs. If the queue mode is not
        engaged, the timer flags C3F-C0F are set the same way as for
        TFMOD = 0.
PACMX -- 8-bit pulse accumulator maximum count bit
    0 = normal operation. When the 8-bit pulse accumulator has reached
        the value $FF, with the next active edge, it will be incremented
        to $00.
    1 = When the 8-bit pulse accumulator has reached the value $FF, it
        will not be incremented further. The value $FF indicates a count
        of 255 or more.
BUFEN -- IC buffer enable bit
    0 = input capture and pulse accumulator holding registers are
        disabled.
    1 = input capture and pulse accumulator holding registers are
        enabled.
LATQ -- input capture latch or queue mode select bit
    The BUFEN bit should be set to enable IC and pulse accumulator's
    holding registers. Otherwise, LATQ latching mode is disabled.
    0 = queue mode of input capture is enabled
    1 = latch mode is enabled. Latching function occurs when modulus
        down-counter reaches 0 or a 0 is written into the count register
        MCCNT. With a latching event, the contents of IC registers and
        8-bit pulse accumulators are transferred to their holding registers.
        The 8-bit pulse accumulators are cleared.

Figure 8.44 ■ Input control system control register (ICSYS)

DELAY COUNTER CONTROL REGISTER (DLYCT)

If enabled, after detection of a valid edge on input capture pin, the delay counter counts the pre-selected number of P clock cycles (the same frequency as the E clock but it differs 90° in phase), and then it will generate a pulse to latch the TCNT value into the input-capture register (TCx). The pulse will be generated only if the level of input signal, after the preset delay, is the opposite of the level before the transition. This will avoid reactions to narrow pulses caused by noise.

After counting, the counter will be cleared automatically. Delay between the two active edges of the input signal period should be longer than the selected counter delay.

The contents of the DLYCT register are shown in Figure 8.45.

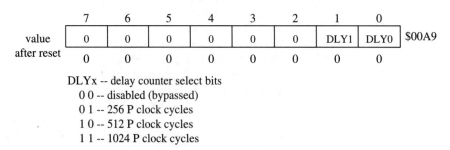

Figure 8.45 ■ Delay counter control register (DLYCT)

The ECT timer has a 16-bit *modulus down counter* (MCCNT) that is used to control the latching of the input-capture register (and its associated 8-bit pulse accumulator) into the holding register. This modulus down counter has a 16-bit load register to hold the reload value for the counter and the counter itself. The operation of this modulus counter is controlled by the MCCTL register (shown in Figure 8.46) and the MCFLG register (shown in Figure 8.47). The clock source of the modulus counter is the M clock, prescaled by a factor programmable from 1 to 16. The M clock has the same frequency as the E clock.

7	6	5	4	3	2	1	0	
MCZI	MODMC	RDMCL	ICLAT	FLMC	MCEN	MCPR1	MCPR0	$00A
0	0	0	0	0	0	0	0	

value after reset

MCZI -- modulus counter underflow interrupt enable bit
  0 = modulus counter underflow interrupt is disabled
  1 = modulus counter underflow interrupt is enabled
MODMC -- modulus mode enable bit
  0 = the counter counts once from the value written to it and will stop at $0000.
  1 = modulus mode is enabled. When the counter reaches $0000, the counter is loaded with the latest value written into to the modulus count register.
RDMCL -- read modulus down-counter load bit
  0 = reads of the modulus count register will return the present value of the count register.
  1 = reads of the modulus count register will return the contents of the load register (i.e., the reload value is returned)
ICLAT -- input capture force latch action bit
  When input capture latch mode is enabled (LATQ and BUFEN bit in ICSYS are set), writing 1 to this bit immediately forces the contents of the input capture registers TC0 to TC3 and their corresponding 8-bit pulse accumulators to be latched into the associated holding registers. The pulse accumulators will be cleared when the latch action occurs. Writing 0 to this bit has no effect. Read of this bit always will return 0.
FLMC -- force load register into the modulus counter count register bit
  This bit is active only when the modulus down-counter is enabled (MCEN = 1). Writing a 1 into this bit loads the load register into the modulus counter count register. This also resets the modulus counter prescaler. Writing 0 to this bit has no effect.
MCEN -- modulus down-counter enable bit
  0 = modulus counter is disabled and preset to $FFFF
  1 = modulus counter is enabled
MCPR1 & MCPR0 -- modulus counter prescaler select bits
  0 0 = prescale rate is 1
  0 1 = prescale rate is 4
  1 0 = prescale rate is 8
  1 1 = prescale rate is 16

Figure 8.46 ■ Modulus down-counter control register (MCCTL)

7	6	5	4	3	2	1	0	
MCZF	0	0	0	POLF3	POLF2	POLF1	POLF0	$00A7
0	0	0	0	0	0	0	0	

value after reset

MCZF -- modulus counter underflow interrupt flag
  This flag is set when the modulus down-counter reaches 0. Writing 1 to this bit clears the flag.
POLF3-POLF0 -- first input capture polarity status bits
  These are read-only bits. Writing to these bits has no effect. Each status bit gives the polarity of the first edge which has caused an input capture to occur after capture latch has been read.
  0 = The first input capture has been caused by a falling edge.
  1 = The first input capture has been caused by a rising edge.

Figure 8.47 ■ Modulus down-counter flag register (MCFLG)

Like the RTI module, the modulus down counter can also be programmed to generate periodic interrupts. This can be achieved by selecting the modulus mode and enabling its interrupt.

8.10.4 Applications of the Enhanced Capture Function

Two examples will be given to illustrate the applications of the enhanced input-capture function. Beware that these programs can only be run on a demo board that has one of the 68HC12 members that has an enhanced capture timer module.

Example 8.18

Modify example 8.2 to take advantage of the queue mode of the enhanced input-capture function.

Solution:

```
#include "d:\miniide\hc12.inc"
        org      $800
period  rmb      2                       ; memory to store the period
        org      $1000
        movb     #$90,TSCR               ; enable timer counter and fast timer flags clear
        bclr     TIOS,$01                ; select input capture zero
        movb     #$44,TMSK2              ; disable TCNT overflow interrupt, set prescale
                                         ; factor to 16
        movb     #$01,TCTL4              ; choose to capture the rising edge of PT0 pin
        movb     #$0A,ICSYS              ; enable timer flag-setting, IC buffer, and queue
                                         ; mode
        clr      DLYCT                   ; disable delay counter
        bset     ICOVW,$01               ; no input capture overwrite for IC0
        ldd      TC0                     ; empty the input capture register TC0
        ldd      TC0H                    ; empty the holding register TC0H
        brclr    TFLG1,$FE,*             ; wait for the arrival of the second rising edge
        ldd      TC0
        subd     TC0H                    ; subtract the first edge from the second edge
        std      period
        swi
        end
```

The program in C is as follows:

```
#include <hc12.h>
void main( )
{
        unsigned int period;

        TSCR = 0x90;        /* enable timer counter, enable fast flag clear*/
        TIOS &= 0xFE;       /* select input capture zero */
/* disable TCNT overflow interrupt, set prescale factor to 16 */
        TMSK2 = 0x44;
        TCTL4 = 0x01;       /* capture the rising edge of PT0 pin */
/* enable timer flag-setting mode, IC buffer, and queue mode */
        ICSYS = 0x0A;
        DLYCT = 0x00;       /* disable delay counter */
        ICOVW |= 0x01;      /* disable input capture overwrite */
        period = TC0;       /* empty TC0 and clear the COF flag */
```

```
        period = TCOH;        /* empty the TCOH register */
/* wait for the arrival of the second rising edge */
        while (!(TFLG1 & 0x01));
        period = TC0 - TCOH;
        asm ("swi");
}
```

▲

Example 8.19

▼

Suppose you want to measure the pulse width of a signal connected to the PT0 pin in a noisy environment. Write a program to perform the operation. Ignore any noise pulse shorter than 256 P clock cycles.

Solution: Since the range of the pulse width is unknown, we need to consider the timer overflow. The program in Example 8.3 is modified to perform the measurement as follows:

```
#include "d:\miniide\hc12.inc"
setuservector equ    $F69A
tov_vec_no equ    23

               org        $800
edge1          rmb        2
overflow       rmb        2
pulse_width rmb    2
               org        $1000
; the following 7 instructions set up TOV interrupt jump vector
               ldd        #tov_isr
               pshd
               ldab       #tov_vec_no
               clra
               ldx        setuservector
               jsr        0,x
               leas       2,sp

               ldd        #0
               std        overflow
               movb       #$90,TSCR
; disable TCNT overflow interrupt,  set prescale factor to 16
               movb       #$44,TMSK2
               bclr       TIOS,$01           ; select input capture zero
               movb       #$01,DLYCT         ; select delay count to 256 P cycles
               movb       #$01,ICOVW         ; prohibit overwrite to TC0 register
               movb       #$0,ICSYS          ; disable queue mode
; capture the rising edge on PT0 pin
               movb       #$01,TCTL4
               bclr       TFLG1,$FE          ; clear COF flag
               brclr      TFLG1,$01,*        ; wait for the arrival of the first rising edge
               bclr       TFLG2,$7F          ; clear TOF flag
               bset       TMSK2,$80          ; enable TCNT overflow interrupt
               cli                           ;          "
               ldd        TC0                ; clear COF flag and save the captured first edge
               std        edge1
```

```
; capture the falling edge on PT0 pin
                    ldaa        #$02
                    staa        TCTL4
                    brclr       TFLG1,$01,*        ; wait for the arrival of the falling edge
                    ldd         TC0
                    subd        edge1
                    std         pulse_width
                    bcc         next
; second edge is smaller, so decrement overflow count by one
                    ldx         overflow
                    dex
                    stx         overflow
next                swi
tov_isr             bclr        TFLG2,$7F          ; clear TOF flag
                    ldx         overflow
                    inx
                    stx         overflow
                    rti
                    end
```

The modification to the C program in Example 8.3 is also minor and will be left for you as an exercise.

▲

8.11 Pulse Accumulators Associated with Enhanced Capture Timer

An 8-bit pulse accumulator is associated with each of the buffered input-capture channels. Each of these pulse accumulators has an associated holding register. The block diagram of these four 8-bit pulse accumulators is shown in Figure 8.48. The user can prevent 8-bit pulse accumulators from counting further than $FF by the PACMX bit in the ICSYS register. In this case, a value of $FF means that 255 counts or more have occurred. Among these four pulse accumulators, only two (PACN1 and PACN3) of them can generate interrupts when they overflow.

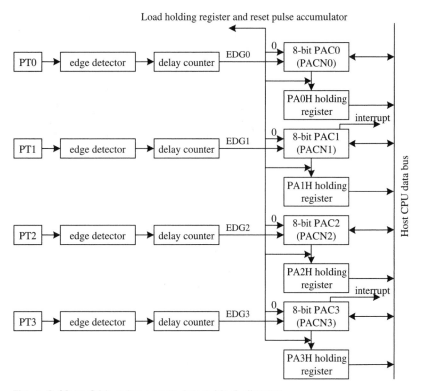

Figure 8.48 ■ 8-bit pulse accumulators block diagram

Each pair of pulse accumulators can be used as a 16-bit pulse accumulator. These pulse accumulators also have two operation modes: *latch* and *queue* mode. The diagram for 16-bit pulse accumulator is shown in Figure 8.49.

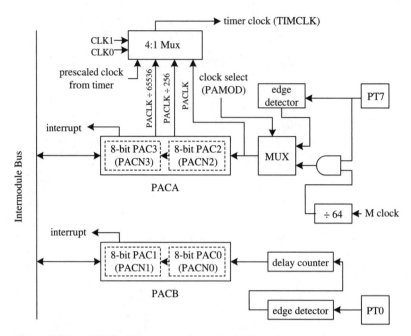

Figure 8.49 ■ 16-bit pulse accumulator block diagram

8.11.1 Operations of the Enhanced Pulse Accumulators

As shown in Figure 8.48, an 8-bit pulse accumulator counts the number of active edges at the input of its channel. As in enhanced input-capture function, the detected edge may be delayed for making sure it is not caused by noise. The delay time is set in the same way as for its associated input-capture channel.

Two pulse accumulators may generate interrupts when they overflow from $FF to $00. These two pulse accumulators are PAC1 and PAC3. When PAC3 overflows, it sets the PAOVF flag (bit 1) of the PAFLG register. When PAC1 overflows, it sets the PBOVF (bit 1) flag of the PBFLG register. These two flags can be cleared by writing a one to them. When the timer fast flag feature is enabled (by setting the TFFCA bit in the TSCR register), any access to the PACN3 and PACN2 will clear the PAOVF flag, whereas any access to the PACN1 and PACN0 will clear the PBOVF flag.

In an 8-bit configuration, pulse accumulators can be configured in latch mode and queue mode. The latch or queue mode is selected by programming the ICSYS register.

In latch mode, the value of the pulse accumulator is transferred to its holding register when the modulus down counter reaches zero, a write of $0000 to the modulus down counter, or when the force latch control bit ICLAT is written. At the same time, the pulse accumulator is cleared.

In queue mode, reads of an input-capture holding register will transfer the contents of the associated pulse accumulator to its holding register. At the same time, the pulse accumulator is cleared.

To concatenate PAC3 and PAC2 into a 16-bit pulse accumulator, set the PAEN bit (bit 6) of the PACTL register (shown in Figure 8.25). After being configured into a 16-bit pulse accumulator, its function and programming are identical to the 16-bit pulse accumulator of the standard timer module. Pin 7 becomes the input to this 16-bit pulse accumulator. The PAFLG register records the interrupt flags PAOVF and PAIF, as shown in Figure 8.26.

PAC1 and PAC0 can be concatenated into a 16-bit pulse accumulator B. This is done by setting the PBEN bit (bit 6) of the PBCTL register. The contents of the PBCTL and the status register PBFLG are shown in Figure 8.50.

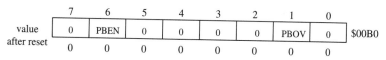

PBEN -- pulse accumulator B system enable bit
    0 = 16-bit pulse accumulator disabled. Eight-bit PAC1 and PAC0 can
        be enabled when their related enable bits in ICPACR are set.
    1 = pulse accumulator B system enabled.
PBOV -- pulse accumulator B overflow interrupt enable bit
    0 = interrupt inhibited.
    1 = interrupt requested if PBOVF is set

(a) Pulse accumulator B control register (PBCTL)

PBOVF -- pulse accumulator B overflow flag
This bit is set when the 16-bit pulse accumulator B overflows from
$FFFF to $0000 or when 8-bit accumulator 1 (PAC1) overflows from
$FF to $00. It is clearded by writing 1 to it or by accessing PACN1
and PACN0 when the TFFCA bit in the TSCR register is set.

(b) Pulse accumulator B flag register (PBFLG)

Figure 8.50 ■ Pulse accumulator B control and flag register

Eight-bit pulse accumulators PAC3 and PAC2 can be enabled when the PAEN bit in the PACTL register is cleared. PAC1 and PAC0 can be enabled when the PBEN bit in the PBCTL register is zero. Eight-bit pulse accumulators are enabled by setting the corresponding bit in the ICPACR register. The contents of this register are shown in Figure 8.51.

PAxEN -- 8-bit pulse accumulator x enable bit
    0 = pulse accumulator x disabled
    1 = pulse accumulator x enabled

Figure 8.51 ■ Input control pulse accumulator control register (ICPACR)

Since these eight-bit pulse accumulators are associated with the enhanced input-capture channels, their active edge inputs are specified using the TCTL4 register. In 16-bit configuration, the active edge of the pulse accumulator A is specified using the PACTL register, whereas the active edge (of pin PT0) of the pulse accumulator B is specified using the TCTL4 register (bits 1 and 0).

Example 8.20

▼

Write an instruction sequence to enable the 8-bit pulse accumulator PAC1 and PAC3 and let them increase on the rising edge of their associated pins. Disable their overflow interrupts.

Solution:

To enable PAC1 and PAC3, we need to clear the PAEN and PBEN bits and set the PA3EN and PA1EN bits. To select the rising edge as their active edge, we need to write the value $44 into the TCTL4 register and also configure the PT3 and PT1 pin for input capture. To disable overflow interrupt, clear the PAOVI and PBOV bits.

The following instruction sequence will perform the required configuration:

```
bclr    PACTL,$42      ; disable 16-bit PA, disable overflow interrupt
bclr    PBCTL,$42      ; disable 16-bit PB, disable overflow interrupt
movb    #$44,TCTL4     ; select the rising edges as active edge
bset    ICPACR,$0A     ; enable PAC3 and PAC1
bclr    DDRT,$81       ; configure PT7 and PT0 for input
```

The applications and programming of these pulse accumulators are similar to those discussed in Section 8.10, and will be left for you to do as an exercise.

▲

8.12   Summary

Many applications require a dedicated timer. Without a timer the following application features will become very difficult or even impossible to implement:

- the measurement of pulse width, frequency, period, duty cycle, and phase difference
- the detection of certain events
- the creation of time delays
- the generation of waveforms

The earlier members of the 68HC12, including the 812A4, the 912B32, and the 912BC32, implemented a standard timer module (TIM) that incorporated eight channels of input-capture/output-compare functions in addition to real-time interrupt, 16-bit pulse accumulator, and pulse-width modulation subsystems (812A4 does not have PWM). Later members, including the 912BE32, the 912D60, the 912DG128, and the 912DT128, implemented an enhanced capture timer (ECT) module, which has the features in the standard timer module enhanced by additional functions.

The heart of the timer system is the 16-bit main timer TCNT. This timer must be enabled in order to run. Its clock signal is the P clock, prescaled by a factor. The prescale factor can be from 1 to 32 in the standard timer module and can be from 1 to 128 in the enhanced capture timer module.

The input-capture function can be programmed to latch the main timer value (TCNT) into the input-capture register on the arrival of an active edge, and optionally generate an interrupt. The input-capture function is often used to measure the period, pulse width, duty cycle, and phase shift. It can also be used as a time reference and count the events that occurred during certain intervals.

There are eight channels (IC0 to IC7) of input-capture function. The input-capture function of TIM has several limitations that make it unsuitable for higher frequency applications and noisy environments. The ECT module adds a holding register to each of the four input-capture channels (IC0 to IC3). This enhancement adds the following capabilities:

- *Interrupt the CPU after two edges have been captured instead of each edge.* This capability enables the user to measure much shorter pulses or periods.

- *Ignore the short pulse.* This capability enables the user to capture event arrival times and measure pulse width or periods in a noisy environment.

- *Selective no-overwrite.* This capability allows the user to do measurements in very high frequency and will not miss the true events because the CPU is busy with other chores.

The 68HC12 has eight output-compare channels (OC0 to OC7). Each output-compare channel has a 16-bit register, a 16-bit comparator, and an output-compare action pin. The output-compare function is often used to create a time delay, generate a waveform, and so on. To use the output-compare function, we make a copy of the main timer, add a delay to this copy, and store the sum into an output-compare register. The 16-bit comparator compares the contents of the main timer with that of the output-compare register. When they are equal, the corresponding timer flag will be set and an optional action on the associated signal pin will be triggered: pull to high, pull to low, or toggle.

An output-compare operation can be forced to take effect immediately by writing a one into the corresponding bit in the CFORC register. This action will not set the timer flag and will not generate an interrupt either.

The output compare channel 7 can control up to eight output-compare channels at the same time. This capability allows us to use two output-compare channels to control the same signal pin.

The RTI function can be used to generate periodic interrupts to remind the CPU to perform routine tasks without unduly delay. The time-out period of RTI is programmable.

Using the output-compare function to generate periodic waveforms is an important application. However, it involves too much overhead. Four 8-bit pulse-width-modulation (PWM) channels are added for this purpose. The PWM function can generate digital waveforms with a wide range of frequencies and duty cycles. After setting up the duty and period values, the CPU can work on other tasks until the period and/or duty cycle needs to be changed. When necessary, each pair of PWM channels can be concatenated into a 16-bit PWM channel.

The TIM module has a 16-bit pulse accumulator module. The PA function has two operation modes: *event counting* and *gated time accumulation*. This function has been used to generate an interrupt after *n* events have occurred, measure the frequency of a signal, count events, measure the pulse width, and so on.

The ECT module adds four buffered 8-bit pulse accumulators. Each pulse accumulator has an 8-bit counter that will increment when an active edge arrives. Each pair of the pulse accumulator can be concatenated into a 16-bit pulse accumulator.

8.13 Exercises

Assume the E clock frequency of the 68HC12 is 8 MHz for the following questions unless it is specified otherwise.

E8.1 Write a program to configure all Port T pins to be used in capturing event arrival times. Use the interrupt-driven approach. Stay in a wait loop after completing the configuration. Exit the wait loop when all eight channels have arrived. Store the arrival times in memory locations starting from $800.

E8.2 Use input-capture channel 1 to measure the duty cycle of a signal. Write an assembly and a C program (in a subroutine format) to do the measurement.

E8.3 Assume that two signals having the same frequency are connected to the pins PT1 and PT0. Write an assembly and a C program to measure their phase difference.

E8.4 Write an assembly and a C program to generate a 2 KHz, 70% duty cycle waveform from the PT6 pin. You must use an OC function.

E8.5 Write an assembly and a C program to generate a 4 KHz, 80% duty cycle waveform from the PT5 pin. You must use an OC function.

E8.6 Write a subroutine that uses the PWM0 to generate a waveform with a programmable frequency and duty cycle. The duty cycle can be from 0% to 100% in 10% steps. The frequency can vary from several ranges:

1. 10 Hz to 90 Hz in 10 Hz steps
2. 100 Hz to 900 Hz in 100 Hz steps
3. 1KHz to 9KHz in 1KHz steps
4. 10KHz to 90KHz in 10KHz steps
5. 100KHz to 900KHz in 100KHz steps

The duty cycle and frequency are passed in the stack.

E8.7 Write a subroutine that can generate a time delay from 1 to 100 seconds using both the polling and the interrupt-driven methods. The number of seconds is passed to the subroutine in accumulator A.

E8.8 What would be the output frequency of the PT0 signal generated by the following program segment?

```
clr       PORTT
movb      #$01,DDRT
bset      TIOS,$81
movb      #$01,OC7M
movb      #$01,OC7D
movb      #$90,TSCR
movb      #$02,TCTL2
movb      #$0,TMSK1
movb      #$08,TMSK2
ldd       #$0
std       TC0
ldd       #$1
std       TC7
```

E8.9 Write a program to configure the main timer of the ECT module properly and use output channel 7 to generate an interrupt every second and output the message "n seconds has passed."

E8.10 Write a program to wait for an event (rising edge) to arrive at the PT0 pin. After that, the program will wait for 100 ms and trigger a pulse 20 ms wide on the PT6 pin.

E8.11 In Example 8.3, we are using the polling method to check for the arrival of edges. Write a program using the interrupt-driven approach to measure the period of an unknown signal. There will be two interrupts related to the PT0 active edges and zero or more TCNT overflow interrupts to be dealt with.

E8.12 Write a program to generate a waveform with 2 KHz frequency and 60% duty cycle for half of the time and 30% duty cycle the other half of the time.

E8.13 Write a program to generate 10 pulses from the PT6 pin. Each pulse has 60 ms high time and 40 ms low time.

E8.14 Write a program to generate an interrupt to the 68HC12 20 ms after the rising edge on the pin PT2 has been detected.

E8.15 Suppose the contents of the TCTL1 and TCTL2 registers are $79 and $9B, respectively. The contents of the TFLG1 are $00. What would occur on pins PT7 to PT0 on the next clock cycle if the value $7F is written into the CFORC register?

E8.16 What is the slowest clock signal that can be generated from the PWM output?

E8.17 Write the assembly language version of the program for Example 8.8.

E8.18 Write an instruction sequence to configure the modulus down counter so that it generates periodic interrupts to the microcontroller every 100 ms.

8.14 Lab Exercises & Assignments

L8.1 *Frequency measurement.* Use the pulse accumulator function to measure the frequency of an unknown signal. The procedure is as follows:

Step 1
Set the function generator output to be a square wave and adjust the output to between 0 and 5V. Connect the signal to the PAI (PT7) pin.

Step 2
Connect the signal to an oscilloscope or a frequency counter. This is for verification purposes.

Step 3
Output the message: "Do you want to continue to measure the frequency? (y/n)".

Step 4
Set up the appropriate frequency of the signal to be measured and enter **y** or **n** to inform the microcontroller if you want to continue the measurement.

Step 5
Your program would read in the answer from the user. If the character read in is *n*, then stop. If the answer is *y*, then repeat the measurement. If the character is something else, then repeat the same question.

Step 6
Perform the measurement and display the frequency in Hz in decimal format on the screen and go back to step 3. Use as many digits as necessary. The output format should look like:

The signal frequency is xxxxx Hz.

Crank up the frequency until the measurement becomes inaccurate. What is the highest frequency that you can measure?

L8.2 *Pulse width measurement.* Use the input-capture function to measure the pulse width. The procedure is as follows:

Step 1
Set the function generator output to be a square wave and adjust the output to between 0 and 5V. Connect the signal to the PAI (PT7) pin.

Step 2
Connect the signal to an oscilloscope or a frequency counter. This is for verification purposes.

Step 3
Output the message: "Do you want to continue to measure the pulse width? (y/n)".

Step 4

Set up the appropriate period (frequency) of the signal to be measured and enter **y** or **n** to inform the microcontroller if you want to continue the measurement.

Step 5

Your program would read in the answer from the user. If the character read in is *n*, then stop. If the answer is *y*, then repeat the measurement. If the character is something else, then repeat the same question.

Step 6

Perform the measurement and display the period in μs in decimal format on the screen and go back to Step 3. Use as many digits as necessary. The output format should look like:

The signal period is xxxxx microseconds.

Crank up the frequency until the measurement becomes inaccurate. What is the shortest period that you can measure?

L8.3 *Waveform generation.* Use the PWM function to generate a digital waveform with a certain frequency and duty cycle. The procedure is as follows:

Step 1

Connect the selected PWM channel pin (say PP3) to the oscilloscope.

Step 2

Output a message to ask the user to enter the frequency of the waveform to be generated. The frequency should be as follows:

10 to 90 Hz (9 steps)
100 to 900 Hz (9 steps)
1000 to 9000 Hz (9 steps)
10000 to 90000 Hz (9 steps)
100000 to 900000 Hz (9 steps)

If an illegal value is entered, ask the user to reenter.

Step 3

Output a message to request the user to enter the duty cycle of the waveform. The duty cycle could be from 10% to 90% (9 steps only). If an illegal value is entered, ask the user to reenter.

Step 4

Read in the value and start the waveform.

9

Serial Interface – SCI & SPI

9.1 Objectives

After completing this chapter, you should be able to:

- explain the four aspects of RS-232 standard

- explain the errors occurred in data transmission

- establish a null-modem connection

- explain the operation of the SCI subsystem

- wire the SCI pins to the RS-232 connector

- program the SCI subsystem to perform data transmission

- describe the operation of the 68HC12 SPI subsystem

- interface an SPI master with one or more SPI slave devices

- use SPI to interface with the shift register HC595

- use SPI to interface with the 8-digit LED display driver MAX7221

- use SPI to interface with the serial EEPROM X25138

9.2 Fundamental Concepts of Serial Communications

So far we have only dealt with parallel data transfer using parallel ports. While using parallel data transfer can satisfy the data transfer requirements for these devices, there are a few drawbacks:

- Parallel data transfer requires many I/O pins. This requirement prevents the microcontroller from interfacing with as many devices as desired in the application.

- Many I/O devices do not have a high enough data rate to justify the use of parallel data transfer.

- Data synchronization for parallel transfer is difficult to achieve over a long distance. This requirement is one of the reasons that data communications are always using serial transfer.

- The cost is higher.

The *serial communication interface* (SCI) was designed to transfer data in an asynchronous mode that utilizes the industrial standard EIA-232 protocol. The EIA-232 was originally called RS-232 due to the fact that it is a recommended standard. You have been using this interface to communicate with and download programs onto the demo board for execution. Only two wires are used by the SCI function.

The *serial peripheral interface* (SPI) was proposed by Motorola to facilitate the data exchange between the microcontrollers designed by Motorola and peripheral devices. This interface is synchronous due to the fact that it requires a clock signal to synchronize the data transfer between the master and the slave devices. In the SPI protocol, a *master* is a device that can initiate data transfer and a *slave* is a device that can only respond to the data transfer requests. Four pins are used by the SPI function.

9.3 The RS-232 Standard

The RS-232 standard was established in 1960 by the Electronic Industry Association (EIA) for interfacing between a computer and a modem. It has experienced several revisions since then. The latest revision, RS-232-E, was published in July 1991. In this revision, EIA has replaced the prefix RS with the prefix EIA. This change represents no change in the standard, but was made to allow users to identify the source of the standard. We will refer to the standard as both RS-232 and EIA-232 throughout this chapter. In data communication terms, both computers and terminals are called *data terminal equipment* (DTE), whereas modems, bridges, and routers are called *data communication equipment* (DCE).

There are four aspects to the EIA-232 standard:

1. Electrical specifications—specifies the voltage level, rise time and fall time of each signal, achievable data rate, and the distance of communication

2. Functional specifications—specifies the function of each signal

3. Mechanical specifications—specifies the number of pins and the shape and dimensions of the connectors

4. Procedural specifications—specifies the sequence of events for transmitting data, based on the functional specifications of the interface

9.3.1 EIA-232 Electrical Specifications

The following electrical specifications of the EIA-232-E are of interest to us:

- The interface is rated at a signal rate less than 20 kbps. With good design, however, we can achieve a higher data rate.

- The signal can transfer correctly within 15 meters. Greater distance can be achieved with good design.

- Driver maximum output voltage (when in open circuit) is –25V to +25V.

- Driver minimum output voltage (loaded output) is –25V to –5V or +5V to +25V.

- Receiver input voltage range is –25V to +25V.

- A voltage more negative than –3V at the receiver's input is interpreted as logic one.

- A voltage more positive than +3V at the receiver's input is interpreted as logic zero.

9.3.2 EIA-232-E Functional Specifications

The EIA-232-E interface provides one primary channel and one secondary channel. Table 9.1 summarizes the functional specification of each circuit. Because the secondary channel is rarely used, only the functions of signals in the primary channel will be discussed. There is one data circuit in each direction, so full-duplex operation is possible. One ground lead is for protective isolation; the other serves as the return circuit for both data leads. Hence transmission is unbalanced, with only one active wire. The timing signals provide clock pulses for synchronous transmission. When the DCE is sending data over circuit BB, it also sends 1-0 and 0-1 transitions on circuit DD, with transitions timed to the middle of each BB signal element. When the DTE is sending data, either the DTE or DCE can provide timing pulses, depending on the circumstances. The control signals are explained in the procedural specifications.

Pin No.	Circuit	Description
1	-	Shield
2	BA	Transmitted data
3	BB	Received data
4	CA/CJ	Request to send/ready for receiving[1]
5	CB	Clear to send
6	CC	DCE ready
7	AB	Signal common
8	CF	Received line signal detector
9	-	(reserved for testing)
10	-	(reserved for testing)
11	-	unassigned[3]
12	SCF/CI	Secondary received line signal detection/data rate selector (DCE source)[2]
13	SCB	Secondary clear to send
14	SBA	Secondary transmitted data
15	DB	Transmitter signal element timing (DCE source)
16	SBB	Secondary received data
17	DD	Receiver signal element timing
18	LL	Local loopback
19	SCA	Secondary request to send
20	CD	DTE ready
21	RL/CG	Remote loopback/signal quality detector
22	CE	Ring indicator
23	CH/CI	Data signal rate selector (DTE/DCE source)[2]
24	DA	Transmitter signal element timing (DTE source)
25	TM	Test mode

1. When hardware flow control is required, circuit CA may take on the functionality of circuit CJ. This is one change from the former EIA-232.
2. For designs using interchange circuit SCF, interchange circuits CH and CI are assigned to pin 23. If SCF is not used, CI is assigned to pin 12.
3. Pin 11 is unassigned. It will not be assigned in future versions of EIA-232. However, in international standard ISO 2110, this pin is assigned to select transmit frequency.

Table 9.1 ■ Functions of EIA-232-E signals

9.3.3 EIA-232-E Mechanical Specifications

The EIA-232-E uses a 25-pin D-type connector, as shown in Figure 9.1. The 25-pin connector is available from many vendors. The exact dimensions can be found in the standard documentation published by EIA.

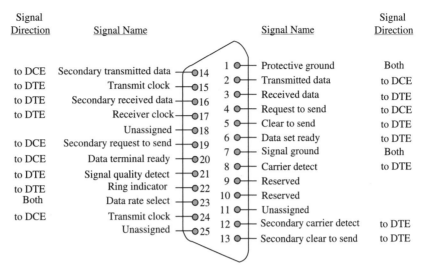

Signal Direction	Signal Name			Signal Name	Signal Direction
to DCE	Secondary transmitted data	14	1	Protective ground	Both
to DTE	Transmit clock	15	2	Transmitted data	to DCE
to DTE	Secondary received data	16	3	Received data	to DTE
to DTE	Receiver clock	17	4	Request to send	to DCE
	Unassigned	18	5	Clear to send	to DTE
to DCE	Secondary request to send	19	6	Data set ready	to DTE
to DCE	Data terminal ready	20	7	Signal ground	Both
to DTE	Signal quality detect	21	8	Carrier detect	to DTE
to DTE	Ring indicator	22	9	Reserved	
Both	Data rate select	23	10	Reserved	
to DCE	Transmit clock	24	11	Unassigned	
	Unassigned	25	12	Secondary carrier detect	to DTE
			13	Secondary clear to send	to DTE

Figure 9.1 ■ EIA-232-E connector and pin assignment

9.3.4 EIA-232-E Procedural Specifications

The sequence of events that occurs during data transmission using the EIA-232-E is easier to understand by studying examples. We will use two examples to explain the procedure.

In the first example, two DTEs are connected with a point-to-point link using a modem. The modem requires only the following circuits to operate:

- Signal ground (AB)
- Transmitted data (BA)
- Received data (BB)
- Request to send (CA)
- Clear to send (CB)
- Data set ready (CC)
- Carrier detect (CF)

Before the DTE can transmit data, the *data-set-ready* (DSR) circuit must be asserted to indicate that the modem is ready to operate. This signal should be asserted before the DTE attempts to make a request to send data. The DSR pin can simply be connected to the power supply of the DCE to indicate that it is switched on and ready to operate. When a DTE is ready to send data, it asserts the *request-to-send* (RTS) signal. The modem responds, when ready, with *clear-to-send* (CTS), indicating that data may be transmitted over circuit BA. If the arrangement is half-duplex, then the RTS also inhibits the receive mode. The DTE sends data to the local modem bit-serially. The local modem modulates the data into the carrier signal and transmits the resultant signal over the dedicated communication lines. Before sending out modulated data, the local modem sends out a carrier signal to the remote modem so that the remote modem is ready to receive the data. The remote modem detects the carrier and asserts the *data-carrier-detect* (DCD) signal. The assertion of the DCD signal tells the remote DTE that the local modem is transmitting. The remote modem receives the modulated signal, demodulates it to recover the data, and sends it to the remote DTE over the received-data pin. The circuit connections are illustrated in Figure 9.2.

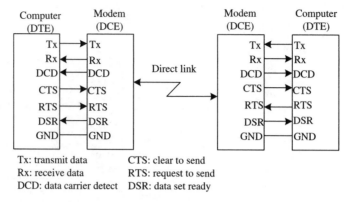

Tx: transmit data   CTS: clear to send
Rx: receive data   RTS: request to send
DCD: data carrier detect   DSR: data set ready

Figure 9.2 ■ Point-to-point asynchronous connection

The next example involves two computers exchanging data through a public telephone line. One of the computers (initiator) must dial the phone (automatically or manually) to establish the connection, just like people talking over the phone. Two additional leads are required for this application:

- Data terminal ready (DTR)
- Ring indicator (RI)

The data transmission in this setting can be divided into three phases:

PHASE 1

Establishing the connection. The following events occur in this phase:

1. The transmitting computer asserts the *data terminal ready* (DTR) signal to indicate to the local modem that it is ready to make a call.

2. The local modem opens the phone line and dials the destination number. The number can be stored in the modem or transmitted to the modem by the computer via the transmit-data pin.

3. The remote modem detects a ring on the phone line and asserts the *ring indicator* (RI) to inform the remote computer that a call has arrived.

4. The remote computer asserts the DTR signal to accept the call.

5. The remote modem answers the call by sending a carrier signal to the local modem via the phone line. It also asserts the DSR signal to inform the remote computer that it is ready for data transmission.

6. The local modem asserts both DSR and DCD signals to indicate that the connection is established and it is ready for data communication.

7. For full-duplex data communication, the local modem also sends a carrier signal to the remote modem. The remote modem then asserts the DCD signal.

PHASE 2

Data Transmission. The following events occur during this phase:

1. The local computer asserts the RTS signal when it is ready to send data.

2. The local modem responds by asserting the CTS signal.

3. The local computer sends data bit-serially to the local modem over the transmit-data pin. The local modem then modulates its carrier signal to transmit the data to the remote modem.

4. The remote modem receives the modulated signal from the local modem, demodulates it to recover the data, and sends it to the remote computer over the received-data pin.

PHASE 3

Disconnection. Disconnection requires only two steps:

1. When the local computer has finished the data transmission, it drops the RTS signal.

2. The local modem then de-asserts the CTS signal and drops the carrier (equivalent to hanging up the phone).

The circuit connection for this example is shown in Figure 9.3.

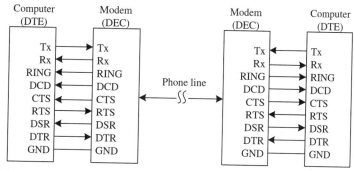

Figure 9.3 ■ Asynchronous connection over public phone line

A timing signal is not required in an asynchronous transmission.

9.3.5 Data Format

In the asynchronous data transfer, data are transferred character by character. Each character is preceded by a start bit (a low), followed by seven or eight data bits, and terminated by one to two stop bits. The data format of a character is shown in Figure 9.4.

Start bit	0	1	2	3	4	5	6	7	Stop bit 1	Stop bit 2

Figure 9.4 ■ The format of a character

As shown in Figure 9.4, the least significant bit is transmitted first, and the most significant bit is transmitted last. Stop bit(s) is(are) high. The start bit and stop bit identify the start and end of a character.

Since there is no clock information in the asynchronous format, the receiver uses a clock signal with a frequency that is a multiple (usually 16) of the data rate to sample the incoming data in order to detect the arrival of the start bit and determine the logical value of each data bit. A clock, with a frequency that is 16 times the data rate, can tolerate a difference of about 5% in the clocks at the transmitter and receiver.

The method for detecting the arrival of a start bit is similar among all microcontrollers:

When the RxD pin is *idle* (*high*) for at least three sampling times and then is followed by a low voltage, the SCI will look at the third, fifth, and seventh samples after the first low sample (these are called verification samples) to determine if a valid start bit has arrived. This process is illustrated in Figure 9.5. If the majority of these three samples are low, then a valid start bit is detected. Otherwise, the SCI will restart the process. After detecting a valid start bit, the SCI will start to shift in the data bits.

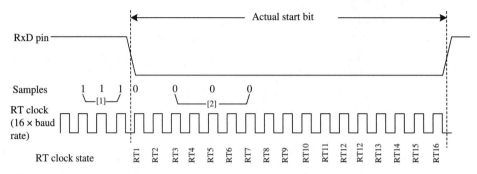

1. A 0 following three 1s.
2. Majority of samples 3, 5, and 7 are 0s

Figure 9.5 ■ Detection of start bit (ideal case)

The method for determining the data bit value is also similar for most microcontrollers:

The SCI uses a clock with frequency about 16 times (most often) that of the data rate to sample the RxD pin. If the majority of the eighth, ninth, and tenth samples are ones, then the data is determined to be one. Otherwise, the data bit is determined to be zero.

The stop bit is high. In older designs, a character can be terminated by one, one-and-a-half, or two stop bits. In newer designs, only one stop bit is used. Using this format, it is possible to transfer data character by character without any gaps.

The term *baud rate* is defined as the number of bit changes per second. Since the RS-232 standard uses a *non-return-to-zero* (NRZ) encoding method, baud rate is identical to bit rate.

Example 9.1

▼

Sketch the output of the letter *g* when it is transmitted using the format of one start bit, 8 data bits, and 1 stop bit.

Solution: Letters are represented in ASCII code. The ASCII code of letter *g* is $67 (= 01100111). Remember that the least significant bit goes out first. The format of the output of letter *g* is shown in Figure 9.6.

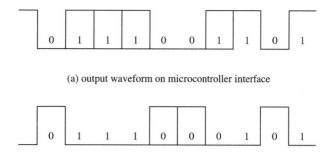

(a) output waveform on microcontroller interface

(b) output waveform on EIA-232-E interface

Figure 9.6 ■ Data format for letter *g*

9.3.6 Data Transmission Errors

The following errors may occur during the data transfer process using asynchronous serial transmission:

- *Framing error.* A framing error occurs when a received character is improperly framed by the start and stop bits; it is detected by the absence of the stop bit. This error indicates a synchronization problem, faulty transmission, or a break condition.

- *Receiver overrun.* One or more characters in the data stream were received but were not read from the buffer before subsequent characters were received.

- *Parity error.* A parity error occurs when an odd number of bits change value. It can be detected by a parity error detecting circuit.

9.3.7 Null Modem Connection

One of the most popular applications of the EIA-232 interface in today's microprocessor laboratory is to connect the PC to a single-board computer (also called a demo board). Both the PC and the single board computer are DTEs, but the EIA standard doesn't allow for a direct connection between two DTEs. In order to make this scheme to work, a *null modem* is needed—the null modem interconnects leads in such a way as to fool both DTEs into thinking that they are connected to modems. The null modem connection is shown in Figure 9.7.

In Figure 9.7, the transmitter timing and receiver timing signals are not needed in asynchronous data transmission. The ring indicator (RI) signal is not needed either, because the transmission is not made through a public phone line.

Pin	Circuit name	DTE	DTE
22	Ring indicator	CE	CE
20	Data terminal ready	CD	CD
8	Data carrier detect	CF	CF
6	Data set ready	CC	CC
5	Clear to send	CB	CB
4	Request to send	CA	CA
3	Received data	BB	BB
2	Transmitted data	BA	BA
24	Transmitter timing	DA	DA
17	Receiver timing	DD	DD
7	Signal ground	AB	AB

Figure 9.7 ■ Null Modem connection

9.4 The 68HC12 Serial Communication Interface

A block diagram of the 68HC12 multiple serial interfaces is shown in Figure 9.8. All except B family members have two identical SCI modules (SCI0 and SCI1). The 68HC12 SCI interface uses a data format of one start bit, eight or nine data bits, and one stop bit. When the SCI is configured to use nine data bits, one of the bits can be used as the parity bit. The collection of the start bit, data bits, and the stop bit is called a *frame*.

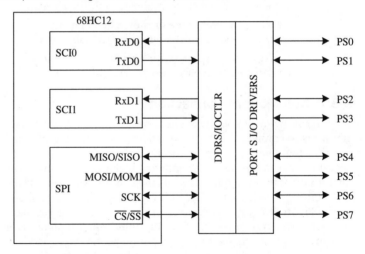

Note. B family members (912B32, 912BC32, and 912BE32) do not have SCI1

Figure 9.8 ■ Multiple serial interface (SCIs and SPI) block diagram

One SCI channel uses two signal pins from the port S: The channel SCI0 uses pins PS1 and PS0 whereas the channel SCI1 uses pins PS3 and PS2. The PS0 pin (PS2) is used for received data whereas the PS1 pin (PS3) is used for transmitted data. The SCI function has the capability to send a break to attract the attention of the other party of data communication. A *break* is defined as the transmission or reception of logic zero for one frame or longer time.

The SCI function supports hardware parity for transmission and reception. When enabled, a parity bit is generated in hardware for transmitted data and received data. Received parity errors are flagged in hardware.

9.4.1 SCI Baud Rate Generation

As we stated earlier, the computer system uses a clock signal with frequency equal to an integral multiple (for example, 16 times) of the data rate to detect the arrival of the start bit and determine the logic value of data bits. The 68HC12 SCI module uses a 13-bit counter to generate this clock signal. This circuit is called a *baud rate generator*.

This baud rate generator divides down the P clock to derive the clock signal for reception and transmission. The divide factor for the baud rate generator is listed in Table 9.2.

Desired SCI baud rate	Baud rate divisor for P = 4.0 MHz	Baud rate divisor for P = 8.0 MHz
110	2273	4545
300	833	1667
600	417	833
1200	208	417
2400	104	208
4800	52	104
9600	26	52
14,400	17	35
19,200	13	26
38,400	--	13

Table 9.2 ■ Baud rate generation

9.4.2 Registers Associated with SCI

The baud rate clock signal is generated by the 16-bit SCI baud rate control register. Its contents are shown in Figure 9.9.

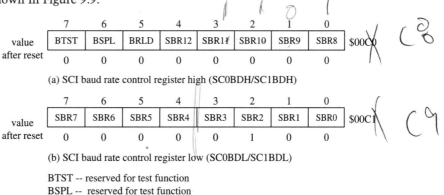

	7	6	5	4	3	2	1	0	
value	BTST	BSPL	BRLD	SBR12	SBR11	SBR10	SBR9	SBR8	$00C0
after reset	0	0	0	0	0	0	0	0	

(a) SCI baud rate control register high (SC0BDH/SC1BDH)

	7	6	5	4	3	2	1	0	
value	SBR7	SBR6	SBR5	SBR4	SBR3	SBR2	SBR1	SBR0	$00C1
after reset	0	0	0	0	0	1	0	0	

(b) SCI baud rate control register low (SC0BDL/SC1BDL)

BTST -- reserved for test function
BSPL -- reserved for test function
BRLD -- reserved for test function

Figure 9.9 ■ SCI baud rate control register

The low byte of the SCI baud rate control register must be written in order for change to take place. The 13-bit value is referred to as *BR*. This value determines the baud rate of the SCI function. The desired baud rate is determined by the following formula:

SCI baud rate = $f_p \div (16 \times BR)$ ---- (9.1)

where f_p is the P clock frequency and BR is the value written to bits SBR12-SBR0 to establish the baud rate.

The SCI0 (SCI1) has two other control registers that set up other parameters: SC0CR1 (SC1CR1) and SC0CR2 (SC1CR2). Their contents are shown in Figures 9.10 and 9.11, respectively.

The SCI can be configured to operate in loop mode (when bit 7 = 1) or normal mode (bit 7 = 0). The loop mode releases the RxD pin to be used as a general-purpose I/O pin. There are several variations in the loop mode, as shown in Table 9.3.

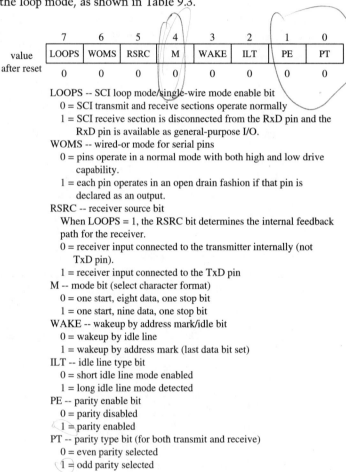

	7	6	5	4	3	2	1	0	
value	LOOPS	WOMS	RSRC	M	WAKE	ILT	PE	PT	$00C2/$00CA
after reset	0	0	0	0	0	0	0	0	

LOOPS -- SCI loop mode/single-wire mode enable bit
    0 = SCI transmit and receive sections operate normally
    1 = SCI receive section is disconnected from the RxD pin and the
        RxD pin is available as general-purpose I/O.
WOMS -- wired-or mode for serial pins
    0 = pins operate in a normal mode with both high and low drive
        capability.
    1 = each pin operates in an open drain fashion if that pin is
        declared as an output.
RSRC -- receiver source bit
    When LOOPS = 1, the RSRC bit determines the internal feedback
    path for the receiver.
    0 = receiver input connected to the transmitter internally (not
        TxD pin).
    1 = receiver input connected to the TxD pin
M -- mode bit (select character format)
    0 = one start, eight data, one stop bit
    1 = one start, nine data, one stop bit
WAKE -- wakeup by address mark/idle bit
    0 = wakeup by idle line
    1 = wakeup by address mark (last data bit set)
ILT -- idle line type bit
    0 = short idle line mode enabled
    1 = long idle line mode detected
PE -- parity enable bit
    0 = parity disabled
    1 = parity enabled
PT -- parity type bit (for both transmit and receive)
    0 = even parity selected
    1 = odd parity selected

Figure 9.10 ■ SCI control register 1 (SC0CR1/SC1CR1)

	7	6	5	4	3	2	1	0	
value	TIE	TCIE	RIE	ILIE	TE	RE	RWU	SBK	$00C3/$00CB
after reset	0	0	0	0	0	0	0	0	

TIE -- transmit interrupt enable bit
    0 = TDRE interrupt disabled
    1 = SCI interrupt requested when TDRE status flag is set.
TCIE -- transmit complete interrupt enable bit
    0 = TC interrupt disabled
    1 = SCI interrupt requested when TC flag is set
RIE -- receiver interrupt enable bit
    0 = RDRF and OR interrupts disabled
    1 = SCI interrupt requested when RDRF status flag or OR status
       flag is set
ILIE -- idle line interrupt enable bit
    0 = IDLE interrupt disabled
    1 = SCI interrupt requested when IDLE status flag is set
TE -- transmit enable bit
    0 = transmitter disabled
    1 = SCI transmit logic is enabled and the TxD pin is dedicated to
       the transmitter. The TE bit can be used to queue an idle
       preamble
RE -- receiver enable
    0 = receiver disabled
    1 = enable the SCI receive circuitry
RWU -- receiver wakeup control bit
    0 = normal SCI receiver
    1 = enables the wakeup function and inhibits further receiver
       interrupts. Normally, hardware wakes the receiver by
       automatically clearing this bit.
SBK -- send break bit
    0 = break generator off
    1 = generate a break code, at least 10 or 11 contiguous 0s. As long
       as SBK remains set, the transmitter sends 0s.

Figure 9.11 ■ SCI control register 2 (SC0CR2/SC1CR2)

LOOPS	RSRC	DDRS1	WOMS	Function of Port S bit 1/3
0	x	x	x	Normal operation
1	0	0	0/1	Loop mode without TxD output (= high impedance)
1	0	1	0	Loop mode with TxD output (CMOS)
1	0	1	1	Loop mode with TxD output (open drain)
1	1	0	x	Single-wire mode without TxD output (pin is used as receiver input only, TxD = high impedance)
1	1	1	0	Single-wire mode without TxD output (pin is used as receiver input only, TxD = high impedance)
1	1	1	1	Single-wire mode for the receiving and transmitting (open drain)

Table 9.3 ■ Loop mode functions

The SCI interface allows one or more 68HC12 microcontrollers and other microcontrollers to be connected in a *multi-drop* environment in which one microcontroller is designated as the *master* while other microcontrollers are *slaves*. An address is assigned to all microcontrollers so that they can specify which microcontroller they intend to communicate with. Any microcontroller that intends to communicate must obtain the consent of the master. A communication protocol must be established for all microcontrollers in the network to follow in order for the communication to proceed correctly.

In a multi-drop environment, software for each SCI receiver evaluates the first character (which represents an address) of each message. If the message is intended for a different receiver, the SCI puts itself in a sleep mode so that the rest of the message will not generate requests for service. Whenever a new message is started, logic in the sleeping receivers causes them to wake up so they can evaluate the initial character(s) of the new message.

The 68HC12 allows the user to choose between two methods to wake up the SCI interface:

- *Idle-line wakeup.* Whenever the RxD line becomes idle for one frame time or longer, the SCI circuit will be awakened. Systems using this type of wakeup method must provide at least one character (frame) time of idleness between messages in order to wake up sleeping receivers, but they must not allow any idle time between characters within a message.

- *Address mark wake-up.* The *most significant bit* (MSB) of the character is used to indicate whether the character is an address (1) or a data (0) character. The sleeping receiver will wake up whenever an address mark character is received. Systems using this method to wake up set the MSB of the first character in each message and leave it clear for all other characters in the message. Idle periods may occur within messages, and no idle time is required between messages for this wakeup method.

The bit 2 of the SCI control register 1 (SC0CR1 or SC1CR1) determines when the receiver starts counting logic 1 as idle character bits. The counting begins either after the start bit or after the stop bit. If the count begins after the start bit, then a string of ones preceding the stop bit may cause false recognition of an idle character. Beginning the count after the stop bit avoids false idle character recognition, but requires properly synchronized transmissions.

To use parity checking, we need to enable (set the PE bit) parity checking and also specify the type of parity (odd or even) to use.

The SCI control register 2 (SC0CR2 and SC1CR2) is used to enable/disable four SCI-related interrupts, enable or disable transmission and reception, control the receiver wakeup, and send out break characters.

Each SCI channel has two status registers to record the status of SCI operation. These two registers are SC0SR1 (SC1SR1) and SC0SR2 (SC1SR2). Their contents are shown in Figures 9.12 and 9.13.

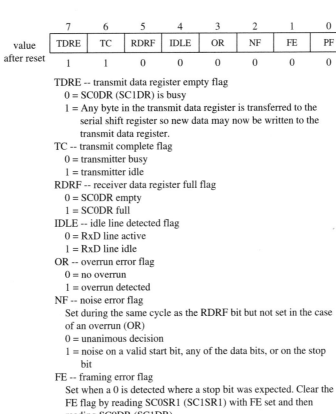

7	6	5	4	3	2	1	0	
TDRE	TC	RDRF	IDLE	OR	NF	FE	PF	$00C4/$00CC

value after reset: 1 1 0 0 0 0 0 0

TDRE -- transmit data register empty flag
  0 = SC0DR (SC1DR) is busy
  1 = Any byte in the transmit data register is transferred to the
     serial shift register so new data may now be written to the
     transmit data register.
TC -- transmit complete flag
  0 = transmitter busy
  1 = transmitter idle
RDRF -- receiver data register full flag
  0 = SC0DR empty
  1 = SC0DR full
IDLE -- idle line detected flag
  0 = RxD line active
  1 = RxD line idle
OR -- overrun error flag
  0 = no overrun
  1 = overrun detected
NF -- noise error flag
  Set during the same cycle as the RDRF bit but not set in the case
  of an overrun (OR)
  0 = unanimous decision
  1 = noise on a valid start bit, any of the data bits, or on the stop
    bit
FE -- framing error flag
  Set when a 0 is detected where a stop bit was expected. Clear the
  FE flag by reading SC0SR1 (SC1SR1) with FE set and then
  reading SC0DR (SC1DR).
  0 = stop bit detected
  1 = zero detected rather than a stop bit
PF -- parity error flag
  0 = parity correct
  1 = incorrect parity detected

Figure 9.12 ■ SCI status register 1 (SC0SR1/SC1SR1)

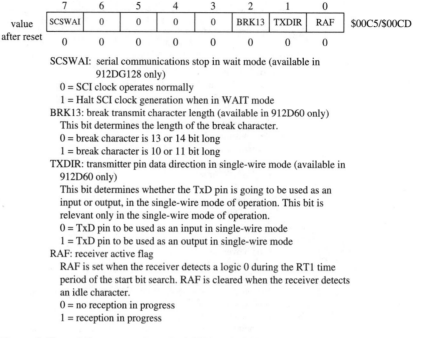

	7	6	5	4	3	2	1	0	
value	SCSWAI	0	0	0	0	BRK13	TXDIR	RAF	$00C5/$00CD
after reset	0	0	0	0	0	0	0	0	

SCSWAI: serial communications stop in wait mode (available in
912DG128 only)
0 = SCI clock operates normally
1 = Halt SCI clock generation when in WAIT mode
BRK13: break transmit character length (available in 912D60 only)
This bit determines the length of the break character.
0 = break character is 13 or 14 bit long
1 = break character is 10 or 11 bit long
TXDIR: transmitter pin data direction in single-wire mode (available in
912D60 only)
This bit determines whether the TxD pin is going to be used as an
input or output, in the single-wire mode of operation. This bit is
relevant only in the single-wire mode of operation.
0 = TxD pin to be used as an input in single-wire mode
1 = TxD pin to be used as an output in single-wire mode
RAF: receiver active flag
RAF is set when the receiver detects a logic 0 during the RT1 time
period of the start bit search. RAF is cleared when the receiver detects
an idle character.
0 = no reception in progress
1 = reception in progress

Figure 9.13 ■ SCI status register 2 (SC0SR2/SC1SR2)

The TDRE flag and TC flag represent different situations. When the byte in the SC0DRL (or SC1DRL) register is transferred to the transmit shift register, the TDRE flag is set to one and allows a new byte to be written into the SC0DRL register. The TC flag will be set only when the transmitter is idle.

The NF flag will be set whenever one of the three samples is different from the other two used to detect the start bit, or determine the stop bit or value of the data bit. The setting of the NF flag may not be an error condition. However, the setting of any one of the OR, FE, or PF flags is an error condition.

The receive-related flag bits in SC0SR1 (or SC1SR1) are cleared by a read of the SC0SR1 register followed by a read of the transmit/receive data register low byte. The transmit-related bits in SC0SR1 (TDRE and TC) are cleared by a read of the SC0SR1 register followed by a write to the transmit/receive data register low byte.

The byte to be transmitted is written into the SCI transmit data register low. The incoming data byte is shifted in and transferred to the received data register low. These two registers share the same address (in each SCI channel). Two SCI data registers are implemented to support the 9-bit data format. When the 8-bit data format is selected, we only need to deal with the SC0DRL (SC1DRL) register. These two registers are shown in Figures 9.14 and 9.15.

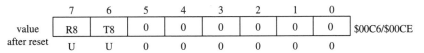

	7	6	5	4	3	2	1	0	
value	R8	T8	0	0	0	0	0	0	$00C6/$00CE
after reset	U	U	0	0	0	0	0	0	

Bit 6 to 0 cannot be written but will be read as 0s

Figure 9.14 ■ SCI data register high (SCODRH/SC1DRH)

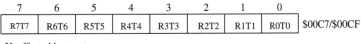

7	6	5	4	3	2	1	0	
R7T7	R6T6	R5T5	R4T4	R3T3	R2T2	R1T1	R0T0	$00C7/$00CF

Unaffected by reset

Figure 9.15 ■ SCI data register low (SCODRL/SC1DRL)

When the 9-bit format is used, the received bit 8 is stored in the R8 bit of the SCI data register high. The bit 8 to be transmitted is written into the T8 bit. When using the 9-bit data format, the T8 bit does not have to be written for each data byte. The same value is transmitted as the ninth bit until this bit is written.

Example 9.2
▼

Write an instruction sequence to configure the SCI channel 0 to operate with the following parameters:

- 9600 baud (P clock is 8 MHz)
- 1 start bit, 8 data bits, and 1 stop bit format
- no interrupt
- address mark wakeup
- disable wakeup initially
- long idle line mode
- enable receive and transmit
- no loop back
- disable parity

Solution: The following instruction sequence will achieve the desired configuration:

```
movb    #$00,SCOBDH    ; set up baud rate
movb    #52,SCOBDL     ;      "
movb    #$0C,SCOCR1
movb    #$0C,SCOCR2
```

The equivalent C statements are as follows:

```
SCOBDH = 0x00;
SCOBDL = 52;
SCOCR1 = 0x0C;
SCOCR2 = 0x0C;
```

▲

9.5 Interfacing SCI with EIA-232

Because the SCI circuit uses 0V and 5V to represent logic zero and one, respectively, it cannot be connected to the EIA-232 interface circuit directly. A voltage translation circuit, which is called the *EIA-232 transceiver*, is needed to translate the voltage levels of the SCI signals (RxD and TxD) to and from those of the corresponding EIA-232 signals.

EIA-232 transceiver chips are available from many vendors. The LT1080/1081 from Linear technology, the ST232 from SGS Thompson, the ICL232 from Intersil, the MAX232 from MAXIM, and the DS14C232 from National Semiconductor are EIA-232 transceiver chips that can operate with a single 5V power supply and generate EIA-232-compatible outputs. These chips are also pin-compatible.

In this section, we will discuss the use of the DS14C232 chip from National Semiconductor to perform the voltage translation. The pin assignment and the use of each pin are shown in Figure 9.16.

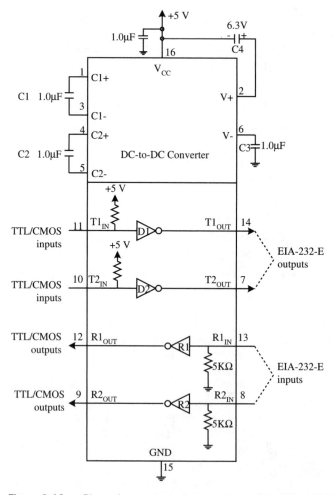

Figure 9.16 ■ Pin assignments and connections of the DS14C232

Adding an EIA-232 transceiver chip will then allow the 68HC12 to use the SCI interface to communicate with the EIA-232 interface circuit. An example of such a circuit is shown in Figure 9.17. A null-modem wiring is followed in this circuit.

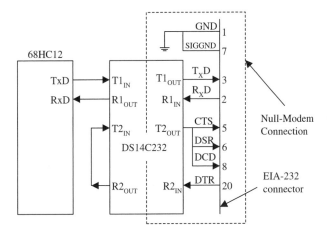

Figure 9.17 ■ Diagram of the SCI and EIA-232 circuit connection

Example 9.3

Write a subroutine to send a break to the communication port controlled by the SCI0 interface. The duration of the break is approximately 40,000 E clock cycles, or 5 ms at 8 MHz.

Solution: A break character is represented by 10 or 11 consecutive zeros and can be sent out by the bit 0 of the SC0CR2 register. As long as bit 0 of the SC0CR2 register remains set, the SCI will keep sending out the break character. The example assembly subroutine is as follows:

```
#include "d:\miniide\hc12.inc"

sendbrk    bset    SC0CR2,$01      ; turn on send break
           jsr     wait_5ms
           bclr    SC0CR2,$01      ; turn off send break
           rts
wait_5ms   movb    #$90,TSCR       ; enable TCNT and fast flag clear
           movb    #$00,TMSK2      ; set prescale factor to one
           movb    #$01,TIOS       ; select OC0 function
           ldd     TCNT
           addd    #40000
           std     TC0             ; start an OC0 operation
           brclr   TFLG1,$01,*     ; wait for 5 ms
           rts
```

The C language version of the function is as follows:

```
void send_break (void)
{
     SC0CR2 |= 0x01;          /* start to send break /
     TSCR = 0x90;             /* enable TCNT and fast flag clear */
     TMSK2 = 0x00;            /* set prescale factor to one */
     TIOS = 0x01;             /* select OC0 function */
```

```
TC0 = TCNT + 40000u;      /* start an OC0 operation */
while (!(TFLG1 & 0x01));   /* wait for 5 ms */
SC0CR2 &= 0xFE;           /* stop sending break */
}
```

Example 9.4

Write a subroutine to output the character in accumulator A to the SCI0 channel using the polling method.

Solution: A new character should be sent out only when the transmit data register is empty. When the polling method is used, the subroutine will wait until the bit 7 of the SC0SR1 register is set before sending out the character in accumulator A. The assembly language function is as follows:

```
#include "d:\miniide\hc12.inc"

putc_sc0      brclr      SC0SR1,$80,*    ; wait for TDRE to be set
              staa       SC0DRL          ; output the character
              rts
```

The C language version of the function is as follows:

```
void putc_sc0 (char cx)
{
      while (!(SC0SR1 & 0x80));
      SC0DRL = cx;
}
```

Example 9.5

Write a subroutine to read a character from the SCI channel 0 using the polling method. The character will be returned in accumulator A.

Solution: Using the polling method, the subroutine will wait until the RDRF bit (bit 5) of the SC0SR1 register becomes set and then read the character held in the SC0DRL register. The subroutine that reads a character from SCI channel 0 is as follows:

```
#include "d:\miniide\hc12.inc"

getc_sc0      brclr      SC0SR1,$20,*    ; wait until RDRF bit is set
              ldaa       SC0DRL          ; read the character
              rts
```

The C language version of the function is as follows:

```
char  getc_sc0 (void)
{
      while(!(SC0SR1 & 0x20));
      return (SC0DRL);
}
```

Example 9.6

Write a subroutine that outputs a string pointed to by index register X to the SCI channel 0 using the polling method.

Solution: The subroutine will call *putc_sc0()* repeatedly until all characters of the string have been output.

```
puts_sc0    ldaa    1,x+          ; get a character and move the pointer
            beq     done          ; is this the end of the string?
            jsr     putc_sc0
            bra     puts_sc0
done        rts
```

The C language version of the program is as follows:

```
void    puts_sc0 (char *cx)
{
        while (!(*cx++))
            putc_sc0(*cx);
}
```

Example 9.7

Write a subroutine that inputs a string from the SCI channel 0. The string is terminated by the carriage return character and must be stored in a buffer pointed to by index register X.

Solution: The subroutine will call *getc_sc0()* repeatedly until the carriage return character is input.

```
gets_sc0    jsr     getc_sc0
            cmpa    #CR           ; is the character a carriage return?
            beq     exit
            staa    1,x+          ; save the character in the buffer pointed to by X
            bra     gets_sc0      ; continue
exit        clr     0,x           ; terminate the string with a NULL character
            rts
```

The C language version of the function is as follows:

```
void gets_sc0 (char *buf)
{
        while ((*buf++ = getc_sc0()) != CR);
        *buf = 0;  /* terminate the string with a NULL character */
}
```

The above subroutines can also be implemented using the interrupt approach, as shown in Example 9.8.

Example 9.8

Use the interrupt-driven approach and write a subroutine to output a string pointed to by index register X. The program is to be executed on a board that contains the D-Bug12 monitor. Write an instruction sequence to test your subroutine.

Solution: The subroutine will enable the SCI interrupt and stay in a wait loop to wait for interrupt (caused by transmit data register empty) to occur and output the string. The interrupt service simply outputs a character. The assembly language version of the subroutine is as follows:

```
puts_sc0   bset    SC0CR2,$80    ; enable TDRE interrupt
           cli                   ; enable global interrupt
loop       ldaa    0,x           ; wait for TDRE interrupt to occur
           bne     loop          ; "
           bclr    SC0CR2,$80    ; disable TDRE interrupt
           sei
           rts


sc0_isr    staa    SC0DRL
           inx
           stx     3,sp
           rti
```

The instruction sequence to test this subroutine is as follows:

```
#include "d:\miniide\hc12.inc"
sc0_vec_no equ       11
setuservector equ    $F69A
CR         equ       $0D
LF         equ       $0A

           org       $1000
; the following 7 instructions set up SC0 interrupt jump vector
           ldd       #sc0_isr
           pshd
           ldab      #sc0_vec_no
           clra
           ldx       setuservector
           jsr       0,x
           leas      2,sp

           jsr       init_sc0
           ldx       #msg
           jsr       puts_sc0
           swi
msg        db        "This is the test message!",CR,LF,0


; ************************************************************
; The init_sc0 routine configures the SC0 channel with 9600 baud, no parity, enable
; transmit and receive, select long idle mode.
; ************************************************************
init_sc0   movb      #$00,SC0BDH
           movb      #52,SC0BDL
           movb      #$0C,SC0CR1
```

```
                    movb        #$0C,SC0CR2
                    rts
```

The C language version of the program is as follows:

```
#include <hc12.h>
void sc0_isr (void);
void init_sc0 (void);
void puts_sc0 (char *cx);
char *cx;
void main (void)
{
        char msg[] = "This is the test message!";

        init_sc0();
        asm ("ldd #_sc0_isr");
        asm ("pshd");
        asm ("ldab #11");
        asm ("clra");
        asm ("ldx $F69A");
        asm ("jsr 0,x");
        asm ("leas 2,sp");
        cx = &msg[0];
        puts_sc0 (cx);
        asm("swi");
}
void init_sc0(void)
{
        SC0BDH = 0x00;
        SC0BDL = 52;
        SC0CR1 = 0x0C;
        SC0CR2 = 0x0C;
}
void puts_sc0(char *cx)
{
        SC0CR2 |= 0x80;         /* set the TIE bit */
        INTR_ON();              /* enable transmit data register empty interrupt */
        while (!(*cx));
        SC0CR2 &= 0x7F;         /* disable transmit data register empty interrupt */
        INTR_OFF();
}
#pragma interrupt_handler sc0_isr
void sc0_isr (void)
{
        SC0DRL = *cx++;
}
```

The interrupt-driven version of the *gets_sc0()* subroutine can be written in a similar manner and will therefore be left for you as an exercise problem.

9.6 The SPI Function

The *serial peripheral interface* (SPI) allows the 68HC12 to communicate synchronously with peripheral devices and other microcontrollers. The SPI system in the 68HC12 can operate as a master or as a slave. The SPI is also capable of interprocessor communications in a multiple-master system. All 68HC12 members have one SPI subsystem.

When the SPI is enabled, all pins that are defined by the configuration as inputs will be inputs regardless of the state of the DDRS bits for those pins. All pins that are defined as SPI outputs will be outputs only if the DDRS bits for those pins are set. Any SPI output whose corresponding DDRS bit is cleared can be used as a general-purpose input.

The SPI subsystem is mainly used in interfacing with peripherals such as TTL shift registers, LED/LCD display drivers, phase-locked loop (PLL) chips, memory components with serial interface, or A/D and D/A converter chips that do not need a very high data rate.

The SPI function on the 68HC12 has been modified slightly from that in the 6HC11 with the expectation that it will improve its applicability.

9.6.1 SPI Signal Pins

Four of the port S pins are also used as SPI pins: SDI/MISO (PS4), SDO/MOSI (PS5), SCK (PS6), and $\overline{CS}/\overline{SS}$ (PS7).

- **SDI/MISO**: *master-in-slave-out* (serial data input). This pin is configured as an input in a master device and as an output in a slave device. It is one of the two lines that transfer serial data in one direction with the most significant bit sent first. The MISO line of a slave device is placed in a high-impedance state if the slave device is not selected.

- **SDO/MOSI**: *master-out-slave-in* (serial data output). The MOSI pin is configured as an output in a master device and as an input in a slave device. It is the second of the two lines that transfer serial data in one direction with the most significant bit sent first.

- **SCK**: *serial clock*. The serial clock is used to synchronize data movement both in and out of the device through its SDO/MOSI and SDI/MISO pins. The master and slave devices are capable of exchanging one byte of information during a sequence of eight SCK clock cycles. Since the SCK signal is generated by the master device, this line becomes an input on a slave device.

- $\overline{CS}/\overline{SS}$: *chip select* (or *slave select*). The slave select input line is used to select a slave device. For a slave device, it has to be low prior to data transactions and must stay low for the duration of the transaction. When configured as an input, the $\overline{SS}$ line on the master device must be tied high. If it goes low, the mode error flag in the SPI status register (SP0SR) will be set. When configured as an output and enabled (by setting the SSOE bit in the SP0CR1 register and the DDS7 bit of the DDRS register) in the master mode, this pin can be connected to the SS input pin of a slave. This pin will automatically go low for each transmission to select the external device and go high during each idling state to deselect the external device. The function of this pin is determined by the combination of the DDS7 and SSOE bits as illustrated in Table 9.4.

DDS7	SSOE	Master mode	Slave mode
0	0	$\overline{SS}$ input with MODF feature	$\overline{SS}$ input
0	1	Reserved	$\overline{SS}$ input
1	0	General-purpose output	$\overline{SS}$ input
1	1	$\overline{SS}$ output	$\overline{SS}$ input

Table 9.4 ■ SS output selection

9.6.2 SPI Operation

In the SPI system, the 8-bit data register in the master and the 8-bit data register in the slave are linked together to form a distributed 16-bit register. When a data transfer operation is performed, this 16-bit register is serially shifted eight bit positions by the SCK clock from the master so that the data are effectively exchanged between the master and the slave. Data written into the SP0DR register of the master becomes the output data for the slave, and data read from the SP0DR register of the master after a transfer operation is the input data from the slave.

There are four clock formats to be used by the SPI system. The combinations of the CPOL and CPHA bits of the SP0CR1 register determine these four formats. The CPOL bit determines whether the clock (SCK) is idle low (normal) or high (inverted). The CPHA bit is used to accommodate two fundamentally different protocols by shifting the clock by one half cycle or no phase shift. These four formats are shown in Figure 9.18.

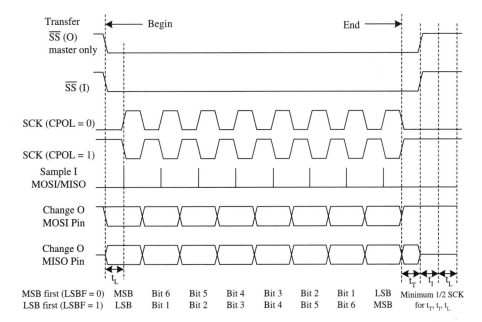

Figure 9.18a ■ SPI Clock format 0 (CPHA = 0)

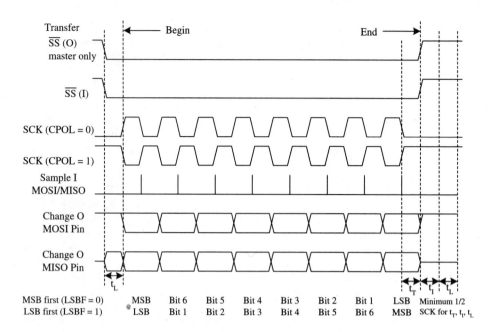

Figure 9.18b ■ SPI Clock format 1 (CPHA = 1)

Sometimes it is more convenient to specify the clock edge for data transfer. The 00 and 11 combinations of CPOL and CPHA are for data transfers using the rising edge of the SCK clock, whereas the 01 and 10 combinations are for data transfers using the falling edge of the SCK clock.

9.6.3 Bidirectional Mode (MOMI or SISO)

The SPI subsystem can be used in bi-directional mode. In this mode, the SPI uses only one serial data pin for an external device interface. The MSTR bit of the SP0CR1 register decides which pin to be used. The MOSI pin becomes a serial data I/O (MOMI) pin for the master mode, and the MISO pin becomes a serial data I/O (SISO) pin for the slave mode. The direction of each serial I/O pin depends on the corresponding DDRS bit. The possible combinations are shown in Figure 9.19.

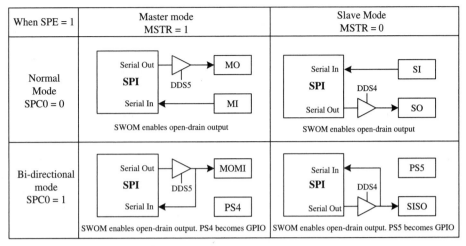

Figure 9.19 ■ Normal mode and bidirectional mode

9.7 Registers Related to the SPI Subsystem

Most of the SPI operational parameters are set by two SPI control registers: SP0CR1 and SP0CR2. Their contents are shown in Figures 9.20 and 9.21.

	7	6	5	4	3	2	1	0	
value	SPIE	SPE	SWOM	MSTR	CPOL	CPHA	SSOE	LSBF	$00D0
after reset	0	0	0	0	0	1	0	0	

SPIE: SPI interrupt enable bit
   0 = SPI interrupts are inhibited
   1 = SPI interrupt is requested every time the SPIF or MODF status
      flag is set.
SPE: SPI system enable bit
   0 = SPI hardware is initialized but is in the disabled state
   1 = SPI enabled and pins PS4-PS7 are dedicated to SPI function
SWOM: Port S wired-OR mode bit
   Controls only pins PS4-PS7.
   0 = PS4-PS7 output buffers operate normally
   1 = PS4-PS7 output buffers behave as open-drain outputs
MSTR: SPI master/slave mode select bit
   0 = slave mode
   1 = master mode
CPOL and CPHA: SPI clock polarity, clock phase bits
   Functions are shown in Figure 9.18a and 9.18b
SSOE: slave select output enable bit
   The $\overline{SS}$ output feature is enabled only in master mode by asserting the
   SSOE and DDRS7.
LSBF: SPI LSB first enable bit
   0 = data is transferred most-significant bit first
   1 = data is transferred least-significant bit first

Figure 9.20 ■ SPI control register 1 (SP0CR1)

7	6	5	4	3	2	1	0	
0	0	0	0	PUPS	RDS	SPSWA	SPC0	$00D1
0	0	0	0	1	0	0	0	

value after reset

PUPS: Pullup port S enable bit (not available in 912D60)
    0 = no internal pullups on port S
    1 = All port S input pins have an active pullup device. If a pin is
        programmed as output, the pullup device becomes inactive.
RDS: Reduce drive of port S bit (not available in 912D60)
    0 = Port S output drivers operate normally
    1 = All port S output pins have reduced drive capability for lower
        power and less noise.
SPSWAI: serial interface stop in wait mode (available only in 912D60,
    912DG128, 912DT128, M9S12DP256)
    0 = Serial port S clock operate normally
    1 = Halt serial interface clock generation in Wait mode
SPC0: Serial pin control 0 bit
    This bit decides serial pin configurations with MSTR control bit. All
    possible combinations are shown in Table 9.5.

Figure 9.21 ■ SPI control register 2 (SP0CR2)

Pin mode		SPC0[1]	MSTR	MISO[2]	MOSI[3]	SCK[4]	SS[5]
#1	normal	0	0	slave out	slave in	SCK in	SS in
#2			1	master in	master out	SCK out	SS I/O
#3	bidirectional	1	0	slave I/O	general-purpose I/O	SCK in	SS in
#4			1	general-purpose I/O	master I/O	SCK out	SS I/O

1. The serial pin control 0 bit enables bidirectional configurations.
2. Slave output is enabled if DDRS4 = 1, SS = 0, and MSTR = 0, (#1, #3)
3. Master output is enabled if DDRS5 = 1 and MSTR = 1, (#2, #4)
4. SCK output is enabled if DDRS6 = 1 and MSTR = 1, (#2, #4)
5. SS output is enabled if DDRS7 = 1, SSOE = 1, and MSTR = 1, (#2, #4)

Table 9.5 ■ MISO and MOSI pin configurations

The SPI must be enabled before it can start data transfer. Setting the bit 6 of the SP0CR1 register will enable the SPI subsystem.

The bit 1 (SSOE) of the SP0CR1 register allows the user to use the SS pin to select the slave device for data transfer automatically. However, this feature is not useful if you want to use the SPI to interface with multiple slave devices at the same time.

Some slave devices may transfer data with the least significant bit first. The bit 0 (LSBF) of the SP0CR1 allows the user this flexibility.

The bit 0 of the SPI control register 2 (SP0CR2), along with the MSTR and DDRS bits, allows the user to have more flexibility in using the SPI pins. This is illustrated in Table 9.5.

The SPI data shift rate (also called the *baud rate*) is programmable. The baud rate is programmed by setting the lowest three bits of the SP0BR register to appropriate values. The contents of the SP0BR register are shown in Figure 9.22.

	7	6	5	4	3	2	1	0	
value	0	0	0	0	0	SPR2	SPR1	SPR0	$00D2
after reset	0	0	0	0	0	0	0	0	

Bit 7 to 3 cannot be written but will be read as 0s

SPR2-SPR0: SPI clock (SCK) rate select bits
The baud rate selection is shown in Table 9.6.

Figure 9.22 ■ SPI baud rate register (SP0BR)

SPR2	SPR1	SPR0	E clock divisor	Frequency at E Clock = 4 MHz	Frequency at E Clock = 8 MHz
0	0	0	2	2.0 MHz	4.0 MHz
0	0	1	4	1.0 MHz	2.0 MHz
0	1	0	8	500 KHz	1.0 MHz
0	1	1	16	250 KHz	500 KHz
1	0	0	32	125 KHz	250 KHz
1	0	1	64	62.5 KHz	125 KHz
1	1	0	128	31.3 KHz	62.5 KHz
1	1	1	256	15.6 KHz	31.3 KHz

Table 9.6 ■ SPI clock rate selection

The SPI has a status register that records the progress of data transfer and errors as shown in Figure 9.23. The application program can check the bit 7 of the SP0SR register or wait for the SPI interrupt to find out if the SPI transfer has completed. When transferring data in higher frequency using the SPI format, using the polling method is more efficient due to the overhead involved in interrupt handling.

	7	6	5	4	3	2	1	0	
value	SPIF	WCOL	0	MODF	0	0	0	0	$00D3
after reset	0	0	0	0	0	0	0	0	

SPIF: SPI interrupt request bit
SPIF is set after the eight SCK cycles in a data transfer, and it is cleared by reading the SP0SR register (with SPIF set) followed by an access to the SPI data register
WCOL: write collision
0 = no write collision
1 = Indicates that a serial transfer is in progress when the microcontroller tried to write new data into the SP0DR register.
MODF: mode error interrupt status flag
This bit is set if the MSTR control bit is set and the slave select input pin becomes low. This condition is not permitted in normal operation. This flag is cleared by a read of the SP0SR register followed by a write to the SP0CR1 register.

Figure 9.23 ■ SPI status register (SP0SR)

The SPI subsystem has an 8-bit data register (SP0DR). This register is both the input and output register for SPI data. When the CPU writes a byte into this register, eight clock signals are generated to shift out the 8-bit data written into this register. If the user wants to shift in a byte of data from the SPI slave, he will still write a byte into the SP0DR register. At the end of an SPI transfer, the slave shifts eight bits of data into this register.

All port S output pins can be configured to have reduced drive capability. All port S input pins could be programmed to have pull-up devices. The configuration for reduced drive and pull-up devices can be accomplished by programming the PURDS register. The contents of the PURDS register are shown in Figure 9.24.

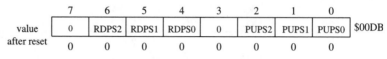

Bit 7 and 3 cannot be written but will be read as 0s

RDPS2: reduce drive PS7-PS4
  0 = Port S output drives for bits 7 to 4 operate normally
  1 = Port S output pins for bits 7 to 4 have reduced drive capability
      for lower power and less noise
RDPS1: reduce drive of PS3 and PS2
  0 = Port S output drives for bits 3 and 2 operate normally
  1 = Port S output pins for bits 3 and 2 have reduced drive
      capability for lower power and less noise
RPDS0: reduce drive of PS1 and PS0
  0 = Port S output drives for bits 1 and 0 operate normally
  1 = Port S output pins for bits 1 and 0 have reduced drive
      capability for lower power and less noise
PUPS2: Pullup port S enable PS7-PS4
  0 = No internal pullups on port S bits 7 to 4
  1 = Port S input pins for bits 7 to 4 have an active pullup device. If a
      pin is programmed as output, the pullup device becomes
      inactive.
PUPS1: Pullup port S enable PS3 and PS2
  0 = No internal pullups on port S bits 3 and 2
  1 = Port S input pins for bits 3 and 2 have an active pullup device.
PUPS0: Pullup port S enable PS1 and PS0
  0 = No internal pullups on port S bits 1 and 0
  1 = Port S input pins for bits 1 and 0 have an active pullup device.

Figure 9.24 ■ Pullup and reduced drive register for port S (PURDS)

For those port S pins that are not used by the SPI transfer, they can be used as general-purpose I/O pins. You can read or write into the port S data register (PORTS) in order to access these pins. Since they are general-purpose I/O pins, you will need to use the port S data direction register DDRS to configure their directions. The contents of DDRS are shown in Figure 9.25.

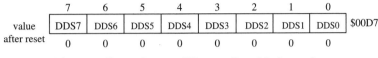

	7	6	5	4	3	2	1	0	
value	DDS7	DDS6	DDS5	DDS4	DDS3	DDS2	DDS1	DDS0	$00D7
after reset	0	0	0	0	0	0	0	0	

After reset, all general-purpose I/O are configured for input only.
    0 = configure the corresponding I/O pin for input
    1 = configure the corresponding I/O pin for output
DDS2 & DDS0 -- Data direction for port S bit 2 and 0
    If the SCI receiver are configured for 2-wire SCI operation,
    corresponding port S pins are input regardless of the state of these bits.
DDS3 & DDS1 -- Data direction for port S bit 3 and 1
    If the SCI receiver are configured for 2-wire SCI operation,
    corresponding port S pins are output regardless of the state of these
    bits.
DDS6-DDS4: Data direction for port S bits 6 to 4
    If the SPI is enabled and expects the corresponding port S pin to be an
    input, it will be an input regardless of the state of the DDRS bit. If the
    SPI is enabled and expects the bit to be an output, it wil be an output
    only if the DDRS bit is set.
DDS7 -- Data direction for port S bit 7
    In SPI slave mode, DDS7 has no effect; the PS7 pin is dedicated as the
    SS input. In SPI master mode, DDS7 determines whether PS7 is an
    error detect input to the SPI or a general-purpose or slave select output
    line.

Figure 9.25 ■ Port S data direction register (DDRS)

Example 9.9

Suppose there is an SPI-compatible peripheral output device that has the following characteristics:

- Has a *clk* input pin that is used as the data shifting clock signal
- Has a *data_in* pin to shift in data on the rising edge of the clk input
- Has a $\overline{CS}$ pin, which enables the chip to shift in data when it is low
- The highest data-shifting rate is 2 MHz
- Most significant bit shifted in first

Describe how to connect the SPI pins of the 68HC12BC32 and this peripheral device and write an instruction sequence to configure the SPI subsystem properly for data transfer. Assume the E clock frequency is 8 MHz.

Solution: The pin connection is shown in Figure 9.26.

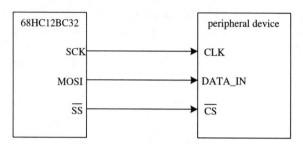

Figure 9.26 ■ Circuit connection between the 68HC12 SPI and the peripheral device

For pin directions, we need to:

- Configure the pins SS, SCK, and MOSI for output
- Configure the PS1 (TxD) pin for output and the PS0 (RxD) pin for input

For this configuration, write the value $E2 into the DDRS register.

The data transfer rate is 2 MHz for the 8 MHz E clock. So, set the SPI clock divide factor to four. For this, write the value $01 into the SP0BR register.

The SP0CR1 register should be configured as follows:

- SPI interrupt disabled
- SPI subsystem enabled
- Normal SPI pins
- Master mode enabled
- Rising edge to shift data out
- SS pin enabled
- Data is transferred most significant bit first

For this configuration, write the value $52 into the SP0CR1 register.

To configure the SP0CR2 register, we will:

- Enable Port S pins internal pullup
- Select normal port S output drive
- Normal MOSI pin

For this configuration, write the value $00 into the SP0CR2 register.

Write the value $01 into the PURDS register to choose the normal drive for all port S output pins and enabled PS0 pin pull-up devices.

Therefore, the following instruction sequence will achieve the desired configuration:

```
movb        #$E2,DDRS
movb        #$01,SP0BR
movb        #$52,SP0CR1
movb        #$00,SP0CR2
movb        #$01,PURDS
```

In C, this can be achieved by the following statements:

```
DDRS = 0xE2;
SP0BR = 0x01;
SP0CR1 = 0x52;
SP0CR2 = 0x00;
PURDS = 0x01;
```

To send a byte into this chip, simply write this byte into the SP0DR register. For example, the following instruction sequence will send the value $39 into this device:

```
movb    #$39,SP0DR
brclr   SP0SR,$80,*        ; wait until eight bits have been shifted out
```

In C, this can be achieved by the following statements:

```
SP0DR = 0x39;
while (!(SP0SR & 0x80));
```

▲

9.8 SPI Circuit Connection

In a system that uses the SPI subsystem, one device (normally a microcontroller) is configured as the master and the other devices are configured as slaves. Either a peripheral chip or a 68HC12 can be configured as a slave device. The master SPI device controls the data transfer and can control one or more SPI slave devices simultaneously.

In a single-slave configuration, the circuit would be connected as shown in Figure 9.27. If the $\overline{SS}$ output enable feature is set, then the $\overline{SS}$ pin will automatically go low to enable the slave device before the data transfer is started.

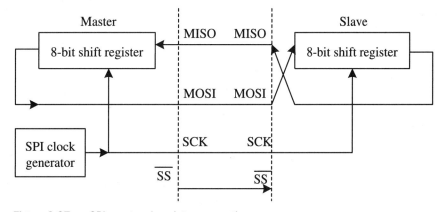

Figure 9.27 ■ SPI master-slave interconnection

There could be several connection methods in a multi-slave SPI environment. One possibility is shown in Figure 9.28. In this connection method, the 68HC12 can choose any peripheral device for data transfer. In this figure, we use port P pins to drive the $\overline{SS}$ inputs for peripheral devices. Any other unused general-purpose I/O pins can be used for this purpose.

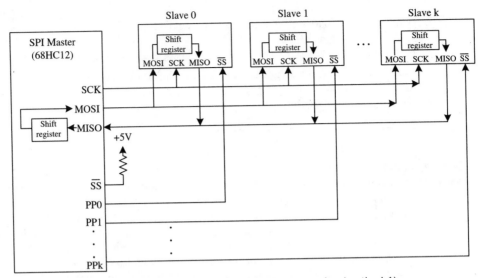

Figure 9.28 ■ Single-master and multiple-slave device connection (method 1)

If we don't need the capability of selecting an individual peripheral device for data transfer, we can save the port P pins by using the connection shown in Figure 9.29. Figure 9.29 differs from Figure 9.28 in the following ways:

1. The MISO pin of each slave is wired to the MOSI pin of the slave device to its right. The MOSI pins of the master and slave 0 are still wired together.

2. The MISO pin of the master is wired to the same pin of the last slave device.

3. The $\overline{SS}$ inputs of all slaves are tied to ground to enable all slaves.

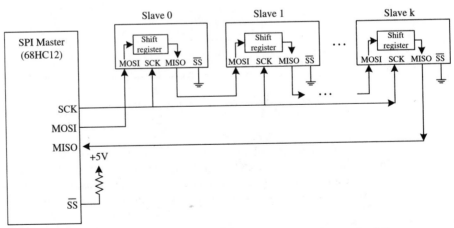

Figure 9.29 ■ Single-master and multiple-slave device connection (method 2)

Thus the shift registers of the SPI master and slaves become a ring. The data of slave k are shifted to the master SPI, the data of the master are shifted to the slave 0, the data of slave 0 are shifted to the slave 1, and so on. In this configuration, a minimal number of pins control a large number of peripheral devices. However, the master does not have the freedom to select an arbitrary slave device for data transfer without going through other slave devices.

This type of configuration is often used to extend the capability of the SPI slave. For example, suppose there is an SPI-compatible seven-segment display driver/decoder that can drive only four digits. By using this configuration, up to $4 \times k$ digits can be displayed when k driver/decoders are cascaded together.

Depending on the capability and role of the slave device, either the MISO or MOSI pin may not be used in the data transfer. Many SPI-compatible peripheral chips do not have the MISO pin.

9.9 SPI-Compatible Chips

The SPI is a protocol proposed by Motorola to interface peripheral devices to a microcontroller. As long as a peripheral device supports the SPI interface protocol, it can be used with any microcontroller that implements the SPI subsystem. Many semiconductor manufacturers produce SPI-compatible peripheral chips. A partial list is given in Appendix F. The Motorola SPI protocol is compatible with the National Semiconductor Microwire protocol. Therefore, any peripheral device that is compatible with the SPI can also be interfaced with the Microwire protocol.

In the following sections, we will look at a few SPI-compatible peripheral chips.

9.10 The HC595 Shift Register

As shown in Figure 9.30, the HC595 consists of an 8-bit shift register and an 8-bit D-type latch with 3-state parallel outputs. The shift register accepts serial data and provides a serial output. The shift register also provides parallel data to the 8-bit latch. The shift register and the latch have different clock sources. This device also has an asynchronous reset input. The maximum data shift rate is 6 MHz.

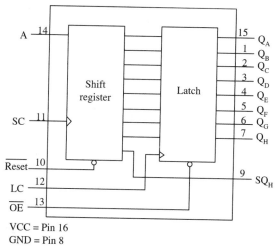

Figure 9.30 ■ HC595 block diagram and pin assignment

The functions of the pins in Figure 9.30 are as follows:

- A: *Serial data input.* The data on this pin is shifted into the 8-bit shift register.
- SC: *Shift clock.* A low-to-high transition on this input causes the data at the serial input pin to be shifted into the 8-bit shift register.
- $\overline{\text{Reset}}$. A low on this pin resets the shift register portion of this device only. The 8-bit latch is not affected.
- LC: *Latch clock.* A low-to-high transition on this pin loads the contents of the shift register into the output latch.
- $\overline{\text{OE}}$: *Output enable.* A low on this pin allows the data from the latches to be presented at the outputs. A high on this pin forces the outputs (Q_A to Q_H) into the high-impedance state. This pin does not affect the serial output.
- Q_A to Q_H: Non-inverted, tri-state, latch outputs.
- SQ_H: *Serial data output.* This is the output of the eighth stage of the 8-bit shift register. This output does not have tri-state capability.

One application of the HC595 is to expand the number of parallel output ports of the 68HC12. Many parallel output ports can be added to the 68HC12 by using the SPI subsystem and multiple HC595s. One way to connect the SPI subsystem and multiple HC595s is illustrated in Figure 9.31.

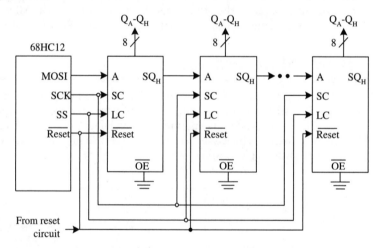

Figure 9.31 ■ Serial connection of multiple HC595s to the SPI

The procedure for outputting multiple bytes using this connection is as follows:

Step 1
Program the DDRS register to configure each SPI pin and the TxD and RxD pins properly. Write the value $E2 into the DDRS register.

Step 2
Program the SP0CR1 register to enable the SPI function, using the rising edge of the SCK signal to shift data in and out, select master mode, normal port S pins, and set the data rate to 4 Mbits/sec, configure the $\overline{\text{SS}}$ pin as a general-purpose output pin, and shift data out most significant bit first. Write the value $50 into the SP0CR1 register and write the value $00 into the SP0BR register. Write the value $00 into the SP0CR2 register.

Step 3
Choose normal mode for all port S pins. Write the value $01 into the PURDS register.

Step 4
Write a byte into the SP0DR register to trigger eight pulses from the SCK pin.

Step 5
Repeat step 4 as many times as needed.

Step 6
Set the $\overline{SS}$ pin to low and then pull it to high to load the data in the shift register of each HC595 into the output latch. After this step, the QA through QH pins of each HC595 contains valid data.

Example 9.10

Write a program to output the contents of 8 bytes that are stored from $800 to the HC595s in Figure 9.31.

Solution: The program is as simple as follows:

```
#include "d:\miniide\hc12.inc"

            org       $800
buf         db        1,2,3,4,5,6,7,8
            org       $1000
            jsr       spi_init          ; initialize the SPI function
            ldx       #$800             ; use X as the array pointer
            ldab      #8                ; use B as the loop counter
loop        ldaa      1,x+              ; read one byte and move the pointer
            staa      SP0DR
            brclr     SP0SR,$80,*       ; wait for the SPI transfer to complete
            dbne      b,loop
            swi
;****************************************************************
; The following function initializes the SPI subsystem
;****************************************************************
spi_init    movb      #$E2,DDRS
            movb      #$50,SP0CR1
            movb      #$00,SP0CR2
            movb      #$00,SP0BR
            movb      #$01,PURDS
            rts
            end
```

The C language version of the program is as follows:

```c
#include <hc12.h>
void spi_init (void);
void main (void)
{
        char data [8] = {1, 2, 3, 4, 5, 6, 7, 8};
        int i;
        spi_init ();
        for (i = 0; i < 8; i++) {
```

```
                    SPODR = data[i];          /* start an SPI transfer */
                    while(!(SPOSR & 0x80));    /* wait until SPI transfer is complete */
            }
            asm("swi");
    }
    void spi_init (void)
    {
            SPOCR1 = 0x50;
            SPOCR2 = 0x00;
            SPOBR = 0x00;
            DDRS = 0xE2;
            PURDS = 0x01;
    }
```

Another method of expanding the number of parallel output ports is to connect multiple HC595s in parallel to the SPI subsystem, as shown in Figure 9.32. With this circuit connection, we can selectively output data to any HC595 by generating a rising edge on the corresponding port P pin after the data has been shifted out. The procedure is as follows:

Step 1
Program the DDRS register to configure SPI pin directions.

Step 2
Program SPI registers to enable the SPI subsystem, set the data rate, select the rising edge of the SCK signal to shift data out, select master mode, disable interrupt, and so on.

Step 3
Write a byte to the SP0DR register to trigger SPI transfer.

Step 4
Set the PP*i* pin to low and then pull it to high to load the byte from the shift register of the *i*th HC595 into the output latch.

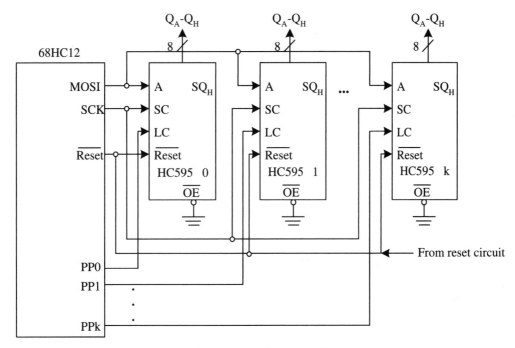

Figure 9.32 ■ Parallel connection of multiple HC595s to the SPI

The programming of this circuit is very straightforward and will be left for you as an exercise problem.

9.11 The MAX7221 Seven-Segment Display Driver

The MAX7221 from MAXIM is an 8-digit seven-segment display driver with a serial interface that is compatible with SPI and Microwire. In addition to driving seven-segment displays, it can also be used to drive bar-graph displays, industrial controllers, panel meters, and LED matrix displays.

When being used to drive seven-segment displays, the MAX7221 can address individual digits and updates without rewriting the entire display. Only common cathode seven-segment displays can be driven directly. The user is allowed to select code-B decoding or no decoding for each digit.

The MAX7221 has a scan limit register that allows the user to display from one to eight digits. It also has a test mode that forces all LEDs on.

9.11.1 MAX7221 Pins

The MAX7221 has 23 signals and is housed in a 24-pin DIP package. Its pin assignment is shown in Figure 9.33. The function of each pin is as follows:

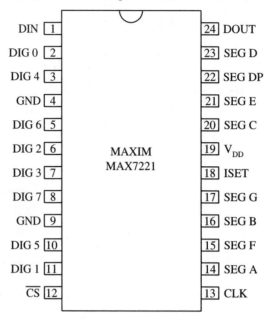

Figure 9.33 ■ Pin assignment of MAX7221

- DIN: *Serial data input.* Data is loaded into the internal 16-bit shift register on CLK's rising edge.
- DIG 0 – DIG 7: *Eight-digit drive lines.* These lines sink current from the display common cathode. The MAX7221's digit drivers are in high-impedance when they are turned off.
- GND: Ground.
- $\overline{CS}$: *Chip-select input.* Serial data is loaded into the shift register while $\overline{CS}$ is low. The last 16 bits of serial data are latched on $\overline{CS}$'s rising edge.
- CLK: *Serial-clock input.* 10 MHz maximum data shift rate. Data is shifted into the internal shift register on CLK's rising edge and shifted out of the DOUT pin on CLK's falling edge.
- SEG A – SEG G, DP: *Seven-segment drive and decimal point drive.* These signals are in a high-impedance state when they are turned off.
- ISET: *Current setting resistor.* This pin is connected to the VDD through a resistor (R_{SET}) to set the peak segment current.
- V_{DD}: Positive power supply.
- DOUT: *Serial-data out.* This pin is used to daisy chain several MAX7221s and is never in a high-impedance state.

9.11.2 MAX7221 Functioning

The functional diagram of the MAX7221 is shown in Figure 9.34.

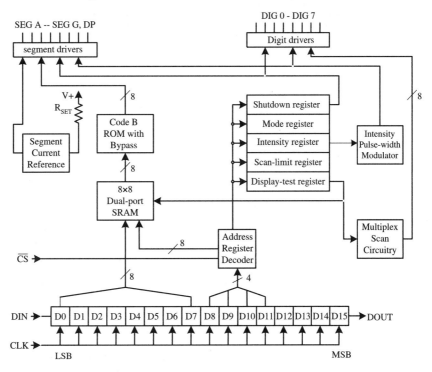

Figure 9.34 ■ MAX7221 functional diagram

The function of every block is as follows:

SEGMENT DRIVERS

This block provides the current to drive the segment pattern outputs SEG A, .., SEG G, and DP. The source of segment patterns can come from the Code B ROM (in decode mode) or dual-port SRAM (in no decode mode). The dual-port SRAM consists of eight digit registers. The intensity of the segment pattern is modulated by the *intensity pulse-width modulator*. The magnitude of the current is controlled by the *segment current reference* block.

CODE B ROM WITH BYPASS

This ROM stores the segment pattern of decimal digits 0 to 9 and characters -, E, H, L, and P. A 4-bit address presented by the dual-port SRAM is used to select the segment pattern from this ROM. The output of this ROM will be used to drive the segment outputs SEG A., .., DP. The intensity of the segment pattern is under the control of the output intensity register.

If you don't like the patterns provided by the Code B ROM, you can bypass them and supply your own. To do this, you need to choose the *no decode mode*. In no decode mode, you need to store the desired patterns (corresponding to eight digits) in the dual port SRAM and these patterns will be used to drive the segment outputs. You can choose to bypass from one to all eight digits.

The segment pattern stored in the ROM is shown in Table 9.7.

7-segment character	Register Data (output from SRAM)						On segment = 1							
	D7*	D6-D4	D3	D2	D1	D0	DP*	A	B	C	D	E	F	G
0		x	0	0	0	0	1	1	1	1	1	1	1	0
1		x	0	0	0	1	0	1	1	0	0	0	0	0
2		x	0	0	1	0	1	1	0	1	1	0	1
3		x	0	0	1	1	1	1	1	1	0	0	1
4		x	0	1	0	0	0	1	1	0	0	1	1
5		x	0	1	0	1	1	0	1	1	0	1	1
6		x	0	1	1	0	1	0	1	1	1	1	1
7		x	0	1	1	1	1	1	1	0	0	0	0
8		x	1	0	0	0	1	1	1	1	1	1	1
9		x	1	0	0	1	1	1	1	1	0	1	1
-		x	1	0	1	0	0	0	0	0	0	0	1
E		x	1	0	1	1	1	0	0	1	1	1	1
H		x	1	1	0	0	0	1	1	0	1	1	1
L		x	1	1	0	1	0	0	0	1	1	1	0
P		x	1	1	1	0	1	1	0	0	1	1	1
blank		x	1	1	1	1	0	0	0	0	0	0	0

* The decimal point is set by bit D7 = 1

Table 9.7 ■ Code B pattern

DIGIT DRIVERS & MULTIPLEX SCAN CIRCUITRY

The MAX7221 uses the time-multiplexing technique to display eight digits. When DIG i (i = 0 to 7) is asserted, the corresponding pattern is sent to the segment driver from either ROM or SRAM. Since only one digit output is asserted, only one digit will be displayed at a time. However, the multiplex scan circuitry will only allow one digit to be lighted for a short period of time (slightly over 1 ms) before it switches to display the next digit. In this way, all eight digits are lighted in turn many times in a second. Due to the effect of *persistence of vision*, all eight digits appear to be lighted at the same time.

MODE REGISTER

The mode register sets BCD code B (0-9, E, H, L, P, and -) or no decode operation for each digit. Each bit in the register corresponds to one digit. A logic high selects code B decoding while low bypasses the decoder. Examples of the decode mode control-register format are shown in Table 9.8.

Decode mode	Register data								Hex Code
	D7	D6	D5	D4	D3	D2	D1	D0	
No decode for digits 7-0	0	0	0	0	0	0	0	0	00
Code B decode for digit 0 No decode for digits 7-1	0	0	0	0	0	0	0	1	01
Code B decode for digits 3-0 No decode for digits 7-4	0	0	0	0	1	1	1	1	0F
Code B decode for digits 7-0	1	1	1	1	1	1	1	1	FF

Table 9.8 ■ Decode-mode register examples

When the code B decode mode is used, the decoder looks only at the lower nibble of the data in the digit registers (D3-D0), disregarding bits D4-D6. D7, which sets the decimal point (SEG DP), is independent of the decoder and is positive logic (D7 = 1 turns on the decimal point). Digit registers are the dual port SRAM.

INTENSITY CONTROL & INTERDIGIT BLANKING

The MAX7221 allows display brightness to be controlled with an external resistor (R_{SET}) connected between V_{DD} and ISET. The peak current sourced from the segment drivers is nominally 100 times the current entering ISET. This resistor can either be fixed or variable to allow brightness adjustment from the front panel. Its minimum value should be 9.53 KΩ, which typically sets the segment current at 40 mA. Display brightness can also be controlled digitally by using the intensity register.

Digital control of display brightness is provided by an internal pulse-width modulator, which is controlled by the lower nibble of the intensity register. The modulator scales the average segment current in 16 steps from a maximum of 15/16 down to 1/16 of the peak current set by R_{SET}. Table 9.9 lists the intensity register format. The minimum interdigit blanking time is set to 1/32 of a cycle.

Duty cycle (min on)	D7	D6	D5	D4	D3	D2	D1	D0	Hex code
1/16	x	x	x	x	0	0	0	0	x0
2/16	x	x	x	x	0	0	0	1	x1
3/16	x	x	x	x	0	0	1	0	x2
4/16	x	x	x	x	0	0	1	1	x3
5/16	x	x	x	x	0	1	0	0	x4
6/16	x	x	x	x	0	1	0	1	x5
7/16	x	x	x	x	0	1	1	0	x6
8/16	x	x	x	x	0	1	1	1	x7
9/16	x	x	x	x	1	0	0	0	x8
10/16	x	x	x	x	1	0	0	1	x9
11/16	x	x	x	x	1	0	1	0	xA
12/16	x	x	x	x	1	0	1	1	xB
13/16	x	x	x	x	1	1	0	0	xC
14/16	x	x	x	x	1	1	0	1	xD
15/16	x	x	x	x	1	1	1	0	xE
15/16 (max on)	x	x	x	x	1	1	1	1	xF

Table 9.9 ■ Intensity register format

SCAN-LIMIT REGISTER

The scan-limit register sets how many digits are displayed, from one to eight. They are displayed in a multiplexed manner with a typical scan rate of 800 Hz with 8 digits displayed. If fewer digits are displayed, the scan rate is $8f_{OSC}/N$, where N is the number of digits scanned. Since the number of scanned digits affects the display brightness, the scan limit register should not be used to blank portions of the display (such as leading zero suppression). Table 9.10 lists the scan-limit register format.

Scan limit	Register data								Hex code
	D7	D6	D5	D4	D3	D2	D1	D0	
Display digit 0 only	x	x	x	x	x	0	0	0	x0
Display digits 0 & 1	x	x	x	x	x	0	0	1	x1
Display digits 0 to 2	x	x	x	x	x	0	1	0	x2
Display digits 0 to 3	x	x	x	x	x	0	1	1	x3
Display digits 0 to 4	x	x	x	x	x	1	0	0	x4
Display digits 0 to 5	x	x	x	x	x	1	0	1	x5
Display digits 0 to 6	x	x	x	x	x	1	1	0	x6
Display digits 0 to 7	x	x	x	x	x	1	1	1	x7

Table 9.10 ■ Scan-Limit register format

SHUTDOWN MODE REGISTER

When the MAX7221 is in shutdown mode, the segment drivers are in a high-impedance state. Data in the digit and control registers remain unaltered. Shutdown can be used to save power or as an alarm to flash the display by successively entering and leaving shutdown mode. For the minimum supply current in shutdown mode, logic inputs should be at ground or V_{DD} (CMOS-logic levels).

Typically it takes less than 250 µs for the MAX7221 to leave shutdown mode. The display driver can be programmed while in shutdown mode, and shutdown mode can be overridden by the display-test function.

When the bit 0 of the shutdown register is zero, the MAX7221 is in shutdown mode. Otherwise, the chip is in normal mode.

DISPLAY TEST REGISTER

The display test register operates in two modes: normal and display test. The display test mode turns all LEDs on by overriding, but alternating, all controls and digit registers (including the shutdown register). In display test mode, eight digits are scanned and the duty cycle is 15/16. When the bit of the display test register is zero, the MAX7221 is in normal operation mode. Otherwise it is in display test mode.

SERIAL SHIFT REGISTER

At the bottom of Figure 9.34 is a 16-bit serial shift register. The meaning of the contents of this register is illustrated in Table 9.11. When the CS signal is low, data will be shifted in from the DIN pin on the rising edge of the CLK signal. On the rising edge of CS, the lower 8 bits of the shift register will be transferred to a destination specified by the value represented by bits D11 to D8 of the shift register. The mapping of these four address bits is shown in Table 9.12. The upper four bits of this register are not used.

D15	D14	D13	D12	D11	D10	D9	D8	D7	D6	D5	D4	D3	D2	D1	D0
x	x	x	x		Address			MSB				Data			LSB

Table 9.11 ■ Serial data format

In Table 9.12, Digit 0 to Digit 7 are registers in dual port SRAM that are used to hold the decimal digits or the patterns (when no decode mode is selected) to be displayed. Whenever the register *Digit i* (i = 0..7) is selected to supply a segment pattern (directly (in no decode mode) or indirectly (in decode mode)) to the segment drivers, the associated digit driver pin *DIG i* is also asserted.

Register	Register data					
	D15-D12	D11	D10	D9	D8	Hex code
No op	x	0	0	0	0	x0
Digit 0	x	0	0	0	1	x1
Digit 1	x	0	0	1	0	x2
Digit 2	x	0	0	1	1	x3
Digit 3	x	0	1	0	0	x4
Digit 4	x	0	1	0	1	x5
Digit 5	x	0	1	1	0	x6
Digit 6	x	0	1	1	1	x7
Digit 7	x	1	0	0	0	x8
Decode Mode	x	1	0	0	1	x9
Intensity	x	1	0	1	0	xA
Scan limit	x	1	0	1	1	xB
Shutdown	x	1	1	0	0	xC
Display test	x	1	1	0	1	xF

Table 9.12 ■ Register address map

The register name in the first row of the table is *no op*. This register is used when cascading multiple MAX7221s. To cascade multiple MAX7221s, connect all devices' CS inputs together and connect DOUT to DIN on adjacent devices. For example, if four MAX7221s are cascaded, then to write to the fourth chip, set the desired 16-bit word, followed by three no-op codes (hex *X0XX*, msb first). When CS goes high, data is latched in all devices. The first three chips receive no-op commands, and the fourth receives the intended data.

CHOOSING THE VALUE FOR R_{SET}

Appropriate values must be chosen for the R_{SET} resistor in order to supply enough current to the LED segments. Table 9.13 shows us how to choose the R_{SET} value for a given LED segment current (I_{SEG}) and the LED forward voltage drop (V_{LED}). Any other combinations of I_{SEG} and V_{LED} that are not in the table can be obtained by using interpolation. For example, suppose V_{LED} is 2.3 V when I_{SEG} is 10 mA, then the R_{SET} value can be calculated as follows:

$$(2.3 - 2.0) / (2.5 - 2.0) = (63.7 - R_{SET}) / (63.7 - 59.3)$$

The value of R_{SET} is solved to be 61 KΩ, which is not a standard resistor. The closest standard resistor 62 KΩ will be used instead.

I_{SEG} (mA)	V_{LED} (V)				
	1.5	2.0	2.5	3.0	3.5
40	12.2 KΩ	11.8 KΩ	11.0 KΩ	10.6 KΩ	9.69 KΩ
30	17.8 KΩ	17.1 KΩ	15.8 KΩ	15.0 KΩ	14.0 KΩ
20	29.8 KΩ	28.0 KΩ	25.9 KΩ	24.5 KΩ	22.6 KΩ
10	66.7 KΩ	63.7 KΩ	59.3 KΩ	55.4 KΩ	51.2 KΩ

Table 9.13 ■ R_{SET} vs. segment current and LED forward voltage drop

9.11.3 Using the MAX7221

In this section, we will use examples to illustrate the application of the MAX7221.

Example 9.11

Use a MAX7221 and seven seven-segment displays to display the value –126.5°F. Show the circuit connection and write a program to display the given value. Assume enough brightness can be obtained when there is 10 mA current flowing through the LED and the V_{LED} at this current is 2.3V.

Solution: The circuit connection is shown in Figure 9.35. The R_{SET} value is chosen to be 62 KΩ to satisfy the V_{LED} and I_{SEG} requirements.

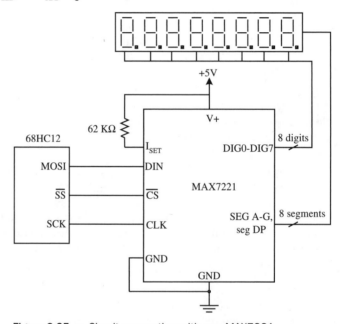

Figure 9.35 ■ Circuit connection with one MAX7221

In this circuit, we need to send 12 16-bit data to the MAX7221 in order to complete the set up. The values to be written are as follows:

Scan Limit Register
We want to display digits 0, 1, 2, 3, 4, 5, 6. We need to write the value $06 into the scan limit register. To achieve this, we can write the value $0B06 to the MAX7221. The hex value B is the address of the scan limit register.

Decode Mode Register
Among the seven characters (–126.5°F) to be displayed, the degree character and letter F cannot be decoded. We will use the *decode mode* for the leading five characters and the *no decode mode* for the degree character and letter F. Therefore, the value to be stored in the decode register is $FC. To achieve this, we need to send the value $09FC to the MAX7221, where 9 is the address of the decode mode register.

Digit 0 Register (to display letter F)

We have chosen no decode mode for the letter F. To display the letter F, we need to turn on segments A, E, F, and G, and turn off segments B, C, D, and decimal point DP. The value to be stored in digit 0 is $47. The address of this register is 1. We need to send the value $0147 to the MAX7221.

Digit 1 Register (to display degree character)

The degree character is displayed in no decode mode. The seven-segment pattern for a degree character is %1100011 (from segment a to g). Since a decimal point is not associated with this character, the value to be stored in digit 1 is $63. Therefore, we need to send the value $0263 to the MAX7221.

Digit 2 Register (to display the value 5)

This digit can be displayed in decode mode. A decimal point is not associated with it. The address for this register is 3. Therefore, the value to be stored in digit 2 is $05. To achieve this, we need to send the value $0305 to the MAX7221.

Digit 3 Register (to display the value 6)

This digit can be displayed in decode mode and a decimal point is associated with this digit. The value to be stored at this register is %10000110 (from segment DP, A to G). The address of this register is 4. Therefore, the value to be written to the MAX7221 is $0486.

Digit 4 Register (to display the value 2)

This digit can be displayed in decode mode and a decimal point is not associated with it. The value to be stored at this register is $02. The address of this register is 5. Therefore, the value to be written to the MAX7221 is $0502.

Digit 5 Register (to display the value 1)

This digit can be displayed in decode mode and a decimal point is not associated with it. The value to be stored in this register is $01. The address of this register is 6. Therefore, we need to write the value $0601 to the MAX7221.

Digit 6 Register (to display negative sign –)

This character can be displayed in decode mode and the decimal point is not associated with it. The value to be stored in this register is $0A. The address of this register is 7. Therefore, we need to send the value $070A to the MAX7221.

Intensity Register

We will use the maximum intensity to display the value –126.5°F. Therefore, the value to be stored in this register is $0F. The address of this register is $A. Therefore, we need to send the value $0A0F to the MAX7221.

Shutdown Register

The chip should perform in normal operation. Therefore, the value to be stored in this register is $01 (bit 0 must be one). The address of this register is $C. Therefore, the value to be sent to the MAX7221 is $0C01.

Display Test Register

The chip should perform in normal operation. Therefore, the value to be stored in this register is $00 (bit 0 must be zero). The address of this register is $F. Therefore, the value to be sent to the MAX7221 is $0F00.

The SPI function must be configured properly:

1. Pin directions: The SCK and MOSI pins must be configured as SPI output, whereas the SS pin must be configured as general output. Write the value $CA into the DDRS register. This value sets all TxD pins for output and all RxD pins for input.

2. SP0CR1 register: Disable SPI input, enable SPI, choose normal port S pin, set master mode, choose rising edge to shift out data, set SS pin as general output, data transfer MSB first. Write the value $50 into the SP0CR1 register.

3. SP0CR2 register: Select port S input pullup, normal port S drive, and normal direction. Write the value $08 into SP0CR2.

4. SP0BR register: Select 2 as the E clock divide factor. Write the value $00 into the SP0BR register. This will set the data shift rate to 4 MHz and will be within the maximum MAX7221 transfer rate.

5. PURDS register: Select normal drive and enable port S input pins to pull-up. Write the value $07 into the PURDS register.

The following subroutine will configure the SPI function accordingly:

```
#include     "d:\miniide\hc12.inc"
spi_init     movb     #$50,SP0CR1
             movb     #$08,SP0CR2
             movb     #$00,SP0BR
             movb     #$CA,DDRS
             movb     #$07,PURDS
             rts
```

We will use the following assembler directives to define the value to be written into the MAX7221 device:

```
display_dat  db     $0B,$06     ; value scan limit register
             db     $09,$FC     ; value for decode mode register
             db     $0A,$0F     ; value for intensity register
             db     $0C,$01     ; value for shutdown register
             db     $0F,$00     ; value for display test register
             db     $01,$47     ; value for Digit 0 register
             db     $02,$63     ; value for Digit 1 register
             db     $03,$05     ; value for Digit 2 register
             db     $04,$86     ; value for Digit 3 register
             db     $05,$02     ; value for Digit 4 register
             db     $06,$01     ; value for Digit 5 register
             db     $07,$0A     ; value for Digit 6 register
```

The following subroutine will transfer the appropriate data to the MAX7221:

```
; ***********************************************************
; The following subroutine transfers the array of 16-bit data to the MAX7221.
; The starting address of the data is passed in index register X and the number
; of 16-bit words is passed in accumulator B.
; ***********************************************************
transfer     bclr     PORTS,$80      ; pull SS pin to low
             ldaa     1,x+           ; get one byte and also move the pointer
             staa     SP0DR          ; start an SPI transfer
             brclr    SP0SR,$80,*    ; wait until the SPI transfer is complete
             ldaa     1,x+           ; get one byte and also move the pointer
```

```
        staa      SP0DR
        brclr     SP0SR,$80,*
        bset      PORTS,$80       ; pull SS Pin to high and transfer data to appropriate
;                                 ; register
        dbne      b,transfer      ; continue until data is complete
        rts
```

The C language version of this subroutine will be left as an exercise.

Cascading multiple MAX7221s is easy. The procedure is as follows:

- Tie the $\overline{CS}$ signal of all MAX7221s together and connect them to an output pin of the microcontroller.

- Tie the SCK signal of all MAX7221s together and connect them to the SCK pin of the microcontroller.

- Wire the DIN pin of the first MAX7221 to the MOSI pin of the 68HC12, connect the DOUT pin of the first MAX7221 to the DIN pin of its adjacent MAX7221, and so on.

The circuit for cascading two MAX7221s is shown in Figure 9.36. In this example, we use the $\overline{SS}$ pin as a general-purpose output pin to drive the $\overline{CS}$ pin. Again we use the same R_{SET} to achieve the same I_{SEG} and V_{LED} values.

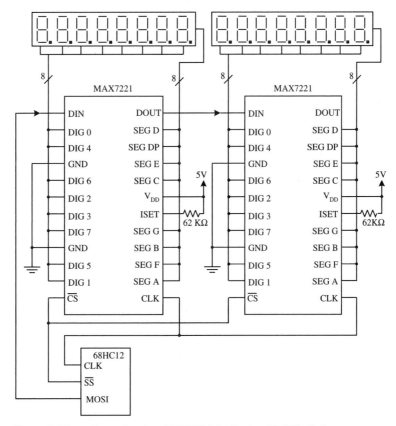

Figure 9.36 ■ Cascading two MAX7221 to display 16 BCD digits

When multiple MAX7221s are cascading together, we need to shift data to all devices before transferring data (by pulling the $\overline{CS}$ signal to high) to the destination register of each MAX7221.

▲

Example 9.12

▼

Write a program to configure the SPI subsystem and display the value:
38.9°C 02 13 04 20 01

on the 16 seven-segment displays. Assume the E clock frequency is 8 MHz.

Solution: We can use the same configuration for the SPI subsystem in this circuit. We will divide the digit string into two halves so that each MAX7221 displays one half as follows:
38.9°C 02
13 04 20 01

The first MAX7221 will display seven characters, whereas the second MAX7221 will display eight digits. We will pair these two substrings as follows for data transfer purposes:

digit	7	6	5	4	3	2	1	0	
		3	9	8	°	C	0	2	◄ first MAX7221
	1	3	0	4	2	0	0	1	◄ second MAX7221

Figure 9.37 ■ Data arrangement for transfer

In this circuit configuration, we need to send 13 32-bit data to perform the data setup. The values to be transferred are as follows:

Scan Limit Register
The first MAX7221 will display seven digits and the second MAX7221 will display eight digits. For this setup, we need to send the value $0B06 to the first MAX7221 and send the value $0B07 to the second MAX7221.

Decode Mode Register
For the first MAX7221, we will use no decode mode for digit 2 and 3 and decode mode for other digits. For the second MAX7221, all digits will be displayed in decode mode. For this setup, we will send the value $09F3 to the first MAX7221 and send the value $09FF to the second MAX7221.

Intensity Register
For this example, we will use the maximum intensity to display the values. For this setup, we will send the value $0A0F to both MAX7221s.

Shutdown Register
Both chips should perform normally. Therefore the value to be stored in this register is $01, and the value to be sent to both MAX7221 is $0C01.

Display Test Register
Both chips should perform in normal operation. Therefore the value to be stored in this register is $00 (bit 0 must be zero). We need to send the value $0F00 to both MAX7221s.

Digit 7 Register

The first MAX7221 won't display anything on digit 7 whereas the second MAX7221 will display the value 1 on digit 7. For this value, we can send $0000 to the first MAX7221 and send the value $0801 to the second MAX7221.

Digit 6 Register

Both MAX7221s will display 3 on this digit. Therefore, send the value $0703 to both MAX7221s.

Digit 5 Register

The first MAX7221 will display 9 on digit 5 whereas the second MAX7221 will display 0 on digit 5. Therefore, we will send the value $0609 to the first MAX7221 and send the value $0600 to the second MAX7221.

Digit 4 Register

The first MAX7221 will display 8 on digit 4 whereas the second MAX7221 will display 4 on digit 4. Therefore we will send the value $0508 to the first MAX7221 and send the value $0504 to the second MAX7221.

Digit 3 Register

The first MAX7221 will display º on digit 3 whereas the second MAX7221 will display 2 on digit 3. Remember the first MAX7221 selected no decode mode for this digit. Therefore we will send the value $0463 to the first MAX7221 and send the value $0402 to the second MAX7221.

Digit 2 Register

The first MAX7221 will display C on digit 2 whereas the second MAX7221 will display 0 on digit 2. Remember the first MAX7221 selected no decode mode for this digit. Therefore we will send the value $034E to the first MAX7221 and send the value $0300 to the second MAX7221.

Digit 1 Register

Both MAX7221s will display 0 on this digit. Therefore we should send the value $0200 to both MAX7221s.

Digit 0 Register

The first MAX7221 will display 2 on digit 0 whereas the second MAX7221 will display 1 on this digit. Therefore we will send the value $0102 to the first MAX7221 and send the value $0101 to the second MAX7221.

We can use the same configuration as in Example 9.11 for the SPI subsystem in this example. The data to be transferred to the MAX7221s are defined as follows:

```
;*****************************************************************
; The data to be sent to two MAX7221s are paired together. The data for the second
; MAX7221 are listed in the front because they should be sent out first.
;*****************************************************************
max_data    db      $0B,$07,$0B,$06     ; data for scan limit registers
            db      $09,$FF,$09,$F3     ; data for decode mode registers
            db      $0A,$0F,$0A,$0F     ; data for intensity registers
            db      $0C,$01,$0C,$01     ; data for shutdown registers
            db      $0F,$00,$0F,$00     ; data for display test registers
            db      $08,$01,$00,$00     ; data for digit 7 registers
            db      $07,$03,$07,$03     ; data for digit 6 registers
            db      $06,$00,$06,$09     ; data for digit 5 registers
            db      $05,$04,$05,$08     ; data for digit 4 registers
```

```
        db      $04,$02,$04,$63     ; data for digit 3 registers
        db      $03,$00,$03,$4E     ; data for digit 2 registers
        db      $02,$00,$02,$00     ; data for digit 1 registers
        db      $01,$01,$01,$02     ; data for digit 0 registers
```

The following subroutine will transfer the appropriate data to two MAX7221s:

```
; ****************************************************************
; The following subroutine transfers the array of 32-bit data to the MAX7221.
; The starting address of the data is passed in index register X and the number
; of 32-bit words is passed in accumulator B.
; ****************************************************************
transfer    bclr    PORTS,$80          ; pull SS pin to low
            ldaa    1,x+               ; send the first byte of the second MAX7221
            staa    SP0DR              ;   "
            brclr   SP0SR,$80,*        ;   "
            ldaa    1,x+               ; send the second byte of the second MAX7221
            staa    SP0DR              ;   "
            brclr   SP0SR,$80,*        ;   "
            ldaa    1,x+               ; send the first byte of the first MAX7221
            staa    SP0DR              ;
            brclr   SP0SR,$80,*        ;
            ldaa    1,x+               ; send the second byte of the first MAX7221
            staa    SP0DR              ;   "
            brclr   SP0SR,$80,*        ;   "
            bset    PORTS,$80          ; pull SS Pin to high and transfer data to appropriate
                                       ; register
;
            dbne    b,transfer         ; continue until data is complete
            rts
```

The C language version of this subroutine will be left for you as an exercise.

9.12 Using the Serial E²PROM X25138

Some applications require data logging to be done in the field where the system must be powered by the battery. In order to prevent the loss of data, using nonvolatile EEPROM with a serial interface is a good design option. Building large lookup tables is another application of the serial EEPROM. Using serial EEPROM does not occupy the microprocessor memory space.

The Xicor X25138 is a 128K-bit serial E²PROM internally organized as $16K \times 8$ bits. As shown in Figure 9.38, the X25138 has six signals in addition to the power and the ground.

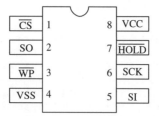

Figure 9.38 ■ X25138 pin assignment in DIP package

The X25138 features an SPI interface and software protocol allowing operation on a simple three-wire bus. The bus signals include a *clock input* (SCK), plus separate *data in* (SI) and *data out* (SO) lines. Access to the device is controlled through a *chip select* ($\overline{CS}$) input, allowing any number of devices to share the same bus.

The *hold input* (HOLD) forces the X25138 to ignore transitions on its inputs, thus allowing the host microcontroller to service higher priority interrupts. The $\overline{WP}$ input can be used to disable write attempts to the status register, thus providing a mechanism for limiting end user capability of altering zero, one-quarter, one-half, or all of the memory.

The X25138 can shift data at the clock rate of 5 MHz. Similar to parallel EEPROM, a write cycle will take about 5 ms in the X25138. Data retention time can reach 100 years. The X25138 can endure 100,000 write cycles.

9.12.1 Principles of Operation

The X25138 contains an 8-bit instruction register. It is accessed via the SI input, with data being clocked in on the rising edge of the SCK input. The $\overline{CS}$ input must be low and the $\overline{HOLD}$ and $\overline{WP}$ inputs must be high during the entire operation.

Table 9.14 contains a list of the instructions and their opcodes. All instructions, addresses, and data are transferred MSB first.

Instruction name	Instruction format	Operation
WREN	0000 0110	Set the Write Enable Latch (enable write operation)
WRDI	0000 0100	Reset the Write Enable Latch (disable write operation)
RDSR	0000 0101	Read Status Register
WRSR	0000 0001	Write Status Register
READ	0000 0011	Read data from memory array beginning at selected address
WRITE	0000 0010	Write data to memory array beginning at selected address (1 to 32 bytes)

Table 9.14 ■ X25138 instruction set

Data input is sampled on the first rising edge of SCK after $\overline{CS}$ goes low. The SCK signal is static, allowing the user to stop the clock and then resume operations. If the clock line is shared with other peripheral devices on the SPI bus, the user can assert the $\overline{HOLD}$ input to place the X25138 into a *"PAUSE"* condition. After releasing $\overline{HOLD}$, the X25138 will resume operation from the point that the $\overline{HOLD}$ was first asserted.

WRITE-ENABLE LATCH

The X25138 contains a write-enable latch. This latch must be set before a write operation can be completed internally. The WREN instruction will set the latch and the WRDI instruction will reset the latch. This latch is automatically reset upon a power-on condition and after the completion of a byte, page, or status register write cycle.

STATUS REGISTER

The RDSR instruction provides access to the *status register*. The status register may be read at any time, even during a write cycle. The contents of the status register are shown in Figure 9.39.

7	6	5	4	3	2	1	0
WPEN	x	x	x	BL1	BL0	WEL	WIP

Figure 9.39 ■ The X25138 status register

In Figure 9.39, bits WPEN, BL0, and BL1 are set by the WRSR instruction. Bits WEL and WIP are read-only and are automatically set by other operations.

The *write-in-process* (WIP) bit indicates whether the X25138 is busy with a write operation. When set to a one, a write is in progress, when set to zero, no write is in progress. During a write, all other bits are set to one.

The WEL bit indicates the status of the write enable latch. When set to one, the latch is set; when set to a zero, the latch is reset.

The block lock bits (BL1 and BL0) are nonvolatile and allow the user to select one of four levels of protection. The X25138 is divided into four 32Kbits segments. One, two, or all four segments may be protected. That is, the user may read the segments but will be unable to alter data within the selected segments. The partitioning is controlled as illustrated in Table 9.15.

Status register bits BL1	BL0	Array Addresses protected
0	0	none
0	1	$3000-$3FFF
1	0	$2000-$3FFF
1	1	$0000-$3FFF

Table 9.15 ■ The X25138 memory partitioning

The WPEN bit stands for *Write Protection Enable*. The WPEN and WEL bits, along with the $\overline{\text{WP}}$ input, provide write protection to the memory array and the status register as shown in Table 9.16.

WPEN	$\overline{\text{WP}}$	WEL	Protected blocks	Unprotected blocks	Status register
0	x	0	protected	protected	protected
0	x	1	protected	writable	writable
1	low	0	protected	protected	protected
1	low	1	protected	writable	protected
x	high	0	protected	protected	protected
x	high	1	protected	writable	writable

Table 9.16 ■ X25138 memory array and status register protection

HARDWARE WRITE PROTECTION

The *write protect* ($\overline{\text{WP}}$) pin and the nonvolatile WPEN bit in the status register control the hardware write-protect feature. Hardware write-protection is enabled when the $\overline{\text{WP}}$ pin is low, and the WPEN bit is one. Hardware write protection is disabled when either the $\overline{\text{WP}}$ pin is high or the WPEN bit is zero. When the chip is hardware write protected, nonvolatile writes are disabled to the status register, including the block lock bits and the WPEN bit itself, as well as the block-protected sections in the memory array. Only the sections of the memory array that are not block-protected can be written.

IN CIRCUIT PROGRAMMABLE ROM MODE

Since the WPEN bit is write protected, it cannot be changed back to a low state, so write protection is enabled as long as the $\overline{\text{WP}}$ pin is held low. Thus an in-circuit programmable ROM function can be implemented by hardwiring the $\overline{\text{WP}}$ pin to V_{SS}, writing to and block-locking the desired portion of the array to be ROM, and then programming the WPEN bit high.

CLOCK AND DATA TIMING

Data input on the SI pin is latched on the rising edge of SCK. Data is output on the SO pin on the falling edge of SCK.

READ SEQUENCE

To read from the EEPROM memory array, the signal $\overline{CS}$ is first pulled low to select the device. The 8-bit READ instruction is transmitted to the X25138, followed by the 16-bit address of which the last 14 are used. After the READ opcode and address are sent, the data stored in the memory at the selected address is shifted out on the SO line. The data stored in memory at the next address can be read sequentially by continuing to provide clock pulses. The address is automatically incremented to the next higher address after each byte of data is shifted out. When the highest address is reached ($3FFF), the address counter rolls over to address $0000 allowing the read cycle to be continued indefinitely. The read operation is terminated by taking the signal $\overline{CS}$ high. The EEPROM array read operation sequence is illustrated in Figure 9.40.

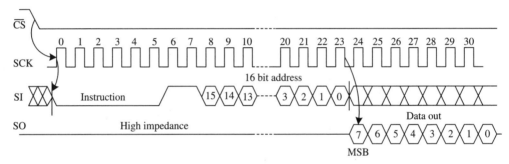

Figure 9.40 ■ Read EEPROM array operation sequence

To read the status register, the $\overline{CS}$ signal is first pulled low to select the device, followed by the 8-bit RDSR instruction. After the RDSR opcode is sent, the contents of the status register are shifted out on the SO pin. Figure 9.41 illustrates the read status register sequence.

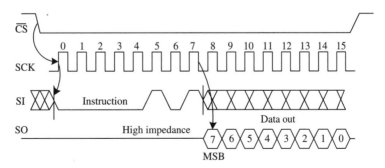

Figure 9.41 ■ Read status register sequence

WRITE SEQUENCE

Prior to any attempt to write data into the X25138, the *write-enable* (WREN) latch must first be set by issuing the WREN instruction. As shown in Figure 9.42, the $\overline{CS}$ signal is first pulled to low, then the WREN instruction is clocked into the X25138. After all eight bits of the instruction are transmitted, the $\overline{CS}$ signal must be pulled high. If the user continues the write operation without taking the $\overline{CS}$ signal high after issuing the WREN instruction, the write operation will be ignored.

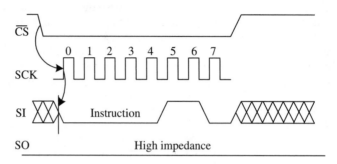

Figure 9.42 ■ Write enable latch sequence

To write data to the EEPROM memory array, the user issues a WRITE instruction, followed by the address and then the data to be written (shown in Figure 9.43). This operation takes at least 32 clock cycles. The $\overline{CS}$ signal must go low and remain low for the duration of the write operation. The microcontroller may continue to write up to 32 bytes of data to the X25138. The only restriction is that the 32 bytes must reside on the same page. If the address counter reaches the end of the page and the clock continues, the counter will roll over to the first address of the page and overwrite any data that may have been written. The timing diagram of an N-byte write sequence is shown in Figure 9.44.

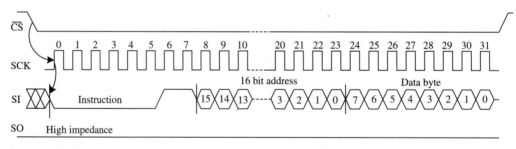

Figure 9.43 ■ Byte write operation sequence

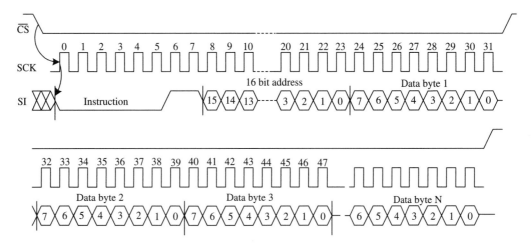

Figure 9.44 ■ Page write operation sequence

For the WRITE operation to be completed, the $\overline{CS}$ signal can only be brought high after bit 0 of data byte N is clocked in. If it is brought high at any other time, the write operation will not be completed.

To write to the status register, the WRSR instruction is followed by data to be written. Data bits 0, 1, 4, 5, and 6 must be zero. Figure 9.45 illustrates this sequence.

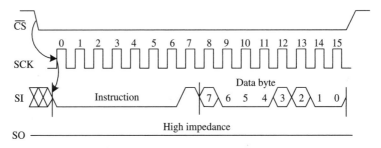

Figure 9.45 ■ Write status register operation sequence

While the write is in progress following a status register or EEPROM write sequence, the status register may be read to check the WIP bit. During this time the WIP bit will be high.

HOLD OPERATION

The $\overline{HOLD}$ input should be high under normal operation. If a data transfer is to be interrupted, $\overline{HOLD}$ can be pulled to low to suspend the transfer until it can be resumed. The only restriction is that the SCK input must be low when $\overline{HOLD}$ is first pulled low and SCK must also be low when $\overline{HOLD}$ is released.

The $\overline{HOLD}$ input may be tied high either directly to V_{CC} or tied to V_{CC} through a resistor.

9.12.2 Using SPI to Interface with the X25138

Adding an X25138 to the 68HC12 using the SPI interface is straightforward. A circuit connection that does not provide the data protection function is shown in Figure 9.46.

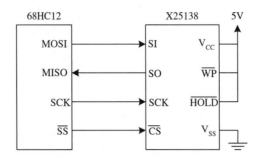

Figure 9.46 ■ Interfacing X25138 to the 68HC12

The same SPI configuration used in Example 9.12 can be used in this circuit. In order to write and read data to and from the X25138, we need to invoke a set of subroutines that provide the basic memory operations. These subroutines are:

1. *EE_wren*: This routine sets the write-enable latch.

2. *EE_wrdi*: This routine clears the write-enable latch.

3. *rd_sr*: This routine returns the contents of the EEPROM status register in accumulator A.

4. *wr_sr*: This routine writes the contents of accumulator A into the EEPROM status register.

5. *rd_byte*: This routine reads a byte from EEPROM at the address specified by index register X. The contents of the EEPROM byte are returned in accumulator A.

6. *rd_block*: This routine reads a block of EEPROM locations starting from the address in index register X and stores the bytes in the buffer pointed to by index register Y.

7. *wr_byte*: This routine writes the byte in accumulator A to the EEPROM location pointed to by index register X.

8. *wr_page*: This routine writes up to a page (32) of bytes into the EEPROM, starting from the EEPROM location specified in X. The data buffer to be output is pointed to by index register Y.

9. *ackpoll*: This routine polls the WIP bit until it is cleared.

The assembly language versions of these subroutines follow:

```
#include    "d:\miniide\hc12.inc"
WREN    equ     $06             ; set the write-enable latch
WRDI    equ     $04             ; reset the write-enable latch
RDSR    equ     $05             ; read status register
WRSR    equ     $01             ; write status register
READ    equ     $03             ; read data from memory array
WRITE   equ     $02             ; write data to memory array
```

```
; ************************************************************
;
; This routine sets the EEPROM write enable latch.
; ************************************************************
;
EE_wren      jsr        ackpoll              ; wait until EEPROM is ready for write operation
             bclr       PORTS,$80            ; enable the X25138
             ldaa       #WREN                ; send out write-enable latch command
             staa       SP0DR                ;         "
             brclr      SP0SR,$80,*          ;         "
             bset       PORTS,$80            ; disable the X25138
             rts
; ************************************************************
; This routine resets the EEPROM write-enable latch.
; ************************************************************
;
EE_wrdi      jsr        ackpoll              ; wait until EEPROM is ready for write operation
             bclr       PORTS,$80            ; enable the X25138
             ldaa       #WRDI                ; send out disable write latch command
             staa       SP0DR                ;         "
             brclr      SP0SR,$80,*          ;         "
             bset       PORTS,$80            ; disable the X25138
             rts
; ************************************************************
;
; This routine reads the EEPROM status register and returns it in A.
; ************************************************************
;
rd_sr        bclr       PORTS,$80
             ldaa       #RDSR                ; send out RDSR command
             staa       SP0DR                ;         "
             brclr      SP0SR,$80,*          ;         "
             staa       SP0DR                ; shift in status register
             brclr      SP0SR,$80,*          ;         "
             ldaa       SP0DR                ; place the X25138 status register contents in A
             bset       PORTS,$80            ; disable X25138
             rts
; ************************************************************
; This routine writes into the X25138 status register. The byte to be written into the
; status register is passed in accumulator A.
; ************************************************************
;
wr_sr        jsr        ackpoll              ; wait until EEPROM is ready for write operation
             jsr        EE_wren              ; set write-enable latch
             bclr       PORTS,$80            ; enable X25138
             ldab       #WRSR                ; send out WRSR command
             stab       SP0DR                ;         "
             brclr      SP0SR,$80,*          ;         "
             staa       SP0DR                ; send out new status register value
             brclr      SP0SR,$80,*          ;         "
             bset       PORTS,$80            ; disable X25138
             rts


; ************************************************************
; This routine reads a byte from a memory location with address passed in X.
; ************************************************************
;
rd_byte      bclr       PORTS,$80            ; enable the X25138
             ldaa       #READ                ; send out READ command
             staa       SP0DR                ;         "
             brclr      SP0SR,$80,*          ;         "
```

```
            xgdx                               ; place address in D
            staa        SP0DR                  ; send out the high byte of address
            brclr       SP0SR,$80,*            ;         "
            stab        SP0DR                  ; send out the low byte of address
            brclr       SP0SR,$80,*            ;         "
            staa        SP0DR                  ; shift back the byte
            brclr       SP0SR,$80,*            ;         "
            ldaa        SP0DR                  ; put the EEPROM byte contents in A
            bset        PORTS,$80              ; disable X25138
            rts
; ****************************************************************
; This routine reads a block of data in EEPROM starting from a certain address. The
; starting address of EEPROM is passed in X and the starting address of the buffer
; to hold the data is passed in Y. The number of bytes to read is passed in B.
; ****************************************************************
;
rd_block    bclr        PORTS,$80              ; enable the X25138
            ldaa        #READ                  ; send out READ command
            staa        SP0DR                  ;         "
            brclr       SP0SR,$80,*            ;         "
            pshb                               ; save byte count in stack
            xgdx                               ; put address in D
            staa        SP0DR                  ; send the high byte of the EEPROM address
            brclr       SP0SR,$80,*            ;         "
            stab        SP0DR                  ; send the low byte of EEPROM address
            brclr       SP0SR,$80,*            ;         "
            pulb                               ; retrieve byte count
loop1       staa        SP0DR                  ; shift in the data
            brclr       SP0SR,$80,*            ;         "
            ldaa        SP0DR                  ; get the data byte
            staa        1,y+                   ; save data byte in buffer and move pointer
            dbne        b,loop1                ; have we finished yet?
            rts
; ****************************************************************
; This subroutine writes a byte in accumulator A into the EEPROM at the address
; contained in index register X.
; ****************************************************************
;
wr_byte     psha                               ; save data byte in stack
            jsr         ackpoll                ; wait until EEPROM is ready for write operation
            jsr         EE_wren                ; set write-enable latch
            bclr        PORTS,$80              ; enable X25138
            ldaa        #WRITE                 ; send out WRITE command
            staa        SP0DR                  ;         "
            brclr       SP0SR,$80,*            ;         "
            xgdx                               ; transfer address to D
            staa        SP0DR                  ; send out the high byte of address
            brclr       SP0SR,$80,*            ;         "
            stab        SP0DR                  ; send out the low byte of address
            brclr       SP0SR,$80,*            ;         "
            pula                               ; retrieve data byte
            staa        SP0DR                  ; write out the data byte
            brclr       SP0SR,$80,*            ;         "
            bset        PORTS,$80              ; disable X25138
            rts
```

```
;*****************************************************************
; This routine writes a block of data up to N bytes. The starting address of EEPROM
; is passed in X and the starting address of data to be written is passed in Y.
; The number of bytes to be written is passed in B. If N > 32, then do nothing.
;*****************************************************************
wr_page     cmpb      #32               ; is byte count greater than page size?
            bhi       done              ; byte count greater than page size is an error
            jsr       ackpoll           ; wait until EEPROM is ready for write operation
            jsr       wren              ; set write-enable latch
            bclr      PORTS,$80         ; enable X25138
            ldaa      #WRITE            ; send out WRITE command
            staa      SP0DR             ;       "
            brclr     SP0SR,$80,*       ;       "
            pshb                        ; save the byte count
            xgdx                        ; put EEPROM address in D
            staa      SP0DR             ; send out the high byte of EEPROM address
            brclr     SP0SR,$80,*       ;       "
            stab      SP0DR             ; send out the low byte of EEPROM address
            brclr     SP0SR,$80,*       ;       "
            pulb                        ; retrieve the byte count
loop        ldaa      1,y+              ; get one byte and move the pointer
            staa      SP0DR             ; write one byte into EEPROM
            brclr     SP0SR,$80,*       ;       "
            dbne      b,loop            ; have we finished yet?
            bset      PORTS,$80         ; disable X25138
done        rts

;*****************************************************************
; This routine verifies if the EEPROM is ready for a new command. If the EEPROM is
; not ready, this routine will wait until EEPROM becomes ready.
;*****************************************************************
ackpoll     bclr      PORTS,$80         ; enable X25138
            ldaa      #RDSR             ; send the read status register command
            staa      SP0DR             ;            "
            brclr     SP0SR,$80,*       ;            "
            staa      SP0DR             ; start an SPI transfer to shift in the status
            brclr     SP0SR,$80,*       ; register contents
            bset      PORTS,$80         ; disable X25138
            ldaa      SP0DR
            asra                        ; shift out the WIP bit
            bcs       ackpoll           ; keep polling until the WIP bit is clear
            rts
```

The C language versions of these routines are as follows:

```
#include <hc12.h>
#define     WREN      0x06
#define     WRDI      0x04
#define     RDSR      0x05
#define     WRSR      0x01
#define     READ      0x03
#define     WRITE     0x02
void EE_wren (void);
void EE_wrdi (void);
char rd_sr (void);
```

```c
void wr_sr (char cx);
char rd_byte (int addr);
void rd_block (int addr, char *buf, char n);
void wr_byte (int addr, char cx);
void wr_page (int addr, char *buf, char n);
void ackpoll (void);

void EE_wren (void)
{
        ackpoll ();    /* wait until EEPROM is ready for write operation */
        PORTS &= 0x7F;              /* enable X25138 */
        SP0DR = WREN;              /* set write-enable latch */
        while (!(SP0SR & 0x80));
        PORTS |= 0x80;            /* disable X25138 */
}

void EE_wrdi (void)
{
        ackpoll ();                /* wait until EEPROM is ready for write operation */
        PORTS &= 0x7F;            /* enable X25138 */
        SP0DR = WRDI;            /* reset write-enable latch */
        while (!(SP0SR & 0x80));
        PORTS |= 0x80;          /* disable X25138 */
}

char rd_sr (void)
{
        PORTS &= 0x7F;
        SP0DR = RDSR; /* send out RDSR command */
        while (!(SP0SR & 0x80));
        SP0DR = 0x00;                  /* start an SPI transfer to shift in the status register */
        while (!(SP0SR & 0x80));
        PORTS |= 0x80;            /* disable X25138 */
        return SP0DR;
}

void wr_sr (char cx)
{
        ackpoll ();
        EE_wren();                /* set write-enable latch */
        PORTS &= 0x7F;            /* enable X25138 */
        SP0DR = WRSR;            /* send out WRSR command */
        while (!(SP0SR & 0x80));
        SP0DR = cx; /* write new value to EEPROM status register */
        while (!(SP0SR & 0x80));
         PORTS |= 0x80;            /* disable X25138 */
}
char rd_byte (int addr)
{
        PORTS &= 0x7F;              /* enable X25138 */
        SP0DR = READ;              /* send out READ command */
        while (!(SP0SR & 0x80));
        SP0DR = addr/256;            /* send out high byte of EEPROM address */
        while (!(SP0SR & 0x80));
```

```
                    SPODR = addr % 256;              /* send out low byte of EEPROM address */
                    while (!(SP0SR & 0x80));
                    SPODR = 0x00;                    /* perform an SPI read */
                    while (!(SP0SR & 0x80));
                    PORTS |= 0x80;                   /* disable X25138 */
                    return      SPODR;
}
void rd_block (int addr, char *buf, char n)
{
            int i;
            PORTS &= 0x7F;                           /* enable X25138 */
            SPODR = READ;                            /* send out READ command */
            while (!(SP0SR & 0x80));
            SPODR = addr / 256;                      /* send out the high byte of address */
            while (!(SP0SR & 0x80));
            SPODR = addr % 256;                      /* send out the low byte of address */
            while (!(SP0SR & 0x80));
            for (i = 0; i < n; i++) {
                    SPODR = 0x00;        /* trigger an SPI transfer to shift in one byte */
                    while (!(SP0SR & 0x80));
                    *buf++ = SPODR;      /* save EEPROM byte in buffer */
            }
            PORTS |= 0x80;                           /* disable X25138 */
}


void wr_byte (int addr, char cx)
{
            ackpoll ();                              /* wait until EEPROM is ready for write operation */
            EE_wren ( );                             /* set write-enable latch */
            PORTS &= 0x7F;                           /* enable X25138 */
            SPODR = WRITE;                           /* send out WRITE command */
            while (!(SP0SR & 0x80));
            SPODR = addr / 256;                      /* send out the high byte of address */
            while (!(SP0SR & 0x80));
            SPODR = addr % 256;                      /* send out the low byte of address */
            while (!(SP0SR & 0x80));
            SPODR = cx;                              /* send out data byte */
            while (!(SP0SR & 0x80));
            PORTS |= 0x80;                            /* disable X25138 */
}

void wr_page (int addr, char *ptr, char n)
{
            char        i;
            if (n > 32) return;                      /* byte count greater than page size is an error */
            ackpoll ();                              /* wait until EEPROM is ready for write operation */
            EE_wren ( );                             /* set write-enable latch */
            PORTS &= 0x7F;                           /* enable X25138 */
            SPODR = WRITE;                           /* send out WRITE command */
            while (!(SP0SR & 0x80));
            SPODR = addr / 256;                      /* send out the high byte of address */
            while (!(SP0SR & 0x80));
            SPODR = addr % 256;                      /* send out the low byte of address */
```

```
                        while (!(SPOSR & 0x80));
                        for (i = 0; i < n; i++) {              /* write N bytes into EEPROM */
                                    SPODR = *ptr++;
                                    while(!(SPOSR & 0x80));
                        }
                        PORTS |= 0x80;
        }

        void ackpoll (void)
        {
                    do {
                                PORTS &= 0x7F;        /* enable X25138 */
                                SPODR = RDSR;         /* send out RDSR command */
                                while (!(SPOSR & 0x80));
                                SPODR = 0x00;         /* start an SPI transfer to shift in the status register */
                                while (!(SPOSR & 0x80));
                    } while (SPODR & 0x01);           /* wait while WIP bit is still set */
        }
```

9.13 Summary

When high-speed data transfer is not necessary, using serial data transfer enables us to make the most use of the limited number of I/O pins available to us. Serial data transfer can be performed asynchronously or synchronously. The SCI interface is an asynchronous interface that is designed to be compatible with the EIA-232 standard. The EIA-232 standard has four aspects: electrical, mechanical, procedural, and functional. For a short distance, modems are not needed for two computers to communicate by using a connection called a *null modem*. This is achieved by connecting signals in such a way as to fool the two computers into thinking that they are connected through modems.

Since there is no common clock signal for data transfer synchronization, users of the EIA-232 standard follow a common data format: Each character is framed with a start bit and a stop bit. A character can have eight or nine data bits. The ninth bit is often used as a parity bit for error checking. The receiver uses a clock signal with a frequency that is 16 times the data rate to sample the RxD pin to detect the start bit and determine the logic values of data bits.

Errors could happen during the data transmission process. The most common errors include framing, receiver overrun, and parity errors. A framing error occurs when the start and stop bits improperly frame a received character. A framing error is detected by a missing stop bit. A receiver-overrun error occurs when one or multiple characters are received but not read by the processor. A parity error occurs when an odd number of bits change value.

All except B series 68HC12 members have two identical *serial communication interface* (SCI) subsystems. Since the EIA-232 standard uses a voltage level different from those to represent logic one and zero, a transceiver is required to perform the voltage translation so that the SCI subsystem can interface with the EIA-232 circuit. Due to the widespread use of the EIA-232 standard, transceiver chips are available from many vendors. The DS14C232 from National Semiconductor is used as an example to illustrate the SCI hardware interfacing.

The SPI subsystem is a synchronous serial interface. This interface uses three wires (excluding the $\overline{SS}$ pin) to perform data transfer. In the SPI protocol, data transfer is initiated by the master device. Other slave devices can only respond. The master supplies the clock signal for data transfer synchronization.

The SPI interface has been supported by many peripheral device vendors. They produce SPI-compatible LED and LCD display drivers, A/D converters, D/A converters, alarm timer chips, shift-registers, EEPROMs, SRAMs, phase-lock loop chips, and so on.

The HC595 is a serial-to-parallel shift register that can be used to expand the number of parallel output ports to a microcontroller. The MAX7221 is an eight-digit seven-segment display driver chip. Multiple MAX7221s can be cascaded to display more than eight BCD digits.

EEPROM chips with serial interface are very suitable for data logging and table lookup applications. The Xicor X25138 is a 128Kbits EEPROM chip organized as 16K × 8 bits. The X25138 has an instruction register to hold incoming commands. Six commands are implemented. This chip supports single-byte read, block read, single-byte write, and page write operations. Data protection is available for applications that require high security.

9.14 Exercises

E9.1 Sketch the output of the letters *k* and *p* when they are transmitted, using the format of 1 start bit, 8 data bit, and 1 stop bit.

E9.2 Write an instruction sequence to configure the SCI channel 1 to operate with the following parameters:

- 19,200 baud (P clock is 8 MHz)
- 1 start bit, 8 data bits, and 1 stop bit format
- enable both transmit and receive interrupts
- idle line wakeup
- disable wakeup initially
- long idle line mode
- enable receive and transmit
- no loop back
- enable parity

E9.3 Write a subroutine to send a break to the communication port controlled by SCI channel 1. The duration of the break must be approximately 100,000 E clock cycles.

E9.4 Use the interrupt-driven approach and write a subroutine that inputs a string from SCI interface channel 0. The string is terminated by the carriage return character and is to be stored in a buffer pointed to by index register X. The program is to be executed on a board that contains the D-Bug12 monitor. Write an instruction sequence to test your subroutine.

E9.5 Write a subroutine to output the contents of accumulator A as two hex digits to channel SCI0.

E9.6 Modify Example 9.4 so that the *putchar* routine will expand the CR character into the CR/LF pair and expand the LF character into the LF/CR pair. You can add a flag to indicate whether the expansion should be performed.

E9.7 Add an echo flag (1 byte) and modify Example 9.5 so that the received character will be echoed back to the SCI0 when the flag is one. Otherwise, no echo will be performed.

E9.8 Write a subroutine to output the contents of accumulator A as two hex digits to SCI channel 0.

E9.9 Write a subroutine to input two hex digits from the SCI channel 0 and echo them back to SCI channel 0. When this routine is run on the CME-12BC demo board, it will allow you to input two hex digits from the keyboard and echo them on the terminal screen.

E9.10 Modify the *putchar* routine that you wrote in problem E9.6 so that it will handle the backspace character like this:

- Check if the input character is the backspace character (ASCII code $08).
- If the entered character is the backspace, then echo it, output a space character, and output another backspace character.

Why would you do this?

E9.11 Suppose there is an SPI-compatible peripheral output device that has the following characteristics:

- Has a CLK input pin that is used as the data shifting clock signal
- Has an SI pin to shift in data on the falling edge of the CLK input
- Has a $\overline{\text{CE}}$ pin, which enables the chip to shift in data when it is low
- The highest data-shifting rate is 500 KHz
- The most significant bit is shifted in first

Describe how to connect the SPI pins for the 68HC12BC32 and this peripheral device, and write an instruction sequence to configure the SPI subsystem properly for data transfer. Assume the E clock frequency is 8 MHz.

E9.11 The HC589 is another SPI-compatible shift register. This chip has both serial and parallel inputs and is often used to expand the number of parallel input ports. The block diagram of the HC589 is shown in Figure 9E.11. The functions of these pins are as follows:

- A,..,H: *Parallel data inputs*. Data on these inputs are loaded in the data latch on the rising edge of the LC signal.
- SA: *Serial data input*. Data on this pin are shifted into the shift register on the rising edge of the SC signal if the SS/$\overline{\text{PL}}$ pin is high. Data on this pin are ignored when the SS/$\overline{\text{PL}}$ signal is low.
- SS/$\overline{\text{PL}}$: *Serial shift/parallel load*. When a high level is applied to this pin, the shift register is allowed to shift data serially. When a low level is applied, the shift register accepts parallel data from the data latch.
- SC: *Serial shift clock*. A low-to-high transition on this pin shifts data on the serial data input into the shift register; data on stage H are shifted out from Q_H, where it is replaced by the data previously stored in stage G. The highest shift clock rate is 5 MHz.
- LC: *Latch clock*. A low-to -high transition on this pin loads the parallel data on inputs A-H into the data latch.
- $\overline{\text{OE}}$: *Output enable*. A high level on this pin forces the Q_H output into a high-impedance state. A low level enables the output. This control does not affect the state of the input latch or the shift register.
- Q_H: *Serial data output*. This is a three-state output from the last stage of the shift register.

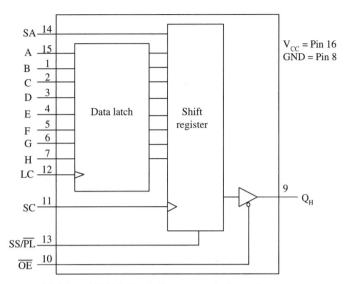

Figure 9E.11 ■ HC589 block diagram and pin assignment

Like the HC595, the HC589 can be cascaded. To cascade, the Q_H output is connected to the SA input of its adjacent HC589. All HC589s should share the same shift clock SC. Suppose we want to use two HC589s to interface with two DIP switches so that we can input 4 hex digits into the 68HC12. Describe the circuit connection and write an instruction sequence to read the data from these two DIP switches.

E9.12 Suppose you are going to use four HC595s to drive four seven-segment displays using the circuit shown in Figure 9.31. Write an instruction sequence to display the number 1982 on these four seven-segment displays.

E9.13 Write a program to display the numbers 1, 2, ..., 8 on those eight seven-segment displays in Figure 9.35. Display one digit at a time and turn off the other seven digits. Each digit is displayed for one second. Perform this operation continuously.

E9.14 Write the C language version of the program for Example 9.11.

E9.15 Modify the solution to Example 9.12 so that it can display the value 103.5°F 10 48 10 30 01.

E9.16 Write the C language version of the program for Example 9.12.

E9.17 Write a program to store the first 100 prime numbers in the X25138 starting from memory location $0000. Each prime number will occupy 2 bytes.

E9.18 The Motorola MC14489 is a five-digit seven-segment display driver chip. The data sheet is in the accompanying CD-ROM (also available in Motorola's Web site). Describe how to interface one MC14489 with the HC12 using the SPI subsystem. Write a program to display the value 12345 on the seven-segment displays driven by the MC14489.

E9.19 Write an interrupt-driven version of *gets_sc0*.

9.15 Lab Exercises & Assignment

L9.1 Write a program to be run on the CME-12BC or other demo board. This program and the user interact as follows:

1. The program outputs the message *Please enter your age:* and waits for the user to enter his/her age.

2. The user enters his/her age and the program reads it.

3. The program outputs the message *Please enter your height in inches:* and waits for the user to enter his/her height.

4. The user enters his/her height in inches and the program reads it.

5. The program outputs the message *Please enter your weight in lbs:* and waits for the user to enter his/her weight.

6. The user enters his/her weight in lbs and the program reads it.

7. The program outputs the following messages on the terminal screen and exit:
   You are *xxx* years old.
   You are *kk* ft *mm* inches tall.
   You weigh *zzz* lbs.

L9.2 Use the SPI to interface with four HC595s to drive four seven-segment displays. Write a program to display the number 1289 on them. Light one digit at a time for one second. Repeat the same operation continuously.

L9.3 Use the SCI I/O and SPI together to display the current time of day. Use one MAX7221 to drive six seven-segment displays. Write a program to perform the following operation:

1. Output the prompt *Please enter the current time of day: hh mm ss* to the SCI channel 0 (this would be displayed on the PC monitor screen).

2. The user enters the time and your program reads it and saves it in memory.

3. The program uses the timer output compare function to create a one second delay to update the time and display.

10

Analog-to-Digital Converter

10.1 Objectives

After completing this chapter, you should be able to:

- explain the A/D conversion process

- describe the resolution, the various channels, and the operation modes of the 68HC12 A/D converter

- interpret A/D conversion results

- describe the procedure for using the 68HC12 A/D converter

- configure the A/D converter for the application

- use the temperature sensor LM34 made by National Semiconductor

- measure the barometric pressure using the BPT sensor from Sensortechnics

- use an external A/D converter

10.2 Fundamental Concepts of the A/D Converter

Many microcontroller applications deal with non-electric quantities such as weight, humidity, pressure, mass or air flow, temperature, light intensity, speed, and so on. These quantities are *analog* values because they have a continuous set of values over a given range, in contrast to the discrete values of digital signals. To enable the microcontroller to process these quantities, we need to represent these quantities in digital form; thus an *analog-to-digital (A/D) converter* is required.

An A/D converter can only deal with electrical voltage. A nonelectric quantity must be converted into a voltage before A/D conversion can be performed. Conversion from a nonelectric quantity to a voltage requires the use of a *transducer*. In general, a transducer is a device that converts the quantity from one form to another. For example, a *temperature sensor* is a transducer that can convert the temperature into a voltage. A *load cell* is the transducer that can convert a weight into a voltage.

A transducer may not generate an output voltage in the range suitable for A/D conversion. A voltage *scaler* (or *shifter*) are often needed to transform the transducer output voltage into a range that can be handled by the A/D converter. The circuit that performs the scaling and shifting of the transducer output is called a *signal-conditioning circuit*. The overall A/D process is illustrated in Figure 10.1.

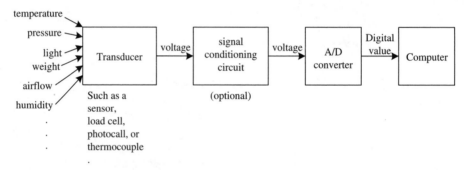

Figure 10.1 ■ The A/D conversion process

The accuracy of an A/D converter is dictated by the number of bits used to represent the analog quantity. The greater the number of bits, the better the accuracy. All 68HC12 members (except the 812A4) implement one or two 10-bit A/D converters. The 812A4 implements an 8-bit A/D converter.

There are three major A/D conversion methods in use today: the *flash*, *sigma-delta*, and *successive approximation* methods. All 68HC12 members use the charge-redistribution–based successive-approximation method to implement the A/D converter.

10.3 Successive Approximation Method

The block diagram of a successive-approximation A/D converter is shown in Figure 10.2. An A/D converter based on the successive-approximation method approximates the analog signal to n-bit code in n steps. It first initializes the *successive-approximation register* (SAR) to zero and then performs a series of assumptions, starting with the most significant bit and proceeding toward the least significant bit. The algorithm of the successive-approximation method is illustrated in Figure 10.3. It assumes that the SAR register has n bits. For every bit of the SAR, the algorithm:

- Assumes the bit to be a one
- Converts the value of the SAR to an analog voltage
- Compares the D/A output with the analog input and clears the bit to zero if the D/A output is larger (which indicates that the guess is wrong)

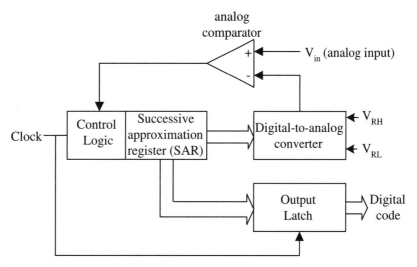

Figure 10.2 ■ Block diagram of a successive approximation A/D converter

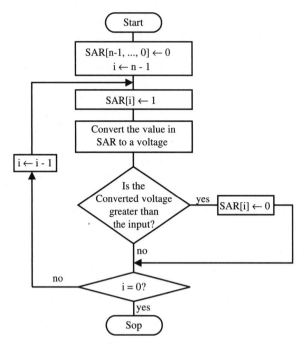

Figure 10.3 ■ Successive approximation A/D conversion method

10.4 Signal Conditioning Circuits

Not all transducer outputs are appropriate for an A/D converter. An A/D converter often needs a circuit to scale and/or shift the transducer output to fit its input range. This section introduces a circuit for voltage scaling and also a circuit for shifting and scaling.

10.4.1 Optimal Voltage Range for the A/D Converter

An A/D converter requires a *low-reference voltage* (V_{LREF}) and a *high-reference voltage* (V_{HREF}) to perform the conversion. The low-reference voltage is often set to 0V and the high-reference voltage is often set to the power supply V_{DD}. Most A/D converters are ratiometric, in other words:

- A 0V analog input is converted to the digital value zero.
- A V_{DD} analog input is converted to the digital value $2^n - 1$.
- A k volt input will be converted to the digital value $k \times (2^n - 1) \div V_{DD}$.

where n is the number of bits the A/D converter uses to represent a conversion result. The value n is also called the *resolution* of the A/D converter.

Since the A/D converter is ratiometric, the optimal voltage range for the A/D converter is $0\text{~}V_{DD}$ as long as the V_{LREF} and V_{HREF} are set to 0V and V_{DD}, respectively. The A/D conversion result k corresponds to an analog voltage given by the following equation:

$$V_k = V_{LREF} + (\text{range} \times k) \div (2^n - 1) \qquad \text{--- (10.1)}$$

where V_k is the analog voltage corresponding to the converted result k and range = $V_{HREF} - V_{LREF}$.

Example 10.1

Suppose there is a 12-bit A/D converter with $V_{LREF} = 0.5V$ and $V_{HREF} = 3.5V$. Find the corresponding voltage values for A/D conversion results of 20, 100, 800, 1,200, 2,400, and 3,600.

Solution: range $= V_{HREF} - V_{LREF} = 3.5V - 0.5V = 3V$

The voltage corresponding to the results of 20, 100, 800, 1,200, 2,400, and 3,600 are:

$0.5V + (20 \times 3) \div (2^{12} - 1) = 0.515V$
$0.5V + (100 \times 3) \div (2^{12} - 1) = 0.573V$
$0.5V + (800 \times 3) \div (2^{12} - 1) = 1.086V$
$0.5V + (1,200 \times 3) \div (2^{12} - 1) = 1.379V$
$0.5V + (2,400 \times 3) \div (2^{12} - 1) = 2.258V$
$0.5V + (3,600 \times 3) \div (2^{12} - 1) = 3.137V$

10.4.2 Voltage Scaling Circuit

There are situations in which the transducer output voltages are in the range of $0 \sim V_Z$, where $V_Z < V_{DD}$ (power supply). Because V_Z sometimes can be much smaller than V_{DD}, the A/D converter cannot take advantage of the available full dynamic range, and therefore conversion results can be very inaccurate. The voltage scaling circuit can be used to improve the accuracy because it allows the A/D converter to utilize its full range. The diagram of a voltage scaling circuit is shown in Figure 10.4. Because the OP AMP has an (almost) infinite input impedance, the current that flows through resistor R2 will be the same as the current that flows through R1. Therefore, the voltage gain of this circuit is given by the following equation:

$A_V = V_{OUT} \div V_{IN} = (R_1 + R_2) \div R_1 = 1 + R_1/R_2$ (8.2)

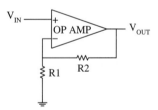

Figure 10.4 ■ A voltage scaler

Example 10.2

Suppose the transducer output voltage ranges from 0V to 100 mV. Choose the appropriate values of R_1 and R_2 to scale this range to 0~5V.

Solution:

$5V \div 100\,mV = 50$
$\therefore R_2/R_1 = 49$

By choosing 330 KΩ for R_2 and 6.8KΩ for R_1, we obtain a R_2/R_1 ratio of 48.53, which is accurate to within 3%.

10.4.3 Voltage Shifting/Scaling Circuit

There are transducers whose outputs are in the range of $V_1 \sim V_2$ (V_1 can be negative and V_2 can be smaller than V_{DD}) instead of in the range of $0V \sim V_{DD}$. The accuracy of A/D conversion can be improved by using a circuit that shifts and scales the transducer output so that it falls in the full range of $0V \sim V_{DD}$.

An OP AMP circuit that can shift and scale the transducer output is shown in Figure 10.5c. This circuit consists of a summing (Figure 10.5a) and an inverting circuit (Figure 10.5b). The voltage V_{IN} comes from the transducer output, whereas V_1 is an adjusting voltage. By choosing appropriate values for V_1 and the resistors, the desired voltage shifting and scaling can be achieved.

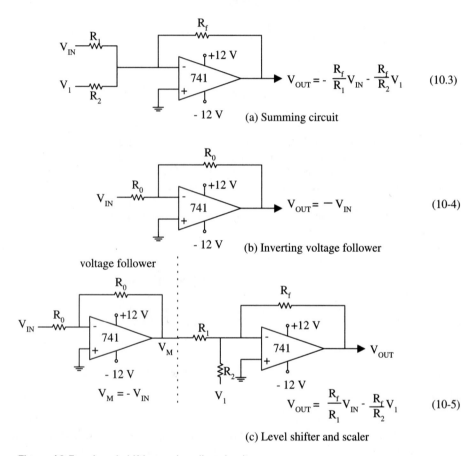

$$V_{OUT} = -\frac{R_f}{R_1}V_{IN} - \frac{R_f}{R_2}V_1 \qquad (10.3)$$

(a) Summing circuit

$$V_{OUT} = -V_{IN} \qquad (10\text{-}4)$$

(b) Inverting voltage follower

$$V_M = -V_{IN}$$

$$V_{OUT} = \frac{R_f}{R_1}V_{IN} - \frac{R_f}{R_2}V_1 \qquad (10\text{-}5)$$

(c) Level shifter and scaler

Figure 10.5 ■ Level shifting and scaling circuit

Example 10.3

Choose the appropriate values of resistors and the adjusting voltage so that the circuit in Figure 10.5c can shift the voltage from $-1.5V \sim 3.5V$ to $0V \sim 5V$.

Solution: Applying Equation 10.5:

$$0 = -1.5 \times (R_f/R_1) - (R_f/R_2) \times V_1$$
$$5 = 3.5 \times (R_f/R_1) - (R_f/R_2) \times V_1$$

By choosing $R_f = R_1 = R_0 = 15K\Omega$, $R_2 = 120K\Omega$, and setting $V_1 = -12V$, we can shift and scale the voltage to the desired range.

From this example, you can see that the selection of resistor values and the voltage V_1 is a trial-and-error process at best.

10.5 The 68HC12 A/D Converter

The 68HC12 members have either one or two 8-channel A/D converters. The 812A4 has an 8-channel, 8-bit A/D converter. All B family members have an 8-channel, 10-bit A/D converter. The 912D60, the 912DG128, the 912DT128, and the 912DT128 have two identical 8-channel, 10-bit A/D converters.

After the A/D converter has been enabled, the A/D conversion will be started when the ATDCTL5 register is written. Once started, the 68HC12 A/D converter will perform either a single conversion sequence (in non-scan mode) or an unlimited number of conversion sequences (in scan mode). A single conversion sequence consists of either four or eight conversions, depending on the state of the selected 8-channel mode (S8CM) bit when the ATDCTL5 register is written.

Except for the 812A4, all other 68HC12 members can be programmed to perform either the 8-bit or 10-bit A/D conversion depending on the need of the applications. The clock source to the A/D converter is the P-clock divided by a prescale factor. The maximum conversion frequency is 2 MHz and the lowest conversion frequency is 500 KHz. Users must keep this requirement in mind when choosing the prescale factor.

The overall block diagram of the A/D converter is shown in Figure 10.6. The pin V_{DDA} is the analog power supply input pin. This voltage must be set to $5V \pm 10\%$. The V_{SSA} pin is the analog ground input and must be connected to 0V. The V_{RHx} (x = 0 or 1) and V_{RLx} are the high and low reference voltages, respectively. For those 68HC12 members that have two A/D converters, we need to add a zero and one to specify the pins at either the ATD0 or the ATD1 subsystem. The high reference voltage must be between $V_{DDA}/2$ and V_{DDA}. The low reference voltage must be between V_{SSA} and $V_{DDA}/2$. However, Motorola recommends that the high and the low reference voltages should be set to V_{DDA} and V_{SSA}, respectively, in order to achieve the best accuracy.

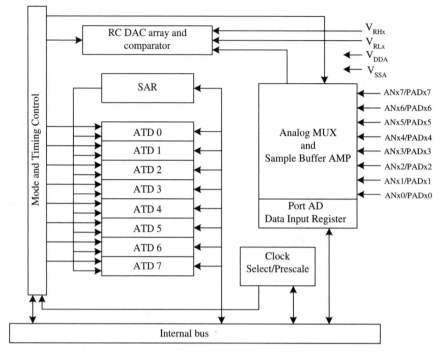

Figure 10.6 ■ 68HC12 ATD block diagram

10.6 ATD Registers

Each of the two 68HC12 A/D converters has six control registers for configuring the A/D parameters. Among these six registers, the first two are not implemented so we will ignore them in this text. For the 812A4 and B series members, the mnemonics of these six control registers are ATDCTL0, ..., ATDCTL5. For those members that have two A/D converters, the control registers associated with ATD0 are ATD0CTL0, ..., ATD0CTL5, whereas those associated with ATD1 are ATD1CTL0, ..., ATD1CTL5. You need to use appropriate names for these control registers when writing programs. The functions of these control registers are described in this section. The A/D registers in ATD0 are mapped to the same memory locations as their corresponding registers when there is only one A/D converter. For example, ATDCTL2 and ATD0CTL2 are mapped to the same address.

10.6.1 ATD Control Register 2 (ATDCTL2, ATD0CTL2 & ATD1CTL2)

The contents of this register are shown in Figure 10.7. The most critical bit in this register is the ADPU bit. This bit must be set in order to make the A/D converter work. Like timer applications, enabling the fast flag clear feature is an efficient way to use the A/D conversion. The A/D converter interrupt can be enabled (by setting the ASCIE bit) to inform the CPU that the conversion has completed.

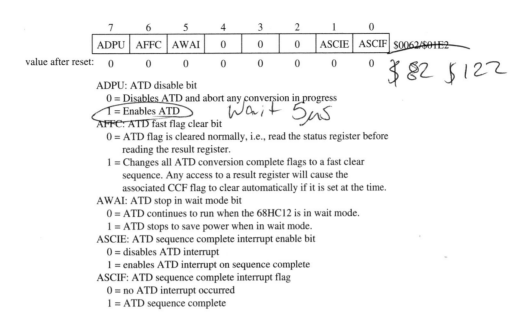

ADPU: ATD disable bit
  0 = Disables ATD and abort any conversion in progress
  1 = Enables ATD
AFFC: ATD fast flag clear bit
  0 = ATD flag is cleared normally, i.e., read the status register before reading the result register.
  1 = Changes all ATD conversion complete flags to a fast clear sequence. Any access to a result register will cause the associated CCF flag to clear automatically if it is set at the time.
AWAI: ATD stop in wait mode bit
  0 = ATD continues to run when the 68HC12 is in wait mode.
  1 = ATD stops to save power when in wait mode.
ASCIE: ATD sequence complete interrupt enable bit
  0 = disables ATD interrupt
  1 = enables ATD interrupt on sequence complete
ASCIF: ATD sequence complete interrupt flag
  0 = no ATD interrupt occurred
  1 = ATD sequence complete

Figure 10.7 ■ ATD control register 2 (ATDCTL2/ATD1CTL2)

The A/D converter needs at least 5 μs to stabilize after setting the ADPU bit. We must wait at least this amount of time before starting an A/D conversion.

10.6.2 ATD Control Register 3 (ATDCTL3, ATD0CTL3 & ATD1CTL3)

As shown in Figure 10.8, this register is used to select the power-up mode, interrupt control, and freeze control.

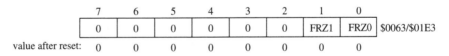

7	6	5	4	3	2	1	0	
0	0	0	0	0	0	FRZ1	FRZ0	$0063/$01E3

value after reset: 0 0 0 0 0 0 0 0

FRZ1 and FRZ0: background debug (freeze) enable bit
  00: contiue conversions in active background mode
  01: reserved
  10: finish current conversion, then freeze
  11: freeze when background mode is active

Figure 10.8 ■ ATD control register 3 (ATDCTL3/ATD1CTL3)

10.6.3 ATD Control Register 4 (ATDCTL4, ATD0CTL4 & ATD1CTL4)

The ATD control register 4 selects the A/D resolution and sample times, and sets up the prescaler. The 68HC12 A/D converter allows the user to select the conversion resolution depending on the needs of the application. Using a lower resolution can save some conversion time. Writes to the ATD control registers initiate a new conversion sequence. If a write occurs while a conversion is in progress, the conversion is aborted and ATD activity halts until a write to ATD control register 5 occurs. The contents of this register are shown in Figure 10.9.

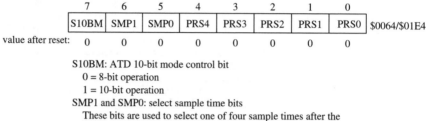

S10BM: ATD 10-bit mode control bit
  0 = 8-bit operation
  1 = 10-bit operation
SMP1 and SMP0: select sample time bits
  These bits are used to select one of four sample times after the
  buffered sample and transfer has occurred. See Table 10.1.
PRS4--PRS0: select divide-by factor for ATD P-clock prescaler bits
  The P clock is divided by this value plus one, and then fed into a
  divide-by-two circuit to generate the ATD module signal.
  Table 10.2 shows the divide-by operation and the appropriate
  range of system clock frequencies. The ATD conversion frequency
  must be between 500KHz and 2 MHz.

Figure 10.9 ■ ATD control register 4 (ATDCTL4/ATD1CTL4)

SMP1	SMP0	Final sample time	Total 8-bit conversion time	Total 10-bit conversion time
0	0	2 ATD clock periods	18 ATD clock periods	20 ATD clock periods
0	1	4 ATD clock periods	20 ATD clock periods	22 ATD clock periods
1	0	8 ATD clock periods	24 ATD clock periods	26 ATD clock periods
1	1	16 ATD clock periods	32 ATD clock periods	34 ATD clock periods

Table 10.1 ■ Final sample time selection

Prescale value	Total Divisor	Max P clock[1]	Min P clock[2]
00000	2	4 MHz	1 MHz
00001	4	8 MHz	2 MHz
00010	6	8 MHz	3 MHz
00011	8	8 MHz	4 MHz
00100	10	8 MHz	5 MHz
00101	12	8 MHz	6 MHz
00110	14	8 MHz	7 MHz
00111	16	8 MHz	8 MHz
01xxx		Do not use	
1xxxx			

1. Maximum conversion frequency is 2 MHz. Maximum P clock divisor value becomes maximum conversion rate that can be used on this ATD module.
2. Minimum conversion frequency is 500 KHz. Minimum P clock divisor value becomes minimum conversion rate that this ATD can perform.

Table 10.2 ■ Clock prescale values

Like any other A/D converter, the 68HC12 A/D converter needs a clock signal to operate. The clock frequency must be between 500 KHz and 2 MHz. When selecting the prescale factor, you must make sure that the resultant clock frequency is in this range.

10.6.4 ATD Control Register 5 (ATDCTL5, ATD0CTL5 & ATD1CTL5)

The ATD control register 5 is used to select the conversion modes, the conversion channel(s), and initiate conversions. A write to ATDCTL5 initiates a new conversion sequence. If a conversion is in progress when a write occurs, that sequence is aborted and the SCF and CCF bits are reset.

The contents of ATD control register 5 are shown in Figure 10.10. The A/D converter allows you to select a single channel, a group of four channels, or a group of eight channels to perform the conversion. The A/D converter can be programmed to perform a single sequence of conversion or perform continuously.

	7	6	5	4	3	2	1	0	
	0	S8CM	SCAN	MULT	CD	CC	CB	CA	$0065/$01E5
value after reset:	0	0	0	0	0	0	0	0	

S8CM: *select 8-channel mode bit*
    0 = conversion sequence consists of four conversions
    1 = conversion sequence consists of eight conversions
SCAN: *enable continuous channel scan bit*
    0 = single conversion sequence
    1 = continuous conversion sequences (scan mode)
MULT: *enable multichannel conversion bit*
    0 = ATD sequencer runs all four or eight conversions on a single
        input channel selected via the CD, CC, CB, and CA bits
    1 = ATD sequencer runs each of the four or eight conversions on
        sequential channels in a specific group. Refer to Table 10.3.
CD, CC, CB, and CA: *channel select for conversion bits*
    The channel selection is shown in Table 10.3.

Figure 10.10 ■ ATD control register 5 (ATDCTL5/ATD1CTL5)

S8CM	CD	CC	CB	CA	Channel Signal	Result in ADRx if MULT = 1
0	0	0	0	0	AN0	ADR0
			0	1	AN1	ADR1
			1	0	AN2	ADR2
			1	1	AN3	ADR3
0	0	1	0	0	AN4	ADR0
			0	1	AN5	ADR1
			1	0	AN6	ADR2
			1	1	AN7	ADR3
0	1	0	0	0	Reserved	ADR0
			0	1	Reserved	ADR1
			1	0	Reserved	ADR2
			1	1	Reserved	ADR3
0	1	1	0	0	VRH	ADR0
			0	1	VRL	ADR1
			1	0	$(V_{RH} + V_{RL})/2$	ADR2
			1	1	Test/reserved	ADR3
1	0	0	0	0	AN0	ADR0
		0	0	1	AN1	ADR1
		0	1	0	AN2	ADR2
		0	1	1	AN3	ADR3
		1	0	0	AN4	ADR4
		1	0	1	AN5	ADR5
		1	1	0	AN6	ADR6
		1	1	1	AN7	ADR7
1	1	0	0	0	Reserved	ADR0
		0	0	1	Reserved	ADR1
		0	1	0	Reserved	ADR2
		0	1	1	Reserved	ADR3
		1	0	0	VRH	ADR4
		1	0	1	VRL	ADR5
		1	1	0	$(V_{RH} + V_{RL})/2$	ADR6
		1	1	1	Test/reserved	ADR7

Shaded bits are "don't care" if the MULT bit = 1 and the entire block of four or eight channels makes up a conversion sequence. When MULT = 0, all four bits (CD, CC, CB, & CA) must be specified and a conversion sequence consists of four or eight consecutive conversions of the single specified channel.

Table 10.3 ■ Multichannel mode result register assignment

10.6.5 ATD Status Registers (ATDSTAT0, ATDSTAT1, ATD0STAT0, ATD0STAT1, ATD1STAT0 & ATD1STAT1)

Each A/D channel has two status registers for indicating whether a sequence of conversion is complete. The contents of these two registers are shown in Figure 10.11.

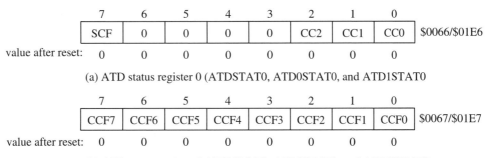

(a) ATD status register 0 (ATDSTAT0, ATD0STAT0, and ATD1STAT0)

(b) ATD status register 1 (ATDSTAT1, ATD0STAT1, and ATD1STAT1)

SCF: sequence complete flag
> This bit is set at the end of the conversion sequence when in the single conversion sequence mode and is set at the end of the first conversion sequence when in the continuous conversion mode. When AFFC = 0, SCF is cleared when the ATD control register 5 is being written to initiate a new conversion sequence. When AFFC = 1, SCF is cleared after the first result register is read.

CC2--CC0: conversion counter bits for current 4 or 8 conversions
> This 3-bit value reflects the contents of the conversion counter pointer in a four or eight count sequence. This value also reflects which result register is written next, indicating which channel is currently being converted.

CCF7--CCF0: conversion complete flag
> Each CCF bit is associated with an individual ATD result register. For each register, this bit is set at the end of conversion for the associated ATD channel and remains set until that ATD result register is read.

Figure 10.11 ■ ATD status register

10.6.6 ATD Test Registers (ATDTESTH/L, ATD0TESTH/L & ATD1TESTH/L)

Each of the 68HC12 A/D converters has a 16-bit test register to facilitate the factory testing operation. End users do not need to pay much attention to this register. The contents of this register are shown in Figure 10.12.

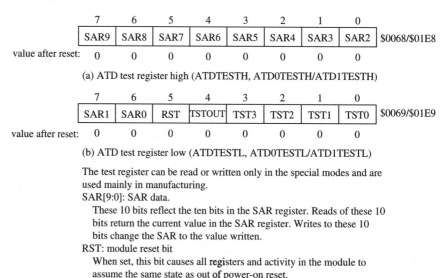

	7	6	5	4	3	2	1	0	
	SAR9	SAR8	SAR7	SAR6	SAR5	SAR4	SAR3	SAR2	$0068/$01E8
value after reset:	0	0	0	0	0	0	0	0	

(a) ATD test register high (ATDTESTH, ATD0TESTH/ATD1TESTH)

	7	6	5	4	3	2	1	0	
	SAR1	SAR0	RST	TSTOUT	TST3	TST2	TST1	TST0	$0069/$01E9
value after reset:	0	0	0	0	0	0	0	0	

(b) ATD test register low (ATDTESTL, ATD0TESTL/ATD1TESTL)

The test register can be read or written only in the special modes and are used mainly in manufacturing.
SAR[9:0]: SAR data.
    These 10 bits reflect the ten bits in the SAR register. Reads of these 10 bits return the current value in the SAR register. Writes to these 10 bits change the SAR to the value written.
RST: module reset bit
    When set, this bit causes all registers and activity in the module to assume the same state as out of power-on reset.
TSTOUT: multiplex output of TST[3:0] (factory use)
TST[3:0]: test bits 3 to 0
    Selects one of 16 reserved factory testing modes.

Figure 10.12 ■ ATD test registers

10.6.7 Port AD Data Input Register (PORTAD & PORTAD0/PORTAD1)

When the A/D converter is not enabled, the associated port can be used as an input port. The values of these pins can be read from the port AD data register. The 812A4 and all members in the B series should use the PORTAD register to access the PORTAD pins, whereas other members should use either the PORTAD0 or the PORTAD1 register to access the pins associated with PORTAD0 or PORTAD1.

10.6.8 ATD Result Registers (ADR0H/L..ADR7H/L, ADR00H/L..ADR07H/L, & ADR10H/L..ADR17H/L)

Each ATD channel has eight result registers to hold the conversion result. Each result register is 16-bit and consists of a high and a low result register. For the 812A4 and members of the B series, the mnemonics ADR0H/L..ADR7H/L are used to access the high and low result registers of channel 0 to 7. For other members, the mnemonics ADR00H/L..ADR07H/L are used to access the high and low result registers of channel 0 to 7 for ATD0, whereas ADR10H/L..ADR17H/L are used to access the high and low result registers of channel 0 to 7 for ATD1.

All except the 812A4 can choose to use either the 8-bit or 10-bit conversion mode. When the 8-bit mode is chosen, the conversion result is stored in the high result register. In 10-bit mode, the high result register holds the upper eight bits of the conversion, whereas the highest two bits (bit 7 and 6) of the low result register hold the lowest two bits of the conversion result.

10.7 The Procedure for Using the A/D Converter

The procedure for using the A/D converter is as follows:

Step 1

Connect the hardware properly. The A/D-related pins must be connected as follows:

V_{DDA}: connect to 5V

V_{SSA}: connect to 0V

V_{RH} (or V_{RHx}, x = 0 or 1): 5V

V_{RL} (or V_{RLx}, x = 0 or 1): 0V

If the transducer output is not in the appropriate range, then we should use a signal conditioning circuit to shift and scale it to between V_{RL} and V_{RH}.

Step 2

Configure ATD control registers 2 to 4 properly and wait for the ATD to stabilize (need to wait for 5 μs).

Step 3

Select the appropriate channel(s) and operation modes by programming the ATD control register 5. Writing into the ATD control register 5 also starts the A/D conversion.

Step 4

Wait until the SCF flag of the status register ATDSTAT0 (ATD0STAT0 or ATD1STAT0 for members having two ATD modules) is set, then collect the A/D conversion results and store them in memory.

Example 10.4

▼

Write a subroutine to initialize the ATD converter for the 912BC32 and start the conversion with the following parameters:

- nonscan mode
- select channel 7
- fast flag clear all
- stop ATD in wait mode
- disable interrupt
- finish current conversion then freeze when BDM becomes active
- 10-bit operation and 2 A/D clock periods of sample time
- choose 2 MHz as the conversion frequency for the 8-MHz P-clock

Solution: The setting of the ATD control registers 2 to 5 are as follows:

The setting of the *ATD control register 2*:

- enable ATD (set bit 7 to one)
- select fast flag clear all (set bit 6 to one)
- stop ATD when in wait mode (set bit 5 to one)
- disable ATD interrupt (set bit 1 to zero)
- all other bits are cleared
- write the value $E0 into this control register

The setting of the *ATD control register 3*:

- when BDM becomes active, complete the current conversion then freeze
- write the value $02 into this control register

The setting of the *ATD control register 4*:

- select 10-bit operation (set bit 7 to one)
- 2 A/D clock periods for sample time (set bits 6 and 5 to 00)
- max P clock is 8 MHz and conversion frequency is 2 MHz (set bits 4 to 0 to 00001 to set prescale factor to four)
- write the value $81 to this control register

The setting of the *ATD control register 5*:

- conversion sequence consists of four conversions (set bit 6 to zero)
- nonscan mode (set bit 5 to zero)
- single channel mode (set bit 4 to zero)
- select channels 0 to 3 (set bits 3..0 to 0111)
- write the value $07 to this control register

The following subroutine will perform the desired ATD configuration:

```
#include      "d:\miniide\hc12.inc"
ATD_init      movb        #$E0,ATDCTL2
              ldaa        #10
; the next two instructions create 5 µs delay
wait          deca                              ; 1 E-clock cycle execution time
              bne         wait                  ; 3/1 E-clock cycles execution time
              movb        #$02,ATDCTL3
              movb        #$81,ATDCTL4
              rts
```

This routine does not write into the ATDCTL5 register because that will start the conversion. We should write into the ATDCTL5 register only when we want to perform the conversion.

The C language version of the subroutine is as follows:

```
void ATD_init (void)
{
        int i;
        ATDCTL2 = 0xE0;
        for (i = 0; i < 40; i++) {
                asm ("nop");
        }
        ATDCTL3 = 0x02;
        ATDCTL4 = 0x81;

}
```

Example 10.5

Write a subroutine to initialize the ATD0 module of the 912DG128 with the following parameters:
- nonscan mode
- channels 0 to 7
- fast flag clear all
- stop ATD in wait mode
- disable interrupt
- finish current conversion then freeze when BDM becomes active
- 10-bit operation and 4 A/D clock periods of sample time
- choose 2 MHz as the conversion frequency for the 8-MHz P-clock

Solution: The setting of ATD control registers 2 to 5 are as follows:
The setting of the *ATD control register 2*:
- enable ATD (set bit 7 to one)
- select fast flag clear all (set bit 6 to one)
- stop ATD when in wait mode (set bit 5 to one)
- disable ATD interrupt (set bit 1 to zero)
- all other bits are cleared
- write the value $E0 into this control register

The setting of the *ATD control register 3*:
- when BDM becomes active, complete the current conversion then freeze
- write the value $02 into this control register

The setting of the *ATD control register 4*:
- select 10-bit operation (set bit 7 to one)
- 4 A/D clock periods for sample time (set bits 6 and 5 to 01)
- max P clock is 8 MHz and conversion frequency is 2 MHz (set bit 4 to 0 to 00001 to set prescale factor to four)
- write the value $A1 to this control register

The setting of the *ATD control register 5*:
- conversion sequence consists of eight conversions (set bit 6 to one)
- nonscan mode (set bit 5 to zero)
- multiple channel mode (set bit 4 to one)
- select channels 0 to 7 (set bits 3..0 to 0000)
- Write the value $50 to this control register

The following subroutine will perform the desired ATD configuration:

```
#include     "d:\miniide\hc12.inc"
ATD0_init    movb     #$E0,ATDCTL2
             ldaa     #10
```

```
                              ; the next two instructions create 5 ?s delay
              wait0          deca                                      ; 1 E-clock cycle execution time
                            bne          wait                         ; 3/1 E-clock cycles execution time
                            movb         #$02,ATDCTL3
                            movb         #$A1,ATDCTL4
                            rts
```

This routine doesn't write into the ATDCTL5 register because that will start the A/D conversion.

The C language version of the subroutine is as follows:

```c
void ATD0_init (void)
{
        int i;
        ATD0CTL2 = 0xE0;
        for (i = 0; i < 40; i++) {
                asm ("nop");
        }
        ATD0CTL3 = 0x02;
        ATD0CTL4 = 0xA1;
}
```

Example 10.6

Write a program to perform an A/D conversion on the analog signal connected to the AN7 pin. Collect 20 A/D conversion results and store them at memory location starting from $800. Use the same configuration as in Example 10.4.

Solution: To collect 20 conversion results, we need to write into the ATDCTL5 register 5 times. Each time we will wait for the SCF flag to be set and collect four results. The program is as follows:

```
              #include   "d:\miniide\hc12.inc"

              org        $1000
              ldx        #$800            ; use index register X as a pointer to the buffer
              jsr        ATD_init         ; initialize the ATD converter
              ldy        #5
       loop5  movb       #$07,ATDCTL5     ; start an A/D conversion sequence
              brclr      ATDSTAT0,$80,*
              movw       ADR0H,2,x+       ; collect and save the conversion result (left-justified)
              movw       ADR1H,2,x+       ; post-increment the pointer by two
              movw       ADR2H,2,x+       ;          "
              movw       ADR3H,2,x+       ;          "
              dbne       y,loop5
              swi
              end
```

The C language version of the program is as follows:

```c
#include <hc12.h>
void ATD_init (void);
int buf[20];
void main (void)
```

```
{
        int  i;

        ATD_init();
        for (i = 0; i < 5; i++) {
                ATDCTL5 = 0x07;              /* start an A/D conversion */
                while (!(ATDSTAT0 & 0x80)); /* wait for the A/D conversion to complete */
/* store the result right-justified */
                buf[4*i + 0] = ADR0H * 4 + ADR0L/64;
                buf[4*i + 1] = ADR1H * 4 + ADR1L/64;
                buf[4*i + 2] = ADR2H * 4 + ADR2L/64;
                buf[4*i + 3] = ADR3H * 4 + ADR3L/64;
        }
        asm ("swi");
}
```

10.8 Measuring the Temperature

The LM34 is a precision Fahrenheit temperature sensor made by National Semiconductor. The voltage output of the LM34 is linearly proportional to the Fahrenheit temperature. No external calibration or trimming is required; the LM can achieve an accuracy of ±0.5°F at room temperature and ±1.5°F over a full –50 to +300°F temperature range. It can be used with single power supplies or with plus and minus supplies.

The LM34 draws about 75 μA of current from the supply and hence has very low self-heating, less than 0.2°F in still air. The LM34 series is available in TO-46, TO-92, and SO-8 packages. It has three signal pins: V_{OUT} for output voltage, V_s for supply voltage, and GND for ground. Two typical circuit connections for the LM34 are shown in Figure 10.13. These two circuits will sense the ambient temperature of the LM34 and provide accurate readings.

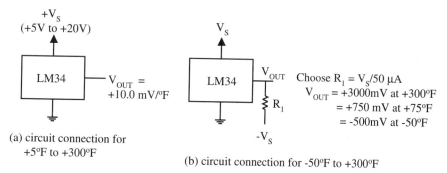

(a) circuit connection for
+5°F to +300°F

(b) circuit connection for -50°F to +300°F

Figure 10.13 ■ Circuit connection for the LM34

Example 10.7

Digital Fahrenheit Thermometer. Use the circuit in Figure 10.13 as a building block in a system to measure the room temperature. Display the result in three integral and one fractional digits using seven segment displays driven by the MAX7221. Measure and display the temperature in the range of –44°F to 212°F.

Solution: The voltage output from Figure 10.13 will be –440 mV at –44°F and 2120 mV at 212°F. Higher accuracy can be obtained from the 68HC12 A/D converter if the voltage output corresponding to 212°F is scaled to 5V and the voltage output at –44°F could be scaled and shifted to 0V. The circuit shown in Figure 10.14 will scale and shift the temperature sensor output to the range of 0V ~ 5V. Since the 68HC12 ATD module has 10-bit resolution, the ambient temperature is equal to the conversion result divided by 4 and then subtracted by 44.

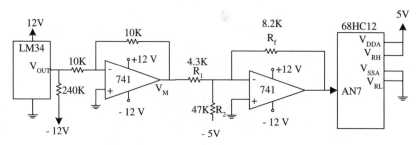

Figure 10.14 ■ Circuit connection between the LM34 and the 68HC12

In this example, the current temperature will be displayed on six seven-segment displays driven by the MAX7221. The circuit connection between the 68HC12 and the MAX7221 is shown in Figure 10.15.

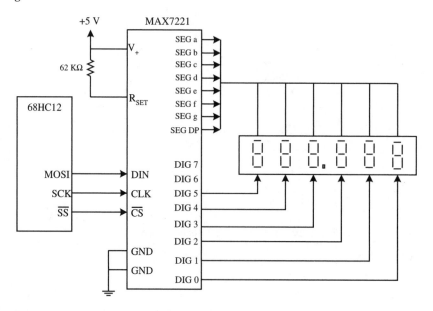

Figure 10.15 ■ Digital thermometer display circuit

Because ambient temperature does not change very fast, it should be adequate to measure and update the temperature once every second. The characters ° and F will be displayed on digits 1 and 0, respectively. The negative sign will be displayed when the temperature is negative. The negative sign will be shown in DIG 5 when the temperature is equal or lower than –10°F. When the temperature is between 0°F and –10°F, the negative sign will be displayed in DIG 5 and DIG 4 is blanked. For nonnegative temperatures, we will blank the leading zeros.

There are three modules that need to be configured: SPI, MAX7221, and the ATD converter.

The SPI function can be configured using the routine *spi_init* of Example 9.11. The MAX7221 should be configured as follows:

Scan limit register
We need to display digits DIG 5..DIG 0. The value to set this up is $0B05.

Decode mode register
Among the six characters to be displayed, the degree character ° and letter F cannot be decoded. The leading four characters should be displayed using the decode mode, whereas the lowest two digits should be displayed using the no-decode mode. The value to set this up is $09FC.

Intensity register
Temperature value will be displayed in the maximum intensity. The value to set this up is $0A0F.

Shutdown register
The chip is supposed to perform normal operation. Write the value $0C01 to the MAX7221.

Display test register
The MAX7221 is used to perform normal operation. Write the value $0F00 to the MAX7221.

Digit 0 register
Letter F is to be displayed in this digit using the no-decode mode. To display letter F, segments a, e, f, and g should be turned on, whereas segments b, c, d, and decimal point DP should be turned off. The value to set this up is $0147.

Digit 1 register
The degree character is to be displayed in this digit in no-decode mode. The seven-segment pattern for the degree character is %1100011 (from segment a to g). The value to set this up is $0263.

Digit 2 register
This digit displays the fractional digit in decode mode. The value of this digit could be from 0 to 9. The value to set this up is $030x (where x is the value of the fractional digit).

Digit 3 register
This digit shows the one's digit in decode mode. The value of this digit could be from 0 to 9. The value to set this up is $048x (x = 0..9). This digit displays the decimal point.

Digit 4 register
This digit shows the ten's digit in decode mode. The value of this digit could be from 0 to 9. The value to be sent to the MAX7221 is $050x (x = 0..9). If the temperature is above 0 and below 10, then this digit should be blanked. To blank this digit, send the value $050F to the MAX7221.

Digit 5 register

There are three possible uses for this digit: (1) Display the hundred's digit (send the value $060x, where, x = 1 or 2); (2) display the negative sign (send the value $060A); or (3) be blanked when the temperature is above 0 and below 100 (send the value $060F).

The configuration of the A/D converter is identical to that in Example 10.4. The procedure for measuring temperature and updating the display is as follows:

Step 1

Initialize the SPI module.

Step 2

Initialize the MAX7221 chip.

Step 3

Configure the ATD module.

Step 4

Start A/D conversion and read the conversion result.

Step 5

Divide the conversion result by 4 and then subtract the quotient by 44 to obtain the current temperature. Separate each temperature digit by performing repetitive division by 10.

Step 6

Transfer temperature data to the MAX7221 to update the display.

Step 7

Wait for one second and then go to step 4 again.

The following program will perform the configuration, start the A/D conversion, and display the temperature on seven-segment displays:

```
              #include   "d:\miniide\hc12.inc"
              org        $800
scan_limit    db         $0B,$05              ; data to set up scan limit register
decode_mode   db         $09,$FC              ; data to set up decode mode register
intensity     db         $0A,$0F              ; data to set up intensity register
shutdown      db         $0C,$01              ; data to set up shut down register
display_test  db         $0F,$00              ; data to set up display test register
digit0        db         $01,$47              ; data to update digit 0
digit1        db         $02,$63              ; data to update digit 1
digit2        rmw        1                    ; data to update digit 2
digit3        rmw        1                    ; data to update digit 3
digit4        rmw        1                    ; data to update digit 4
digit5        rmw        1                    ; data to update digit 5
f_is_zero     rmb        1                    ; flag to indicate if fractional digit is zero
              org        $1000
              lds        #$8000               ; set up the stack pointer
              jsr        spi_init             ; configure SPI function
              jsr        m7221_init           ; configure MAX7221
              jsr        atd_init             ; configure ATD module
              clr        f_is_zero            ; fractional digit is not zero
forever       jsr        init_disp            ; initialize digits 3 to 5
              movb       #$07,ATDCTL5         ; start an A/D conversion sequence
              brclr      ATDSTAT0,$80,*       ; wait until conversion is done
              ldd        ADR0H                ; read the conversion result
              ldx        #64                  ; right-justify the conversion result
```

```
                idiv                            ;     "
                xgdx                            ;     "
                ldx        #4                   ; convert to Fahrenheit temperature
                idiv                            ;     "
; compute the value for displaying fractional digit to be sent to MAX7221
                cmpb       #0
                bne        check_1
                movw       #$0300,digit2        ; fractional digit is zero
                movb       #1,f_is_zero
                bra        ones_digit
check_1         cmpb       #1
                bne        check_2
                movw       #$0303,digit2        ; fractional digit is three
                bra        ones_digit
check_2         cmpb       #2
                bne        is_3
                movw       #$0305,digit2        ; fractional digit is five
                bra        ones_digit
is_3           movw       #$0308,digit2        ; fractional digit is eight
ones_digit     xgdx
                subd       #44                  ; shift by 44 degree
                bpl        above_zero           ; is temperature above zero?
                bmi        below_zero
                movb       #$0F,digit4+1        ; temperature is 0, so blank digit 4
                movb       #$0F,digit5+1        ; & blank digit 5
                jmp        update_T
below_zero     movw       #$060A,digit5        ; set up the negative sign
                negb                            ; convert to positive temperature
                tst        f_is_zero
                bne        no_op                ; is fractional digit zero?
                incb                            ; add 1 to negative temperature
                ldaa       #10                  ; take 10's complement of the
                suba       digit2+1             ; fractional digit
                staa       digit2+1             ;     "
no_op          clra                            ; clear A
                ldx        #10                  ; separate one's digit
                idiv                            ;     "
                stab       digit3+1             ; fix the digit3 value
                xgdx                            ; place ten's digit in B
                tstb                            ; is ten's digit zero?
                bne        not_zero
                ldab       #$0F
                stab       digit4+1             ; blank ten's digit
                jmp        update_T
not_zero       stab       digit4+1             ; fix ten's digit's value
                jmp        update_T
above_zero clra
                cmpb       #9
                bhi        hi_than_9            ; is temperature equal or above 10 degrees?
                stab       digit3+1             ; between 0 and 9
                movb       #$0F,digit4+1        ; blank digit 4
                movb       #$0F,digit5+1        ; blank digit 5
                jmp        update_T
hi_than_9      cmpb       #99                  ; is temperature between 10 and 100?
```

```
                       bhi       three_dig
                       ldx       #10
                       idiv
                       stab      digit3+1              ; set the one's digit value
                       xgdx
                       stab      digit4+1              ; set the ten's digit value
                       movb      #$0F,digit5+1         ; blank the most significant digit
                       jmp       update_T
three_dig              ldx       #10
                       idiv
                       stab      digit3+1              ; set the one's digit value
                       xgdx
                       ldx       #10
                       idiv
                       stab      digit4+1              ; set the ten's digit value
                       xgdx
                       stab      digit5+1              ; set the hundred's digit value
update_T               ldx       #digit2
                       ldab      #4
                       jsr       transfer              ; send data to MAX7221 to update temperature
                       jsr       delay_1s
                       jmp       forever
;****************************************************************************
;
; The following routine initializes the SPI function:
;****************************************************************************
;
spi_init               movb      #$50,SP0CR1           ; enable SPI, shift data on rising edge, master mode
                       movb      #$08,SP0CR2           ; enable internal pull-up, normal drive
                       movb      #$00,SP0BR            ; shift rate set to 4 MHz
                       movb      #$CA,DDRS
                       movb      #$07,PURDS            ; all port S pins have pull up
                       rts
;****************************************************************************
;
; The following routine initializes the MAX7221 chip, including degree and letter F:
;****************************************************************************
m7221_init ldx         #scan_limit
                       ldab      #7
                       jsr       transfer
                       rts
;****************************************************************************
;
; The following routine transfer data to the MAX7221 chip:
;****************************************************************************
transfer               bclr      PORTS,$80
                       ldaa      1,x+
                       staa      SP0DR
                       brclr     SP0SR,$80,*           ; wait for SPI transfer to complete
                       ldaa      1,x+
                       staa      SP0DR
                       brclr     SP0SR,$80,*
                       bset      PORTS,$80             ; transfer data to appropriate register
                       dbne      b,transfer
                       rts
;****************************************************************************
;
```

```
; The following function initializes the ATD module:
;*********************************************************************
;
atd_init        movb        #$E0,ATDCTL2
                ldaa        #10
; the next two instructions create 5 ?s delay
wait            deca
                bne         wait
                movb        #$02,ATDCTL3
                movb        #$81,ATDCTL4
                rts
;*********************************************************************
; The following function initializes display data:
;*********************************************************************
;
init_disp       movw        #$0300,digit2
                movw        #$0480,digit3
                movw        #$0500,digit4
                movw        #$0600,digit5
                rts
;*********************************************************************
; The following function creates a one-second time delay:
;*********************************************************************
;
delay_1s        pshx
                movb        #$90,TSCR           ; enable TCNT & fast flags clear
                movb        #$03,TMSK2          ; configure prescale factor to 8
                movb        #$01,TIOS           ; enable OC0
                ldx         #20                 ; prepare to perform 20 OC actions
                ldd         TCNT
again           addd        #50000              ; start an output compare operation
                std         TC0                 ; with 50 ms time delay
                brclr       TFLG1,$01,*
                ldd         TC0
                dbne        x,again
                pulx
                rts
                end
```

In C, the program is as follows:

```
#include <hc12.h>
void    spi_init (void);
void    m7221_init(void);
void    atd_init (void);
void    transfer (char *ptr, char n);
void    delay_1s (void);                        /* create one-second time delay */
void    init_disp(void);                        /* initialize display data */
/* max7221 configuration data for registers and digits 0 and 1*/
char max_conf [14] = {0x0B,0x05,0x09,0xFC,0x0A,0x0F,0x0C,0x01,0x0F,0x00,0x01,0x47,0x02,0x63};
char temp_dat [8];                              /* four leading temperature digits data */
void main ()
{
        int  temp;
        char xx, is_zero;
        spi_init ();
        m7221_init();
```

```
atd_init ();
while (1) {
    init_disp();                            /* set up common temperature digit data */
    ATDCTL5 = 0x07;                         /* start an A/D conversion */
    while (!(ATDSTAT0 & 0x80));             /* wait for A/D conversion to complete */
    temp = ADR0H * 4 + ADR0L/64;           /* combine the A/D result register */
    xx = temp % 4;                          /* get the fractional digit value */
    temp = temp / 4;                        /* convert to temperature (not shifted by 44 degrees yet ) */

    switch (xx) {
        case 0: temp_dat[1] = 0x00;        /* fractional digit is 0 */
            is_zero = 1;
            break;
        case 1: temp_dat[1] = 0x03;        /* fractional digit is 3 */
            is_zero = 0;
            break;
        case 2: temp_dat[1] = 0x05;        /* fractional digit is 5 */
            is_zero = 0;
            break;
        case 3: temp_dat[1] = 0x08;        /* fractional digit is 8 */
            is_zero = 0;
            break;
        default: temp_dat[1] = 0x00;
            break;
    }
    if (is_zero && temp < 35) {             /* temperature is -44 ~ -10 F */
        temp_dat[7] = 0x0A;                 /* set up negative sign */
        temp = abs (temp - 44);
        temp_dat[3] = temp % 10;            /* compute one's digit */
        temp_dat[5] = temp / 10;            /* compute ten's digit */
    }
    else if (!is_zero && temp < 35) {
        temp_dat[7] = 0x0A;                 /* set up negative sign */
        temp_dat[1] = 10 - temp_dat[1];     /* complement the fractional digit */
        temp = abs(temp - 43);
        temp_dat[3] = temp % 10;
        temp_dat[5] = temp / 10;
    }
    else if (is_zero && temp < 45) {        /* temperature is between 0 and -9 */
        temp_dat[7] = 0x0A;                 /* set negative sign */
        temp_dat[3] = temp - 44;
        temp_dat[5] = 0x0F;                 /* blank leading zeros */
    }
    else if (!is_zero && temp < 45) {  /* temperature is -9 ~ 0 F */
        temp_dat[1] = 10 - temp_dat[1];     /* complement the fractional digit */
        temp_dat[3] = abs(temp - 43);       /* compute the one's digit */
        temp_dat[5] = 0x0F;                 /* blank the ten's digit because it is zero */
        temp_dat[7] = 0x0A;                 /* set up negative sign */
    }
    else if (temp < 54) {                   /* temperature is 1~9 F */
        temp_dat[3] = temp - 44;            /* compute the one's digit */
        temp_dat[5] = 0x0F;                 /* blank the leading zeros */
        temp_dat[7] = 0x0F;
```

```
                                  }
                else if (temp < 144) {                      /* temperature is 10 ~ 99 F */
                        temp_dat[3] = (temp - 44) % 10;   /* compute ones digit */
                        temp_dat[5] = (temp - 44) / 10;   /* compute ten's digit */
                        temp_dat[7] = 0x0F;               /* blank hundred digit */
                        }
                else {
                        temp_dat[3] = (temp - 44) % 10;   /* compute one's digit */
                        temp = (temp - 44) / 10;
                        temp_dat[5] = temp % 10;          /* compute ten's digit */
                        temp_dat[7] = temp / 10;          /* compute hundred's digit */
                        };
                transfer (&temp_dat[0], 4);                 /* update display data */
                delay_1s ();                                /* wait for one second */
        }
}
void spi_init(void)
{
        SP0CR1 = 0x50;
        SP0CR2 = 0x08;
        SP0BR = 0x00;
        DDRS = 0xCA;
        PURDS = 0x07;
}
void m7221_init (void)
{
        transfer (&max_conf[0],7);
}
/*  transfer n bytes of data to MAX7221  */
void transfer (char *ptr, char n)
{
        char i;
        for (i = 0; i < n; i++) {
            PORTS &= 0x7F;                                  /* set pin 7 to low */
                SP0DR = *ptr++;
                while (!(SP0SR & 0x80));
                SP0DR = *ptr++;
                while (!(SP0SR & 0x80));
                PORTS |= 0x80;                              /* pull pin 7 to high */
        }
}
/* this function initializes the digit registers of the MAX7221 driver chp */
void init_disp (void)
{
        temp_dat[0] = 0x03;                                 /* fractional digit */
        temp_dat[1] = 0x00;
        temp_dat[2] = 0x04;                                 /* one's digit */
        temp_dat[3] = 0x80;
        temp_dat[4] = 0x05;                                 /* ten's digit */
        temp_dat[5] = 0x00;
        temp_dat[6] = 0x06;                                 /* hundred's digit */
        temp_dat[7] = 0x00;
}
void atd_init (void)
{
```

```
        int i;
        ATDCTL2 = 0xE0;
        for (i = 0; i < 40; i++)                    /* delay for 5 us */
            asm ("nop");
        ATDCTL3 = 0x02;
        ATDCTL4 = 0x81;
}
/* this function uses OC0 function to create a one-second time delay */
void delay_1s (void)
{
        int i;
        TSCR = 0x90;                                /* enable TCNT & fast flag clear */
        TMSK2 = 0x03;                               /* set the clock prescale factor to 8 */
        TIOS = 0x01;                                /* select OC0 function */
        TC0 = TCNT + 50000u;
        i = 20;
        while (i) {
            while (!(TFLG1 & 0x01));                /* wait for 50 ms */
            TC0 = TC0 + 50000u;
            i−;
        }
}
```

Adding a signal conditioning circuit increases the chip count and power consumption to our applications. If the PCB size is very critical, we could consider using a chip that integrates the temperature sensor and the A/D converter on the same chip. Chips that integrate a temperature sensor, an A/D converter, and a serial interface on one chip are available from several vendors. The LM74 from National Semiconductor and the DS1620 from Dallas Semiconductor are two available chips. Using these chips, the microcontroller needs only shift in the temperature from the chip.

▲

10.9 Measuring Barometric Pressure

Barometric pressure refers to the air pressure existing at any point within the earth's atmosphere. This pressure can be measured as an absolute pressure (with reference to absolute vacuum), or can be referenced to some other value or scale. The meteorology and avionics industries traditionally measure the absolute pressure, and then reference it to a sea level pressure value. This complicated process is used in generating maps of weather systems.

Mathematically, atmospheric pressure is exponentially related to altitude. Once the pressure at a particular location and altitude is measured, the pressure at any other altitude can be calculated.

Several different units have been used to measure the barometric pressure: in-Hg, kPa, mbar, or psi. A comparison of barometric pressure using four different units at sea level up to 15,000 ft is shown in Table 10.4.

Altitude (ft)	Pressure (in-Hg)	Pressure (mbar)	Pressure (kPa)	Pressure (psi)
0	29.92	1013.4	101.4	14.70
500	29.38	995.1	99.5	14.43
1000	28.85	977.2	97.7	14.17
6000	23.97	811.9	81.2	11.78
10000	20.57	696.7	69.7	10.11
15000	16.86	571.1	57.1	8.28

Table 10.4 ■ Altitude versus pressure data

Sensortechnics Inc. manufactures a *barometric pressure transducer* (BPT) kit that has three external connectors: $+V_S$. GND, and V_{OUT}. The operating parameters are listed in Table 10.5. This product can be found in the *Electronic Components Catalog* prepared by Farnell Inc.

Parameter	Value
Reference conditions	V_S = 9.0 V, T (ambient) = 25°C, R_L = 100 KΩ
Supply voltage	7-24V DC
Operating pressure	800-1100mbar
Breakdown pressure	2 bar
Voltage output (span)	5.0 ± 500 mV
Operating temperature	-40 to 85°C
Compensated range	-10 to 60°C
Nonlinearity and hysteresis	0.1% FSO (max)
Repeatability	0.2% FSO (typical)
Temperature shift (-10 to 60°C)	0.3% FSO/10°C (max)
Response time	1 msec (typical)
Long term stability	0.1% FSO (typical)

Manufacturer's Type No. 144SC0811-BARO.

Table 10.5 ■ BPT Parameters

The BPT is a calibrated and signal-conditioned transducer that provides a true 4.5~5.5V output in the barometric pressure ranging from 800 to 1100 mbar. Internal voltage regulation allows the device to operate on a power supply between 7 and 24V. Applications include barometry, weather stations, and absolute pressure compensation in sensitive equipment. The transducer is designed for use in noncorrosive, nonionic media; in other words, in dry air and gasses.

Since the voltage output of BPT is from 4.5 to 5.5V, a voltage shifting and scaling circuit is needed to get the best conversion result from the 68HC12 A/D converter.

Example 10.8

Design a digital barometer that incorporates a BPT and the 68HC12 with seven-segment displays. Write a program that updates the barometric pressure once every second.

Solution: The circuit that scales and shifts the BPT output to the range of 0~5V is shown in Figure 10.16. The values of R_1, R_2, R_f, and V_1 must be calculated by trial and error. One possible combination is shown in the figure.

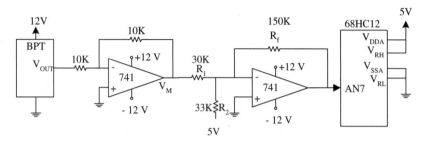

Figure 10.16 ■ Barometric pressure sensor output scaling and shifting circuit

The circuit in Figure 10.15 can be used to display the barometric pressure. The only modification that needs to be made is eliminating two of the seven-segment displays that are not needed.

The pressure data will be displayed in four BCD digits. Since the A/D conversion result 0 represents the pressure of 800 mbar, while the conversion result 1023 represents the pressure of 1100 mbar, the translation from the A/D result to pressure is not exact. The barometric pressure for the A/D conversion result **y** can be computed by the following equation:

$$P_y = 800 + (y \times 300 \div 1023) = 800 + (y \times 100 \div 341) \quad (10.6)$$

Both the A/D converter and the SPI module of the 68HC12 can be configured using the same subroutines as Example 10.7.

The MAX7221 should be configured as follows:

Scan limit register
We need to display digits DIG 3..DIG 0. The value to set this up is $0B03.

Decode mode register
All four digits can be displaced using the decode mode. The value to set this up is $090F.

Intensity register
We will use the maximum intensity to display the barometric pressure. The value to set this up is $0A0F.

Shutdown register
The chip is supposed to perform a normal operation. Write the value $0C01 to the MAX7221.

Display test register
The MAX7221 is supposed to perform a normal operation. Write the value $0F00 to the MAX7221.

Digit 0 register
This digit displays the one's digit in decode mode. The value to set this up is $010x (x = 0..9).

Digit 1 register
This digit displays the ten's digit in decode mode. The value to set this up is $020x (x = 0..9).

Digit 2 register
This digit displays the hundred's digit in decode mode. The value to set this up is $030x (x = 0..9).

Digit 3 register
This digit shows the thousand's digit in decode mode. The value to set this up is $040x (x = 0..1).

The procedure for measuring barometric pressure and updating the display is as follows:

Step 1
Initialize the SPI module.

Step 2
Initialize the MAX7221 chip.

Step 3
Configure the ATD module.

Step 4
Start A/D conversion and read the conversion result.

Step 5
Multiply the conversion result by 100, then divide the result by 341, and add 800 to the result to obtain the barometric pressure. Separate each pressure digit by performing repeated divisions by 10.

Step 6
Transfer temperature data to the MAX7221 to update the display.

Step 7
Wait for one second and then go to step 4.

The following assembly program takes the pressure sample, performs A/D conversion, collects the conversion result, prepares display data, and transfers data to the MAX7221 for display:

```
              #include  "d:\miniide\hc12.inc"
              org       $800
scan_limit    db        $0B,$03        ; data to set up scan limit register
decode_mode   db        $09,$0F        ; data to set up decode mode register
intensity     db        $0A,$0F        ; data to set up intensity register
shutdown      db        $0C,$01        ; data to set up shut down register
display_test  db        $0F,$00        ; data to set up display test register
digit0        rmw       1              ; data to update digit 0
digit1        rmw       1              ; data to update digit 1
digit2        rmw       1              ; data to update digit 2
digit3        rmw       1              ; data to update digit 3

              org       $1000
              jsr       spi_init       ; configure SPI function
              jsr       m7221_init     ; configure MAX7221
```

```
              jsr       atd_init              ; configure ATD modeul

forever       jsr       init_disp             ; initialize digit 0 to 3
              movb      #$07,ATDCTL5          ; start an A/D conversion sequence
              brclr     ATDSTAT0,$80,*        ; wait until conversion is done
              ldd       ADR0H                 ; read the conversion result
              ldx       #64                   ; right-justify the conversion result
              idiv                            ;     "
              xgdx                            ;     "
; *************************************************************
; Use the equation pressure = 800 + (A/D result ×100 / 341) to obtain
; barometric pressure.
; *************************************************************
;
              ldy       #100
              emul                            ; multiply conversion result by 100
              ldx       #341                  ; compute result * 100 / 341
              ediv                            ;     "
              cpd       #171                  ; check the remainder
              blo       less_half
              iny                             ; round up the result
less_half     xgdy                            ; place the pressure in D
              addd      #800
; *************************************************************
; Use repeated division by 10 to separate each digit:
; *************************************************************
;
              ldx       #10
              idiv                            ; separate the one's digit
              stab      digit0+1              ; and put it in buffer
              xgdx
              ldx       #10
              idiv                            ; separate the ten's digit
              stab      digit1+1              ; and put it in buffer
              xgdx
              ldx       #10
              idiv                            ; separate the hundred's digit
              stab      digit2+1              ; and put it in buffer
              xgdx
              tstb
              beq       update_T              ; if thousand digit is zero, blank it
              stab      digit3+1              ; store the thousand's digit

update_T      ldx       #digit0
              ldab      #4
              jsr       transfer              ; send data to MAX7221 to update temperature
              jsr       delay_1s
              jmp       forever
; *************************************************************
; The following routine initializes the SPI function:
; *************************************************************
;
spi_init      movb      #$50,SP0CR1           ; enable SPI, disable interrupt , master mode
              movb      #$08,SP0CR2
              movb      #$00,SP0BR            ; set prescale factor to two
              movb      #$CA,DDRS
```

```
                    movb      #$07,PURDS
                    rts
; ****************************************************************************
; The following routine initializes the MAX7221 chip:
; ****************************************************************************
m7221_init          ldx       #scan_limit
                    ldab      #5
                    jsr       transfer
                    rts
; ****************************************************************************
; The following routine transfers data to the MAX7221 chip:
; ****************************************************************************
transfer            bclr      PORTS,$80
                    ldaa      1,x+
                    staa      SP0DR
                    brclr     SP0SR,$80,*      ; wait for SPI transfer to complete
                    ldaa      1,x+
                    staa      SP0DR
                    brclr     SP0SR,$80,*
                    bset      PORTS,$80        ; transfer data to appropriate register
                    dbne      b,transfer
                    rts
; ****************************************************************************
; The following function initializes the ATD module:
; ****************************************************************************
atd_init            movb      #$E0,ATDCTL2
                    ldaa      #10
; the next two instructions create 5 ?s delay
wait                deca
                    bne       wait
                    movb      #$02,ATDCTL3
                    movb      #$81,ATDCTL4
                    rts
; ****************************************************************************
; The following function initializes display data. One's, ten's, and hundred's digits are set to
; zero:
; ****************************************************************************
init_disp           movw      #$0100,digit0
                    movw      #$0200,digit1
                    movw      #$0300,digit2
                    movw      #$040F,digit3    ; blank thousand's digit initially
                    rts
; ****************************************************************************
; The following function creates one-second time delay:
; ****************************************************************************
delay_1s            pshx
                    movb      #$90,TSCR        ; enable TCNT & fast flags clear
                    movb      #$03,TMSK2       ; configure prescale factor to 8
                    movb      #$01,TIOS        ; enable OC0
                    ldx       #20              ; prepare to perform 20 OC actions
                    ldd       TCNT
again               addd      #50000           ; start an output compare operation
                    std       TC0              ; with 50 ms time delay
```

```
brclr     TFLG1,$01,*
ldd       TC0
dbne      x,again
pulx
rts
end
```

The C language version of this program is straightforward and will be left for you as an exercise.

▲

10.10 Using the External A/D Converter AD7894

Some of our applications require a higher resolution (for example 14-bit) A/D converter than that provided by the 68HC12. Unless we choose a microcontroller with higher A/D resolution, we would need to use an external A/D.

The AD7894 is a 14-bit, successive approximation A/D converter that operates from a single +5V power supply. The A/D conversion result is output via a serial port. The chip accepts an analog input range of ±10V (AD7894-10), ±2.5V (AD7894-3), or 0 to 2.5V (AD7894-2). The typical power consumption is 20 mW.

When operating in automatic power-down mode, the chip powers down itself once the conversion is complete and wakes up before the next conversion cycle. This feature makes it suitable for battery powered or portable applications.

The block diagram and pin assignment are shown in Figure 10.17.

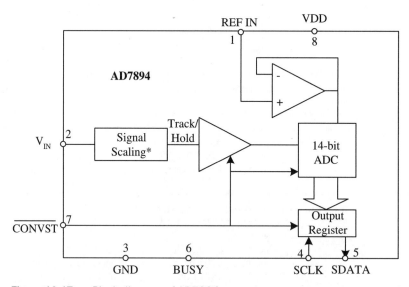

Figure 10.17 ■ Block diagram of AD7894

10.10.1 Signals

The function of each signal is as follows:

REF IN: *Voltage reference input.* An external reference source should be connected to this pin to provide the reference voltage for the AD7894's conversion process. The nominal reference voltage for correct operation of the AD7894 is +2.5V.

VIN: *Analog input channel.* The analog input range is ± 10V (AD7894-10), ± 2.5V (AD7894-3), and 0 to 2.5V (AD7894-2).

GND: *Analog ground.* Ground reference for the whole chip.

SCLK: *Serial clock input.* An external serial clock is applied to this pin to obtain serial data from the AD7894. A new serial data bit is clocked out on the falling edge of this serial clock. Data is guaranteed to be valid for 10 ns after this falling edge so data can be accepted on the falling edge when a fast serial clock is used. The serial clock should be taken low at the end of the serial data transmission.

SDATA: *Serial data output.* Serial data output from the AD7894 is provided at this pin. Serial data is clocked out by the falling edge of SCLK, but the data can also be read at the falling edge of SCLK. Sixteen bits of serial data are provided as two leading zeros followed by 14 bits of conversion data. On the 16th falling edge of SCLK, the SDATA line is held for the data hold time and then disabled (three-stated). Output data coding is two's complement for AD7894-10 and AD7894-3, and is straight binary for the AD7894-2.

BUSY: This pin is used to indicate when the part is doing a conversion. The Busy pin goes high on the falling edge of CONVST and returns low when the conversion is complete.

$\overline{\text{CONVST}}$**:** *Conversion start.* On the falling edge of this input, the track/hold circuit goes into its hold mode and conversion is initiated. If $\overline{\text{CONVST}}$ is low at the end of conversion, the AD7894 will go into power-down mode. In this case, the rising edge of $\overline{\text{CONVST}}$ will cause the chip to wake up.

V_{DD}**:** Positive supply voltage, +5V ± 5%.

10.10.2 Chip Functioning

In high sampling mode, A/D conversion is initiated on the falling edge of the $\overline{\text{CONVST}}$ input. The clock signal required by the conversion is generated internally using a laser-trimmed clock oscillator circuit. Conversion takes 5 µs in the high sampling mode and 10 µs for the automatic power-down mode. Analog input track/hold acquisition time is 0.35 µs. To obtain optimum performance from the device, the read operation should not occur during the conversion or during 250 ns prior to the next conversion. This allows the AD7894 to operate at throughput rates up to 160 kHz.

TRACK/HOLD AMPLIFIER

The track/hold amplifier on the analog input of the AD7894 allows the ADC to accurately convert an input sine wave of full-scale amplitude to 14-bit accuracy. The track/hold amplifier acquires an input signal in less than 0.35 µs.

In high sampling-operation mode, the track/hold amplifier goes from its tracking mode to its hold mode at the start of conversion (i.e., the falling edge of $\overline{\text{CONVST}}$). At the end of conversion, the AD7894 goes back to tracking mode. The acquisition time of the track/hold amplifier starts from this moment. For the automatic power-down mode, the rising edge of $\overline{\text{CONVST}}$ wakes up the chip and the track/hold amplifier goes into hold mode 5 µs after the rising edge of $\overline{\text{CONVST}}$. Once again, the chip returns to its tracking mode at the end of conversion when the BUSY signal goes low.

REFERENCE INPUT

The reference input must be set to 2.5V as precisely as possible. Errors in this reference input will add to the specified full-scale errors. Either the AD780 or the AD680 can be used to generate a precise + 2.5V reference voltage.

TIMING AND CONTROL

The timing and control sequence required to obtain optimum performance from the AD7894 is shown in Figure 10.18. As shown in the sequence, conversion is started on the falling edge of $\overline{CONVST}$ and new data from this conversion is available in the output register 5 μs later. Once the read operation has taken place, a further 250 ns should be allowed before the next falling edge of $\overline{CONVST}$ to optimize the settling of the track/hold amplifier before the next conversion is initiated.

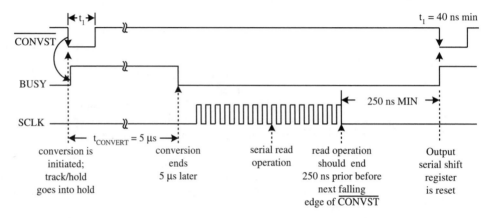

Figure 10.18 ■ Mode 1 timing operation diagram for high sampling performance

With the serial clock frequency at its maximum of 16 MHz, the achievable throughput time for the AD7894 is 5 μs (conversion time) plus 1.0 μs (read time) plus 250 ns (quiet time). This results in a minimum throughput time of 6.25 μs. A serial clock of less than 16 MHz can be used, but the throughput rate will be lower.

MODE 1 OPERATION (HIGH SAMPLING PERFORMANCE)

The timing diagram in Figure 10.18 is for optimum performance in mode 1 operation, where the falling edge of $\overline{CONVST}$ starts the conversion and causes the BUSY signal to go high. The BUSY signal goes low when the conversion is complete (about 5 μs after the falling edge of $\overline{CONVST}$). Data from this conversion is available in the output register of the AD7894. This data can be read out from the SDATA pin. The read operation must be completed 250 ns before the next conversion is initiated.

MODE 2 OPERATION (AUTO POWER DOWN AFTER CONVERSION)

The timing diagram in Figure 10.19 is for the optimum performance in mode 2 operation, where the converter goes into power-down mode once BUSY goes low, after conversion, and wakes up before the next conversion takes place. This is achieved by keeping $\overline{CONVST}$ low at the end of conversion. The rising edge of $\overline{CONVST}$ wakes up the AD7894. This wakeup time is typically 5 μs and is controlled internally by a monostable circuit. For optimum results, the $\overline{CONVST}$ pulse should be between 40 ns and 2 μs or greater than 6 μs in width. The narrower pulse allows a system to instruct the AD7894 to begin waking up and perform a conversion when ready, whereas a pulse greater than 6 μs will give control over when the sampling instant takes place.

Note that the 10 μs wakeup time shown in Figure 10.19 is for a $\overline{\text{CONVST}}$ pulse less than 2 μs. If a $\overline{\text{CONVST}}$ pulse greater than 6 μs is used, the conversion will not complete for a further 5 μs after the falling edge of $\overline{\text{CONVST}}$. For the fastest serial clock of 16 MHz, the read operation will take 1.0 μs and this must be completed at least 250 ns before the falling edge of the next $\overline{\text{CONVST}}$, to allow the track/hold amplifier to have enough time to settle.

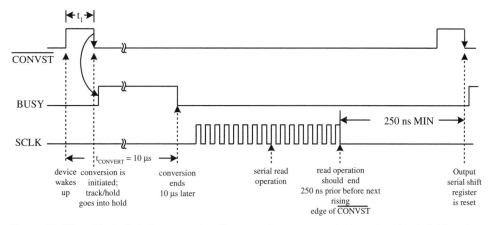

Figure 10.19 ■ Mode 2 timing operation diagram where automatic sleep function is initiated

SERIAL INTERFACE

The serial interface to the AD7894 consists of just three wires: a *serial clock input* (SCLK), a *serial data output* (SDATA), and a *conversion status output* (BUSY). Figure 10.20 shows the timing diagram for the read operation of the AD7894. Serial data is clocked out on the falling edge of SCLK and is valid on both the rising and falling edges of SCLK.

The *most significant bit* (MSB) of the conversion is shifted out first, whereas the least significant bit is shifted out last. On the 16th falling edge of SCLK the LSB will be valid for a specified time to allow the bit to be read on the falling edge of the SCLK, and then the SDATA line is disabled (three-stated). After the last data bit is shifted out, the SCLK input should return to low and remain low until the next serial data read operation. The serial clock input does not have to be continuous during the serial read operation. The 16 bits of data can be read from the AD7894 in 2 bytes; for example, in two SPI transfers.

$t_2 = t_3 = 31.25$ ns MIN, $t_4 = 60$ ns MAX, $t_5 = 10$ ns MIN, $t_6 = 20$ ns MAX @ 5V, A, B versions

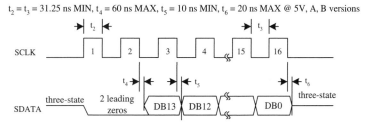

Figure 10.20 ■ Data read operation

10.10.3 Interfacing the AD7894 with the 68HC12

The AD7894 is fully compatible with the 68HC12 SPI system. To use the AD7894, set the 68HC12 SPI in master mode. The circuit connection between the 68HC12 and the AD7894 is shown in Figure 10.21.

The AD680 is a precision voltage reference IC that generates a 2.5V from V_{OUT}. Capacitors are used to filter out the high and low frequency power supply noise caused by sudden high current drawn by the AD680 and AD7894.

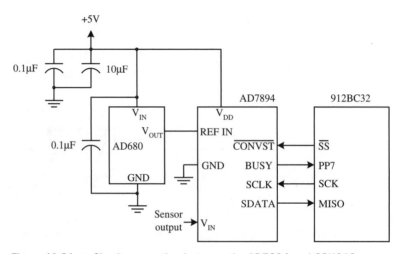

Figure 10.21 ■ Circuit connection between the AD7894 and 68HC12

The SS pin of the 68HC12 should be configured as a general output pin. The procedure for starting an A/D conversion and reading back the conversion result (choose *high performance sampling* mode) is as follows:

Step 1
Configure the SPI function properly. For this application, we can configure the SPI to transfer data at 4 MHz, configure the SS pin for general output, and use the falling edge to shift in data (CPHA = 1).

Step 2
Pull the SS pin to high, then pull down to low and then pull to high to generate a low going pulse. This pulse will start the A/D conversion.

Step 3
Wait until the A/D conversion is complete and perform two SPI transfers to read in the conversion result and store it in memory.

Step 4
Perform other processing.

Example 10.9

Write a subroutine to initialize the SPI function and a subroutine to initiate an A/D conversion operation for the circuit shown in Figure 10.21, and return the result in double accumulator D.

Solution:

The routines are as follows:

```
#include "d:\miniide\hc12.inc"

spi_init     movb     #$54,SP0CR1
             movb     #$08,SP0CR2
             movb     #$00,SP0BR      ; set highest transfer rate
             movb     #$EA,DDRS       ; set port S pin directions
             rts

get_sample   bset     PORTS,$80       ; pull SS pin to high
             bclr     PORTS,$80       ; initiate A/D conversion
             bset     PORTS,$80       ; mode 1 operation
             brset    PORTP,$80,*     ; wait until A/D conversion is complete
             staa     SP0DR           ; start an SPI transfer to read in upper byte
             brclr    SP0SR,$80,*     ; of conversion result
             ldaa     SP0DR           ; leave the upper byte of A/D result in A
             staa     SP0DR           ; read in the lower byte of the conversion result
             brclr    SP0SR,$80,*     ;   "
             ldab     SP0DR
             rts
```

In C, the programs is as follows:

```c
void spi_init (void)
{
        SP0CR1 = 0x54;            /* enable SPI, disable SPI interrupt, falling edge shift */
        SP0CR2 = 0x08;
        SP0BR = 0x00;            /* choose highest transfer rate */
        DDRS = 0xEA;
}

int get_sample ()
{
        int xx;
        PORTS |= 0x80;           /* pull SS pin to high */
        PORTS &= 0x7F;           /* initiate A/D conversion /
        PORTS |= 0x80;           /* set mode 1 operation */
        while (!(PORTP&0x80));    /* wait until A/D conversion is complete */
        SP0DR = 0x00;            /* start an SPI transfer */
        while (!(SP0SR & 0x80)); /* wait until SPI transfer is complete */
        xx = SP0DR * 256;
        SP0DR = 0x00;            /* read in the lower byte of A/D conversion result /
        while (!(SP0SR&0x80));
        xx += SP0DR;             /* combine the upper and lower bytes */
        return xx;
}
```

10.11 Summary

A data acquisition system consists of four major components: a transducer, a signal conditioning circuit, an A/D converter, and a computer. The transducer converts a nonelectric quantity into a voltage. The transducer may not be appropriate for processing by the A/D converter. The signal conditioning circuit shifts and scales the output from a transducer to a range that can take advantage of the full capability of the A/D converter. The A/D converter converts an electric voltage into a digital value. The computer does the final processing.

The accuracy of an A/D converter is dictated by the number of bits used to represent the analog quantity. The more the number of bits is used, the better the accuracy. The 68HC12 uses the successive approximation algorithm to perform the A/D conversion. All 68HC12 members except the 812A4 implement one or two 8-channel 10-bit A/D converters. The 68HC12 A/D converter has six A/D control registers. Among them, only ATDCTL2 to ATDCTL5 are implemented. The A/D conversion is started when the ATDCTL5 register is written into.

The A/D converter can operate in single-channel or multiple-channel mode. The user has a choice of performing a sequence of four (or eight) conversions or continuously performing four (or eight) conversion sequences. In the single-channel mode, a sequence of conversions is either four or eight conversions on the same channel. In the multiple-channel mode, the ATD sequencer runs each of the four or eight conversions on sequential channels in a specific group.

The LM34 temperature sensor and the BPT barometric pressure sensor are used as examples to demonstrate the A/D conversion process.

Some applications require an A/D converter with resolution higher than 10 bits. The solution is either to use a different microcontroller that has a higher resolution A/D converter, or use an external A/D converter with higher resolution. The AD7894 is a 14-bit A/D converter that has an SPI-compatible serial interface.

10.12 Exercises

E10.1 Explain the difference between digital and analog signals.

E10.2 The successive approximation method can also be used in computing the square root of an integer. Write an assembly program and a C program to compute the square root of a 32-bit non-negative integer using this algorithm.

E10.3 Design a signal conditioning circuit that can scale the voltage from the range of 0~80 mV to 0~5V.

E10.4 Design a signal conditioning circuit that can scale and shift the voltage from the range of –80mV~160mV to the range of 0V~5V.

E10.5 Design a signal conditioning circuit to shift and scale the voltage from the range of –40mV~240mV to the range of 0V~5V.

E10.6 Suppose there is a 10-bit A/D converter with V_{LREF} = 0.5V and V_{HREF} = 3.5V. Find the corresponding voltage values for A/D conversion results of 20, 100, 300, 600, 800, and 960.

E10.7 Suppose there is a 12-bit A/D converter with V_{LREF} = 0.0V and V_{HREF} = 5.0V. Find the corresponding voltage values for A/D conversion results of 100, 200, 360, 512, 1,024, 2,048, and 3,600.

E10.8 Write a subroutine to initialize the ATD converter for the 912BC32 and start the conversion with the following parameters:

- nonscan mode
- select channel 0~7

- fast flag clear all
- stop ATD in wait mode
- disable interrupt
- finish current conversion then freeze when BDM becomes active
- 10-bit operation and 2 A/D clock periods of sample time
- choose 1 MHz as the conversion frequency for 8 MHz P-clock

E10.9 The LM35 is a Centigrade temperature sensor made by National Semiconductor Inc. This device has three external connection pins like the LM34. The pin assignment and circuit connection for converting temperature are shown in Figure 10E.1. Use this device to construct a circuit to display the room temperature in Celsius. Assume the temperature range is from −20°C to 44°C.

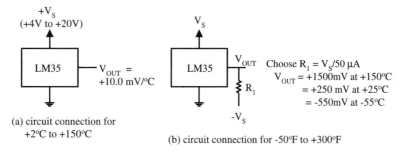

$+V_S$ (+4V to +20V)

LM35 — $V_{OUT} = +10.0$ mV/°C

(a) circuit connection for +2°C to +150°C

V_S

LM35 V_{OUT}

R_1

$-V_S$

Choose $R_1 = V_S/50\ \mu A$
$V_{OUT} = +1500$ mV at +150°C
$= +250$ mV at +25°C
$= -550$ mV at -55°C

(b) circuit connection for -50°F to +300°F

Figure 10E.1 ■ Circuit connection for the LM35

E10.10 The Microbridge AWM3300V is a mass airflow sensor manufactured by Honeywell. The block diagram of the AWM3300V is shown in Figure 10E.2. The Microbridge mass airflow sensor AWM3300V is designed to measure the airflow. Its applications include air-conditioning, medical ventilation/anesthesia control, gas analyzers, gas metering, fume cabinets, and process control. The AWM3300V operates on a single 10V±10mV power supply. The sensor output (from V_{OUT}) corresponding to the airflow rate of 0~1.0 liter/min is 1.0~5.0V. The AWM3300V can operate in the temperature range of −25 to 85°C. It takes 3 ms for the output voltage to settle after power-up. Design a circuit to measure and display the mass airflow using the AWM3300V. Write a program to configure the 68HC12 A/D converter, start the A/D conversion, and display the mass airflow.

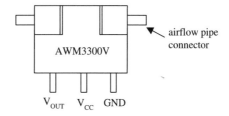

airflow pipe connector

AWM3300V

V_{OUT} V_{CC} GND

Figure 10E.2 ■ Microbridge AWM3300V

10.13 Lab Exercises & Assignments

L10.1 *A/D Converter Setup.* Perform the following steps:

Step 1
Set the function generator output to between 0 and 5V.

Step 2
Connect the ATD channel 7 input pin to the functional generator output. Set the frequency to 10 KHz.

Step 3
Set the V_{RH} and V_{RL} to 5 and 0V respectively.

Step 4
Write a program to perform the following operations:

- Start the A/D conversion.
- Take 64 samples, convert them, and store the conversion results at $800~$83F.
- Compute the average value and store it at $840~$841.

Step 5
Assemble the program and download the program onto the CME-12BC demo board for execution.

Step 6
Run the program for square and triangular waveforms. Do the A/D conversion results look like the selected waveforms?

L10.2. *Digital thermometer.* Use the LM34 temperature sensor, the CA3140 op amp, and the required resistors to construct a digital thermometer. The temperature should be displayed in three integral and one fractional digits. Use the MAX7221 to drive seven-segment displays. The requirements are as follows:

1. The temperature range is from 0°F to 212°F.

2. Use four seven-segment displays to display the temperature reading. Use one display to show the fractional digit. The decimal point must be displayed.

3. Use a water bath to change the ambient temperature to the LM34. You must wrap the LM34 carefully so that it does not get wet.

4. Measure the temperature five times in one second and display the average of the measurements.

5. Update the temperature once every second.

11

Development Support

11.1 Objectives

After completing this chapter, you should be
able to:

- understand the basic debugging concepts and
  tradeoffs of each approach

- explain the development support features
  provided by the 68HC12

- explain the functioning of instruction tagging

- explain BDM breakpoint modes

- understand the BDM hardware and firmware
  commands

- use the AX-BDM12 to perform source-level
  debugging on the CME-BC12 demo board

11.2 Fundamental Concepts of Software Debugging

Software debugging in microcontroller-based product development can be classified into two approaches: software-only and hardware-assisted.

Using a demo board with a resident monitor or simulator to run the program is the software-only approach. In the hardware-assisted approach, we will use either the logic analyzer or an emulator to trace the program execution. No matter which approach is used, the program need to be run in order to find out if there are any errors. When the program execution result is not what we expect, we examine our program to find out if there are any obvious logical errors. Many errors can be identified in this manner. However, there are other errors that are not obvious, even when we look at them.

Most universities that teach the 68HC12 use demo boards with an on-board monitor. The monitor allows us to set breakpoints at places that we suspect errors might exist. The contents of CPU registers will be displayed by the breakpoint handling routine (often implemented by the SWI instruction). Other monitor commands for displaying memory contents are also available. By doing so, we will be able to determine whether our program executes correctly up to the breakpoint. Instruction tracing is also available from most monitors in case we need to pinpoint the exact instructions that caused the error. All input/output functionality can actually be exercised using this approach. The source-level debugger that we discussed in Chapter 3 may be written to communicate with the monitor on the demo board or invoke the simulator to execute our program. Using this approach, you don't need to examine the contents of memory locations in order to find out if the program execution result is correct. Instead, values of program variables are available for examination during the process of program execution. In other words, debugging activity is carried out at the source level.

In the hardware-assisted approach, most designers use the *in-circuit emulator* (ICE) to identify program errors. The in-circuit emulator needs to reconstruct instructions being executed from the data flowing on the system bus of the target prototype (including address, data, and control signals). This approach requires certain built-in hardware support from the microcontroller used in the prototype in order to identify the boundary of instructions. The in-circuit emulator also allows us to set breakpoints at those locations that are suspicious so that program execution results can be examined. This approach is especially useful when a demo board with a resident monitor is not available. An in-circuit emulator usually costs much more than a demo board with a resident monitor.

The 68HC12 provides the following features to support development and debugging of applications and systems:

- *Background debug mode* (BDM). The BDM module uses the BKGD pin to communicate with the host computer (which could be a PC or another 68HC12) that executes the debugging software. The debug host uses this serial interface to send commands to the target system to perform debugging activities. The debug host can use this serial interface to read from or write into any memory location or register of the target system, stop processing application programs, trace application program instructions one at a time, or jump to a certain location to start program execution. The debug software running on the debug host can be written to allow the user to download programs to the target system, set breakpoints, program EEPROM or flash memory, and display register and memory contents at the breakpoints.

- *Instruction tagging.* This feature provides a way of forcing the 68HC12 to stop executing an application program when the tagged instruction reaches the CPU. When the tagged instruction is detected, the 68HC12 CPU enters background debug mode rather than executing the tagged instruction.

■ *Hardware breakpoints.* The 68HC12 implements a hardware circuit that compares actual address and data values to predetermined data in setup registers. A successful comparison places the CPU in background debug mode or initiates a *software interrupt* (SWI).

11.3 Instruction Queue

The 68HC12 has an instruction queue that performs instruction prefetching to boost the CPU performance. The prefetched instruction(s) may not be executed due to the branch instruction. It is still possible to monitor CPU activity on a cycle-by-cycle basis for debugging. The 68HC12 instruction queue provides at least three bytes of program information to the CPU when instruction execution begins. To help external debugging hardware identify what instruction is being executed, the 68HC12 provides information about data movement in the queue and when the CPU begins to execute instructions via status signals IPIPE1 and IPIPE0. Information available on the IPIPE1 and IPIPE0 pins is time multiplexed. External circuitry can latch data movement information on rising edges of the E-clock signal and latch execution start information on falling edges of the E clock. The meaning of data on the pins is shown in Table 11.1.

Data movement — IPIPE[1:0] captured at rising edge of E clock [1]		
IPIPE[1:0]	**Mnemonic**	**Meaning**
0:0	--	No movement
0:1	LAT	Latch data from bus
1:0	ALD	Advance queue and load from bus
1:1	ALL	Advance queue and load from latch
Execution Start — IPIPE[1:0] captured at falling edge of E clock [2]		
IPIPE[1:0]	**Mnemonic**	**Meaning**
0:0	--	No start
0:1	INT	Start interrupt sequence
1:0	SEV	Start even instruction
1:1	SOD	Start odd instruction

1. Refers to data that was on the bus at the previous E falling edge.
2. Refers to bus cycle starting at this E falling edge.

Table 11.1 ■ IPIPE Decoding

Instructions are fetched a few cycles before they are executed by the CPU. To monitor cycle-by-cycle CPU activity, it is necessary to externally reconstruct what is happening in the instruction queue. Internally, the CPU only needs to buffer the data from program fetches. For a system debug it is necessary to keep the data and its associated address in the reconstructed instruction queue. The raw signals required for reconstruction of the queue are ADDR, DATA, R/W, ECLK, and status signals IPIPE1 and IPIPE0.

The instruction queue consists of two 16-bit queue stages and a holding latch on the input of the first stage. To advance the queue means to move the word in the first stage to the second stage and move the word from either the holding latch or the data bus input buffer into the first stage. To start the even (or odd) instruction means to execute the opcode in the high-order (or low-order) byte of the second stage of the instruction queue.

11.4 Background Debug Mode (BDM)

Background debug mode (BDM) is used for system development, in-circuit testing, field-testing, and programming. BDM is implemented in on-chip hardware and provides a full set of debug options.

As shown in Figure 11.1, the external host development system communicates with the BDM control logic via the BKGD pin. The external host development system performs debug activity by sending commands to the BDM module via the BKGD pin. Some of the commands are hardware commands whereas others are firmware commands. Hardware commands are decoded and executed directly by the hardware, whereas each firmware command is implemented by a small program (written in 68HC12 instructions).

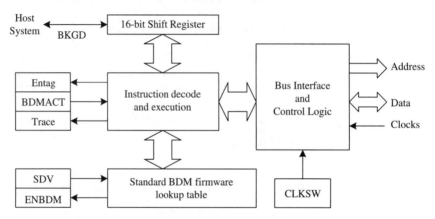

Figure 11.1 ■ BDM block diagram

Because the BDM control logic does not reside in the CPU, BDM hardware commands can be executed while the CPU is operating normally. The control logic generally uses the CPU idle bus cycle to execute these commands (if they need to access memory), but can steal bus cycles when necessary. The execution of firmware commands requires the CPU to be in active background debug mode, so that the ROM that stores the programs that implement all firmware commands becomes available in the standard 64-KB memory map.

Hardware commands are used to read and write target system memory locations and to enter active background debug mode. Target system memory includes all memory accessible by the CPU.

Firmware commands are used to read and write CPU registers and exit from active debug mode. CPU registers include double accumulator D, index register X, index register Y, stack pointer SP, and program counter.

Hardware commands can be executed any time and in any mode excluding a few exceptions (which will be discussed later). Firmware commands can only be executed when the system is in active background debug mode.

These commands can be utilized to download programs into memory, set breakpoints, execute the user program up to the breakpoint, trace the program execution, display values of the CPU and I/O registers, and display the contents of memory locations. Tool developers can take advantage of these commands to implement low cost source-level debuggers.

11.5 BDM Serial Interface

Commands and data are transferred into the BDM module via the BKGD pin serially. One bit takes 16 E-clock cycles to transfer. The BDM serial interface requires the external hardware to generate a falling edge on the BKGD pin to indicate the start of each bit time. The external hardware circuit provides this falling edge whether data is transmitted or received.

The BKGD pin is a pseudo-open–drain pin that can be driven either by an external debugging hardware or by the 68HC12. Data is transferred most significant bit first at 16 E-clock cycles per bit. The interface times out if 512 E-clock cycles occur between falling edges from the host. The hardware clears the command register when this occurs.

The BKGD pin can receive a high or low or transmit a high or low level as shown in figures 11.2, 11.3, and 11.4. Interface timing is synchronous to the 68HC12 but asynchronous to the external host.

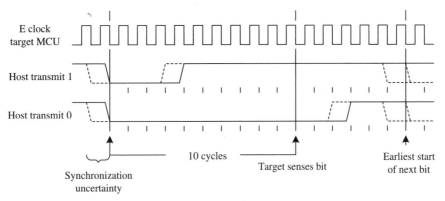

Figure 11.2 ■ BDM host to target serial bit timing

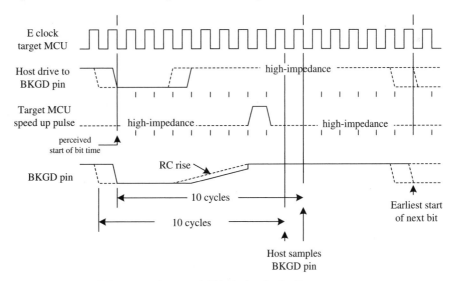

Figure 11.3 ■ BDM target to host serial bit timing (logic 1)

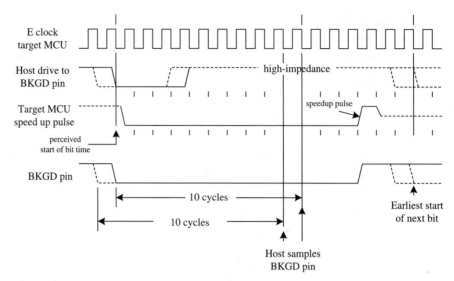

Figure 11.4 ■ BDM target to host serial bit timing (logic 0)

Figure 11.2 shows an external host transmitting a value one or zero to the BKGD pin of the target 68HC12 microcontroller. The host is asynchronous to the target, so there is a zero to one cycle delay from the host-generating falling edge to where the target perceives the beginning of the bit time. Ten target E cycles later, the target senses the bit level on the BKGD pin. Typically, the host actively drives the pseudo-open–drain BKGD pin during host-to-target transmissions to speed up rising edges.

Figure 11.3 shows the host receiving a logic 1 from the target microcontroller, there is a zero to one cycle from the host-generated falling edge on the BKGD pin to the perceived start of the bit time in the target microcontroller. The host holds the BKGD pin low long enough for the target to recognize it (at least two target E cycles). The host then releases the low drive before the target microcontroller drives a brief active high speedup pulse seven cycles after the perceived start of the bit time. The host should sample the bit level about 10 cycles after it starts the bit time.

Figure 11.4 shows the host receiving a logic 0 from the target microcontroller. The host initiates the bit time but the target microcontroller finishes it. Since the target wants the host to receive a logic 0, it drives the BKGD pin low for 13 E-clock cycles, then briefly drives it high to speed up the rising edge. The host also samples the bit level about 10 cycles after starting the bit time.

Pins TAGHI and TAGLO are used for instruction tagging. TAGHI and BKGD share the same pin. TAGLO and LSTRB share the same pin.

The TAGHI pin is used to tag high byte of an instruction. When instruction tagging is on, a logic 0 at the falling edge of the external clock (E) tags the high half of the instruction word being read into the instruction queue.

The TAGLO pin is used to tag the low byte of an instruction. When instruction tagging is on and low strobe is enabled, a logic 0 at the falling edge of the external clock (E) tags the lower half of the instruction word being read into the instruction queue.

11.6 BDM Registers

Seven BDM registers are mapped into the standard 64-KB address space when BDM is active. The mapping is shown in Table 11.2.

Address	Register	Mnemonic
$FF00	BDM instruction register	INSTRUCTION
$FF01	BDM status register	STATUS
$FF02 -- $FF03	BDM shift register	SHIFTER
$FF04 -- $FF05	BDM address register	ADDRESS
$FF06	BDM CCR holding register	CCRSAV

Table 11.2 ■ BDM registers

The content of the instruction register is determined by the type of background command being executed. The status register indicates BDM operating conditions. The shift register contains data being received or transmitted via the serial interface. The address register is temporary storage for BDM commands. The CCR holding register preserves the content of the *condition code register* (CCR) while BDM is active.

The only registers of interest to users are the *status register* and the *CCR holding register*. The other BDM registers are used only by the BDM firmware to execute commands. The registers are accessed by means of the hardware commands READ_BD and WRITE_BD but should not be written during the BDM operation.

11.6.1 BDM Instruction Register

The instruction register is written by the BDM hardware as a result of serial data shifted in on the BKGD pin (this can be done in all operation modes). It is readable and writable in special peripheral mode on the parallel bus. The hardware clears the instruction register if 512 E-clock cycles occur between falling edges (on the BKGD pin) from the host.

When a hardware BDM command is shifted in, the content of the instruction register is interpreted as shown in Figure 11.5. When a firmware command is executed, the meanings of the bits in the BDM instruction register are shown in Figure 11.6.

7	6	5	4	3	2	1	0	
H/F	DATA	R/W	BKGND	W/B	BD/U	0	0	$FF00

value after reset   0   0   0   0   0   0   0   0

H/F: Hardware/Firmware flag
     0 = firmware instruction
     1 = hardware instruction
DATA: data flag
     0 = No data
     1 = Data included in command
R/W: Read/Write flag
     0 = Write
     1 = Read
BKGND: Hardware request to enter active background mode
     0 = Not a hardware background command
     1 = Hardware background command (instruction = $90)
W/B: Word/byte transfer flag
     0 = byte transfer
     1 = word transfer
BD/U -- BDM Map/User map flag
     Indicates whether BDM registers and ROM are mapped to
     addresses $FF00 to $FFFF in the standard 64-KB address space.
     Used only by hardware read/write commands.
     0 = BDM resources not in map
     1 = BDM resources in map (user resources in map)

Figure 11.5 ■ BDM instruction register (hardware command)

7	6	5	4	3	2	1	0	
H/F	DATA	R/W	TTAGO		RNEXT			$FF00

value after reset   0   0   0   0   0   0   0   0

H/F: Hardware/Firmware flag
     0 = firmware control logic
     1 = hardware control logic
DATA: data flag
     0 = No data
     1 = Data included in command
R/W: Read/Write flag
     0 = Write
     1 = Read
TTAGO: Trace, Tag, GO field (shown in Table 11.3)

RNEXT: Register/Next field
     Indicates which register is being affected by a command. In the case
     of a READ_NEXT or WRITE_NEXT command, index register X is
     pre-incremented by two and the word pointed to by X is then read
     or written. The meaning of this field is shown in Table 11.4.

Figure 11.6 ■ BDM instruction register (firmware command)

TTAGO value	Instruction
00	--
01	GO
10	TRACE1
11	TAGGO

Table 11.3 ■ TTAGO decoding

REGN value	Instruction
000	--
001	--
010	READ/WRITE NEXT
011	PC
100	D[1]
101	X[1]
110	Y[1]
111	SP[1]

Note. 1. Not available in 812A4 and B family members

Table 11.4 ■ RNEXT decoding

11.6.2 BDM Status Register

The contents of the BDM status register are shown in Figure 11.7. It is readable and writable in special peripheral mode on the parallel bus.

	7	6	5	4	3	2	1	0
value after reset	ENBDM	BDMACT	ENTAG	SDV	TRACE	CLKSW	0	0
Special single chip:	0	1	0	0	0	0	0	0
Special peripheral:	0	1	0	0	0	0	0	0
All other modes:	0	0	0	0	0	0	0	0

ENBDM: Enable BDM
 0 = BDM disabled
 1 = BDM enabled
BDMACT: BDM active status
 0 = BDM not active
 1 = BDM active
ENTAG: Tagging enable
 0 = Tagging not enabled, or BDM active
 1 = Tagging active. BDM cannot process serial commands while
   tagging is active.
SDV: Shift data valid
 0 = Data phase of command not complete
 1 = Data phase of command is complete
TRACE: TRACE1 BDM firmware command is being executed
 0 = TRACE1 command is not being executed
 1 = TRACE1 command is being executed
CLKSW: Clock switch (not available in 812A4 and B family)
 0 = BDM system operates with BCLK
 1 = BDM system operates with ECLK

Figure 11.7 ■ BDM status register

The *enable BDM bit (ENBDM)* controls whether the BDM is enabled or disabled. When enabled, BDM can be made active to allow firmware commands to be executed. When disabled, BDM cannot be made active but BDM hardware commands are still allowed. This bit is set immediately out of reset in special single chip mode. In secure mode, this bit will not be set by the firmware until after the EEPROM and flash erase verify tests are complete.

The *background mode active status (BDMACT)* bit can only be set by BDM hardware upon entry into BDM. It can only be cleared by the standard BDM firmware lookup table upon exit from BDM active mode.

The *enable tagging (ENTAG)* bit indicates whether instruction tagging is enabled or disabled. It is set when the TAGGO command is executed and cleared when BDM is entered. The serial system is disabled and the tag function is enabled 16 cycles after this bit is set. BDM cannot process serial commands while tagging is active.

The *shift-data-valid (SDV)* bit is set and cleared by the BDM hardware. It is set after data has been transmitted as part of a firmware read command or after data has been received as part of a firmware write command. It is cleared when the next BDM command has been received or BDM is exited. SDV is used by the standard BDM firmware to control program flow execution.

The *TRACE* bit gets set when a BDM TRACE1 firmware command is first recognized. It will stay set as long as continuous back-to-back TRACE1 commands are executed. This bit will get cleared when the next command that is not a TRACE1 command is recognized.

The *BDM clock switch (CLKSW)* bit controls which clock the BDM operates with. It is only writable from a hardware BDM command. The BDM system can operate with the BCLK or ECLK. If ECLK rate is slower than BCLK rate, CLKSW is ignored and BDM system is forced to operate with ECLK. A 150-cycle delay at the clock speed that is active during the data portion of the command will occur before the new clock source is guaranteed to be active. The start of the next command uses the new clock for timing subsequent BDM communications.

11.6.3 BDM CCR Holding Register

When entering background debug mode, the BDM CCR holding register is used to save the contents of the condition code register of the user's program. It is also used for temporary storage in the standard BDM firmware mode. The BDM CCR holding register can be written to modify the CCR value.

11.6.4 BDM Shift Register

The 16-bit BDM shift register contains data being received or transmitted via the serial interface. It is also used by the standard BDM firmware for temporary storage.

11.6.5 BDM Address Register ($FF04–$FF05)

The 16-bit address register is loaded with the address to be accessed by BDM hardware commands.

11.6.6 BDM Internal Register Position Register (BDMINR)

The contents of the BDMINR register are shown in Figure 11.8. The upper five bits of this register show the state of the upper five bits of the base address for the system's relocatable register block. BDMINR is a shadow of the INTRG register that maps the register block to any 2KB space within the first 32KB of the 64KB address space.

	7	6	5	4	3	2	1	0
	REG15	REG14	REG13	REG12	REG11	0	0	0
Value after reset	0	0	0	0	0	0	0	0

Figure 11.8 ■ BDM internal register position (BDMINR)

11.7 BDM Commands

BDM commands are classified into two categories: hardware and firmware.

Hardware commands are used to read and write target system memory locations and to enter active background debug mode. Target system memory includes all memories that are accessible by the CPU such as on-chip RAM, EEPROM, Flash EEPROM, I/O and control registers, and all external memory.

Hardware commands are executed with minimal or no CPU intervention and do not require the system to be in active BDM for execution, although they can be executed in this mode. When executing a hardware command, the BDM sub-block waits for a free CPU bus cycle so that the background access does not disturb the running application program. If a free cycle is not found within 128 clock cycles, the CPU is momentarily frozen so that the BDM can steal a cycle. When the BDM finds a free cycle, the operation does not intrude on normal CPU operation provided that it can be completed in a single cycle. However, if an operation requires multiple cycles, the CPU is frozen until the operation is complete, even though the BDM found a free cycle.

The BDM hardware commands are listed in Table 11.5.

Command	Opcode	Data	Description
BACKGROUND	$90	None	Enter background mode if firmware is enabled.
READ_BD_BYTE	$E4	16-bit address 16-bit data out	Read from memory with standard BDM firmware lookup table in map. Odd address data on low byte.
READ_BD_WORD	$EC	16-bit address 16-bit data out	Read from memory with standard BDM firmware lookup table in map. Must be aligned access.
READ_BYTE	$E0	16-bit address 16-bit data out	Read from memory with standard BDM firmware lookup table out of map. Odd address data on low byte.
READ_WORD	$E8	16-bit address 16-bit data out	Read from memory with standard BDM firmware lookup table out of map. Must be aligned access.
WRITE_BD_BYTE	$C4	16-bit address 16-bit data in	Write to memory with standard BDM firmware lookup table in map. Odd address data on low byte.
WRITE_BD_WORD	$CC	16-bit address 16-bit data in	Write to memory with standard BDM firmware lookup table in map. Must be aligned access.
WRITE_BYTE	$C0	16-bit address 16-bit data in	Write to memory with standard BDM firmware lookup table out of map. Odd address data on low byte.
WRITE_WORD	$C8	16-bit address 16-bit data in	Write to memory with standard BDM firmware lookup table out of map. Must be aligned access.
ENABLE FIRMWARE[1]	$C4	$FF01, 1xxx xxxx (in)	Write byte $FF01, set the ENBDM bit. This allows execution of commands which are implemented in firmware.
STATUS[2]	$E4	$FF01, 00000000 (out)	READ_BD_BYTE $FF01. Running user code. (BGND instruction is not allowed).
		$FF01, 10000000 (out)	READ_BD_BYTE $FF01. BGND instruction is allowed.
		$FF01, 11000000 (out)	READ_BD_BYTE $FF01. Background mode active (waiting for single wire serial command).

1. ENABLE_FIRMWARE command is a specific case of the WRITE_BD_BYTE command.
2. STATUS command is a special case of the READ_BD_BYTE command.

Table 11.5 ■ Hardware BDM commands

The READ_BD and WRITE_BD commands allow access to BDM register locations. These locations are not normally in the system memory map but share addresses with the application in memory. To distinguish between physical memory locations that share the same address, BDM memory resources are enabled unobtrusively, even if the addresses conflict with the application memory map.

Firmware commands are used to access and manipulate CPU resources. The system must be in active BDM mode to execute standard BDM firmware commands. Normal instruction execution is suspended while the CPU executes the firmware located in the standard BDM firmware lookup table.

As the system enters active BDM mode, the standard BDM firmware lookup table and BDM registers become visible in the on-chip memory map at $FF00–$FFFF, and the CPU begins executing the standard BDM firmware. The standard BDM firmware watches for serial commands and executes them as they are received. The firmware commands are shown in Table 11.6.

Command	Opcode	Data	Description
READ_NEXT	$62	16-bit data out	X = X+2; then read the word pointed to by X.
READ_PC	$63	16-bit data out	Read program counter.
READ_D	$64	16-bit data out	Read accumulator D.
READ_X	$65	16-bit data out	Read index register X.
READ_Y	$66	16-bit data out	Read index register Y.
READ_SP	$67	16-bit data out	Read stack pointer.
WRITE_NEXT	$42	16-bit data in	X = X + 2; then write word pointed to by X.
WRITE_PC	$43	16-bit data in	Write program counter.
WRITE_D	$44	16-bit data in	Write to accumulator D.
WRITE_X	$45	16-bit data in	Write into index register X.
WRITE_Y	$46	16-bit data in	Write into index register Y.
WRITE_SP	$47	16-bit data in	Write to stack pointer.
GO	$08	none	Go to user program.
TRACE1	$10	none	Execute one user instruction then return to BDM.
TAGGO	$18	none	Enable tagging and go to user program.

Table 11.6 ■ Firmware BDM commands

Hardware and firmware commands start with an 8-bit opcode followed by a 16-bit address and/or a 16-bit data word depending on the command. All the read commands return 16 bits of data despite the byte or word implication in the command name.

Eight-bit reads return 16-bit data, of which only 1 byte will contain valid data. If reading an even address, the valid data will appear in the MSB. If reading an odd address, the valid data will appear in the LSB.

Sixteen-bit unaligned reads and writes are not allowed. If attempted, the BDM will ignore the least significant bit of the address and will assume an even address from the remaining bits.

For hardware data read commands, the external host must wait for 150 target clock cycles after sending the address before attempting to obtain the read data. This is to be certain that valid data is available in the BDM shift register, ready to be shifted out. For hardware write commands, the external host must wait for 150 target clock cycles after sending the data to be written before attempting to send a new command. This is to avoid disturbing the BDM shift register before the write has been completed. The 150 target clock cycles delay (in both cases) includes the maximum 128-cycle delay that can be incurred as the BDM waits for a free cycle before stealing a cycle.

For firmware read commands, the external host must wait for 32 target clock cycles after sending the command opcode before attempting to obtain the read data. This allows enough time for the requested data to be made available in the BDM shift register, ready to be shifted out. For firmware write commands, the external host must wait for 32 target clock cycles after sending the data to be written before attempting to send a new command. This is to avoid disturbing the BDM shift register before the write has been completed.

The external host should wait for 64 target clock cycles after a TRACE1 or GO command before starting any new serial command. This is to allow the CPU to exit gracefully from the standard BDM firmware lookup table and resume execution of the user code. Disturbing the BDM shift register prematurely may adversely affect the exit from the standard BDM firmware lookup table.

Figure 11.9 represents the BDM command structure. The command blocks illustrate a series of eight bit times starting with a falling edge. The bar across the top of the blocks indicates that the BKGD pin idles in the idle state. The time for an 8-bit command is 8×16 target clock cycles.

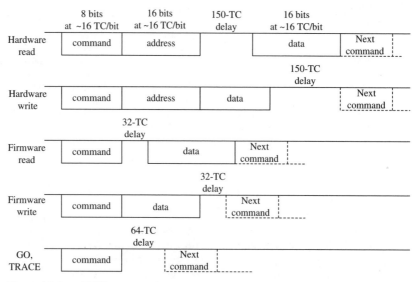

Figure 11.9 ■ BDM command structure

11.8 BDM Operation

The BDM receives and executes commands from a host via a single wire interface BKGD pin. Hardware commands can be executed at any time and in any mode, excluding a few exceptions to be discussed in this section. Firmware commands can only be executed when the system is in active background debug mode.

11.8.1 Executing Firmware Commands

The system must be in active BDM to execute standard BDM firmware commands. BDM can be activated only after being enabled. BDM is enabled by setting the ENBDM bit in the BDM status (BDMSTS) register. The ENBDM bit is set by writing the BDM status register, via the single-wire interface, and using the hardware command WRITE_BD_BYTE.

After being enabled, BDM is activated by one of the following methods:

- Hardware BACKGROUND command
- BDM external instruction tagging mechanism
- CPU BGND instruction
- Breakpoint sub-block's force or tag mechanism (only available in systems that have a breakpoint sub-block)

When BDM is activated, the CPU finishes executing the current instruction and then begins executing the firmware in the standard BDM firmware lookup table. When BDM is activated by the breakpoint sub-block, the type of breakpoint used determines if BDM becomes active before or after execution of the next instruction.

If an attempt is made to activate BDM before being enabled, the CPU resumes normal instruction execution after a brief delay. If BDM is not enabled, any hardware BACKGROUND commands issued are ignored by the BDM and the CPU is not delayed.

In active BDM, the BDM registers and standard BDM firmware lookup table are mapped to addresses $FF00 to $FFFF. BDM registers are mapped to addresses $FF00 to $FF06. The BDM uses these registers that are readable any time by the BDM. These registers are not, however, readable by user programs.

11.8.2 Instruction Tracing

When a TRACE1 command is issued to the BDM in active BDM mode, the CPU exits the standard BDM firmware and executes a single instruction in the user code. Once this has occurred, the CPU is forced to return to the standard BDM firmware and the BDM is active and ready to receive a new command. If the TRACE1 command is issued again, the next user instruction will be executed. This facilitates stepping or tracing through the user code one instruction at a time.

If an interrupt is pending when a TRACE1 command is issued, the interrupt stacking operation occurs but no user instruction is executed. Once back in BDM firmware execution, the program counter points to the first instruction in the interrupt service routine.

11.8.3 Instruction Tagging

The instruction queue and cycle-by-cycle CPU activity are reconstructible in real time or from trace history that is captured by a logic analyzer. However, the reconstructed queue cannot be used to stop the CPU at a specific instruction, because execution has already begun by the time an operation is visible outside the system. A separate instruction tagging mechanism is provided for this purpose.

The tag follows program information as it advances through the instruction queue. When a tagged instruction reaches the head of the queue, the CPU enters active BDM rather than executing the instruction. This is the mechanism by which a development system initiates *hardware breakpoints*. Tagging is disabled when BDM is active, and BDM serial commands are not processed while tagging is active.

Executing the BDM TAGGO command configures two system pins for tagging. The TAGLO signal shares a pin with the LSTRB signal, and the TAGHI signal shares a pin with the BKGD signal. To tag a data value entering the instruction queue, you drive the TAGHI and/or TAGLO pin low before the falling edge of E clock.

The functions of the two tagging pins are shown in Table 11.7. The pins operate independently; that is, the state of one pin does not affect the function of the other. The presence of logic level 0 on either pin at the fall of the external clock (E) performs the indicated function. High tagging is allowed in all modes. Low tagging is allowed only when low strobe (LSTRB) is enabled (LSTRB is allowed only in wide expanded mode and emulation expanded narrow mode).

TAGHI	TAGLO	Tag
1	0	No tag
1	1	low byte
0	0	high byte
0	1	both bytes

Table 11.7 ■ Tag pin function

11.9 BDM & CPU Modes of Operation

BDM is available in all operation modes but must be enabled before firmware commands are executed. Some system peripherals may have a control bit which allows suspending the peripheral function during background debug mode. BDM operates the same in all normal modes.

In special single-chip mode, background operation is enabled and active out of reset. This allows for programming a system with blank memory.

BDM is also active out of special peripheral mode reset and can be turned off by clearing BDMACT bit in the BDM status register. This allows testing of the BDM memory space as well as the user's memory space.

The BDM does not include disable controls that would conserve power during run mode and cannot be used in wait mode if the system disables the clocks to the BDM. The BDM is completely shutdown in stop mode.

11.10 Breakpoints

Breakpoints allow us to stop at the instructions that are suspected to cause the error. Hardware breakpoints are used to debug software on the microcontroller by comparing actual address and data values to predetermined data in setup registers. Depending on the chosen mode, a successful comparison places the CPU in background debug mode or initiates a *software interrupt* (SWI).

The 68HC12 has incorporated the following breakpoints features:

- Mode selection for BDM or SWI generation
- Program fetch tagging for cycle of execution breakpoint
- Second address compare in dual address modes
- Range compare by disabling low byte address
- Data compare in full feature mode for non-tagged breakpoint
- Byte masking for high/low byte data compares
- R/W compare for non-tagged compares
- Tag inhibit on BDM TRACE

11.10.1 Breakpoint Registers

Breakpoint operation consists of comparing data in the *breakpoint address registers* (BRKAH/BRKAL) to the address bus and comparing data in the *breakpoint data registers* (BRKDH/BRKDL) to the data bus. The breakpoint data registers can also be compared to the address bus. The scope of comparison can be expanded by ignoring the least significant byte of address or data matches. The scope of comparison can be limited to program data only by setting the BKPM bit in breakpoint control register to zero.

To trace program flow, setting the BKPM bit causes address comparison of program data only. Control bits are also available that allow checking of read/write matches.

Breakpoint control register 0 and 1 allow us to select breakpoint parameters. Their contents are shown in Figures 11.10 and 11.11.

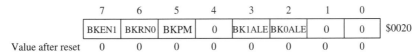

7	6	5	4	3	2	1	0	
BKEN1	BKRN0	BKPM	0	BK1ALE	BK0ALE	0	0	$0020

Value after reset   0   0   0   0   0   0   0   0

    BKEN1, BKEN0: breakpoint mode select enable bits.
       The meaning of these two bits are shown in Table 11.8.
    BKPM: Break on program addresses
       This bit controls whether the breakpoint causes an immediate data
       breakpoint (next instruction boundary) or a delayed program
       breakpoint related to an executable opcode. Data and unexecuted
       opcodes cannot cause a breakpoint if this bit is set. This bit has no
       meaning in SWI dual address mode. The SWI mode only performs
       program breakpoints.
       0 = On match, break at the next instruction boundary
       1 = On match, break if the match is an instruction to be executed.
         This uses tagging as its breakpoint mechanism.
    BK1ALE: Breakpoint 1 range control bit (only valid in dual address mode)
       0 = BRKDL is not used to compare to the address bus
       1 = BRKDL is used to compare to the address bus.
    BK0ALE: Breakpoint 0 range control bit
       0 = BRKAL is not used to compare to the address bus
       1 = BRKAL is used to compare to the address bus

Figure 11.10 ■ Breakpoint control register 0

7	6	5	4	3	2	1	0	
0	BKDBE	BKMBH	BKMBL	BK1RWE	BK1RW	BK0RWE	BK0RW	$0021

Value after reset   0   0   0   0   0   0   0   0

    BKDBE: Enable data bus bit.
       0 = BRKDH/L registers are not used in any comparison
       1 = BRKDH/L registers are used to compare address or data (depending on
         the mode selections BKEN1 and BKEN0)
    BKMBH: Breakpoint mask high bit
       0 = High byte of data bus is compared to BRKDH
       1 = High byte of data bus is not used in comparison
    BKMBL: Breakpoint mask low bit
       0 = low byte of data bus is compared to BRKDL
       1 = low byte of data bus is not used in comparison
    BK1RWE: $R/\overline{W}$ compare enable bit
       Enables the comparison of the $R/\overline{W}$ signal to further specify what causes a
       match. This bit is used in conjunction with a second address in dual address
       mode when BKDBE = 1.
       0 = $R/\overline{W}$ is not used in comparison
       1 = $R/\overline{W}$ is used in comparison
    BK1RW: $R/\overline{W}$ compare value bit
       When BKRWE = 1, this bit determines the type of bus cycle to match.
       0 = A write cycle is matched
       1 = a read cycle is matched
    BK0RWE: $R/\overline{W}$ compare enable bit
       0 = $R/\overline{W}$ is not used in the comparison
       1 = $R/\overline{W}$ is used in comparison
    BK0RW: $R/\overline{W}$ compare value bit
       0 = write cycle is matched
       1 = read cycle is matched

Figure 11.11 ■ Breakpoint control register 1

BKEN1	BKEN0	Mode selected	BRKAH/L usage	BRKDH/L usage	R/W	Range
0	0	Breakpoints off	--	--	--	--
0	1	SWI -- dual address mode	address match	address match	no	yes
1	0	BDM -- full breakpoint mode	address match	data match	yes	yes
1	1	BDM -- dual address mode	address match	address match	yes	yes

Table 11.8 ■ Breakpoint mode control

Table 11.9 gives us a better understanding of the breakpoint range control by showing the address range selected by the bits BK1ALE and BK0ALE.

BK1ALE	BK0ALE	Address range selected
0	0	Upper 8-bit address only for full mode or dual mode BKP0
0	1	Full 16-bit address for full mode or dual mode BKP0
1	0	Upper 8-bit address only for dual mode BKP1
1	1	Full 16-bit address for dual mode BKP1

Table 11.9 ■ Breakpoint address range control

The lower four bits of the breakpoint control register 1 are used in determining whether the R/W signal should be used in breakpoint control. Table 11.10 gives us a better understanding on the breakpoint read/write control by showing the combinations of the lower four bits of the breakpoint control register.

BK1RWE	BK1RW	BK0RWE	BK0RW	Read/Write selected
--	--	0	x	R/$\overline{\text{W}}$ is don't care for full mode or dual mode BKP0
--	--	1	0	R/$\overline{\text{W}}$ is write for full mode or dual mode BKP0
--	--	1	1	R/$\overline{\text{W}}$ is read for full mode or dual mode BKP0
0	x	--	--	R/$\overline{\text{W}}$ is don't care for dual mode BKP1
1	0	--	--	R/$\overline{\text{W}}$ is write for dual mode BKP1
1	1	--	--	R/$\overline{\text{W}}$ is read for dual mode BKP1

Table 11.10 ■ Breakpoint read/write control

11.10.2 Breakpoint Modes

The 68HC12 BDM supports three breakpoint modes:

- *SWI dual address-only breakpoints.* In this mode, dual address-only breakpoints can be set, each of which causes a software interrupt. This is the only breakpoint mode that can force the CPU to execute an SWI instruction. Program tagging is the default in this mode; data breakpoints are not possible. In dual mode, each address breakpoint is affected by the respective BKALE bit. The BKDBE bit becomes an enable bit for the second address breakpoint. The BKxRW, BKxRWE, BKMBH, and BKMBL bits are ignored.

- *Single full-feature breakpoint.* This breakpoint causes the microcontroller to enter background debug mode. Bits BK1ALE, BK1RW, and BK1RWE have no meaning in this mode. The BKBDE bit enables data comparison but has no meaning if BKPM = 1. Bits BKMBH and BKMBL allow masking of high and low byte compares but have no meaning if BKPM = 1. The BK0ALE bit enables comparison of low address byte.

- *BDM dual address-only breakpoints.* This mode has dual address-only breakpoints, each of which causes the microcontroller to enter background debug mode. In this mode, each address breakpoint is affected by the BKPM bit, the BKxALE bits, and the BKxRW and BKxRWE bits. In this mode, the BKDBE becomes an enable signal for the second address breakpoint. The BKMBH and BKMBL bits have no effect when in dual address mode. BDM may be entered by a breakpoint only if an internal signal from the BDM indicates that the background debug mode is enabled. If BKPM = 1, then bits BKxRW, BKxRWE, BKMBH, and BKMBL have no meaning. The restrictions to this mode are as follows:

1. Breakpoints are not allowed if the BDM is already active. Active mode means the CPU is executing out of the BDM ROM.

2. BDM should not be entered from a breakpoint unless the ENABLE bit is set in the BDM. This is important because even if the ENABLE bit in the BDM is negated, the CPU does actually execute the BDM ROM code. It checks the ENABLE and returns if not set. If the BDM is not serviced by the monitor, then the breakpoint would be re-asserted when the BDM returns to normal CPU flow. There is no hardware to enforce the restriction of the breakpoint operation if the BDM is not enabled.

11.11 The BDM-based Debugger

The combination of BDM mode, instruction tagging, and the hardware breakpoint makes the development of a low-cost source level debugger possible. To promote the portability of tools that use BDM mode, Motorola defines a 6-pin connector as shown in Figure 11.12.

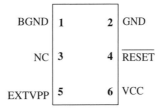

Figure 11.12 ■ BDM connector

Pin 1 should be connected to the BKGD pin of the target microcontroller. Pins 2 and 6 should be connected to the ground and power supply, respectively. Pin 5 should be connected to the voltage required to program EEPROM. Pin 4 allows the external debugger to reset the target microcontroller.

As long as the demo board or prototype board has a BDM connector, then you will be able to use the BDM-based debugger from some vendor. Two components are required to utilize the BDM mode to perform source-level debugging:

1. A BDM pod that has a cable to connect to the BDM connector on the demo board and the parallel (or serial) port on the PC (or workstation). A BDM pod usually contains a microcontroller to communicate with the PC and sends out appropriate BDM commands to the demo board (via the BDM connector) in response to the command received from the PC.

2. A program that runs on the PC and sends commands to the demo board through the BDM pod. This software should have at least the following functions:

   ■ Allow the user to perform some basic configuration.
   ■ Allow the user to download programs onto the demo board (or prototyping board). This may require programming the on-chip EEPROM or flash memory.
   ■ Allow the user to set breakpoints.
   ■ Allow the user to select variables to be watched. This should be done automatically by reading the list file or map file generated by the assembler or compiler because it is not convenient to find out the addresses of variables in C programs.
   ■ Allow the user to trace program to the breakpoint.
   ■ Allow the user to step through the program.
   ■ Update the watched variables and CPU registers automatically after the program execution.

Currently, several vendors have been selling BDM pods for the 68HC12. The usability of these pods is determined by the quality of the PC-based software. You will need to experiment to find out what will suit your needs best. The key consideration is whether the software really achieves the goal of *source-level debugging*.

11.12 Using the Axiom AX-BDM12 Debugger

Axiom manufacturing provides a BDM debugger that works with its 68HC12 demo boards. This debugger consists of a BDM pod and the software (called *hc12bgnd*). The software works fine with the Motorola freeware assembler (*as12*, included on the accompanying CD-ROM) and the ICC12 compiler.

The Axiom BDM pod is attached to the parallel port of the PC. For the CME-12BC demo board, the CONFIG switch should be set as follows:

■ switch 1, 2: off (CME-12BC is configured into single-chip mode out of reset)
■ switch 3: on
■ switch 4, 5: off
■ switch 6: on
■ switch 7, 8: off

After connecting the power plug and the BDM connector, you can start the *hc12bgnd* program. The screen should be similar to that in Figure 11.13.

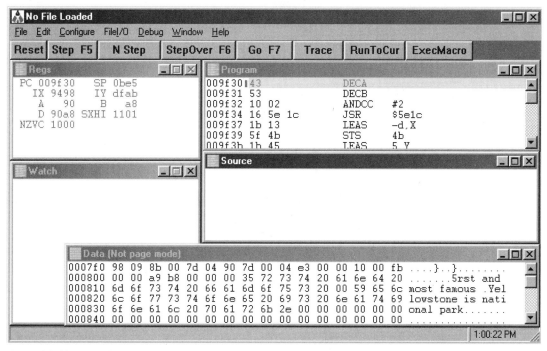

Figure 11.13 ■ hc12bgnd startup screen

The size of the default screen may not be suitable for debugging activities. Adjust the *hc12bgnd* window to an appropriate size that suits your taste.

The function of each menu of the *hc12bgnd* screen is described in the following sections.

CONFIGURE MENU

This is the first thing to start with when you are using the BDM12 debugger for the first time. The contents of the **Configure** menu are shown in Figure 11.14. The contents of the item **Configure** under the **Configure** menu are shown in Figure 11.15. Click on each item from left to right in Figure 11.15 and make the selections as follows:

1. Reset mode: Select **Expanded Wide** for the CME-12BC demo board.

2. Compiler: Only three assemblers and two C compilers are supported. Choose the appropriate assembler or C compiler.

3. Macro files: The popup menu for this choice is shown in Figure 11.16. The appropriate macro files are shown in the figure.

4. Device select: You select the microcontroller to be debugged and select the type of breakpoints to be used. Set Device Select to **912B32** and set Breakpoint Select to either Software or Hardware. The *hc12bgnd* supports 100 software and 2 hardware breakpoints. Choose **Software** breakpoint for this tutorial.

5. Font: Accept the default font. This is not a critical option.

6. Target clock speed: Select **unspecified**.

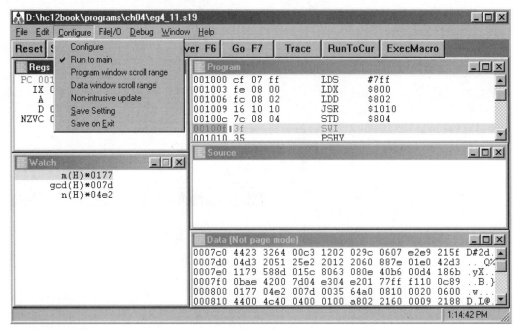

Figure 11.14 ■ Configure menu of the hc12bgnd screen

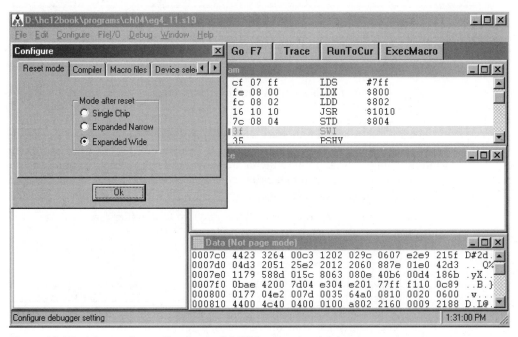

Figure 11.15 ■ Screen for configuring the BDM12 source-level debugger

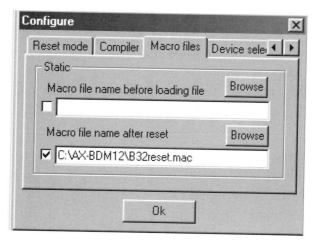

Figure 11.16 ■ Configure the Macro files

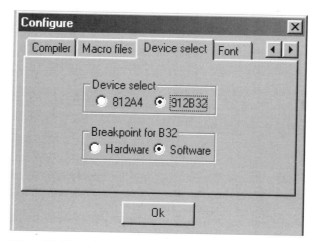

Figure 11.17 ■ Selecting the device to be debugged

FILE MENU

The contents of the file menu are shown in Figure 11.18. The most important operation of this menu is to select a file to be downloaded to the prototype or demo board for debugging. This can be done by selecting the **Load S19 or Hex** option. We will use the program in Example 11 of chapter 4 (file name *eg4_11.a*) to illustrate the debugging process.

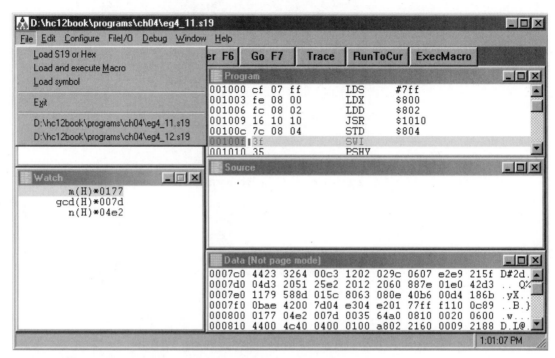

Figure 11.18 ■ File menu of the hc12bgnd debugger

Unless you need to run a special macro, there is no need to do anything to the item **Load and execute Macro**. If you are using the freeware assembler from Axiom, symbols should be loaded into the Watch window automatically. There is no need to manually load any symbol. However, this is not true for the C compiler.

After loading the file *eg4_11.s19*, the window should change and become similar to what is shown in Figure 11.19. The Data window may not be the same as shown in Figure 11.19. The value for gcd (H) in the Watch window may not be zero. The Source window is empty.

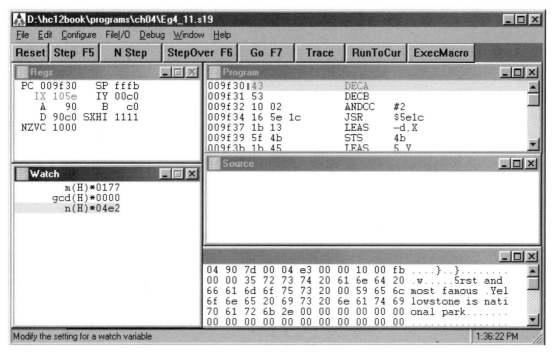

Figure 11.19 ■ Screen after loading the file *eg4_11.s19*

The assembly program *eg4_11.a* is assembled using the Motorola freeware *as12* (modified by Axiom). The debugger automatically adds program variables (declared with assembler directives *dw* and *rmw*) into the Watch window. By default the variable value is displayed in hex. However, you can change it to decimal. To change the number base of a variable, perform the following two actions:

1. Select the variable by clicking the left mouse button on it (try *gcd*). Then press the right mouse to bring up a popup menu, as shown in Figure 11.20.

2. Select **Modify Setting** and the screen will change to what is shown in Figure 11.21. Click on **Decimal** under the column Display as. Perform the same change for variables *m* and *n* and the screen will change to what you see in Figure 11.22.

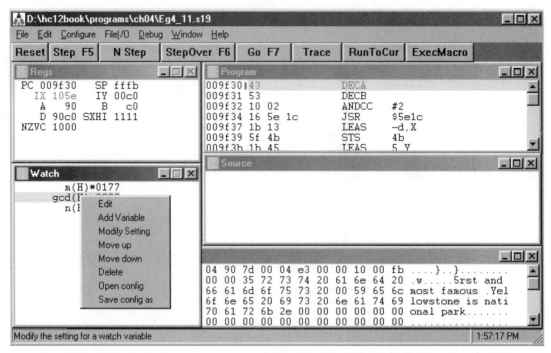

Figure 11.20 ■ Screen after click (left mouse button) on *gcd* and press the right mouse button.

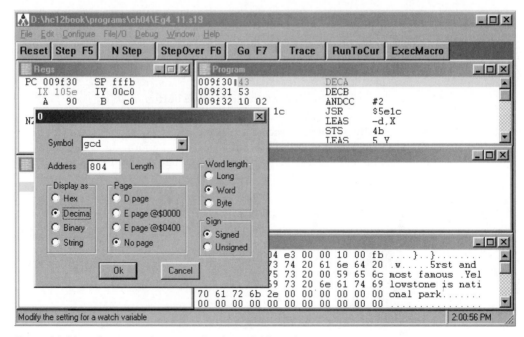

Figure 11.21 ■ Screen to change setting for variable *gcd*.

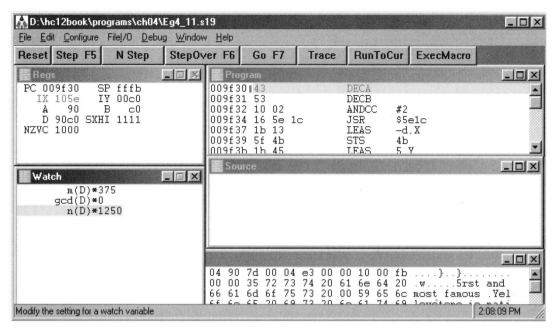

Figure 11.22 ■ Screen after changing the setting of variables *m*, *n*, and *gcd*

You can rearrange the order of variables by clicking on a variable and using the **Move up** and **Move down** commands in Figure 11.20. You can change the value of a variable by selecting the **Edit** command.

The program *eg4_11.a* calls a subroutine to compute the *greatest common divisor* (gcd) of two integers *m* and *n*. Since the Source window is empty, we can close it to make room for the Program window.

To run the program, we need to set the *program counter* (PC) to the start of the program. This can be done by clicking the left mouse on PC in the Regs window and typing in the starting address (try 1000 now) of the program. The Program window will display the instruction located at the content of the program counter ($1000) in the first line.

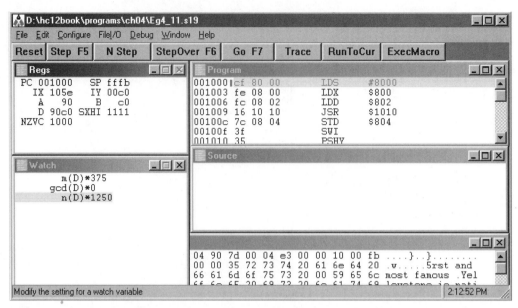

Figure 11.23 ■ Screen after setting PC to the starting address of the program

To debug the program, we need to invoke one or more actions under the **Debug** menu. The actions available in this menu are shown in Figure 11.24.

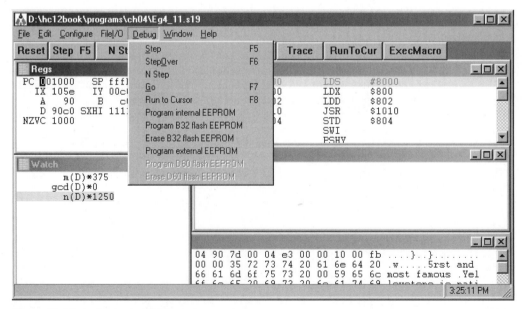

Figure 11.24 ■ Contents of Debug menu

The **Step** entry allows you to execute one instruction and stop. The contents of all affected CPU registers and memory locations will be updated in the Regs and Data windows. The Data window will be useful when you are dealing with arrays or matrices.

If the instruction being executed is not a JSR or BSR instruction, then the effect of the **StepOver** entry is identical to Step. Otherwise, selecting the StepOver entry will step over the subroutine being called. The contents of all CPU registers and memory locations will be updated.

The **N Step** entry under the Debug menu is useful in stepping over a C statement or executing your program one instruction at a time forever. The setup for N Step is shown in Figure 11.25. After executing one instruction, the contents of all windows will be updated.

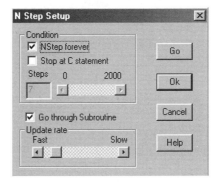

Figure 11.25 ■ Screen for N Step Setup

The **Go** entry will enable the demo board CPU to run the program until a breakpoint is reached or forever. For our example program, the program execution will stop when the SWI instruction is reached. The contents of the watch window will change to what is shown in Figure 11.26. The *gcd* value is 125 and is correct.

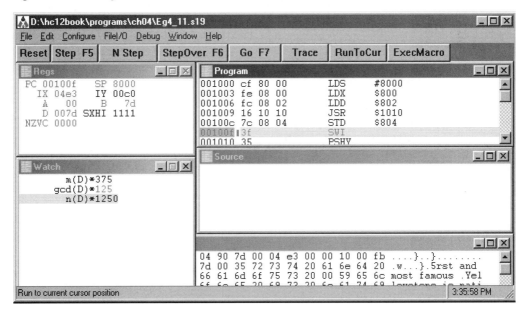

Figure 11.26 ■ Watch window is changed when the SWI instruction is reached.

Setting breakpoints is very simple. You simply click on the address of the instruction at the left column (which represents the address of the instruction) of the Program window. The address will be highlighted with the color purple. For example, try clicking on **001006**; this line will then be highlighted as shown in Figure 11.27. Clicking on the same address one more time will delete the breakpoint.

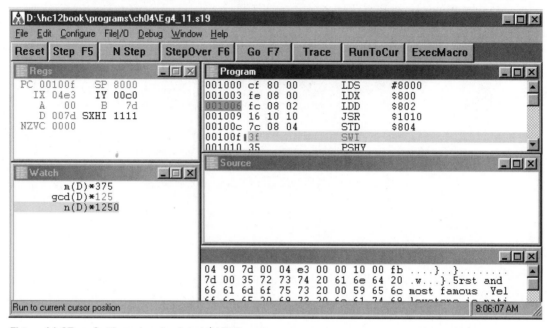

Figure 11.27 ■ Setting a breakpoint at $1006.

The *hc12bngd* debugger does not support 32-bit integers. If you have 32-bit variables to be watched, you need to watch their upper and lower 16 bits separately.

The C language version of the program for computing the *gcd* of two integers *m* and *n* is as follows (file name *bdm_eg1.c*):

```c
#include <hc12.h>
unsigned int m, n, gcd;
unsigned find_gcd (unsigned m, unsigned n);
void swap(unsigned n1, unsigned n2);
void main (void)
{
        m = 375;
        n = 1250;
        gcd = find_gcd(m, n);
        asm ("swi");
}

unsigned find_gcd (unsigned m, unsigned n)
{
        unsigned i, temp;
        temp = 1;
```

```
            if (m > n) swap (&m,&n);              /* make sure m is not greater than n */
            for (i = 2; i <= m; i++)
            {
            if ((m % i) == 0 && (n % i) == 0)     /* check if i can divide both m and n */
                   temp = i;
            }
            return temp;
}

void swap (unsigned *n1, unsigned *n2)
{
            unsigned n3;
            n3 = *n1;
            *n1 = *n2;
            *n2 = n3;
}
```

To use *hc12bgnd* to execute this program, you need to set the compiler to **ImageCraft C Compiler**. In the ICC12 project window (shown in Figure 11.28), select output **format S19 with Source Level Debugging** as the project option. Compile the project and download the program onto the CME-12BC demo board via the AX-BDM12 pod. The screen after loading the *bdm_eg1.s19* file is shown in Figure 11.29.

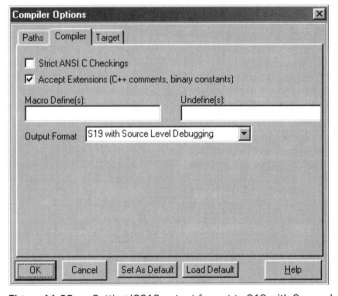

Figure 11.28 ■ Setting ICC12 output format to S19 with Source Level Debugging

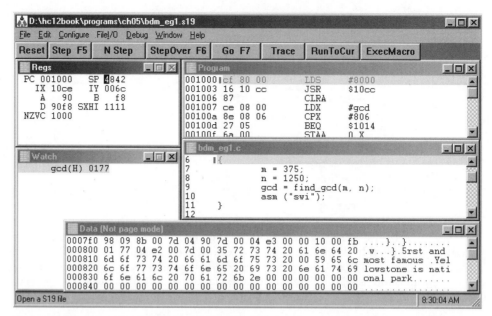

Figure 11.29 ■ Loading *bdm_eg1.s19* onto the CME-12BC demo board using the AX-BMD 12 pod

The AX-BDM12 does not add all the variables into the Watch window automatically. You need to add those variables that you want to watch into the Watch window manually. But before you can do that, you need to find out the addresses of those variables. The ICC12 compiler can optionally generate a map file (in this example, *bdm_eg1.mp*) that contains the starting addresses of all subroutines, labels, and global variables. Local variables are not in the map file. If you want to watch a variable, declare it as a global variable (outside of all functions). The contents (in Courier font) of the file *bdm_eg1.mp* are as follows:

```
Area                             Addr   Size   Decimal Bytes (Attributes)
--------------------------------  ----  ----  --------  -----  ---------
                          text    1000  00D0 =   208. bytes (rel,con)


        Addr   Global Symbol
        -----  -------------------------------
        1000   __start
        1028   _exit
        102A   _main
        1052   _find_gcd
        10AC   _swap
        10CC   __HC12Setup
        10D0   __text_end


Area                             Addr   Size   Decimal Bytes (Attributes)
--------------------------------  ----  ----  --------  -----  ---------
                          bss     0800  0006 =     6. bytes (rel,con)


        Addr   Global Symbol
        -----  -------------------------------
        0800   __bss_start
        0800   _gcd
```

```
0802   _n
0804   _m
0806   __bss_end

Files Linked          [ module(s) ]

C:\ICC\lib\crt12.o    [  crt12.s ]
bdm_eg1.o             [ bdm_eg1. ]
<library>             [  setup.s ]

User Global Definitions

init_sp = 0x8000

User Base Address Definitions

text = 0x1000                      ; starting address of program
data = 0x800                       ; starting address of data region
```

The ICC12 compiler adds an underscore character as a prefix to every variable declared in your C program. Therefore, _m, _n, and _gcd correspond to variables *m*, *n*, and *gcd*, respectively. From the map file, we find that the addresses of variables _m, _n, and _gcd are $804, $802, and $800, respectively. Add them into the watch window and run the program, and the result should be similar to what is shown in Figure 11.30. As you see in Figure 11.30, we have rearranged the order of variables *m*, *n*, and *gcd*. We also set the number base to decimal.

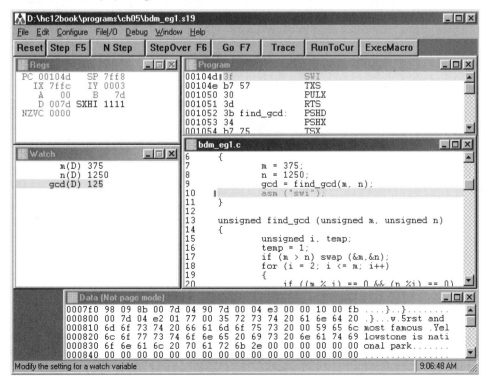

Figure 11.30 ■ Screen after running the program *bdm_egl.c*.

11.13 Summary

In the past, both monitor-assisted and in-circuit emulator approaches have been used in debugging application programs for microcontroller-based designs. Each approach has its own advantages and disadvantages.

The monitor-based approach has the advantage of being inexpensive and easy to use. However, it also has several disadvantages:

1. It cannot be used when a microcontroller has no demo board that contains a monitor.

2. The SWI instruction (for the 68HC11 and 68HC12) cannot be used to test other interrupt service routines because this instruction has been used to implement breakpoints. Some interrupt service routines require hardware circuitry to be built before they can be tested. The SWI instruction can be used to test other interrupt service routines without building the hardware circuitry for generating the interrupts.

The most important advantage for the in-circuit emulator is that it allows us to debug our software even before the hardware circuitry is available. The major disadvantage is its high cost.

The BDM mode of the 68HC12 provides a hardware- and software-integrated approach for debugging the software.

The BDM mode uses the BKGD pin to send commands to be executed by the BDM module. The same pin is also used to read/write the contents of CPU registers and memory locations. Data are transferred in and out serially via the BKGD pin. Each bit takes 16 E-clock cycles to shift in or out of the BDM module. Some of the BDM commands are hardware commands and are executed directly by the BDM hardware (BDM hardware is not in the CPU), so they don't slow down the normal instruction execution. When memory access is needed, the BDM module waits for an idle bus cycle. If no idle bus cycle can be found within 128 E-clock cycles, it then steals a cycle from the CPU (by freezing the CPU operation). Other commands are software commands and are implemented in firmware. The execution of software BDM commands requires the BDM mode to be enabled. The CPU then looks up the firmware ROM to execute the instruction sequence that implements the BDM software command.

The 68HC12 supports instruction tagging that enables hardware breakpoints to be implemented. Signal pins TAGHI and TAGLO are used for instruction tagging.

Motorola defines a 6-pin connector BDM port to promote the compatibility of BDM-based developing tools. Software to be run on a 68HC12 prototype board that provides this port will be able to be debugged using the BDM capability.

BDM allows the tool venders to develop low-cost source-level debuggers. This approach requires a BDM pod hardware to be built, which will be connected to the BDM connector and to either the serial or parallel port of the PC or workstation. A PC-hosted software must be written to send commands to the BDM module to download programs to the prototype board, set up breakpoints, step through the program, or run the program until a breakpoint is reached. The quality of the software is the key to the success of this approach. A tutorial on using the Axiom BDM-12 debugger is provided at the end of the chapter.

11.14 Lab Exercises & Assignments

L11.1 Use the example program in Example 4.12 to walk through the process of using the *hc12bgnd* debugger. Go through the following procedures:

Step 1
Start the *hc12bgnd* program by clicking on its icon.

Step 2

Perform the following configuration:

1. Set reset mode to **Expanded wide**.
2. Select **Free assembler** as the compiler.
3. Choose **C:\AX-BDM12\B32reset.mac** as the macro file to be executed after reset.
4. Click on **912B32** as the device select and choose **Hardware breakpoint**.
5. Set the target clock speed to **unspecified**.

Step 3

Load the *s19* file into the demo board by choosing **Load s19** or **HEX** in the File menu. If you get any warning about inconsistent file formats, ignore them.

Step 4

Set the PC value (in Regs window) to **$1000** (the starting point of the program).

Step 5

If variables *quo_hi*, *quo_lo*, *rem_hi*, and *rem_lo* are not in the Watch window yet, add them into the window. Add the following variables to the Watch window:

1. Add symbol quo_hi at address $800 and set length to 2.
2. Add symbol quo_lo at address $802 and set length to 2.
3. Add symbol rem_hi at address $804 and set length to 2.
4. Add symbol rem_lo at address $806 and set length to 2

The screen at this point should be similar to that shown in Figure L11.1.

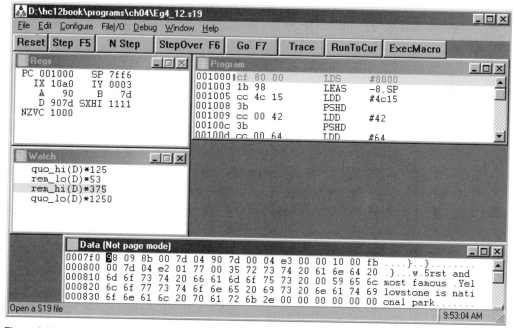

Figure L11.1 ■ Screen after loading the *eg4_12.s19* file.

Step 6

Set a breakpoint at the first SWI instruction after $1000 by clicking on that address.

Step 7

Execute the program until a breakpoint is reached. What are the values for variables in the watch window?

L11.2 Write an assembly program to find three prime numbers closest to but smaller than 5,000 and store them at memory locations $800, $802, and $804, labeling them as first, second, and third, respectively. Follow the procedure described in the previous exercise problem.

L11.3 Write a C program to generate four prime numbers closest to but larger than 9,000, and assign their values to variables *prime_1*, *prime _2*, *prime _3*, and *prime _4*. Run the program using the *hc12bgnd* debugger.

12

Controller Area
Network (CAN)

12.1 Objectives

After completing this chapter, you should be able to:

- describe the layers of the CAN protocol
- describe CAN's error detection capability
- describe the formats of CAN messages
- describe CAN message filtering
- explain CAN error handling
- describe CAN fault confinement
- explain CAN message bit timing
- explain CAN synchronization issue and methods
- describe the CAN message structures
- compute timing parameters to meet the requirements of your applications
- write routines to configure the MSCAN12 module
- write programs to transfer data over the CAN bus

12.2 Overview of Controller Area Network (CAN)

The *Controller Area Network* (CAN) is a serial communication protocol that supports distributed real-time control applications. Though conceived and defined by BOSCH in Germany for automotive applications, CAN is not restricted to that industry. The CAN protocol fulfills the communication needs of a wide range of applications, from high-speed networks to low-cost multiplex wiring. The description of the CAN protocol in this chapter is based on the CAN specification 2.0 published in September 1991 by BOSCH.

CAN is divided into *data link* and *physical* layers. Data link layers are further divided into LLC and MAC sublayers:

- The *LLC* sublayer deals with message filtering, overload notification, and recovery management.

- The *MAC* sublayer represents the kernel of the CAN protocol. It presents messages received to the LLC sublayer and accepts messages to be transmitted by the LLC sublayer. The MAC sublayer is responsible for message framing, arbitration, acknowledgement, error detection, and signaling. The MAC sublayer is supervised by a self-checking mechanism, called *fault confinement*, which distinguishes short disturbances from permanent failures.

The *physical layer* defines how signals are actually transmitted, dealing with the descriptions of bit timing, bit encoding, and synchronization. The physical layer is not defined in the specification, as it will vary according to the requirements of individual applications (for example, transmission medium and signal level implementation).

The layer structure of CAN is illustrated in Figure 12.1. The CAN protocol has gone through several revisions. The latest is revision 2.0. In this chapter, we will cover the latest specification 2.0 as well as the Motorola implementation of CAN on the 68HC12 microcontrollers.

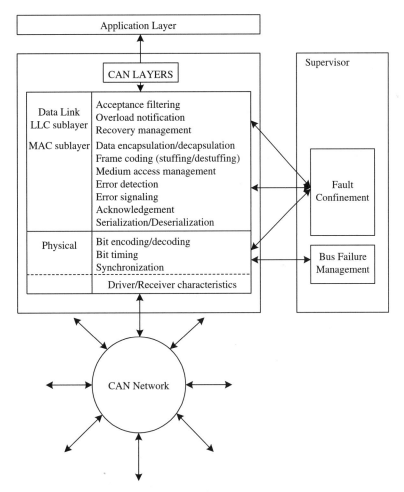

Figure 12.1 ■ CAN layers

Information on the CAN bus is sent in fixed formats of different but limited length. When the bus is free, any connected node may start to transmit a new message.

The CAN system does not specify node addresses in a message. Instead, it uses an identifier to describe the content of a message. The identifier does not indicate the destination of the message, but describes the meaning of the data, so that all nodes in the network are able to decide by message filtering whether the data is to be acted upon by them or not. In addition, the identifier defines a static message priority during bus access.

As a consequence of the concept of message filtering, any number of nodes may receive and act simultaneously upon the same message. Within a CAN network, it is guaranteed that a message is accepted simultaneously either by all nodes or by none of them. Thus data consistency is a property of the system achieved by the concepts of multicasting and error handling.

The speed of CAN may be different in different systems. However, in a given system the bit-rate is uniform and fixed.

A remote node can request another node to send data frames by sending a *remote frame*. The data frames sent by another node and the remote frame would have the same identifier.

When the CAN bus is free, any node may start to transmit a message. The node with the message of the highest priority to be transmitted gains the access. If two or more nodes start transmitting messages at the same time, the bus access conflict is resolved by bit-wise arbitration using the identifier. The mechanism of arbitration guarantees that neither information nor time is lost. If a data frame and a remote frame with the same identifier are initiated at the same time, the data frame prevails over the remote frame. During arbitration every transmitter compares the level of the bit transmitted with the level that is monitored on the bus. If these levels are equal the node may continue to send. When a *recessive level* is sent, but a *dominant level* is monitored, the node has lost arbitration and must withdraw without sending any further bits.

CAN implements the following error-detection measures in each node:

- Monitoring (each transmitter compares the bit levels detected on the bus with the bit levels being transmitted)
- Cyclic redundancy check (CRC)
- Bit-stuffing
- Message frame check

Each node can detect all global errors and all local errors at the transmitters. The implementation of CRC can detect:

- Up to five randomly transmitted errors within a sequence
- Burst errors of length less than 15 in a message
- Errors of any odd number of bits in a sequence

Corrupted messages are flagged by any node that detects an error. Such messages are aborted and are retransmitted automatically. The recovery time of detecting an error until the start of the next message is at most 29 bit times, provided there are no further errors.

CAN nodes can distinguish between short disturbances and permanent failures. Defective nodes are switched off.

A CAN bus consists of a single bidirectional channel that carries data bits. The CAN standard does not specify how the CAN bus should be implemented. Therefore, a CAN bus can be a single wire (plus ground), two differential wires, an optical fiber, and so on.

The CAN bus can have two complementary values: *dominant* or *recessive*. During simultaneous transmission of dominant and recessive bits, the resulting bus value will be dominant. For example, in the case of a wired-AND implementation of the bus, the dominant level would be represented by a logical zero and the recessive level by a logical one. Voltage values that represents the logical levels are not given in the CAN bus specification.

All receivers check the consistency of the message being received and acknowledge a consistent message or flag an inconsistent message automatically.

To reduce the system's power consumption, a CAN device may be set into *sleep* mode, in which there is no internal activity and the bus drivers are disconnected. The sleep mode is finished with a *wakeup* by any bus activity or by internal conditions of the system. Upon wakeup, the internal activity is restarted, although the transfer layer will wait for the system's oscillator to stabilize and then wait until it has synchronized itself to the bus activity (by checking for 11 consecutive recessive bits), before the bus drivers are set to the *on-bus* state again.

12.3 CAN Messages

Message transfer is manifested and controlled by four different frame types:

- *Data frame.* A data frame carries data from a transmitter to the receivers.
- *Remote frame.* A remote frame is used by any node to request the transmission of data frames with the same identifier.
- *Error frame.* An error frame is transmitted by any node that detects an error.
- *Overload frame.* An overload frame is used to provide for an extra delay between the preceding and the succeeding data or remote frames.

Data frames and remote frames are separated from preceding frames by an interframe space.

12.3.1 Data Frames

As shown in Figure 12.2, a data frame is composed of seven different fields: start-of-frame, arbitration, control, data, CRC, ACK, and end-of-frame.

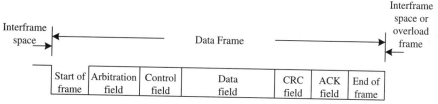

Figure 12.2 ■ CAN data frame

START-OF-FRAME

The *start-of-frame* field marks the beginning of data frames and remote frames. It consists of a single dominant bit. A node is only allowed to start transmission when the bus is idle. All nodes have to synchronize to the leading edge caused by the start-of-frame of the node starting transmission first.

ARBITRATION FIELD

The format of the *arbitration field* is different for the standard format and the extended format frames, as illustrated in Figure 12.3. The identifier's length is 11 bits for the standard format and 29 bits for the extended format.

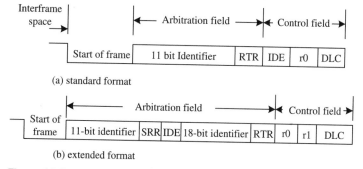

Figure 12.3 ■ Arbitration field

The identifier's length of the standard format is 11 bits and corresponds to the base ID in the extended format. These bits are transmitted most significant bit first. The most significant 7 bits cannot all be recessive.

The identifier's length of the extended format is 29 bits. The format comprises two sections: a base ID with 11 bits and an extended ID with 18 bits. Both the base ID and the extended ID are transmitted most significant bit first. The base ID defines the extended frame's base priority.

The *remote transmission request* (RTR) bit in data frames must be dominant. Within a remote frame, the RTR bit has to be recessive.

The *substitute remote request* (SRR) bit is a recessive bit. The SRR bit of an extended frame is transmitted at the position of the RTR bit in standard frames and therefore substitutes for the RTR bit in the standard frame. As a consequence, collisions between a standard frame and an extended frame, where the base ID of both frames are identical, are resolved in such a way that the standard frame prevails over the extended frame.

The *identifier extension* (IDE) bit belongs to the arbitration field for the extended format and the control field for the standard format. The IDE bit in the standard format is transmitted dominant, whereas in the extended format the IDE bit is recessive.

CONTROL FIELD

The contents of the control field are shown in Figure 12.4. The format of the control field is different for the standard format and the extended format. Frames in standard format include the data length code; the IDE bit, which is transmitted dominant; and the reserved bit $r0$. Frames in extended format include the data length code and two reserved bits, $r0$ and $r1$. The reserved bits must be sent dominant, but the receivers accept dominant and recessive bits in all combinations.

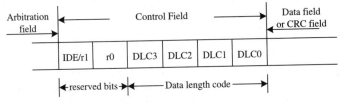

Figure 12.4 ■ Control field

The data length code specifies the number of bytes contained in the data field. Data length can be from zero to eight bytes, as encoded in Table 12.1.

Data length code				Data byte count
DLC3	DLC2	DLC1	DLC0	
d	d	d	d	0
d	d	d	r	1
d	d	r	d	2
d	d	r	r	3
d	r	d	d	4
d	r	d	r	5
d	r	r	d	6
d	r	r	r	7
r	d	d	d	8

d = dominant r = recessive

Table 12.1 ■ CAN data length coding

DATA FIELD

The data field consists of the data to be transmitted within a data frame. It may contain from zero to eight bytes, each of which contains eight bits and are transferred most significant bit first.

CRC FIELD

The CRC field contains the CRC sequence followed by a CRC delimiter, as shown in Figure 12.5.

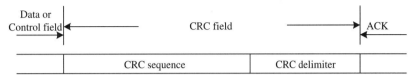

Figure 12.5 ■ CRC field

The frame-check sequence is derived from a cyclic redundancy code best suited to frames with bit counts less than 127.

The CRC sequence is calculated by performing a polynomial division. The coefficients of the polynomial are given by the de-stuffed bit-stream, consisting of the start of frame, arbitration field, control field, data field (if present), and 15 zeros. This polynomial is divided (the coefficients are calculated using modulo-2 arithmetic) by the generator polynomial:

$$X^{15} + X^{14} + X^{10} + X^8 + X^7 + X^4 + X^3 + 1$$

The remainder of this polynomial division is the CRC sequence.

In order to implement this function, a 15-bit shift register CRC_RG (14:0) is used. If $nxbit$ denotes the next bit of the bit-stream, given by the de-stuffed bit sequence from the start of frame until the end of the data field, the CRC sequence is calculated as follows:

```
CRC_RG = 0;                         /* initialize shift register */
do {
        crcnxt = nxtbit ^ CRC_RG(14);   /* exclusive OR */
        CRC_RG(14:1) = CRC_RG(13:0);    /* shift left by one bit */
        CRC_RG (0) = 0;
        If crcnxt
                CRC_RG(14:0) = CRC_RG(14:0) ^ 0x4599;
} while (!(CRC SEQUENCE starts or there is an error condition));
```

After the transmission/reception of the last bit of the data field, CRC_RG (14:0) contains the CRC sequence. The *CRC delimiter* is a single recessive bit.

As shown in Figure 12.6, the *ACK field* is two bits long and contains the ACK slot and the ACK delimiter. A transmitting node sends two recessive bits in the ACK field.

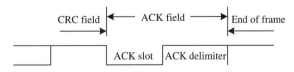

Figure 12.6 ■ ACK field

A receiver that has received a valid message reports this to the transmitter by sending a dominant bit in the ACK slot (i.e., it sends ACK).

A node that has received the matching CRC sequence overwrites the recessive bit in the ACK slot with a dominant bit. This bit will be received by the data frame transmitter and learned that the previously transmitted data frame has been correctly received.

The *ACK delimiter* has to be a recessive bit. As a consequence, the ACK slot is surrounded by two recessive bits (the CRC delimiter and the ACK delimiter).

Each data frame and remote frame is delimited by a flag sequence consisting of seven recessive bits. This seven-bit sequence is the *end-of-frame* sequence.

12.3.2 Remote Frame

A node acting as a receiver for certain data can request the relevant source node to transmit the data by sending a remote frame. The format of a remote frame is shown in Figure 12.7. A remote frame is composed of six fields: *start-of-frame, arbitration, control, CRC, ACK*, and *end-of-frame*.

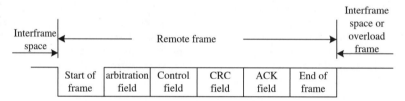

Figure 12.7 ■ Remote frame

The polarity of the RTR bit in the arbitration field indicates whether a transmitted frame is a *data frame* (RTR bit dominant) or a *remote frame* (RTR bit recessive).

12.3.3 Error Frame

The *error frame* consists of two distinct fields. The first field is given by the superposition of error flags contributed from different nodes. The second field is the error delimiter. The format of the error frame is shown in Figure 12.8. In order to terminate an error frame correctly, an *error-passive* node may need the bus to be idle for at least three bit times (if there is a local error at an error-passive receiver). Therefore the bus should not be loaded to 100%. An error-passive node has an error count greater than 127. An error-active node has an error count less than 127.

There are two forms of error flags: an *active-error* flag and a *passive-error* flag.

- An *active-error* flag consists of six consecutive dominant bits.
- A *passive-error* flag consists of six consecutive recessive bits unless it is overwritten by dominant bits from other nodes.

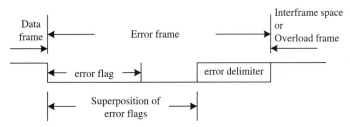

Figure 12.8 ■ Error frame

An *error-active* node signals an error condition by transmitting an *active-error* flag. The error flag's form violates the law of bit stuffing (to be discussed shortly) and applies to all fields from start-of-frame to CRC delimiter, or destroys the fixed form ACK field or end-of-frame field. As a consequence, all other nodes detect an error condition and each starts to transmit an error flag. Therefore the sequence of dominant bits, which actually can be monitored on the bus, results from a superposition of different error flags transmitted by individual nodes. The total length of this sequence varies between a minimum of 6 and a maximum of 12 bits.

An *error-passive* node signals an error condition by transmitting a passive-error flag. The error-passive node waits for six consecutive bits of equal polarity, beginning at the start of the passive-error flag. The passive-error flag is complete when these equal bits have been detected.

The *error delimiter* consists of eight recessive bits. After transmission of an error flag each node sends recessive bits and monitors the bus until it detects a recessive bit. Afterwards it starts transmitting seven more recessive bits.

12.3.4 Overload Frame

The *overload frame* contains two bit fields: *overload flag* and *overload delimiter*. There are three different overload conditions that lead to the transmission of an overload frame:

1. The internal conditions of a receiver require a delay of the next data frame or remote frame.

2. At least one node detects a dominant bit during intermission.

3. A CAN node samples a dominant bit at the eighth bit (i.e., the last bit) of an error delimiter or overload delimiter. The error counters will not be incremented.

The format of an overload frame is shown in Figure 12.9. An overload frame resulting from condition 1 is only allowed to start at the first bit time of an expected intermission, whereas an overload frame resulting from overload conditions 2 and 3 starts one bit after detecting the dominant bit.

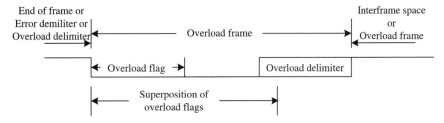

Figure 12.9 ■ Overload frame

No more than two overload frames may be generated to delay the next data frame or remote frame.

The *overload flag* consists of six dominant bits. The format of an overload frame is similar to that of the active error flag. The overload flag's form destroys the fixed form of the *intermission field*. As a consequence, all other nodes also detect an overload condition and each starts to transmit an overload flag. In the event that there is a dominant bit detected during the third bit of *intermission* locally at some node, then it will interpret this bit as the start-of-frame.

The *overload delimiter* consists of eight recessive bits. The overload delimiter is of the same form as the error delimiter. After the transmission of an overload flag, the node monitors the bus until it detects a transition from a dominant to a recessive bit. At this point of time every bus node has finished sending its overload flag and all nodes start transmission of seven more recessive bits in coincidence.

12.3.5 Interframe Space

Data frames and remote frames are separated from preceding frames by a field called *interframe space*. In contrast, overload frames and error frames are not preceded by an interframe space and multiple overload frames are not separated by an interframe space.

For nodes that are not error-passive or have been receivers of the previous message, the interframe space contains the bit fields of *intermission* and *bus idle*, as shown in Figure 12.10. The interframe space of an error-passive node consists of three sub-fields: *intermission*, *suspend transmission*, and *bus idle*, as shown in Figure 12.11.

Figure 12.10 ■ Interframe space for non error-passive nodes or receiver of previous message

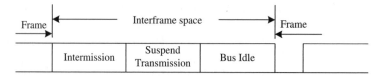

Figure 12.11 ■ Interframe space for error-passive nodes

The *Intermission* subfield consists of three recessive bits. During intermission no node is allowed to start transmission of the data frame or remote frame. The only action permitted is signaling of an overload condition.

The period of bus idle may be of arbitrary length. The bus is recognized to be free, and any node having something to transmit can access the bus. A message, pending during the transmission of another message, is started in the first bit following intermission. When the bus is idle, the detection of a dominant bit on the bus is interpreted as a *start-of-frame*.

After an error-passive node has transmitted a frame, it sends eight recessive bits following intermission, before starting to transmit a further message or recognizing the bus as idle. If, meanwhile, a transmission (caused by another node) starts, the node will become the receiver of this message.

A complete list of CAN frames in standard format is shown in Figure 12.12, and all CAN frames in extended format are given in Figure 12.13.

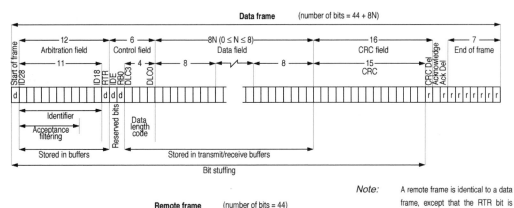

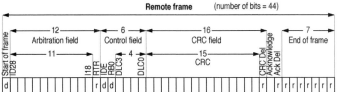

Figure 12.12a ■ CAN frame format–standard format

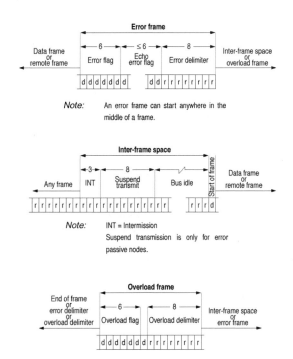

Figure 12.12b ■ Can frame format–standard format

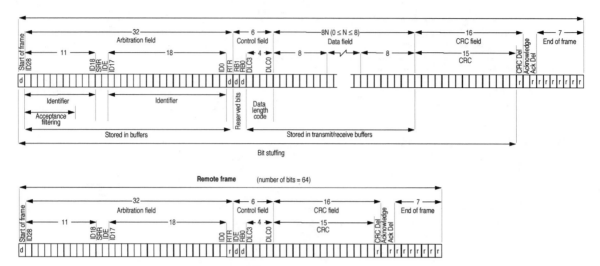

Figure 12.13 ■ CAN extended frame format

12.3.6 Message Filtering

The purpose of message filtering is to control which nodes can receive the message. Message filtering is applied to the whole identifier. Optional mask registers that allow any identifier bit to be set don't care about message filtering, and may be used to select groups of identifiers to be mapped into the attached receive buffers.

If mask registers are implemented, every bit of the mask registers must be programmable; in other words, they can be enabled or disabled for message filtering. The length of the mask register can comprise the whole identifier or only part of it.

12.3.7 Message Validation

The point in time at which a message is taken to be valid is different for the transmitters and receivers of the message.

The message is valid for the transmitter if there is no error until the end of frame. If a message is corrupted, retransmission will follow automatically and according to the rules of prioritization. In order to be able to compete for bus access with other messages, retransmission has to start as soon as the bus is idle.

The message is valid for the receiver if there is no error until the last but one bit of the end of frame.

12.3.8 Bit-Stream Coding

The frame segments, including start-of-frame, arbitration field, control field, data field, and CRC sequence, are coded by bit stuffing. Whenever a transmitter detects five consecutive bits of identical value in the bit-stream to be transmitted, it automatically inserts a complementary bit in the actual transmitted bit-stream.

The remaining bit fields of the data frame or remote frame (CRC delimiter, ACK field, and end of frame) are of fixed form and are not stuffed.

The error frame and overload frame are also of fixed form and are not coded by the method of bit stuffing.

The bit-stream in a message is coded using the *non-return-to-zero* (NRZ) method. This means that during the total bit time the generated bit level is either dominant or recessive.

12.4 Error Handling

There are five types of errors. These errors are not mutually exclusive.

12.4.1 Bit Error

A node that is sending a bit on the bus also monitors the bus. When the bit value monitored is different from the bit value being sent, the node interprets the situation as an error. There are two exceptions to this rule:

- A node that sends a recessive bit during the stuffed bit-stream of the arbitration field or during the ACK slot detects a dominant bit.
- A transmitter that sends a passive-error flag detects a dominant bit.

12.4.2 Stuff Error

A stuff error is detected whenever six consecutive dominant or six consecutive recessive levels occurs in a message field.

12.4.3 CRC Error

The CRC sequence consists of the result of the CRC calculation by the transmitter. The receiver calculates the CRC in the same way as the transmitter. A CRC error is detected if the calculated result is not the same as that received in the CRC sequence.

12.4.4 Form Error

A form error is detected when a fixed-form bit field contains one or more illegal bits. For a receiver, a dominant bit during the last bit of the end-of-frame is not treated as a form error.

12.4.5 Acknowledgement Error

An acknowledgement error is detected whenever the transmitter does not monitor a dominant bit in the ACK slot.

12.4.6 Error Signaling

A node that detects an error condition signals the error by transmitting an error flag. An error-active node will transmit an *active-error* flag; an error-passive node will transmit a *passive-error* flag.

Whenever a node detects a bit error, a stuff error, a form error, or an acknowledgement error, it will start transmission of an error flag at the next bit time.

Whenever a CRC error is detected, transmission of an error flag will start at the bit following the ACK delimiter, unless an error flag for another error condition has already been started.

12.5 Fault Confinement

12.5.1 CAN Node Status

A node in error may be in one of three states: error-active, error-passive, or bus-off.

An error-active node can normally take part in bus communication and sends an active-error flag when an error has been detected.

An error-passive node must not send an active-error flag. It takes part in bus communication, but when an error has been detected only a passive-error flag is sent. After a transmission, an error-passive node will wait before initiating further transmission.

A bus-off node is not allowed to have any influence on the bus.

12.5.2 Error Counts

Two counts are implemented in every bus node to facilitate fault confinement: the transmit error count and the receive error count

These two counts are updated according to the following 12 rules:

1. When a receiver detects an error, the receive error count will be increased by one, except when the detected error was a bit error during the sending of an active-error flag or an overload flag.

2. When a receiver detects a dominant bit as the first bit after sending an error flag, the receive error count will be increased by eight.

3. When a transmitter sends an error flag, the transmit error count is increased by eight. There are two exceptions to this rule: The first exception is when the transmitter is error-passive and the transmitter detects an acknowledgement error because of not detecting a dominant ACK and the transmitter does not detect a dominant bit while sending its passive error flag. The second exception is when the transmitter sends an error flag because a stuff error occurred during arbitration and the stuff bit should have been recessive and the stuff bit has been sent as recessive but is monitored as dominant. The transmit error count is not changed under these two situations.

4. An error-active transmitter detects a bit error while sending an active-error flag or an overload flag, and the transmit error count is increased by eight.

5. An error-active receiver detects a bit error while sending an active-error flag or an overload flag, the receive error count is increased by eight.

6. Any node tolerates up to seven consecutive dominant bits after sending an active error flag or a passive error flag. After detecting the eighth consecutive dominant bit and after each sequence of additional eighth consecutive dominant bits, every transmitter increases its transmit error count by eight and every receiver increases its receive error count by eight.

7. After the successful transmission of a message (getting ACK and no error until the end of frame is finished), the transmit error count is decreased by one, unless it was already zero.

8. After the successful reception of a message (reception without error up to the ACK slot and the successful sending of the ACK bit), the receive error count is decreased by 1, if it was between 1 and 127. If the receive error count was 0, it stays 0, and if it was greater than 127, then it will be set to a value between 119 and 127.

9. A node is *error passive* when the transmit error count equals or exceeds 128, or when the receive error count equals or exceeds 128. An error condition letting a node become error-passive causes the node to send an *active error flag*.

10. A node is *bus-off* when the transmit error count is greater than or equal to 256.

11. An error-passive node becomes *error-active* again when both the transmit error count and receive error count are less than or equal to 127.

12. A node that is bus-off is permitted to become error-active (no longer bus-off) with its error counters both set to 0 after 128 occurrences of 11 consecutive recessive bits have been monitored on the bus.

An error count value greater than roughly 96 indicates a heavily disturbed bus. It may be advantageous to provide the means to test for this condition. If during system start-up only one node is on-line, and if this node transmits some message, it will get no acknowledgement, detect an error, and repeat the message. It can become error-passive but not bus-off for this reason.

12.6 CAN Message Bit Timing

In a CAN environment, the *nominal bit rate* is defined as the number of bits transmitted per second in the absence of resynchronization by an ideal transmitter.

12.6.1 Nominal Bit Time

The inverse of the nominal bit rate is the *nominal bit time*. A nominal bit time can be divided into four non-overlapping time segments, as shown in Figure 12.14.

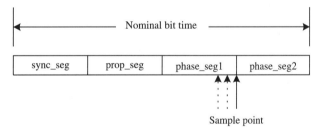

Figure 12.14 ■ Nominal bit time

The *sync_seg* segment is used to synchronize the various nodes on the bus. An edge is expected to lie within this segment.

The *prop_seg* segment is used to compensate for the physical delay times within the network. It is twice the sum of the signal's propagation time on the bus line, the input comparator delay, and the output driver delay.

The *phase_seg1* and *phase_seg2* segments are used to compensate for edge phase errors. These segments can be lengthened or shortened by synchronization.

The *sample point* is the point in time at which the bus level is read and interpreted as the value of that respective bit. The sample point is located at the end of *phase_seg1*. A CAN controller may implement the three samples per bit option in which the majority function is used to determine the bit value. Each sample is separated from the next sample by one time quanta (CAN clock cycle).

The *information processing time* is the time segment starting with the sample point reserved for calculation of the subsequent bit level.

The segments contained in a nominal bit time are represented in units of the *time quantum*. The time quantum is a fixed unit of time that can be derived from the oscillator period. It is expressed as a multiple of a *minimum time quantum*. This multiple is a programmable prescale factor. Thus, the time quantum can have the length of:

time quantum = m × minimum time quantum

where *m* is the value of the prescaler.

12.6.2 Length-of-Time Segments

The segments of a nominal bit time can be expressed in the units of time quantum as follows:

- *sync_seg* is one time quantum long
- *prop_seg* is programmable to be one, two, ..., eight time quanta long
- *phase_seg1* is programmable to be one, two, ..., eight time quanta long
- *phase_seg2* is the maximum of *phase_seg1* and *information processing time*
- The *information processing time* is less than or equal to two time quanta long.

The total number of time quanta in a bit time must be programmable over a range of at least 8 to 25. If three samples are taken, then *phase_seg1* must be at least two time quanta long.

12.7 Synchronization Issues

All CAN nodes must be synchronized while receiving a transmission; in other words, the beginning of each received bit must occur during each node's *sync_seg* segment. This is achieved by synchronization. Synchronization is required due to phase errors between nodes, which may arise due to nodes having slightly different oscillator frequencies, or due to changes in propagation delay when a different node starts transmitting.

Two types of synchronization are defined: *hard synchronization* and *resynchronization*. Hard synchronization is performed only at the beginning of a message frame, when each CAN node aligns the *sync_seg* of its current bit time to the recessive to dominant edge of the transmitted *start-of-frame*. After a hard synchronization, the internal bit time is restarted with *sync_seg*. Resynchronization is subsequently performed during the remainder of the message frame, whenever a change of bit value from recessive to dominant occurs outside of the expected *sync_seg* segment.

12.7.1 Resynchronization Jump Width

There are three possibilities of the occurrence of the incoming recessive to dominant edge:

1. After the *sync_seg* segment but before the *sample point*. This situation is interpreted as a *late edge*. The node will attempt to resynchronize to the bit stream by increasing the duration of its *phase_seg1* segment of the current bit by the number of time quanta by which the edge was late, up to the *resynchronization jump width* limit.

2. After the sample point but before the *sync_seg* segment of the next bit. This situation is interpreted as an *early bit*. The node will attempt to resynchronize to the bit stream by decreasing the duration of its *phase_seg2* segment of the current bit by the

number of time quanta by which the edge was early, up to the *resynchronization jump width* limit. The *sync_seg* segment of the next bit begins immediately.

3. Within the *sync_seg* segment of the current bit time. This is interpreted as no synchronization error.

As a result of resynchronization, *phase_seg1* may be lengthened or *phase_seg2* may be shortened. The amount by which the phase buffer segments may be altered may not be greater than the *resynchronization jump width*. The resynchronization jump width is programmable to be between one and the smaller of four or the *phase_seg1* time quanta.

Clocking information may be derived from transitions from one bit value to the other. The property that only a fixed maximum number of successive bits have the same value provides the possibility of resynchronizing a bus node to the bit-stream during a frame.

The maximum length between two transitions that can be used for resynchronization is 29 bit times.

12.7.2 Phase Error of an Edge

The *phase error* of an edge is given by the position of the edge relative to *sync_seg*, measured in time quanta. The sign of a phase error is defined as follows:

$e < 0$	if the edge lies after the sample point of the previous bit
$e = 0$	if the edge lies within *sync_seg*
$e > 0$	if the edge lies before the sample point

12.7.3 Synchronization Rules

Hard synchronization and resynchronization are two forms of synchronization. They obey the following rules:

- Only one synchronization within one bit time is allowed.

- An edge will be used for synchronization only if the value detected at the previous sample point (previous read bus value) differs from the bus value immediately after the edge.

- Hard synchronization is performed whenever there is a recessive to dominant edge during bus idle.

- All other recessive to dominant edges (and optionally dominant to recessive edges in the case of low hit rates) fulfilling rules 1 and 2 will be used for resynchronization, with the exception that a transmitter will not perform a resynchronization as a result of a recessive to dominant edge with a positive phase error, if only recessive to dominant edges are used for resynchronization.

12.8 Overview of the MSCAN12 Module

The MSCAN12 is the specific implementation of the CAN 2.0A/B specification for the 68HC12 microcontroller family. The 2.0A specification supports the standard identifier format whereas the 2.0B supports the extended identifier format.

The 912BC32 and the 912D60 have one CAN module. The 912DG128 has two identical CAN modules: CAN0 and CAN1. The 912DT128 has three identical CAN modules: CAN0, CAN1, and CAN2.

The MSCAN12 supports both the standard and extended data frames. Each CAN module uses two external pins, one input (RxCAN) and one output (TxCAN). The TxCAN output pin represents the logic level on the CAN: "0" for a *dominant* state, and "1" for a *recessive* state.

A typical CAN system with the MSCAN12 is shown in Figure 12.15. Each CAN station is connected to the CAN bus through a transceiver chip such as the Phillips PCA82C250. The transceiver is capable of driving the large current needed for the CAN bus, and has current protection against defected CAN or defected stations.

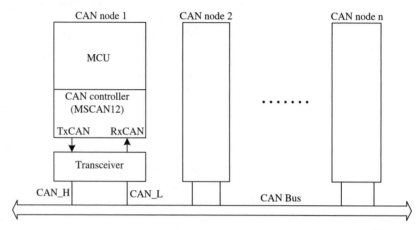

Figure 12.15 ■ A typical CAN system

12.9 MSCAN12 Message Storage

The CAN protocol requires that any CAN node be able to send out a stream of scheduled messages without releasing the bus between two messages. To achieve this goal, the MSCAN12 module implements a triple-buffer transmitter scheme. A two-stage input FIFO is implemented for received messages.

12.9.1 Receive Structures

Received messages are stored in a two-stage input FIFO: the *background receive buffer* (RxBG) and the *foreground receive buffer* (RxFG). Only the foreground receive buffer is addressable by the CPU.

Both buffers have 13 bytes to store the CAN control bits, the identifier (standard or extended) and the data contents. The *receiver full flag* (RXF) in the *CAN receiver flag register* (CRFLG) signals the status of the foreground receive buffer. When the buffer contains a correctly received message with a matching identifier, this flag is set to one.

The MSCAN12 module uses the background receive buffer (RxBG) to receive messages from the CAN bus. If the message passes the filter, the CAN module copies it to the foreground receive buffer, sets the RXF flag, and may optionally generate an interrupt to the CPU.

The user's receive handler must clear the RXF flag and read the received message from RxFG. An *overrun* condition occurs when both the RxFG and RxBG are filled with correctly received messages with accepted identifiers and a new message is being received from the CAN bus. The new message will be discarded and an overrun error interrupt will be requested if it is enabled. While in the overrun situation, the MSCAN12 module still stay synchronized to the CAN bus and is able to transmit messages, but will discard all incoming messages until the error is cleared.

The organization of a receive buffer is shown in Figure 12.16.

address	Register Name
$xx40	Identifier register 0 (IDR0)
$xx41	Identifier register 1 (IDR1)
$xx42	Identifier register 2 (IDR2)
$xx43	Identifier register 3 (IDR3)
$xx44	Data segment register 0 (DSR0)
$xx45	Data segment register 1 (DSR1)
$xx46	Data segment register 2 (DSR2)
$xx47	Data segment register 3 (DSR3)
$xx48	Data segment register 4 (DSR4)
$xx49	Data segment register 5 (DSR5)
$xx4A	Data segment register 6 (DSR6)
$xx4B	Data segment register 7 (DSR7)
$xx4C	Data length register (RxDLR)

Figure 12.16 ■ Receive buffer structure

12.9.2 Transmit Structures

The MSCAN12 has a triple-buffer transmitter that allows up to three messages to be set up in advance to achieve a real-time performance.

All three buffers have a 13-byte data structure similar to the receive buffer. An additional *transmit buffer priority register* (TBPR) associated with each transmit buffer defines the priority of each buffer. The details of all registers will be explained in a later section.

Before transmitting a message, the CPU identifies an available transmit buffer with a set *transmit buffer empty* (TXE) flag in the *transmit flag register* (CTFLG). The CPU then stores the identifier, the control bits, and the data contents into the identified transmit buffer. Finally, the CPU clears the TXE flag to indicate that the buffer is ready for transmission.

The CAN module will then schedule the message for transmission and signal the successful transmission of the buffer by setting the TXE flag. A transmit interrupt may be requested if it is enabled.

In case more than one buffer is scheduled for transmission when the CAN bus becomes available for arbitration, the CAN module uses the local priority setting of the three buffers for prioritization. The lowest value in the priority register is defined to be the highest priority.

When all transmit buffers have been scheduled for transmission and the application software needs to schedule another high-priority message, then it becomes necessary to abort a lower priority message being set up in one of the three transmit buffers. The user requests to abort a scheduled message by setting the corresponding *abort request flag* (ABTRQ) in the *transmission control register* (CTCR). If possible, the CAN module will grant the request by setting the corresponding *abort request acknowledgement* (ABTAK) and the TXE flag in order to release the buffer, and by requesting a transmit interrupt. The transmit interrupt handler can tell if the interrupt is caused by an abort request by checking the ABTAK flag (when it is equal to one).

The organization of a transmit-buffer structure is shown in Figure 12.17.

address	Register Name
$xxb0	Identifier register 0 (IDR0)
$xxb1	Identifier register 1 (IDR1)
$xxb2	Identifier register 2 (IDR2)
$xxb3	Identifier register 3 (IDR3)
$xxb4	Data segment register 0 (DSR0)
$xxb5	Data segment register 1 (DSR1)
$xxb6	Data segment register 2 (DSR2)
$xxb7	Data segment register 3 (DSR3)
$xxb8	Data segment register 4 (DSR4)
$xxb9	Data segment register 5 (DSR5)
$xxbA	Data segment register 6 (DSR6)
$xxbB	Data segment register 7 (DSR7)
$xxbC	Data length register (TxDLR)
$xxbD	Transmit buffer priority register (TBPR)

Note. b is 5, 6, 7 for transmit buffer 0, 1, and 2

Figure 12.17 ■ Transmit buffer structure

12.10 Identifier Acceptance Filter

According to the CAN specification, CAN messages do not contain the destination address. A node decides whether to take action on the incoming message by applying a filtering pattern on the arbitration field of the incoming message.

The CAN filter can be programmed to operate in four different modes:

- *Two identifier acceptance filters.* Each filter is to be applied to the full 29 bits of the identifier and to the RTR, the IDE, and the SRR bits of the CAN frame. This mode implements two filters for a full length extended identifier. Figure 12.18 shows the first 32-bit filter bank (CIDAR0-3, CIDMR0-3) that produces a filter 0 hit. Similarly, the second filter bank (CIDAR4-7, CIDMR4-7) produces a filter 1 hit.

- *Four identifier acceptance filters.* Each filter is to be applied to the 11 bits of the identifier and the RTR bit of the standard identifier or the 14 most significant bits of the extended identifier. Figure 12.19 illustrates how the first 32-bit filter bank (CIDAR0-3, CIDMR0-3) produces filter 0 and 1 hits. Similarly, the second filter bank (CIDAR4-7, CIDMR4-7) produces filter 2 and 3 hits.

- *Eight identifier acceptance filters.* Each filter is to be applied to the first eight bits of the identifier. This mode implements eight independent filters for the first eight bits of a standard identifier. Figure 12.20 shows how the first 32-bit filter bank (CIDAR0-3, CIDMR0-3) produces filter 0 to 3 hits. Similarly, the second filter bank (CIDAR4-7, CIDMR4-7) produces filter 4 to 7 hits.

- *Closed filter.* No CAN message will be copied into the foreground buffer RxFG, and the RXF flag will never be set.

The arbitration field of an incoming message will be loaded into the identifier registers (IDR0-IDR3) in the receive-buffer structure and compared with the *identifier acceptance registers* in Figures 12.18 through 12.20. The identifier acceptance registers (CIDAR0-7) define the acceptable patterns of the standard or the extended identifier (ID10–ID0 or ID28–ID0). Any of these bits can be marked 'don't care' in the *identifier mask registers* (CIDMR0-7). The CAN hardware compares each bit of the identifier registers with the corresponding bit of the acceptance registers excluding those bits that are masked as don't care by identifier mask registers.

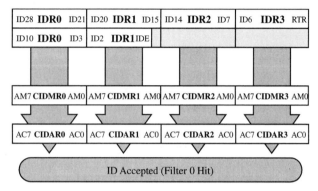

Figure 12.18 ■ 32-bit maskable identifier acceptance filter

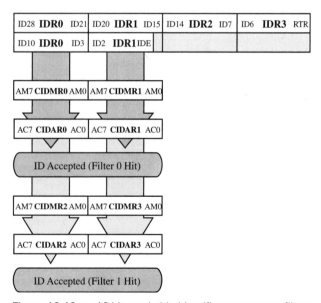

Figure 12.19 ■ 16-bit maskable identifier acceptance filters

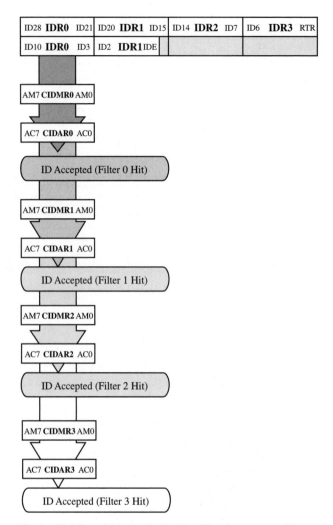

Figure 12.20 ■ 8-bit maskable identifier acceptance filters

The contents of the identifier registers are compared with two, four, or eight acceptance filters simultaneously depending upon which mode is selected. A filter hit is indicated to the application software by a set RXF flag and three bits in the identifier acceptance control register. These identifier hit flags (IDHIT2-0) identify the cause of the receive interrupt. In case more than one hit occurs (two or more filters match), the lower hit number has the priority. A hit will also cause a receive interrupt if it is enabled.

12.11 Interrupts

The MSCAN12 supports four interrupt vectors:

- *Transmit Interrupt.* When one or more of the three transmit buffers is empty, the TXE flag is set.
- *Receive Interrupt.* When a message is successfully received and loaded into the foreground receive buffer, the RXF flag will be set and the receive interrupt will be requested immediately if it has been enabled.
- *Wakeup Interrupt.* A wakeup interrupt will be requested when it is enabled and when an activity on the CAN bus occurred during the MSCAN12 internal sleep mode.
- *Error Interrupt.* An overrun, error, or warning condition occurred. The receive flag register will indicate the following conditions:

1. *Overrun.* This may occur when the CPU is too busy to read the message stored in the receive foreground buffer.
2. *Receiver warning.* The receive error counter has reached the CPU warning limit of 96.
3. *Transmitter warning.* The transmit error counter has reached the CPU warning limit of 96.
4. *Receiver error passive.* The receive error counter has exceeded the error passive limit of 127 and the MSCAN12 has gone to an error-passive state.
5. *Transmitter error passive.* The transmit error counter has exceeded the error passive limit of 127 and the MSCAN12 has gone to an error-passive state.
6. *Bus off.* The transmit error counter has exceeded 255 and MSCAN12 has gone to the bus-off state.

CAN interrupts are directly associated with one or more status flags in either the CAN *receiver flag register* (CRFLG) or the CAN *transmitter control register* (CTCR). These flags can be cleared by writing a "1" to them.

The sources and masks of all CAN interrupts are summarized in Table 12.2.

Function	Source	Local mask	Global mask
Wake-up	WUPIF	WUPIE	
Error	RWRNIF	RWRNIE	
	TWRNIF	TWRNIE	
	RERRIF	RERRIE	
	TERRIF	TERRIE	I bit
	BOFFIF	BOFFIE	
	OVRIF	OVRIE	
Receive	RXF	RXFIE	
Transmit	TXE0	TXEIR0	
	TXE1	TXEIE1	
	TXE2	TXEIE2	

Table 12.2 ■ MSCAN12 interrupt vectors, causes, and masks

12.12. CAN Low-Power Modes

The MSCAN12 has three low-power modes: *sleep*, *soft reset*, and *power down*. In sleep and soft reset modes, power consumption is reduced by stopping all clocks except those needed to access the registers. In power down mode, all clocks are stopped and no power is consumed.

The WAI and STOP instructions put the CPU in low power consumption stand-by mode. The CAN low-power modes are entered by setting the appropriate bits of the CMCR0 register to one. Table 12.3 summarizes the combinations of the MSCAN12 and CPU modes.

MSCAN mode	CPU mode		
	STOP	**WAIT**	**RUN**
Power down	CSWAI = X [1] SLPAK = X SFTRES = X	CSWAI = 1 SLPAK = X SFTRES = X	
Sleep		CSWAI = 0 SLPAK = 1 SFTRES = 0	CSWAI = X SLPAK = 1 SFTRES = 0
Soft reset		CSWAI = 0 SLPAK = 0 SFTRES = 1	CSWAI = X SLPAK = 0 SFTRES = 1
Normal		CSWAI = 0 SLPAK = 0 SFTRES = 0	CSWAI = X SLPAK = 0 SFTRES = 0

(1) 'X' means don't care.

Table 12.3 ■ MSCAN12 vs. CPU operation modes

12.12.1 MSCAN Sleep Mode

The CPU can request the MSCAN12 to enter the sleep mode by setting the SLPRQ bit of the module control register 0 (CMCR0). The time when the MSCAN12 will then enter sleep mode depends on its current activity:

- If it is transmitting, it will continue to transmit until there are no more messages to be transmitted, and then enter sleep mode.
- If it is receiving, it will wait for the end of this message and then enter sleep mode.
- If it is neither transmitting nor receiving, it will immediately enter sleep mode.

Your application software should avoid setting up a transmission (by clearing one or more TXE flags) and then immediately requesting sleep mode (by setting the SLPRQ bit).

During sleep mode, the SLPAK flag is set and the MSCAN12 stops its own clocks and the TxCAN pin will stay in a recessive state. If the RXF flag equals one, the message can be read and the RXF flag can be cleared. However, if the MSCAN12 is in the bus-off state, it stops counting 128×11 consecutive recessive bits due to the stopped clocks. Likewise, copying of RxBG into RxFG will not take place while in sleep mode. It is possible to access the transmit buffers and to clear the TXE flags. No message abort will take place in sleep mode.

The MSCAN12 will leave the sleep mode when either:

- Bus activity occurs
- The CPU clears the SLPRQ bit
- The CPU sets SFTRES

After waking up, the MSCAN12 waits for 11 consecutive recessive bits to synchronize to the bus. As a consequence, if the MSCAN12 is woken up by a CAN frame, this frame will not be received. The receive message buffers (RxBG and RxFG) will contain messages if they were received before the sleep mode was entered. Pending copying of RxBG into RxFG, as well as pending message aborts and pending message transmissions, will now be executed. If the MSCAN12 is still in a bus-off state after sleep mode was left, it continues counting the 128×11 consecutive recessive bits.

12.12.2 MSCAN Soft-Reset Mode

It is likely that a user may accidentally violate the CAN protocol through programming errors. The most likely errors are changing the CAN configuration when the CAN module is in normal operation. The MSCAN12 is designed to prevent the user from changing the contents of the CAN control registers when the CAN module is in normal operation. In order to change any CAN control register, the user must put the CAN module in *soft-reset mode*.

In soft reset mode, the MSCAN12 is stopped. Registers can still be accessed. This mode is used to initialize the module configuration, bit timing, and the CAN message filter.

When setting the SFTRES bit, the MSCAN12 immediately stops all ongoing transmissions and receptions, potentially causing CAN protocol violations. The user is responsible for ensuring that the MSCAN12 is not active when soft reset mode is entered. The recommended procedure is to bring the MSCAN12 into sleep mode before setting the SFTRES bit.

12.12.3 MSCAN Power-Down Mode

The MSCAN12 is in power-down mode when the CPU is in stop mode *or* the CPU is in wait mode and the CSWAI bit is set.

When entering the power-down mode, the MSCAN12 module immediately stops all ongoing transmissions and receptions, potentially causing CAN protocol violations. The user should make sure that the MSCAN12 is not active when the power-down mode is entered. The recommended procedure is to put the CAN module into sleep mode before executing the STOP or the WAI instruction (if CSWAI bit is set).

In power-down mode, the TxCAN pin will be driven into a recessive state and no registers can be accessed.

12.12.4 Wakeup Function

The MSCAN12 can be programmed to apply a low-pass filter function to the RxCAN input line while in sleep mode. This feature can be used to protect the MSCAN12 from waking up due to short glitches on the CAN bus lines. Such glitches can result from electromagnetic interference within noisy environments.

12.13 Timer Link

The MSCAN12 will generate a timer signal whenever a valid frame has been received. Because the CAN specification defines a frame to be valid if no errors occurred before the EOF field has been transmitted successfully, the timer signal will be generated right after the EOF. A pulse of one bit time is generated. As the MSCAN12 receiver also receives the frames being transmitted by itself, a timer signal is also generated after a successful transmission. The generated timer signal can be routed into the on-chip *timer interface module* (TIM/ECT). This signal is connected to the Timer *n* Channel *m* input under the control of the *timer link enable* (TLNKEN) bit in the CMCR0 register. After Timer *n* has been programmed to capture the rising edge events, it can be used under software control to generate 16-bit time stamps that can be stored with the received message.

12.14 Clock System

As shown in Figure 12.21, the MSCAN12 can select the output of the crystal oscillator (EXTALi) or a clock twice as fast as the system clock (ECLK) as its clock source. The clock source has to be chosen such that the tight oscillator tolerance requirements (up to 0.4%) of the CAN protocol are met. For high bus rates (1 Mbps), a 50% duty cycle of the clock is required.

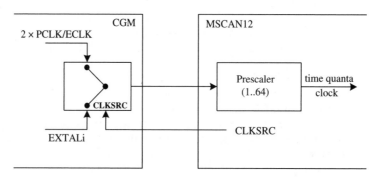

Figure 12.21 ■ Clocking scheme

A programmable prescaler is used to generate the time quanta (T_q) clock. A time quantum is the atomic unit of time handled by the CAN module. A bit time is divided into three segments:

- *Sync_seg.* This segment has a fixed length of one time quantum. Signal edges are expected to happen within this section.
- *Time segment 1.* This segment includes the *prop_seg* and the *phase_seg1* of the CAN standard. It can be programmed by setting the parameter TSEG1 to consist of 4 to 16 time quanta.
- *Time segment 2.* This segment represents the *phase_seg2* of the CAN standard. It can be programmed by setting the TSEG2 parameter to be two to eight time quanta long.

When resynchronization is needed, the CAN module uses the *synchronization jump width* (in the range of one to four time quanta) to readjust the timing of the signal. It is the user's responsibility to make sure that their bit time settings are in compliance with the CAN standard. A summary of the CAN conforming segment settings and related parameter values are displayed in Table 12.4.

Time segment 1	TSEG1	Time segment 2	TSEG2	Synchron. jump width	SJW
5..10	4..9	2	1	1..2	0..1
4..11	3..10	3	2	1..3	0..2
5..12	4..11	4	3	1..4	0..3
6..13	5..12	5	4	1..4	0..3
7..14	6..13	6	5	1..4	0..3
8..15	7..14	7	6	1..4	0..3
9..16	8..15	8	7	1..4	0..3

Table 12.4 ■ CAN standard compliant bit time segment settings

12.15 Registers Associated with MSCAN12

The registers associated with the MSCAN12 module occupy 128 bytes in memory space. The memory map of the MSCAN12 module is shown in Figure 12.22.

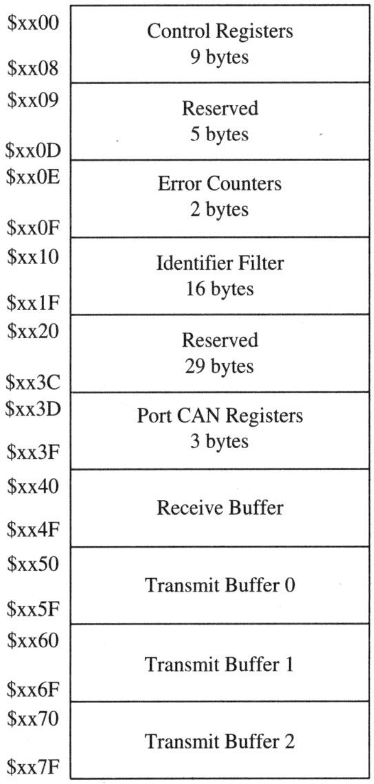

Figure 12.22 ■ MSCAN memory map

Most CAN-related register names start with a letter C. For 68HC12 members that have more than one CAN module, a number (0, 1, or 2) is added after C to indicate which CAN module the register belongs to. For example, the name of the CAN *module control register 0* is:

- CMCR0 in the 912BC32 and 912D60
- C0MCR0 or C1MCR0 in the 912DG128
- C0MCR0, C1MCR1, or C2MCR2 in the 912DT128

By default, each CAN register of the 912BC32/D60 and its corresponding CAN 0 register of the 912DG128/DT128 are mapped to the same address. Unless you remap the register block, you can refer to a CAN 0 register by dropping the character "0" (for example, CMCR0 and C0CMR0 are interchangeable). In the rest of this chapter, we will use CxMCR0 to refer to the CAN module control register 0 where x can be null, zero, one, or two. We will refer to all other CAN registers in the same manner.

The registers associated with the CAN module of the 912BC32 and the 912D60 are mapped to $0100 to $017F. The registers contained in CAN module 0 of the 912DG128 and the 912DT128 are also mapped to the same area.

The registers contained in CAN module 1 of the 912DG128 and the 912DT128 are mapped to the memory space from $0300 to $037F, whereas the registers contained in CAN module 2 of the 912DT128 are mapped to the space from $0200 to $027F.

12.15.1 Message Buffer Registers

The receive buffer and transmit buffer registers are listed in Figures 12.16 and 12.17. When the extended identifier is used in the message, all four identifier registers are used. When the standard identifier is used, only IDR0 and IDR1 have meanings. The contents of these four identifier registers are shown in Figures 12.23 and 12.24, respectively.

bit 7	bit 6	bit 5	bit 4	bit 3	bit 2	bit 1	bit 0	Register Name
ID28	ID27	ID26	ID25	ID24	ID23	ID22	ID21	IDR0
ID20	ID19	ID18	SRR (1)	IDE (1)	ID17	ID16	ID15	IDR1
ID14	ID13	ID12	ID11	ID10	ID9	ID8	ID7	IDR2
ID6	ID5	ID4	ID3	ID2	ID1	ID0	RTR	IDR3

Note. SRR value = 1
IDE value = 1
The contents of IDR0--IDR3 are undefined output of reset

Figure 12.23 ■ Receive/transmit message buffer extended identifier

bit 7	bit 6	bit 5	bit 4	bit 3	bit 2	bit 1	bit 0	Register Name
ID10	ID9	ID8	ID7	ID6	ID5	ID4	ID3	IDR0
ID2	ID1	ID0	RTR	IDE(0)				IDR1
								IDR2
								IDR3

Note. IDE value = 0
The contents of IDR0 -- IDR3 are undefined out of reset.

Figure 12.24 ■ Receive/transmit message buffer standard identifier

When the extended identifier is used, the bit 4 of the IDR1 register is used as the *substitute remote request* (SRR) bit. The user must set this bit to one in the transmission buffer in order to make the message a remote request. This bit position will be interpreted as the SRR bit for the receive buffer when the extended identifier is used.

The bit 3 (IDE flag) of the IDR1 register indicates whether the extended (when set to one) or the standard identifier (when set to zero) format is applied in this buffer. In case of a receive buffer, the flag is set as being received and indicates to the CPU how to process the buffer identifier registers. In case of a transmit buffer, the flag indicates to the MSCAN12 what type of identifier to send.

The RTR bit (bit 4 of the IDR1 register or bit 0 of the IDR3 register, depending upon the type of the identifier) reflects the status of the *remote transmission request* bit in the CAN frame. In case of a receive buffer, the setting of this bit supports the transmission of an answering frame. In case of a transmit buffer, the setting of this flag requests the transmission of a remote frame; otherwise, it indicates that the transmit-buffer stores a data frame.

Only the lowest four bits have meaning in the *data length register* (DLR). In a data frame, these four bits indicate the number of bytes contained in the data frame. The data byte count ranges from zero to eight. The byte count encoding is shown in Table 12.1. In a remote frame, the data length code is transmitted as programmed while the number of transmitted bytes is always zero.

The eight data segment registers contain the data to be transmitted or received.

The contents of the transmit-buffer priority register are shown in Figure 12.25. The contents of this register define the local priority of the associated message buffer. The message with the smallest priority number has the highest priority. In case of more than one buffer having the same lowest priority, the message buffer with the lower index number wins. Only the transmit buffers with the empty (0) TXE flag participate in the prioritization process.

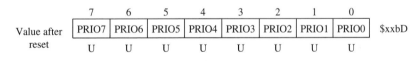

	7	6	5	4	3	2	1	0	
Value after	PRIO7	PRIO6	PRIO5	PRIO4	PRIO3	PRIO2	PRIO1	PRIO0	$xxbD
reset	U	U	U	U	U	U	U	U	

Note. b is 5, 6, and 7 for transmit buffer 0, 1, and 2

Figure 12.25 ■ The Transmit buffer priority register (TBPR)

12.15.2 MSCAN12 Module Control Register 0 (CxMCR0)

The contents of the CxMCR0 register are shown in Figure 12.26. The meanings of most of the bits are straightforward, except for the bit 0. When the CPU sets bit 0, the MSCAN12 immediately enters the soft reset state. Any ongoing transmission or reception is aborted and synchronization to the bus is lost. The following registers will enter and stay in the same state as out of hard reset: CxMCR0, CxRFLG, CxRIER, CxTFLG, and CxTCR. The registers CxMCR1, CxBTR0, CxBTR1, CxIDAC, CxIDAR0-7, and CxIDMR0-7 can only be written by the CPU when the MSCAN12 is in the soft reset state. The values of the error counters are not affected by a soft reset.

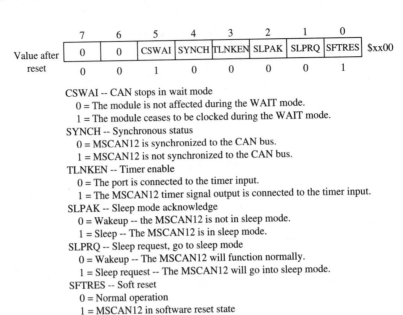

	7	6	5	4	3	2	1	0	
Value after	0	0	CSWAI	SYNCH	TLNKEN	SLPAK	SLPRQ	SFTRES	$xx00
reset	0	0	1	0	0	0	0	1	

CSWAI -- CAN stops in wait mode
  0 = The module is not affected during the WAIT mode.
  1 = The module ceases to be clocked during the WAIT mode.
SYNCH -- Synchronous status
  0 = MSCAN12 is synchronized to the CAN bus.
  1 = MSCAN12 is not synchronized to the CAN bus.
TLNKEN -- Timer enable
  0 = The port is connected to the timer input.
  1 = The MSCAN12 timer signal output is connected to the timer input.
SLPAK -- Sleep mode acknowledge
  0 = Wakeup -- the MSCAN12 is not in sleep mode.
  1 = Sleep -- The MSCAN12 is in sleep mode.
SLPRQ -- Sleep request, go to sleep mode
  0 = Wakeup -- The MSCAN12 will function normally.
  1 = Sleep request -- The MSCAN12 will go into sleep mode.
SFTRES -- Soft reset
  0 = Normal operation
  1 = MSCAN12 in software reset state

Figure 12.26 ■ The MSCAN12 module control register 0 (CxMCR0)

When the SFTRES bit is cleared by the CPU, the CAN module will try to synchronize to the CAN bus: If the CAN module is not in a bus-off state it will be synchronized after 11 recessive bits on the bus; if the CAN module is in a bus-off state it continues to wait for 128 occurrences of 11 recessive bits. Clearing the SFTRES bit and writing to other bits in the CxMCR0 register must occur in separate instructions.

12.15.3 MSCAN12 Module Control Register 1 (CxMCR1)

The contents of this register are shown in Figure 12.27. Bit 2 of the CxMCR1 register activates the loop-back self-test mode. In loop-back self-test mode, the RxCAN input pin is ignored and the TxCAN output goes to the recessive state (1).

The wake-up mode flag (bit 1) of this register defines whether the integrated low-pass filter is applied to protect the MSCAN12 from spurious wake-up.

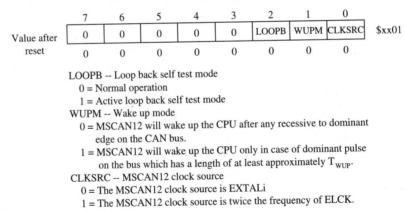

	7	6	5	4	3	2	1	0	
Value after	0	0	0	0	0	LOOPB	WUPM	CLKSRC	$xx01
reset	0	0	0	0	0	0	0	0	

LOOPB -- Loop back self test mode
  0 = Normal operation
  1 = Active loop back self test mode
WUPM -- Wake up mode
  0 = MSCAN12 will wake up the CPU after any recessive to dominant
     edge on the CAN bus.
  1 = MSCAN12 will wake up the CPU only in case of dominant pulse
     on the bus which has a length of at least approximately T_{WUP}.
CLKSRC -- MSCAN12 clock source
  0 = The MSCAN12 clock source is EXTALi
  1 = The MSCAN12 clock source is twice the frequency of ELCK.

Figure 12.27 ■ The MSCAN12 module control register 1 (CxMCR1)

12.15.4 MSCAN12 Bus Timing Register 0 (CxBTR0)

This register defines the synchronization jump width and the prescaler. The contents of this register are shown in Figure 12.28.

7	6	5	4	3	2	1	0
SJW1	SJW0	BRP5	BRP4	BRP3	BRP2	BRP1	BRP0
0	0	0	0	0	0	0	0

Value after reset $xx01

SJW1, SJW0 -- Synchronization jump width
    The synchronization jump width defines the maximum number of time
    quanta (Tq) clock cycles by which a bit may be shortened, or
    lengthened, to achieve resynchronization on data transitions on the bus.
    The number of synchronization jump width is illustrated in Table 12.5.
BRP5 -- BRP0 -- Baud rate prescaler
    These bits determine the time quanta (Tq) clock, which is used to build
    up the individual bit timing, according to Table 12.6.

Figure 12.28 ■ The MSCAN12 bus timing register 0 (CxBTR0)

SJW1	SJW0	Synchronization jump width
0	0	1 Tq clock cycle
0	1	2 Tq clock cycles
1	0	3 Tq clock cycles
1	1	4 Tq clock cycles

Table 12.5 ■ Synchronization jump width

BRP5	BRP4	BRP3	BRP2	BRP1	BRP0	Prescale value
0	0	0	0	0	0	1
0	0	0	0	0	1	2
0	0	0	0	1	0	3
0	0	0	0	1	1	4
--	--	--	--	--	--	--
--	--	--	--	--	--	--
1	1	1	1	1	1	64

Table 12.6 ■ Baud rate prescaler

The clock source selected by the CxMCR1 register divided by the prescale factor becomes the CAN system clock. Each CAN system clock period is referred to as a *time quantum*.

12.15.5 MSCAN12 Bus Timing Register 1 (CxBTR1)

The contents of this register set the *transmit point* and *sample point,* as shown in Figure 12.29.

SAMP -- Sampling

This bit determines the number of samples of the serial bus to be taken per bit time. If set, three samples are taken, the regular one (sample point) and two preceding samples, using a majority rule. For higher bit rates, SAMP should be cleared, which means that only one sample will be taken per bit.

TSEG22 -- TSEG10 -- Time segment

Time segments within the bit time fix the number of clock cycles per bit time, and the location of the sample point. TSEG22 to TSEG20 set the time segment 2 value in a bit time whereas TSEG13 to TSEG10 set the time segment 1 value in a bit time as shown in Table 12.7.

Figure 12.29 ■ The MSCAN12 bus timing register 1 (CxBTR1)

TSEG 13	TSEG 12	TSEG 11	TSEG 10	Time segment 1
0	0	0	0	1 Tq clock cycle
0	0	0	1	2 Tq clock cycles
0	0	1	0	3 Tq clock cycles
0	0	1	1	4 Tq clock cycles
.	.	.	.	.
.	.	.	.	.
1	1	1	1	16 Tq clock cycles

Table 12.7a ■ Time segment values

TSEG 22	TSEG 21	TSEG 20	Time segment 2
0	0	0	1 Tq clock cycle
0	0	1	2 Tq clock cycles
.	.	.	.
.	.	.	.
1	1	1	8 Tq clock cycles

Table 12.7b ■ Time segment values

12.15.6 MSCAN12 Receiver Flag Register (CxRFLG)

All bits of this register are read and clear only. A flag is cleared by writing a one to it. A flag can only be cleared when the condition that caused the setting is no longer valid. Every flag has an associated interrupt enable flag in the CxRIER register. A hard or soft reset will clear the register. The contents of this register are shown in Figure 12.30.

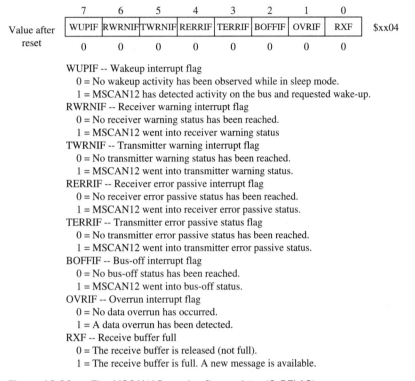

Figure **12.30** ■ The MSCAN12 receive flag register (CxRFLAG)

If the MSCAN12 detects bus activity while it is in sleep mode, it clears the SLPAK bit in the CxMCR0 register and sets the WUPIF flag.

The *receiver warning interrupt flag* will be set when the receive error counter is in the range of 96 to 127 and neither the error interrupt flags nor the bus-off interrupt flag is set.

The *transmitter warning interrupt flag* will be set when the transmit error counter is in the range of 96 to 127 and neither the error interrupt flags nor the bus-off interrupt flag is set.

The *receiver error passive interrupt flag* will be set when the receive error counter exceeds 127 (but is less than 256) and the bus-off interrupt flag is not set.

The *transmitter error passive interrupt flag* will be set when the transmit error counter exceeds 127 (but is less than 256) and the bus-off interrupt flag is not set.

The *bus-off flag* will be set when the transmit error counter has exceeded 255. It cannot be cleared before the MSCAN12 has monitored 128×11 consecutive recessive bits on the bus.

The *RXF* flag is set when a new message is available in the foreground receive buffer. This flag indicates whether the buffer is loaded with a correctly received message.

12.15.6 MSCAN12 Receiver Interrupt Enable Register (CxRIER)

The contents of the MSCAN12 receiver interrupt enable register (CxRIER) are shown in Figure 12.31. Each flag bit in the CxRFLG register has an associated interrupt enable mask in CxRIER.

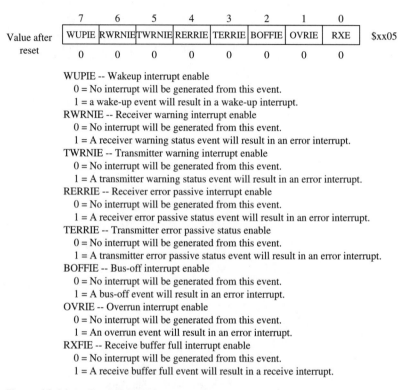

Value after reset

7	6	5	4	3	2	1	0	
WUPIE	RWRNIE	TWRNIE	RERRIE	TERRIE	BOFFIE	OVRIE	RXE	$xx05
0	0	0	0	0	0	0	0	

WUPIE -- Wakeup interrupt enable
    0 = No interrupt will be generated from this event.
    1 = a wake-up event will result in a wake-up interrupt.
RWRNIE -- Receiver warning interrupt enable
    0 = No interrupt will be generated from this event.
    1 = A receiver warning status event will result in an error interrupt.
TWRNIE -- Transmitter warning interrupt enable
    0 = No interrupt will be generated from this event.
    1 = A transmitter warning status event will result in an error interrupt.
RERRIE -- Receiver error passive interrupt enable
    0 = No interrupt will be generated from this event.
    1 = A receiver error passive status event will result in an error interrupt.
TERRIE -- Transmitter error passive status enable
    0 = No interrupt will be generated from this event.
    1 = A transmitter error passive status event will result in an error interrupt.
BOFFIE -- Bus-off interrupt enable
    0 = No interrupt will be generated from this event.
    1 = A bus-off event will result in an error interrupt.
OVRIE -- Overrun interrupt enable
    0 = No interrupt will be generated from this event.
    1 = An overrun event will result in an error interrupt.
RXFIE -- Receive buffer full interrupt enable
    0 = No interrupt will be generated from this event.
    1 = A receive buffer full event will result in a receive interrupt.

Figure 12.31 ■ The MSCAN12 receive interrupt enable register (CxRIER)

12.15.7 MSCAN12 Transmitter Flag Register (CxTFLG)

All bits of this register are read and clear only. A flag can be cleared by writing a one to it. The contents of this register are shown in Figure 12.32.

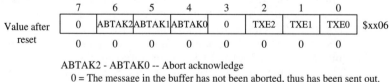

Value after reset

7	6	5	4	3	2	1	0	
0	ABTAK2	ABTAK1	ABTAK0	0	TXE2	TXE1	TXE0	$xx06
0	0	0	0	0	0	0	0	

ABTAK2 - ABTAK0 -- Abort acknowledge
    0 = The message in the buffer has not been aborted, thus has been sent out.
    1 = The message in the buffer has been aborted.
TXE2 - TXE0 -- Transmitter buffer empty
    0 = The associated message buffer is full.
    1 = The associated message buffer if empty.

Figure 12.32 ■ The MSCAN12 transmit flag register (CxTFLG)

The ABTAKx (x = 0..2) flag acknowledges that a message in transmit buffer x has been aborted due to a pending abort request from the CPU.

The transmitter buffer empty flag will be set whenever the MSCAN12 has successfully transmitted a message or aborted a transmission request.

12.15.8 MSCAN12 Transmitter Control Register (CxTCR)

This register allows the CPU to abort a scheduled message buffer (0, 1, or 2) and enables the transmit-buffer empty interrupt. The contents of the MSCAN12 Transmitter Control Register (CxTCR) are shown in Figure 12.33.

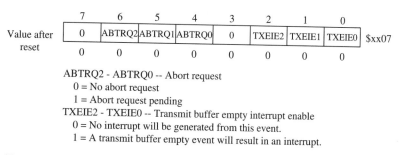

ABTRQ2 - ABTRQ0 -- Abort request
    0 = No abort request
    1 = Abort request pending
TXEIE2 - TXEIE0 -- Transmit buffer empty interrupt enable
    0 = No interrupt will be generated from this event.
    1 = A transmit buffer empty event will result in an interrupt.

Figure 12.33 ■ The MSCAN12 transmit control register (CxTCR)

12.15.9 MSCAN12 Identifier Acceptance Control Register (CxIDAC)

The contents of this register are shown in Figure 12.34. This register is used to set the identifier acceptance mode and indicate the type of acceptance hit.

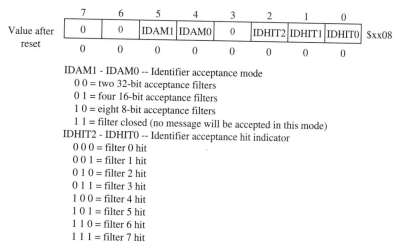

IDAM1 - IDAM0 -- Identifier acceptance mode
    0 0 = two 32-bit acceptance filters
    0 1 = four 16-bit acceptance filters
    1 0 = eight 8-bit acceptance filters
    1 1 = filter closed (no message will be accepted in this mode)
IDHIT2 - IDHIT0 -- Identifier acceptance hit indicator
    0 0 0 = filter 0 hit
    0 0 1 = filter 1 hit
    0 1 0 = filter 2 hit
    0 1 1 = filter 3 hit
    1 0 0 = filter 4 hit
    1 0 1 = filter 5 hit
    1 1 0 = filter 6 hit
    1 1 1 = filter 7 hit

Figure 12.34 ■ The MSCAN12 identifier acceptance control register (CxIDAC)

The IDHIT indicators are always related to the message in the foreground buffer. When a message gets copied from the background to the foreground buffer, the indicators are updated as well.

12.15.10 MSCAN12 Receive Error Counter (CxRXERR)

This read-only register reflects the status of the MSCAN12 receive error counter. The contents of the MSCAN12 receive error counter (CxRXERR) register are shown in Figure 12.35.

	7	6	5	4	3	2	1	0	
	RXERR7	RXERR6	RXERR5	RXERR4	RXERR3	RXERR2	RXERR1	RXERR0	$xx0E
Value after reset	0	0	0	0	0	0	0	0	

Figure 12.35 ■ The MSCAN12 receive error counter (CxRXERR)

12.15.11 MSCAN12 Transmit Error Counter (CxTXERR)

This read-only register reflects the status of the MSCAN12 transmit error counter (CxTXERR). The contents of this register are shown in Figure 12.36.

	7	6	5	4	3	2	1	0	
	TXERR7	TXERR6	TXERR5	TXERR4	TXERR3	TXERR2	TXERR1	TXERR0	$xx0F
Value after reset	0	0	0	0	0	0	0	0	

Figure 12.36 ■ The MSCAN12 transmit error counter (CxTXERR)

12.15.12 MSCAN12 Identifier Acceptance Registers (CxIDAR0-7)

Upon reception, each message is written into the background receive buffer. The CPU will be notified only when the message passes the criteria in the identifier acceptance and identifier mask registers. Otherwise, the message will be overwritten by the next message.

The acceptance registers are applied on the IDR0 to IDR3 registers of incoming messages in a bit-by-bit manner. For extended identifiers, all four acceptance and mask registers are applied. For standard identifiers, only the first two (CxIDAR1, CxIDAR0) are applied. In the latter case it is required to program the lowest three bits of the mask register CxIDMR1 to "don't care".

Identifier acceptance registers are divided into two banks, as shown in Figures 12.37 and 12.38.

	7	6	5	4	3	2	1	0	
CxIDAR0	AC7	AC6	AC5	AC4	AC3	AC2	AC1	AC0	$xx10
CxIDAR1	AC7	AC6	AC5	AC4	AC3	AC2	AC1	AC0	$xx11
CxIDAR2	AC7	AC6	AC5	AC4	AC3	AC2	AC1	AC0	$xx12
CxIDAR3	AC7	AC6	AC5	AC4	AC3	AC2	AC1	AC0	$xx13
Value after reset	U	U	U	U	U	U	U	U	

Figure 12.37 ■ The MSCAN12 identifier acceptance registers (1st bank)

	7	6	5	4	3	2	1	0	
CxIDAR4	AC7	AC6	AC5	AC4	AC3	AC2	AC1	AC0	$xx18
CxIDAR5	AC7	AC6	AC5	AC4	AC3	AC2	AC1	AC0	$xx19
CxIDAR6	AC7	AC6	AC5	AC4	AC3	AC2	AC1	AC0	$xx1A
CxIDAR7	AC7	AC6	AC5	AC4	AC3	AC2	AC1	AC0	$xx1B
Value after reset	U	U	U	U	U	U	U	U	

Figure 12.38 ■ The MSCAN12 identifier acceptance registers (2nd bank)

In Figures 12.37 and 12.38, bits AC7–AC0 comprise a user defined sequence of bits with which the corresponding bits of the *related identifier register* (IDRn) of the receive message buffer are compared. The result of this comparison is then masked with the corresponding identifier mask register.

The CxIDAR0–CxIDAR7 registers can be written only if the SFTRES bit in the CxMCR0 is set.

12.15.13 MSCAN12 Identifier Mask Registers (CxIDMR0-7)

The MSCAN12 identifier mask register (CxIDMR0-7) specifies which of the corresponding bits in the identifier acceptance register are relevant for acceptance filtering. The identifier mask registers are divided into two banks and their contents are shown in Figures 12.39 and 12.40. To receive standard identifiers in 32-bit filter mode, the last three bits (AM2-AM0) in the mask registers CxIDMR1 and CxIDMR5 must be programmed to "don't care". To receive standard identifiers in 16-bit filter mode, the last three bits in the mask registers CxIDMR1, CxIDMR3, CxIDMR5, and CxIDMR7 must be programmed to "don't care".

	7	6	5	4	3	2	1	0	
CxIDMR0	AM7	AM6	AM5	AM4	AM3	AM2	AM1	AM0	$xx14
CxIDMR1	AM7	AM6	AM5	AM4	AM3	AM2	AM1	AM0	$xx15
CxIDMR2	AM7	AM6	AM5	AM4	AM3	AM2	AM1	AM0	$xx16
CxIDMR3	AM7	AM6	AM5	AM4	AM3	AM2	AM1	AM0	$xx17
Value after reset	U	U	U	U	U	U	U	U	

Figure 12.39 ■ The MSCAN12 identifier mask registers (1st bank)

	7	6	5	4	3	2	1	0	
CxIDMR4	AM	AM6	AM5	AM4	AM3	AM2	AM1	AM0	$xx1C
CxIDMR5	AM7	AM6	AM5	AC4	AM3	AM2	AM1	AM0	$xx1D
CxIDMR6	AM7	AM6	AM5	AC4	AM3	AM2	AM1	AM0	$xx1E
CxIDMR7	AM7	AM6	AM5	AC4	AM3	AM2	AM1	AM0	$xx1F
Value after reset	U	U	U	U	U	U	U	U	

Figure 12.40 ■ The MSCAN 12 identifier mask registers (2nd bank)

If a particular bit in an identifier mask register is zero, then the corresponding bit in the identifier acceptance register must be the same as its identifier bit, before a match will be detected. The message will be accepted if all such bits match. If a bit is one, it indicates that the state of the corresponding bit in the identifier acceptance register will not affect whether or not the message is accepted.

The CxIDMR0–CxIDMR7 registers can be written only if the SFTRES bit in CxMCR0 is set.

12.15.14 MSCAN12 Port CAN Control Register (PCTLCAN)

This register configures the CAN port pins that are used for general I/O. As shown in Figure 12.41, only the lowest two bits of this register are implemented.

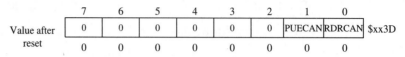

	7	6	5	4	3	2	1	0	
Value after	0	0	0	0	0	0	PUECAN	RDRCAN	$xx3D
reset	0	0	0	0	0	0	0	0	

These two bits control bits 7 through 2 of port CAN.

PUECAN -- pull enable port CAN
  0 = Pull mode disabled for Port CAN
  1 = Pull mode enabled for Port CAN
  The pull mode (pull-up or pull-down) for port CAN is defined in the chip specification.
RDRCAN -- Reduced drive port CAN
  0 = Reduced drive disabled for Port CAN
  1 = Reduced drive enabled for Port CAN

Figure 12.41 ■ The MSCAN12 port CAN control register (PCTLCAN)

12.15.15 MSCAN12 Port CAN Data Register (PORTCAN)

Port CAN pins that are not used as CAN functions are used as general-purpose I/O pins. The MSCAN12 port CAN data register (PORTCAN) allows us to access port CAN pins that are not used by the CAN module. The contents of this register are shown in Figure 12.42.

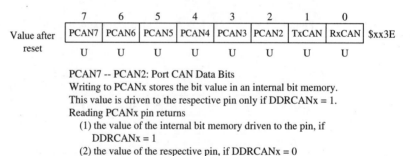

	7	6	5	4	3	2	1	0	
Value after	PCAN7	PCAN6	PCAN5	PCAN4	PCAN3	PCAN2	TxCAN	RxCAN	$xx3E
reset	U	U	U	U	U	U	U	U	

PCAN7 -- PCAN2: Port CAN Data Bits
Writing to PCANx stores the bit value in an internal bit memory.
This value is driven to the respective pin only if DDRCANx = 1.
Reading PCANx pin returns
  (1) the value of the internal bit memory driven to the pin, if
      DDRCANx = 1
  (2) the value of the respective pin, if DDRCANx = 0

Figure 12.42 ■ The MSCAN12 port CAN data register (PORTCAN)

12.15.16 MSCAN12 Port CAN Data Direction Register (DDRCAN)

This register sets the direction of each Port CAN I/O pin. The contents of this register are shown in Figure 12.43.

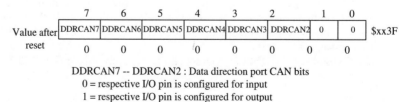

	7	6	5	4	3	2	1	0	
Value after	DDRCAN7	DDRCAN6	DDRCAN5	DDRCAN4	DDRCAN3	DDRCAN2	0	0	$xx3F
reset	0	0	0	0	0	0	0	0	

DDRCAN7 -- DDRCAN2 : Data direction port CAN bits
  0 = respective I/O pin is configured for input
  1 = respective I/O pin is configured for output

Figure 12.43 ■ The MSCAN12 port CAN data direction register (DDRCAN)

12.16 Physical CAN Bus Connection

The CAN standard is designed for data communication over a short distance. It does not specify what medium to use for data transmission. The user can choose optical fiber, shielded cable, or unshielded cable as the transmission medium. Using a shielded or unshielded cable would be more convenient for a short distance communication.

A typical CAN bus system setup using a cable is illustrated in Figure 12.44.

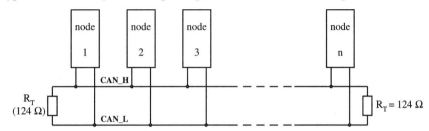

Figure 12.44 ■ A typical CAN bus setup using cable

In Figure 12.44, the resistor R_T is the terminating resistor. A value of 124Ω is recommended for R_T. Each CAN node uses a transceiver to connect to the CAN bus. The CAN bus transceiver is connected to the bus via two bus terminals, CAN_H and CAN_L, which provide differential receive and transmit capabilities. The nominal CAN bus levels are shown in Figure 12.45.

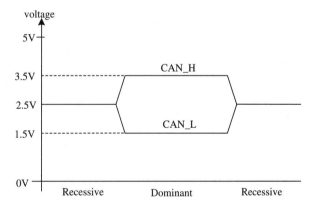

Figure 12.45 ■ Nominal CAN bus levels

12.16.1 The PCA82C250 CAN Transceiver

The Phillips PCA82C250 is one of the most widely used CAN transceivers. It provides differential transmit capabilities to the bus and differential receive capabilities to the CAN controller. A data rate of 1 Mbps can be achieved. Using the PCA82C250 transceiver, at least 110 nodes can be attached to the CAN bus. The block diagram of the PCA82C250 is shown in Figure 12.46.

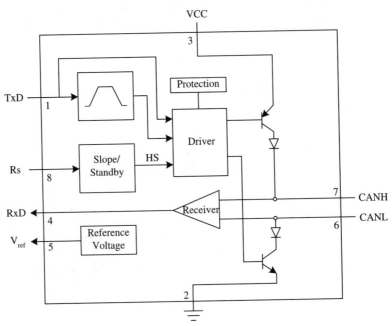

Figure 12.46 ■ The block diagram of PCA82C250

The function of each signal pin is as follows:

V_{CC}	*Power supply input.* This input is normally set to between 4.5 to 5.5 V.
TxD	*Transmit data input* (from the CAN module).
RxD	*Receive data output* (to the CAN module).
V_{REF}	*Reference voltage output.* This pin provides an output voltage of $0.5 \times V_{CC}$ nominal.
R_S	*Slope resistor input.* This pin allows three different modes of operation to be selected: high-speed, slope control, and standby.
CANH	*High-level CAN voltage input/output.*
CANL	*Low-level CAN voltage input/output.*
GND	*Ground input.*

For high-speed operation, the transmitter output transistors are simply switched on and off as fast as possible. In this mode, no measures are taken to limit the rise and fall slopes. Use of a shielded cable is recommended to avoid *radio frequency interference* (RFI) problems. The high-speed mode is selected by connecting pin 8 to ground (with or without a pull-down resistor).

For lower speeds or shorter bus lengths, an unshielded twisted pair or a parallel pair of wires can be used for the bus. To reduce RFI, the rise and fall slopes should be limited. The rise and fall slopes can be programmed with a resistor connected from pin 8 to ground. The slope is proportional to the current output at pin 8.

If a high level is applied to pin 8, the circuit enters a low current standby mode. In this mode, the transmitter is switched off and the receiver is switched to a low current. If dominant bits are detected (differential bus voltage > 0.9V), RxD will be switched to low level. The microcontroller should react to this condition by switching the transceiver back to normal operation (via pin 8). Because the receiver is slow in standby mode, the first message will be lost.

12.16.2 Interfacing PCA82C250 to the 68HC12

A typical method of interfacing the PCA82C250 transceiver to the 68HC12 is shown in Figure 12.47.

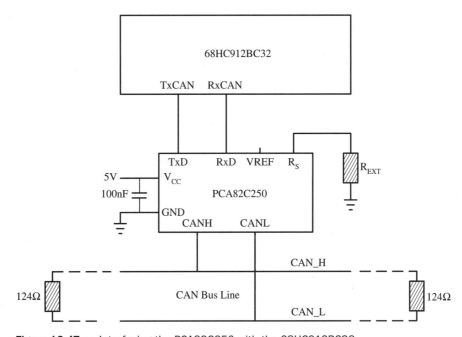

Figure 12.47 ■ Interfacing the PCA82C250 with the 68HC912BC32

The bus line is in a recessive state if no bus node transmits a dominant bit; in other words, all TxD inputs in the network are logic high. Otherwise, if one or more bus nodes transmit a dominant bit (i.e., at least one TxD input is logic low), then the bus line enters the dominant state.

The receiver comparator converts the differential bus signal to a logic level signal that is output at RxD. The serial receive data stream is provided to the CAN bus controller for decoding. The receiver comparator is always active. The PCA82C250 provides a reference output, which may be used to bias one of the inputs of a CAN controller's differential input comparator, and is not needed by the 68HC12 CAN module.

The PCA82C250 provides three different operation modes. Mode control is being provided through the Rs control input.

The first mode is the *high-speed* mode supporting maximum bus speed and/or length. The second mode is called *slope control* mode, which should be considered if unshielded bus wires will be used. In this mode the output slew rate can be decreased intentionally; for example, to reduce electromagnetic emission. The third mode is the *standby* mode, particularly of interest in battery-powered applications, when the system power consumption needs to be very low. In high-speed mode, the value of R_{EXT} should be between 0 and 1.8 KΩ.

An in-depth discussion on the slope control and standby modes is outside of the scope of this text. Interested readers should refer to the Phillips application note AN96116.

The maximum achievable bus length in a CAN bus network is determined by the following physical effects:

1. The loop delay of the connected bus nodes (CAN controller, transceiver) and the delay of the bus line.

2. The differences in bit time quantum length due to the relative oscillator tolerance between nodes.

3. The signal amplitude drop due to the series resistance of the bus cable and the input resistance of bus node.

The resultant equation after taking these three factors into account would be very complicated. As a rule of thumb, the bus length that can be achieved as a function of the bit rate in high-speed mode and with CAN bit timing parameters being optimized for maximum propagation delay is given in Table 12.8.

Bit rate (kbit/s)	Bus length
1000	40
500	100
250	250
125	500
62.5	1000

Table 12.8 ■ CAN bus bit rate/bus length relation

The types and the cross-sections suitable for the CAN bus trunk cable that has more than 32 nodes connected and spans more than 100 meters are listed in Table 12.9.

Bus length/Number of nodes	32	64	100
100 m	0.25 mm2 or AWG 24	0.25 mm2 or AWG 24	0.25 mm2 or AWG 24
250 m	0.34 mm2 or AWG 22	0.50 mm2 or AWG 20	0.50 mm2 or AWG 20
500 m	0.75 mm2 or AWG 18	0.75 mm2 or AWG 18	0.75 mm2 or AWG 18

Table 12.9 ■ Minimum recommended bus wire cross-section for the trunk cable

A standardized CAN bus connection, including the cabling and connectors, is very important to the popularization of the CAN bus. CANopen is an organization that promotes the use of the CAN bus. Its DR-303-1 document published in April 2001 is an attempt to address the issue of cabling and connectors. CANopen sponsors a conference on the CAN bus once a year. Many short courses and presentations are given in the conference. More information about CANopen can be found in the website at www.CANopen.org.

12.16.3 Setting the CAN Timing Parameters

The nominal bit rate of a CAN network is uniform throughout the network and is given by the following equation:

$$f_{NBT} = 1/t_{NBT} \tag{12.1}$$

where t_{NBT} is the *nominal bit time*. As described earlier, a bit time is divided into four separate non-overlapping time segments: *sync_seg, prop_seg, phase_seg1,* and *phase_seg2.*

The purpose of the *sync_seg* segment is to inform all nodes that a bit time is just started. The existence of the propagation delay segment, *prop_seg,* is due to the fact that the CAN protocol allows for non-destructive arbitration between nodes contending for access to the bus and the requirement for *in-frame acknowledgement.* In the case of non-destructive arbitration, more than one node may be transmitting during the arbitration field. Each transmitting node samples data from the bus in order to determine whether it has won or lost the arbitration, and also to receive the arbitration field in case it loses arbitration. When each node samples a bit, the value sampled must be the logical superposition of the bit values transmitted by each of the nodes arbitrating for bus access. In the case of the acknowledge field, the transmitting node transmits a recessive bit but expects to receive a dominant bit; in other words, a dominant value must be sampled at the sample time. The length of the *prop_seg* segment must be selected so that the earliest possible sample of the bit by a node until the transmitted bit values from all the transmitting nodes have reached all of the nodes. When multiple nodes are arbitrating for the control of the CAN bus, the node that transmitted the earliest must wait until the node that transmitted the latest, in order to find out if it won or lost. That is, the worst-case value for t_{PROP_SEG} is given by:

$$t_{PROP\_SEG} = t_{PROP(A, B)} + t_{PROP(B, A)} \tag{12.2}$$

where $t_{PROP(A, B)}$ and $t_{PROP(B,A)}$ are the propagation delays from node A to node B and node B to node A, respectively. In the worst case, node A and node B are at the two ends of the CAN bus.

The propagation delay from node A to node B is given by:

$$t_{PROP(A, B)} = t_{BUS} + t_{Tx} + t_{Rx} \tag{12.3}$$

where t_{BUS}, t_{Tx}, and t_{Rx} are data traveling times on the bus, transmitter propagation delay, and receiver propagation delay, respectively.

Let node A and node B be two nodes at opposite ends of the CAN bus, then the worst case value for t_{PROP_SEG} is:

$$t_{PROP\_SEG} = 2 \times (t_{BUS} + t_{Tx} + t_{Rx}) \tag{12.4}$$

The minimum number of *time quantum* (t_Q) that must be allocated to the *prop_seg* segment is therefore:

$$prop\_seg = round\_up\ (t_{PROP\_SEG} \div t_Q) \tag{12.5}$$

where the *round_up()* function returns a value that equals the argument rounded up to the next integer value.

In the absence of bus errors, bit stuffing guarantees a maximum of 10 bit periods between resynchronization edges (5 dominant bits followed by 5 recessive bits, then followed by a dominant bit). This represents the worst-case condition for the accumulation of phase errors during normal communication. The accumulated phase error must be compensated for by resynchronization and therefore must be less than the programmed resynchronization jump width (t_{RJW}). The accumulated phase error is due to the tolerance in the CAN system clock, and this requirement can be expressed as:

$$(2 \times \Delta f) \times 10 \times t_{NBT} < t_{RJW} \tag{12.6}$$

where Δf is the largest crystal oscillator frequency variation (in a percentage) of all CAN nodes in the network.

Real systems must operate in the presence of electrical noise which may induce errors on the CAN bus. A node transmits an error flag after it has detected an error. In the case of a local error, only the node that detects the error will transmit the error flag. All other nodes receive the error flag and then transmit their own error flags as an echo. If the error is global, all nodes will detect it within the same bit time and will therefore transmit error flags simultaneously. A node can therefore differentiate between a local error and a global error by detecting whether there is an echo after its error flag. This requires that a node can correctly sample the first bit after transmitting its error flag.

An error flag from an error active node consists of six dominant bits, and there could be up to six dominant bits before the error flag, if, for example, the error was a stuff error. A node must therefore correctly sample the 13th bit after the last resynchronization. This can be expressed as:

$$(2 \times \Delta f) \times (13 \times t_{NBT} - t_{PHASE\_SEG2}) < \min(t_{PHASE\_SEG1}, t_{PHASE\_SEG2}) \quad (12.7)$$

where the function *min (,)* returns the smaller of the two arguments.

Thus there are two clock tolerance requirements that must be satisfied. The selection of bit timing values involves consideration of various fundamental system parameters. The requirement of the *prop_seg* value imposes a trade-off between the maximum achievable bit rate and the maximum propagation delay, due to the bus length and the characteristics of the bus driver circuit. The highest bit rate can only be achieved with a short bus length, a fast bus driver circuit, and a high frequency CAN clock source with high tolerance.

The procedure for determining the optimum bit timing parameters that satisfy the requirements for proper bit sampling is as follows:

Step 1
Determine the minimum permissible time for the prop_seg segment using equation 12.4.

Step 2
Choose the CAN system clock frequency. The CAN system clock will be the CPU oscillator output (or ECLK multiplied by two) divided by a prescale factor. The CAN system clock is chosen so that the desired CAN bus nominal bit time is an integer multiple of the time quanta (CAN system clock period) from 8 to 25.

Step 3
Calculate the prop_seg duration. The number of time quanta required for the prop_seg can be calculated by using equation 12.5. If the result is greater than eight, go back to Step 2 and choose a lower CAN system clock frequency.

Step 4
Determine phase_seg1 and phase_seg2. Subtract the prop_seg value and one (for sync_seg) from the time quanta contained in a bit time. If the difference is less than three then go back to Step 2 and select a higher CAN system clock frequency. If the difference is an odd number greater than three then add one to the prop_seg value and recalculate. If the difference is equal to three, then phase_seg1 = 1 and phase_seg2 = 2 and only one sample per bit may be chosen. Otherwise divide the remaining number by two and assign the result to phase_seg1 and phase_seg2.

Step 5
Determine the resynchronization jump width (RJW). RJW is the smaller of four and phase_seg1.

Step 6
Calculate the required oscillator tolerance from equations 12.6 and 12.7. If phase_seg1 > 4, it is recommended that you repeat Steps 2 to 6 with a larger value for the prescaler. Conversely, if phase_seg1 < 4, it is recommended that you repeat

Steps 2 to 6 with a smaller value for the prescaler, as long as prop_seg < 8, as this may result in a reduced oscillator tolerance requirement. If the prescaler is already equal to one and a reduced oscillator tolerance is still required, the only option is to consider using a clock source with a higher frequency.

Example 12.1

▼

Calculate the CAN bit segments for the following system constraints:

Bit rate = 1Mbps
Bus length = 25 m
Bus propagation delay = 5×10^{-9} sec/m
PCA82C250 transceiver plus receiver propagation delay = 150 ns at 85°C
CPU oscillator frequency = 12 MHz

Solution: Let's follow the procedure described earlier:

Step 1
Physical delay of bus = 125 ns
$$t_{PROP\_SEG} = 2 \times (125\ ns + 150\ ns) = 550\ ns$$

Step 2
A prescaler of 1 for a CAN system clock of 12 MHz gives a time quantum of 83.33 ns. This gives 1000/83.33 = 12 time quanta per bit.

Step 3
$$Prop\_seg = round\_up\ (550\ ns \div 83.33\ ns) = round\_up\ (6.6) = 7$$

Step 4
Subtracting 7 for *prop_seg* and 1 for *sync_seg* from 12 time quanta per bit gives 4. Since the result is equal to four and is even, divide it by two (quotient is two) and assign it to *phase_seg1* and *phase_seg2*.

Step 5
RJW is the smaller of four and *phase_seg1* and is two.

Step 6
From equation 12.6:
$$\Delta f < RJW \div (20 \times NBT) = 2 \div (20 \times 12) = 0.83\%$$
From equation 12.7:
$$\Delta f < MIN\ (phase\_seg1, phase\_seg2) \div 2(13 \times NBT - phase\_seg2)$$
$$= 2 \div 308 = 0.65\%$$
The desired oscillator tolerance is the smaller of these values, i.e., 0.65% over a period of 12.83 μs (12.83 bit period).
In summary:

Prescaler	= 1
Nominal bit time	= 12
prop_seg	= 7
sync_seg	= 1
phase_seg1	= 2
phase_seg2	= 2
RJW	= 2
Oscillator tolerance	= 0.65%

▲

Example 12.2

▼

Calculate the bit segments for the following system constraints:

Bit rate = 500Kbps
Bus length = 50 m
Bus propagation delay = 5×10^{-9} sec/m
PCA82C250 transceiver plus receiver propagation delay = 150 ns at 85°C
CPU oscillator frequency = 16 MHz

Solution: Follow the standard procedure to perform the calculation:

Step 1
Physical delay of bus = 250 ns
$$t_{PROP\_SEG} = 2 \times (250 \text{ ns} + 150 \text{ ns}) = 800 \text{ ns}$$

Step 2
A prescaler of 2 for a CAN system clock of 16 MHz gives a time quantum of 125 ns. This gives 2,000/125 = 16 time quanta per bit.

Step 3
$Prop\_seg = round\_up \ (800 \text{ ns} \div 125 \text{ ns}) = round\_up \ (6.4) = 7$

Step 4
Subtract 7 for *prop_seg* and 1 for *sync_seg* from 16 time quanta per bit gives 8.
Since the result is greater than four and is even, divide it by two (quotient is four) and assign it to *phase_seg1* and *phase_seg2*.

Step 5
RJW (= 4) is the smaller of four and *phase_seg1*.

Step 6
From equation 12.6:
$$\Delta f < RJW \div (20 \times NBT) = 4 \div (20 \times 16) = 1.25\%$$
From equation 12.7:
$$\Delta f < MIN \ (phase\_seg1, phase\_seg2) \div 2(13 \times NBT - phase\_seg2)$$
$$= 4 \div 408 = 0.98\%$$
The required oscillator tolerance is the smaller of these values, i.e., 0.98%. Since *phase_seg2* = 4, there is no need to try other prescale values.
In summary:

Prescaler	= 2
Nominal bit time	= 16
prop_seg	= 7
sync_seg	= 1
phase_seg1	= 4
phase_seg2	= 4
RJW	= 4
Oscillator tolerance	= 0.98%

▲

12.16.4 CAN Programming

The programming of the CAN module can be divided into two parts: initialization and data transmission. Initialization is not complicated. However, depending on the application, the program that deals with the CAN data transmission can be very complicated because it depends on how CAN nodes are interacting with each other.

Example 12.3

Write a program to initialize the CAN module of the 912BC32 with the same system constraints as in Example 12.2. Disable all interrupts. Configure the CAN so that it accepts all messages. Use two 32-bit acceptance filters. Stop CAN in wait mode. Use a clock source with a period twice as long as the ECLK.

Solution: To configure register CxMCR1, CxBTR0, CxBTR1, CxIDAC, CxIDAR0-7, and CxIDMR0-7, we need to place the CAN module in soft reset mode.

The following instruction will configure the CAN bit timing as defined in Example 12.2 and take three samples per bit:

```
movw        #$C1B3,CBTR0          ; configure CBTR0 and CBTR1
```

The following instructions will configure the CAN module to accept all messages:

```
movw        #$FFFF,CIDMR0
movw        #$FFFF,CIDMR2
movw        #$FFFF,CIDMR4
movw        #$FFFF,CIDMR6
```

The following instruction will configure the CAN so that a low-pass filter is applied to prevent spurious wakeup, and will choose $2 \times ECLK$ as the clock source:

```
movb  #$03,CMCR1
```

The following instruction will stop the CAN module clock during the wait mode and connect the port to the timer input:

```
movb  #$20,CMCR0
```

The CAN module initialization routine can be obtained by combining all of the above instructions and applying the soft reset:

```
#include "d:\miniide\hc12.inc"
soft_reset   equ         $01                   ; bit position for soft reset
can_init     bset        CMCR0,soft_reset      ; soft reset the CAN module
             movw        #$C1B3,CBTR0          ; configure CBTR0 and CBTR1
             movw        #$FFFF,CIDMR0         ; accept all messages
             movw        #$FFFF,CIDMR2         ;    "
             movw        #$FFFF,CIDMR4         ;    "
             movw        #$FFFF,CIDMR6         ;    "
             movb        #$03,CMCR1
             clr         CRIER                 ; disable all CAN receive interrupts
             clr         CTCR                  ; disable all CAN transmit interrupts
             bclr        CMCR0,soft_reset      ; quit soft reset mode
             movb        #$20,CMCR0
             rts
```

In C language:

```
void can_init (void)
{
        CMCR0 |= 0x01;        /* soft reset CAN */
        CBTR0  = 0xC1;        /* set up bit timing parameters */
        CBTR1  = 0xB3;        /*       "                      */
        CIDMR0 = 0xFF;
        CIDMR1 = 0xFF;
        CIDMR2 = 0xFF;
```

```
                        CIDMR3 = 0xFF;
                        CIDMR4 = 0xFF;
                        CIDMR5 = 0xFF;
                        CIDMR6 = 0xFF;
                        CIDMR7 = 0xFF;
                        CMCR1 = 0x03;
                        CRIER  = 0x00;        /* disable all receive interrupts */
                        CTCR   = 0x00;        /* disable all transmit interrupts */
                        CMCR0 &= 0xFE;        /* quit soft reset mode */
                        CMCR0 = 0x20;
              }
```

▲

Example 12.4
▼

Write a program to initialize the CAN module of the 912BC32 with the same system constraints as in Example 12.1. Disable all interrupts. Configure the CAN so that it accepts only messages with extended identifier S8. Use two 32-bit acceptance filters. Stop CAN in wait mode. Choose a clock source with a period twice as long as the ECLK.

Solution: For the identifier "S8" and the following timing parameters:

```
Prescaler          = 1
Nominal bit time   = 12
prop_seg           = 7
sync_seg           = 1
phase_seg1         = 2
phase_seg2         = 2
RJW                = 2
```

- The values to be written into the CBTR0 and CBTR1 registers are $40 and $11, respectively.
- The values to be written into CIDAR0 to CIDAR3 are $53, $3E, $00, and $00.
- The values to be written into CIDMR0 to CIDMR3 can be $00003FFF. We don't care about what follows the string "S8".

The subroutine that performs the CAN initialization is as follows:

```
#include "d:\miniide\hc12.inc"
soft_reset   equ        $01                   ; bit position for soft reset
can_init     bset       CMCR0,soft_reset      ; soft reset the CAN module
             movw       #$4011,CBTR0          ; configure CBTR0 and CBTR1
             movw       #$533E,CIDAR0         ; setup acceptance identifier
             movw       #$0000,CIDAR2         ;   "
             movw       #$0000,CIDMR0         ; accept the first two bytes and highest two
             movw       #$3FFF,CIDMR2         ; bits of CxIDAR2
             movw       #$FFFF,CIDMR4         ; ignore the second bank of identifier     "
             movw       #$FFFF,CIDMR6         ; acceptance registers            "
             movb       #$00,CIDAC            ; select 32-bit filter
             movb       #$03,CMCR1
             clr        CRIER                 ; disable all CAN receive interrupts
             clr        CTCR                  ; disable all CAN transmit interrupts
             bclr       CMCR0,soft_reset      ; quit soft reset mode
```

```
          movb      #$20,CMCR0
          rts
```

The C language version of the routine is straightforward and will be left for you as an exercise.

▲

Example 12.5

▼

One of the microcontrollers in a manufacturing plant is responsible for monitoring the temperature, pressure, and humidity of all the stations in the plant. The microcontroller collects this information over the CAN bus. The microcontroller uses letters "T", "P", and "H" as identifiers for temperature, pressure, and humidity, respectively. Write an instruction sequence to change the setting of CxIDAR0-3, CxIMAR0-3, CxIDAC so that data frames with standard identifiers "T", "P", and "H" will be accepted.

Solution: We will use the 8-bit filter mode to match the most significant 8 bits. The identifiers will be the ASCII codes of letters T, P, and H.

The following instruction sequences will perform the desired changes:

```
          bset      CMCR0,soft_reset     ; soft reset the CAN module
          movb      #$20,CIDAC           ; choose 8-bit filter mode
          movb      #$54,CIDAR0          ; set identifier "T"
          movb      #$50,CIDAR1          ; set identifier "P"
          movb      #$48,CIDAR2          ; set identifier "H"
          movw      #$0000,CIDMR0        ; set up the mask to compare all 8 bits
          movw      #$00FF,CIDMR2        ;         "
          bclr      CMCR0,soft_reset     ; quit soft reset mode
```

Data transmission could be interrupt-driven or program-driven (polling method). The interrupt-driven approach would be beneficial to applications that have multiple tasks to be performed concurrently. CPU time can be utilized more efficiently.

The data transmit operation can be carried out in two stages:

1. Data preparation. This step includes setting up the priority, identifier, and data in a buffer.
2. Copy data into a transmit buffer and start the transmission (by clearing the TXE flag).

▲

Example 12.6

▼

Write a subroutine that transmits data over the CAN bus. The pointer to the data to be transmitted is passed in index register X. The structure of the buffer that holds the data to be transmitted over the CAN bus is identical to a CAN transmit buffer (see Figure 12.17). The number of data bytes to be transmitted is stored in the data buffer with an offset of 12. A remote frame would have a data length of zero. Use the polling method.

Solution: The following routine will transmit the data over the CAN bus:

```
          #include "c:\miniide\hc12.inc"
tx_can    pshy
          pshd
```

```
; ***************************************************************
; The following 11 instructions identify an empty transmit buffer
; ***************************************************************
tb0        brclr      CTFLG,$01,tb1      ; is transmit buffer 0 available?
           ldy        #Tx0               ; make Y points to Tx0
           clra
           bra        send_it
tb1        brclr      CTFLG,$02,tb2      ; is transmit buffer 1 available?
           ldy        #Tx1
           ldaa       #1
           bra        send_it
tb2        brclr      CTFLG,$04,tb0      ; is transmit buffer 2 available?
           ldy        #Tx2
           ldaa       #2
send_it    movw       2,X+,2,Y+          ; copy identifiers
           movw       2,X+,2,Y+          ;   "
           movw       8,X,8,Y            ; copy data length & transmit buffer priority
           ldab       8,X                ; copy the data byte count into B
           beq        done
loop       movb       1,X+,1,Y+          ; copy one byte of data at a time
           dbne       b,loop
done       cmpa       #0
           bne        chk_1
           bclr       CTFLG,$FE          ; clear TXE0 flag to trigger the transmission
           bra        exit
chk_1      cmpa       #1
           bne        is_2
           bclr       CTFLG,$FD          ; clear TXE1 flag to trigger the transmission
           bra        exit
is_2       bclr       CTFLG,$FB          ; clear TXE2 flag to trigger the transmission
exit       puld
           puly
           rts
```

The C language version of the routine is as follows:

```
#include <hc12.h>
void tx_can (char *ptr)          /* ptr is a pointer to the data buffer to be transmitted */
{
        unsigned char test, mask, i, k;
        unsigned volatile char *pt;
        test = 0;
        while (!test) {
                if (CTFLG & 0x01) {
                        pt = &Tx0;
                        mask = 0x01;
                        test = 1;
                }
                else if (CTFLG & 0x02) {
                        pt = &Tx1;
                        mask = 0x02;
                        test = 1;
                }
                else if (CTFLG & 0x04){
                        pt = &Tx2;
                        mask = 0x04;
```

```
                                test = 1;
                                }
                        else
                                test = 0;
                }
        for (i = 0; i < 4; i++)
                        *pt++ = *ptr++;         /* copy identifiers */
                *(pt+8) = *(ptr+8);             /* copy data byte count */
                k = *(ptr+8);                   /* get the data byte count */
                *(pt+9) = *(ptr + 9);           /* copy transmit buffer priority */
                if (k) {
                        for (i = 0; i < k; i ++)
                                *pt++ = *ptr++;  /* copy data bytes */
                }
                CTFLG = mask;                    /* clear the corresponding transmit buffer empty flag */
        }
```

▲

Example 12.7

▼

Write a subroutine that reads the receive buffer using the polling method. A pointer to the buffer to hold the received frame is passed to this routine (in index register X for the assembly language).

Solution: This routine will continuously checking the RXE flag until it is set and copy the data from receive buffer.

```
        #include "c:\miniide\hc12.inc"
can_rx  pshy
        pshb
        brclr   CRFLG,$01,*       ; wait until received frame is full
        ldy     #RxFG+4           ; set Y to point to the first received data byte
        ldab    8,Y               ; place data length in B
        beq     done              ; is data length equal to 0?
loop    movb    1,Y+,1,X+         ; copy data one byte a time
        dbne    b,loop
done    bclr    CRFLG,$FE         ; clear RXE flag
        pulb
        puly
        rts
```

The C language version of the routine will be left for you as an exercise.

▲

Example 12.8

Suppose a CAN node is expected to transmit the value $204509 over the CAN bus using the highest priority. The identifier to be used is the string "T1". The identifier should be represented in extended format. Depict the contents of the transmit buffer when the message is to be transmitted.

Solution: The meaning and value of each bit in the transmit buffer is shown in Figure 12.48.

Register Name	7	6	5	4	3	2	1	0	Register Contents	Meaning
IDR0	ID28	ID27	ID26	ID25	ID24	ID23	ID22	ID21	0 1 0 1 0 1 0 0	T
IDR1	ID20	ID19	ID18	SRR	IDE	ID17	ID16	ID15	0 0 1 0 1 1 0 0	1[1]
IDR2	ID14	ID13	ID12	ID11	ID10	ID9	ID8	ID7	0 1 0 0 0 0 0 0	
IDR3	ID6	ID5	ID4	ID3	ID2	ID1	ID0	RTR	0 0 0 0 0 0 0 0	
DSR0	D7	D6	D5	D4	D3	D2	D1	D0	0 0 1 0 0 0 0 0	$20
DSR1	D7	D6	D5	D4	D3	D2	D1	D0	0 1 0 0 0 1 0 1	$45
DSR2	D7	D6	D5	D4	D3	D2	D1	D0	0 0 0 0 1 0 0 1	$09
DSR3	D7	D6	D5	D4	D3	D2	D1	D0	x x x x x x x x	
DSR4	D7	D6	D5	D4	D3	D2	D1	D0	x x x x x x x x	
DSR5	D7	D6	D5	D4	D3	D2	D1	D0	x x x x x x x x	
DSR6	D7	D6	D5	D4	D3	D2	D1	D0	x x x x x x x x	
DSR7	D7	D6	D5	D4	D3	D2	D1	D0	x x x x x x x x	
DLR	L7	L6	L5	L4	L3	L2	L1	L0	0 0 0 0 0 0 1 1	3[2]
TBPR	TP7	TP6	TP5	TP4	TP3	TP2	TP1	TP0	0 0 0 0 0 0 0 0	

Note 1. Register IDR1 and the bit 7 and bit 6 of IDR2 forms the ASCII code of "1" and the SRR bit, IDE bit.
2. Data length is 3.

Figure 12.48 ■ Transmit buffer contents

Users of the CAN bus can add commands into data frames to request the receiver of the data frame to perform a certain operation. One example is using the *data segment register 0* (DSR0) to carry the command and other data segment registers to carry arguments to the command.

There is no limit on what can be performed over the CAN bus. The only limitation is our imagination.

12.17 Summary

The CAN specification was initially proposed as a data communication protocol for automotive applications. However, it can also fulfill the communication needs of a wide range of applications, from high-speed networks to low-cost multiplex wiring.

The CAN protocol has gone through several revisions. The latest revision is 2.0A/B. The CAN 2.0A uses standard identifier, while the CAN 2.0B specification accepts extended identifiers. The Motorola MSCAN12 implements the latest CAN specifications.

There are four types of frames: data frames, remote frames, error frames, and overload frames. Users only need to transfer data frames and remote frames. The other two types of frames are only used by the CAN controller to control data transmission and reception.

The CAN protocol allows all nodes on the bus to transmit simultaneously. When there are multiple transmitters on the bus, they arbitrate and the CAN node transmitting a message with the highest priority wins. The simultaneous transmission of multiple nodes will not cause any damage to the CAN bus. CAN data frames are acknowledged in-frame; that is, a receiving node only sets a bit in the acknowledge field of the incoming frame. There is no need to send a separate acknowledge frame.

The CAN bus has two states: *dominant* and *recessive*. The dominant state represents logic 0 and the recessive state represents logic 1 for many CAN implementations. When one node drives the dominant voltage to the bus while other nodes drive the recessive level to the bus, the resultant CAN bus state will be dominant state.

Synchronization is a very critical issue in the CAN bus. Each bit time is divided into four segments: *sync_seg*, *prop_seg*, *phase_seg1*, and *phase_seg2*. The *sync_seg* segment signals the start of a bit time. The sample point is between *phase_seg1* and *phase_seg2*. At the beginning of each frame, every node performs a hard synchronization to align its *sync_seg* segment of its current bit time to the recessive to dominant edge of the transmitted start-of-frame. Resynchronization is performed during the remainder of the frame, whenever a change of bit value from recessive to dominant occurs outside of the expected *sync_seg* segment. Resynchronization is achieved by lengthening the *phase_seg1* segment or shortening the *phase_seg2* segment.

The bus timing registers 0 and 1 (CxBTR0 and CxBTR1) are used to set up the bit timing parameters. Data transmission involves the following steps:

- Set up the contents of identifier registers (IDR0 to IDR3).
- Copy the data to be transmitted into data segment registers (DSR0 to DSR7).
- Load the data byte count into the data length register (DLR).
- Set up the transmit buffer priority register (TBPR).
- Clear the TXEi (i = 0..2) flag to start the transmission.

The CAN module uses a filter to determine whether to accept an incoming frame. The filter can be 32-bit, 16-bit, or 8-bit. As long as there is no error in the incoming frame and the acceptance filter matches, the incoming data frame will be copied into the foreground receive buffer and becomes available to the user.

A CAN module (controller) requires a transceiver such as the Philips PCA82C250 to interface with the CAN bus. The design of the PCA82C250 allows more than 110 nodes to attach to a single CAN bus. The CAN trunk cable could be a shielded cable, unshielded twisted pair, or simply a pair of insulated wires. It is recommended that you use shielded cables for high-speed transfer when there is an RFI problem. Up to 1 Mbps data rate have been achieved over a distance of 40 meters.

Examples are used to illustrate the process of selecting bit timing parameters, CAN control register configuration, and data transmission and reception.

12.18 Exercises

E12.1 Calculate the bit segments for the following system constraints:

Bit rate = 250Kbps
Bus length = 50 m
Bus propagation delay = 5×10^{-9} sec/m
PCA82C250 transceiver plus receiver propagation delay = 150 ns at 85°C
CPU oscillator frequency = 8 MHz

E12.2 Calculate the bit segments for the following system constraints:

Bit rate = 125Kbps
Bus length = 100 m
Bus propagation delay = 5×10^{-9} sec/m
PCA82C250 transceiver plus receiver propagation delay = 150 ns at 85°C
CPU oscillator frequency = 8 MHz

E12.3 Calculate the bit segments for the following system constraints:

Bit rate = 1Mbps
Bus length = 20 m
Bus propagation delay = 5×10^{-9} sec/m
PCA82C250 transceiver plus receiver propagation delay = 150 ns at 85°C
CPU oscillator frequency = 16 MHz

E12.4 Calculate the bit segments for the following system constraints:

Bit rate = 500 Kbps
Bus length = 100 m
Bus propagation delay = 5×10^{-9} sec/m
PCA82C250 transceiver plus receiver propagation delay = 150 ns at 85°C
CPU oscillator frequency = 16 MHz

E12.5 Write a routine to initialize the CAN module of the 912BC32 with the same system constraints as in Example 12.1. Enable receive interrupts but disable transmit interrupt. Configure the CAN so that it accepts only messages with the standard identifier S. Use 8-bit acceptance filters. Stop CAN in wait mode. Use a clock source with a period twice as long as the ECLK.

E12.6 Write a routine to initialize the CAN module of the 912BC32 with the same system constraints as in Example 12.2. Enable all interrupts. Configure the CAN so that it accepts only messages with the extended identifier 32. Use two 32-bit acceptance filters. Stop CAN in wait mode. Use a clock source with a period twice as long as the ECLK.

E12.7 Write an instruction sequence to set up the transmit buffer to transfer the string "T is low" over the CAN bus. Set the identifier to "T2" and set the transmit buffer priority to zero.

E12.8 Write a polling version of the program to call the *can_tx* subroutine to send out a block of data terminated by a null character over the CAN bus.

E12.9 Suppose a CAN node is expected to transmit the value $344508102002 over the CAN bus using the lowest priority. The identifier to be used is the string "T9". The identifier should be represented in extended format. Depict the contents of the transmit buffer when the message is to be transmitted.

E12.10 Suppose a CAN node is expected to transmit the string "time out" over the CAN bus using the highest priority. The identifier to be used is the string "S15". The identifier should be represented in extended format. Depict the contents of the transmit buffer when the message is to be transmitted.

12.19 Lab Exercises & Assignments

L12.1 Practice data transfer over the CAN bus using the following procedures:

Step 1

Use a pair of insulated wires about 20 meters long, connected together at both ends via a 120Ω register.

Step 2

Attach two CME-12BC demo boards to the CAN bus at both ends using appropriate connectors.

Step 3

Write a program to be downloaded onto one (called *board A*) of the demo boards that performs A/D conversion once every second. Use the function generator to generate a triangular waveform to be converted. Send out the A/D conversion result over the CAN bus every 100 ms. Simply use the letter "T" as the identifier of the data frame.

Step 4

Write a program to be downloaded onto another demo board (called *board B*). This program will send out the number of data frames received so far over the CAN bus. This program will use the letter "R" as the identifier. After the number reaches 99, the program will reset the number to 0 and start over again.

Step 5

Board A will display the number received over the CAN bus on two seven-segment displays.

Step 6

Board B will display the received A/D result over the CAN bus in four seven-segment displays.

13

Internal Memory Configuration & External Memory Expansion

13.1 Objectives:

After completing this chapter, you should able to:

- perform the remapping of the register file, on-chip SRAM, on-chip flash memory, and on-chip EEPROM

- explain the mechanism of memory expansion above 64KB

- control the operation, programming, and protection of the on-chip flash memory

- erase and program the on-chip flash memory

- erase and program the on-chip EEPROM

- explain the external memory expansion issue

- make memory space assignment

- design address decoders and memory control circuitry

- explain bus signal waveforms

- perform timing analysis for the memory system

13.2 Overview of the 68HC12 Memory System

Each of the 68HC12 members has a certain amount of on-chip SRAM, EEPROM, and flash memory. The amount of on-chip memories of each 68HC12 member is listed in Table 3.1. These memories are assigned to a certain default memory space after reset. To meet the requirements of different applications, these memories can be remapped to other locations. This can be achieved by programming the appropriate initialization register associated with each of these memories.

Both the EEPROM and the flash memory must be erased before they can be programmed correctly.

Most of the 68HC12-based embedded systems operate in the single chip mode and do not need external memory. Different members of the 68HC12 microcontroller family can satisfy the memory requirements of a wide range of applications. Demo boards that are designed for learning the 68HC12 also provide external SRAM and EEPROM to facilitate the program downloading and testing process. The 68HC12 must be configured to operate in expanded mode in order to access external memory. Two I/O ports become unavailable for I/O operations in expanded mode.

Address and data signals are multiplexed in all 68HC12 members except the 812A4. The 812A4 has separate address and data pins. For other 68HC12 members, the address signals must be latched onto a register such as the HC573 or the HC373 so that address signals remain stable and valid throughout the whole memory access cycle. A memory access cycle is called a *bus cycle*. The inverted E clock signal ECLK can be used to latch the address for de-multiplexing. Use of the ECLK signal is controlled by the NECLK bit of the PEAR register.

When adding external memory to the 68HC12, we must make sure that the timing requirements are satisfied. Each bus cycle takes exactly one E clock cycle to complete. However, the bus cycle can be stretched to support slower memory devices. A bus cycle can be stretched by one to three E clock cycles.

A 16-bit microcontroller like the 68HC12 can only access up to 64 KB of memory space. A special bank-switching technique enables some 68HC12 members to access more than 64 KB memory space. This expanded memory is made available to user programs through the 16KB page window from $8000 to $BFFF. The mapping is done by programming the PPAGE register.

The 812A4 can access up to 4 MB external memory because it has 22 address lines. The 912DG128 and 912DT128 have 128 KB on-chip flash memory, whereas the 9S12DP256 has 256KB on-chip flash memory. These expanded on-chip or external memories are made accessible to the user program via the 16KB window.

13.3 Internal Resource Mapping

The internal register block, SRAM, EEPROM, and flash memory have default locations within the 64 KB standard address space, but may be reassigned to other locations during program execution by setting bits in mapping registers INITRG, INITRM, and INITEE. During normal operation modes these registers can be written once. It is advisable to explicitly establish these resource locations during the initialization phase of program execution, even if default values are chosen, in order to protect the registers from inadvertent modification later.

Writes to the mapping registers go into effect between the cycle that follows the write and the cycle after that. To ensure that there are no unintended operations, a write to one of these registers should be followed by a NOP instruction.

If conflict occurs when mapping resources, the register block will take precedence over other resources; RAM or EEPROM addresses occupied by the register block will not be available for storage. When active, BDM ROM takes precedence over other resources, although a conflict between BDM ROM and register space is not possible. Table 13.1 shows mapping precedence.

Precedence	Resource
1	BDM ROM (if active)
2	Register space
3	SRAM
4	EEPROM
5	flash memory
6	External memory

Table 13.1 ■ Mapping precedence

Only one module will be selected at a time. In the case of more than one module sharing a space, only the module with the highest precedence will be selected. Mapping more than one module to the same location won't damage the system. However, it is not a good idea to map two or more modules to the same location because a significant amount of memories could become unusable.

In expanded modes, all address space not used by internal resources is by default external memory.

13.3.1 Register Block Mapping

After reset, the 512-byte (or 1KB) register block resides at location $0000 but can be reassigned to any 2KB boundary within the 64KB address space. Mapping of internal registers is controlled by five bits in the INITRG register. The contents of this register are shown in Figure 13.1. The bit 7 and bit 0 are always zero for the 9S12DP256.

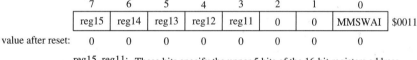

7	6	5	4	3	2	1	0	
reg15	reg14	reg13	reg12	reg11	0	0	MMSWAI	$0011

value after reset: 0   0   0   0   0   0   0   0

reg15..reg11: These bits specify the upper 5 bits of the 16-bit registers address. These five bits can be written only once in normal modes and can be written many times in special modes. There is no restriction on the reading of this register.

MMSWAI: Memory mapping interface stop in wait mode.
0 = Memory mapping interface continues to function during wait mode.
1 = Memory mapping interface access is shut down during wait mode.

Figure 13.1 ■ Contents of the INITRG register

Example 13.1

▼

Write an instruction sequence to remap the register block to $2000.

Solution: The upper five bits of the INITRG register should be set to 00100. The following instruction sequence will achieve the desired assignment:

```
movb      #$20,INITRG
nop                         ; wait for the remap to take effect
```

▲

13.3.2 RAM Mapping

The amount of on-chip SRAM and the default starting addresses of different HC12 members are listed in Table 13.2.

Member	Capacity (KB)	Default start address
812A4	1	$0800
912B32	1	$0800
912BC32	1	$0800
912BE32	1	$0800
912D60	2	$0000
912DG128	8	$2000
912DT128	8	$2000

Table 13.2 ■ 68HC12 on-chip SRAM

Like other internal memory resources, SRAM can be remapped to 2KB (for the 812A4, B family, and the 912D60) or 8KB (for 912DG128 and 912DT128 only) boundary within the 64KB space. The remapping of SRAM is controlled by the INITRM register. The contents of the INITRM are shown in Figure 13.2.

	7	6	5	4	3	2	1	0	
Value after reset:	RAM15	RAM14	RAM13	RAM12	RAM11	0	0	0	$0010
812A4, 912B series	0	0	0	0	1	0	0	0	
912D60	0	0	0	0	0	0	0	0	
912DG128/DT128	0	0	1	0	0	0	0	0	

Note: Only the upper three bits are defined for the 912DG128/DT128

Figure 13.2 ■ RAM initialization register (INITRM)

13.3.3 EEPROM Mapping

The HC12 members have from 768 bytes (B series) up to 4KB (812A4) of on-chip EEPROM. The EEPROM is activated by the EEON bit in the EEPROM initialization register (INITEE).

The EEPROM can be remapped to any 4KB boundary and is controlled by the upper four bits of the INITEE register. After reset EEPROM address space begins at location:

- $1000 for the 812A4
- $0D00 for the B series (912B32, 912BC32, and 912BE32)
- $0C00 for the 912D60 (and the 12D60)
- $0800 for the 912DG128 and the 912DT128

The contents of the INITEE register are shown in Figure 13.3. Bit 0 is forced to 1 in single-chip mode.

	7	6	5	4	3	2	1	0	
	EE15	EE14	EE13	EE12	0	0	0	EEON	$0012
value after reset: 812A4	0	0	0	1	0	0	0	1	
other members:	0	0	0	0	0	0	0	1	

EE15..EE12: These bits specify the upper 4 bits of the 16-bit registers address. These four bits can be written only once in normal modes and can be written many times in special modes. There is no restriction on the reading of this register.

EEON: Internal EEPROM On (Enabled reading).
0 = Removes EEPROM from the map.
1 = Places the on-chip EEPROM in the memory map.

Figure 13.3 ■ Contents of the INITEE register

Example 13.2

Write an instruction sequence to remap the EEPROM of the 912BC32 to start at $4000.

Solution: The upper four bits of the INITEE register must be set to 0100. The following instruction sequence will achieve the desired mapping:

```
movb      #$41,INITEE      ; remap EEPROM and enable it
nop                        ; wait for the remap to take effect
```

13.3.4 Flash Memory Mapping

The 812A4 does not have on-chip flash memory. The B family microcontrollers have 32 KB of flash memory. This memory can be mapped to either the block of $0000~$7FFF or $8000~$FFFF, depending on the setting of the MAPROM bit (bit 1) of the MISC register. When this bit is zero, the flash memory is located from $0000 to $7FFF. Otherwise, it is located from $8000 to $FFFF.

The 912D60 has 60KB of on-chip flash memory. It is divided into 32KB and 28KB blocks. The bit 7 (MAPROM bit) of the MISC register selects the location of the on-chip flash memory. When this bit is zero, the 28KB block is mapped from $1000 to $7FFF and the 32KB block is mapped from $8000 to $FFFF. Otherwise, the 28KB block is mapped from $9000 to $FFFF and the 32KB block is mapped from $0000 to $7FFF.

13.3.5 Program Space Expansion

This section applies to the 912DG128 and 912DT128 only. Both the 912DG128 and 912DT128 have 128KB of on-chip flash memory. It is divided into eight 16KB pages. An application program needs to use a 16KB program space window located at $8000 to $BFFF to access them. The lowest three bits of the PPAGE register select a page to be accessible at a time. The flash memory page space index is shown in Table 13.3. The 16KB page 6 can be accessed at a fixed location from $4000 to $7FFF. The 16KB page 7 can be accessed at a fixed location from $C000 to $FFFF. The rightmost column of Table 13.3 refers to the name of the flash memory array to which the flash memory page belongs. Because flash memory is electrically programmable and erasable, it is also called flash EEPROM by some companies, including Motorola.

Page index 2 (Ppage bit 2)	Page index 1 (Ppage bit 1)	Page index 0 (Ppage bit 0)	16K page no.	Flash array
0	0	0	0	00FEE32K
0	0	1	1	00FEE32K
0	1	0	2	01FEE32K
0	1	1	3	01FEE32K
1	0	0	4	10FEE32K
1	0	1	5	10FEE32K
1	1	0	6*	11FEE32K
1	1	1	7*	11FEE32K

Note: Page 6 and page 7 are also accessible from $4000-$7FFF and $C000-$FFFF, respectively.

Table 13.3 ■ Flash space page index

The 128KB flash memory of the 912DG128 and 912DT128 is divided into four 32KB flash arrays, and a 4-byte register block controls the operation of each array. A register space window is used to access one of the four 4-byte blocks. The bit 2 and bit 1 of the PPAGE register map one register block into the window. The register space window is located from $00F4 to $00F7 after reset. The flash register space page indices are shown in Table 13.4.

Page index 2 (Ppage bit 2)	Page index 1 (Ppage bit 1)	Page index 0 (Ppage bit 0)	Register space page no.	Flash array
0	0	x	$F4-$F7 page 0	00FEE32K
0	1	x	$F4-$F7 page 1	01FEE32K
1	0	x	$F4-$F7 page 2	10FEE32K
1	1	x	$F4-$F7 page 3	11FEE32K

Table 13.4 ■ Flash register space page index

In special mode and for testing purposes only, the 128KB of flash memory for the 912DG128/DT128 can be accessed through a test program space window of 32KB. This window replaces the user's program space window, enabling it to access an entire array. In special mode and with the ROMTST bit set in the MISC register, a program space is located from $8000 to $FFFF. Only two page indices are used to point to one of the four 32KB arrays. These indices can be viewed as the expanded address signals A16 and A15.

Additional flash memory mapping controls are provided by the MISC register. The contents of the MISC register are shown in Figure 13.4.

7	6	5	4	3	2	1	0
0	NDRF	RFSTR1	RFSTR0	EXSTR0	EXSTR1	MAPROM	ROMON

Reset states

| Expanded modes: | 0 | 0 | 0 | 0 | 1 | 1 | 0 | 0 |
| Single-chip modes: | 0 | 0 | 0 | 0 | 1 | 1 | 1 | 1 |

(a) B family MISC register

7	6	5	4	3	2	1	0
MAPROM	NDRF	RFSTR1	RFSTR0	EXSTR0	EXSTR1	ROMON28	ROMON32

Reset states

| Expanded modes: | 0 | 0 | 0 | 0 | 1 | 1 | 0 | 0 |
| Single-chip modes: | 0 | 0 | 0 | 0 | 1 | 1 | 1 | 1 |

(b) 912D60 MISC register

7	6	5	4	3	2	1	0
ROMTST	NDRF	RFSTR1	RFSTR0	EXSTR0	EXSTR1	ROMHM	ROMON

Reset states

| Expanded modes: | 0 | 0 | 0 | 0 | 1 | 1 | 0 | 0 |
| Single-chip modes: | 0 | 0 | 0 | 0 | 1 | 1 | 1 | 1 |

(c) 912DG128/DT128 MISC register

Figure 13.4 ■ MISC register ($0013)

Bits 6 to 2 are common to the B series, the 12D60, and the 912DG128/DT128. The 512-byte memory space following the 512-byte register block is called the *register-following map*. This 512-byte block could be implemented by an external memory chip or peripheral devices. The CME-12BC demo board assigns the LCD device to this area. The NDRF (bit 6) bit enables the narrow bus feature for the 512-byte register-following map. This bit only has effect in expanded wide mode. When it is zero, the register-following map acts the same as an 8-bit external data bus. Otherwise, the register-following map acts as a full 16-bit external data bus.

Bits 5 and 4 determine the amount of clock stretch on accesses to the 512-byte register-following map. In single-chip and peripheral modes, these bits have no meaning or effect. The meanings of these two bits are shown in Table 13.5. The normal 68HC12 bus cycle takes one E clock cycle to complete. After stretching the clock signal, a slower memory chip (with longer access time) can be used and assigned to the 512-byte register following map. Clock stretching will be illustrated later.

RFSTR1 and RFSTR0	E clocks stretched
00	0
01	1
10	2
11	3

Table 13.5 ■ Register-following stretch bit function

Bits 3 and 2 determine the amount of clock stretch on accesses to the external address space. All external memory chips must be stretched by the same number of E clock cycles. In single-chip and peripheral modes, these bits have no meaning or effect. The meanings of these two bits in expanded mode are shown in Table 13.6. By stretching the clock signal, slower but less expensive memory chips can be used to expand the 68HC12's memory capacity. After reset, the bus cycles of external memory devices are stretched by three E clock cycles so that the timing requirements of slow non-volatile memory can be satisfied. Stretching the bus cycle by three E clock cycles allows the reset service routine contained in a slow non-volatile memory (EPROM, EEPROM, or flash memory) to be executed. If external memory chips can operate at a faster rate, the reset routine can reduce the stretched cycles to a smaller number. This is done in the CME-12BC demo board. The CME-12BC demo board stretches the bus cycle by one E clock cycle instead of the three E clock cycles.

EXSTR1 and EXSTR0	E clocks stretched
00	0
01	1
10	2
11	3

Table 13.6 ■ Expanded stretch bit function

Bit 7 of the MISC register for the 912D60 determines the location of flash memory. When it is zero, the 28KB flash memory array is mapped to $1000–$7FFF and the 32KB array is mapped to $8000–$FFFF. Otherwise, the 28KB flash array is mapped to $9000–$FFFF and the 32 KB array is mapped to $0000–$7FFF. Bit 1 and bit 0 of the MISC register enable and disable the 32KB and 28KB flash memory arrays. When set to one, bit 1 and bit 0 enable the 28KB and 32KB flash memory arrays, respectively.

For the 912DG128 and the 912DT128, the bit 7 of the MISC register sets the flash memory window in test mode. When it is zero, the 16KB window is selected and is located from $8000–$BFFF. Otherwise, the 32KB window is selected and is located at $8000–$FFFF. In other modes, this bit is forced to zero. Bit 1 controls the access of flash memory in the first half of the memory map. When it is zero, the 16KB flash memory located at $4000–$7FFF can be accessed. Otherwise, this block cannot be accessed. This bit has no meaning when bit 0 is cleared to zero. Flash memory is enabled when bit 0 is set to one. Otherwise, the whole flash memory cannot be accessed.

For the B series devices, bit 1 (MAPROM) determines the location of the flash memory. When this bit is zero, the flash memory is located at $8000–$FFFF. Otherwise, the flash memory is located at $0000–$7FFF. Bit 0 is the flash memory enable bit. When it is one, the 32KB flash memory is enabled.

13.4 The Flash Memory Operation

The flash memory array is arranged in a 16-bit configuration and may be read as bytes, aligned words, or misaligned words. Access time is one bus cycle for byte and aligned word access and two bus cycles for misaligned word operations. The lowest address bit (A0) of an aligned word access is zero. The address bit A0 equals one for a misaligned word access.

The flash memory module requires an external program/erase voltage (V_{FP}) to program or erase. The external program/erase voltage is provided to the flash memory via the V_{FP} pin. To prevent damage to the flash array, V_{FP} should always be greater than or equal to $V_{DD} - 0.5V$. Programming is by byte or aligned word. The flash memory supports bulk erasure only.

The flash memory has hardware interlocks that protect stored data from accidental corruption. An erase- and program-protected block (1K, 2K, or 8KB) for boot routines is provided in each flash memory array.

13.4.1 Flash Memory Control

Four registers control the operation, programming, and protection of the flash memory. The four flash memory control registers for the B family members are located at $00F4~$00F7. The flash memory control registers for the 32KB/28KB flash memory arrays in the 912D60 are located at $00F4~$00F7 and $00F8~$00FB. Each of the four 32KB arrays of the 912DG128/DT128 flash memory has four control registers. Only one set of them is accessible at a time via the window selected by the bit 2 and bit 1 of the PPAGE register (shown in Table 13.4). The second address attached to each control register applies to the 912D60 only (see Figures 13.4 through 13.7).

The *FEELCK* register enables and disables the write operation to the flash memory. The contents of this register are shown in Figure 13.5.

LOCK: lock register bit
  0 = enable write to FEEMCR register
  1 = disable write to FEEMCR register

Note. This register is denoted as FEE32LCK and FEE28LCK for 32K
        and 28K arrays, respectively.

Figure 13.5 ■ The FEELCK register

A small part (called the *boot block*) of the flash 32KB array can be protected from erasing and programming. The *FEEMCR* register enables and disables the programming and erasing of the boot block of the flash memory array. The contents of this register are shown in Figure 13.6. The boot block is 1KB or 2KB for the B family, and is 8KB for the 912D60 and the 912DG128/912DT128. The address space for the boot block is shown in Table 13.7.

BOOTP: Boot protect
  0 = enable erase and program of boot block
  1 = disable erase and program of boot block

BOOTP cannot be changed when the LOCK control bit in the
FEELCK register is set or if ENPE in the FEECTL register is set.

Note. For the 912D60, this register is denoted as FEE32MCR and
        FEE28MCR for 32K and 28K arrays, respectively.

Figure 13.6 ■ The FEEMCR register

68HC12 member	Address space
B series (1 KB)	$7C00-$7FFF or $FC00-$FFFF
B series (2 KB)	$7800-$7FFF or $F800-$FFFF
912D60	$6000-$7FFF or $E000-$FFFF
912DG128	$E000-$FFFF or $A000-$BFFF
912DT128	$E000-$FFFF or $A000-$BFFF

Note. Only the earliest 912BC32 parts have 1 KB boot block. All later parts have 2 KB boot block.

Table 13.7 ■ Address of boot block

The *FEETST* register is used in testing the flash memory array. The contents of this register are shown in Figure 13.7.

	7	6	5	4	3	2	1	0	
value after	FSTE	GADR	HVT	FENLV	FDISVFP	VTCK	STRE	MWPR	$00F6/$00FA
reset	0	0	0	0	0	0	0	0	

FSTE: Stress test enable
  0 = Disables the gate/drain stress circuitry
  1 = Enables the gate/drain stress circuitry
GADR: Gate/Drain stress test select
  0 = selects the drain stress circuitry
  1 = selects the gate stress circuitry
HVT: Stress test high voltage status
  0 = high voltage not present during stress test
  1 = high voltage present during stress test
FENLV: Enable low voltage
  0 = Disables low voltage transistor in current reference circuit
  1 = Enables low voltage transistor in current reference circuit
FDISVFP: Disable status V_{FP} voltage lock
  When the VFP pin is below normal programming voltage the
  flash module will not allow writing to the LAT bit; the user
  cannot erase or program the flash module. The FDISVFP control
  bit enables writing to the LAT bit regardless of the voltage on the
  V_{FP} pin.
  0 = Enable the automatic lock mechanism if V_{FP} is low.
  1 = Disable the automatic lock mechanism if V_{FP} is low.
VTCK: V_T check test enable
  0 = V_T test disable
  1 = V_T test enable
STRE: Spare test row enable
  0 = LIB accesses are to the flash memory array
  1 = Spare test row in array enabled if SMOD is active
MWPR: Multiple word programming
  0 = Multiple word programming disabled
  1 = Program 32 bits of data
Note. For the 912D60, this register is denoted as FEE32TST and
      FEE28TST for 32K and 28K arrays, respectively.

Figure 13.7 ■ The FEETST register

The *FEECTL* register controls the programming and erasing of the flash memory. The contents of this register are shown in Figures 13.8a and 13.8b.

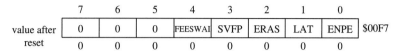

	7	6	5	4	3	2	1	0	
value after	0	0	0	FEESWAI	SVFP	ERAS	LAT	ENPE	$00F7
reset	0	0	0	0	0	0	0	0	

FEESWAI: flash memory stop in wait mode
  0 = Do not halt flash memory clock when the part is in wait mode
  1 = Halt flash memory clock when the 68HC12 is in wait mode
SVFP: Status V_{FP} voltage
  0 = Voltage of V_{FP} pin is below normal programming voltage levels
  1 = Voltage of V_{FP} pin is above normal programming voltage levels
ERAS: Erase control
  0 = Flash memory configured for programming
  1 = Flash memory configured for erasure
LAT: Latch control
  0 = Programming latches disabled
  1 = Programming latches enabled
ENPE: Enable programming/Erase
  0 = Disconnects program/erase voltage to flash memory
  1 = Applies program/erase voltage to flash memory
  ENPE can be asserted only after LAT has been asserted and a write to
  the data and address latches has occurred. If an attempt is made to assert
  ENPE when LAT is negated, or if the latches have not been written to
  after LAT was asserted, ENPE will remain negated after the write cycle
  is complete. The LAT, ERAS, and BOOTP bits cannot be changed when
  ENPE is asserted. A write to FEECTL may only affect the state of
  ENPE. Attempts to read a flash memory array location while ENPE is
  asserted will not return the data addressed. The effects of ENPE, LAT,
  and ERAS on read are shown in Table 13.8.

Figure 13.8a ■ The B family FEECTL register

ENPE	LAT	ERAS	Result of read
0	0	-	Normal read of location addressed
0	1	0	Read of location being programmed
0	1	1	Normal read of location addressed
1	-	-	Read cycle is ignored

Table 13.8 ■ Effects of ENPE, LAT, and ERAS on array reads

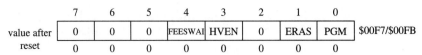

	7	6	5	4	3	2	1	0	
value after	0	0	0	FEESWAI	HVEN	0	ERAS	PGM	$00F7/$00FB
reset	0	0	0	0	0	0	0	0	

FEESWAI: flash memory stop in wait mode
  0 = Do not halt flash memory clock when the part is in wait mode
  1 = Halt flash memory clock when the 68HC12 is in wait mode
HVEN: high-voltage enable
  0 = disables high voltage to array and charge pump
  1 = Enables high voltage to array and charge pump
  This bit can only be set if either PGM or ERAS are set and the proper
  sequence for program or erase is followed.
ERAS: Erase control
  0 = Erase operation is not selected
  1 = Select erase operation
PGM: Program control
  0 = Program operation is not selected.
  1 = Program operation is selected.

Note. For the 912D60, this register is denoted as FEE32CTL and
    FEE28CTL for 32K and 28K arrays, respectively.

Figure 13.8b ■ The FEECTL register for 912D60, 912DG128, and 912DT128

The *FEESWAI* bit allows the user to save more power by stopping the clock input to the flash memory during the wait mode. The status bit *SVFP* indicates whether the programming voltage is below normal programming levels. The *ERAS* bit configures the flash memory for erasing or programming. The *LAT* bit enables or disables the latches for programming. The *ENPE* bit disconnects or applies the programming voltage to the flash memory array.

13.4.2 Erasing the Flash Memory

The flash memory must be erased before it can be programmed correctly. The following sequence demonstrates the recommended procedure for erasing the flash memory for the B family members. The erasing procedures for the 912D60 and the 912DT128/912DT128 are slightly different and can be found in the technical data enclosed in the CD-ROM. The V_{FP} pin voltage must be at the proper level prior to executing Step 4 for the first time.

Step 1
Turn on V_{FP} (apply program/erase voltage to the V_{FP} pin).

Step 2
Set the LAT bit and the ERAS bit of the FEECTL register to configure the flash memory for erasing.

Step 3
Write to any valid address in the flash memory. This allows the erase voltage to be turned on; the data written and the address written are not important. The boot block will be erased only if the control bit BOOTP is negated.

Step 4
Apply erase voltage by setting ENPE.

Step 5
Delay for a single erase pulse (t_{EPULSE}). The duration of this pulse must be at least 5 ms but not longer than 10 ms.

Step 6

Remove erase voltage by clearing ENPE.

Step 7

Delay while high voltage is turning off (t_{VERASE}). The typical value of t_{VERASE} is 1 ms.

Step 8

Read the entire array to ensure that the flash memory is erased.

1. If all of the flash memory locations are not erased, repeat steps 4 through 7 until either the remaining locations are erased, or until the maximum erase pulses have been applied (n_{EP}). Up to five pulses are needed.

2. If all of the flash memory locations are erased, repeat the same number of pulses as required to erase the array. This provides a 100% erase margin.

Step 9

Read the entire array to ensure that the flash memory is erased.

Step 10

Clear the LAT bit.

Step 11

Turn off V_{FP} (reduce voltage on V_{FP} pin to V_{DD}).

The flowchart for the flash memory erasing sequence is illustrated in Figures 13.9a and 13.9b.

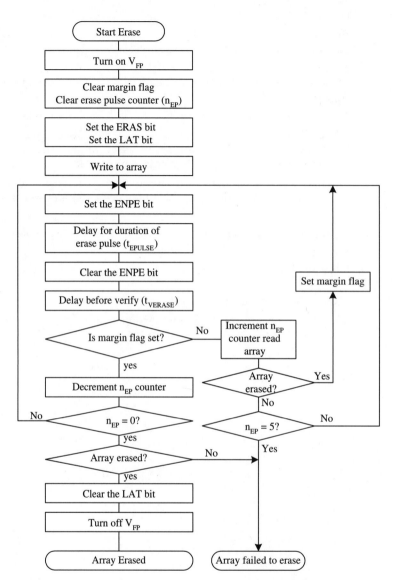

Figure 13.9a ■ Flash memory erase procedure for the 68HC12 B family

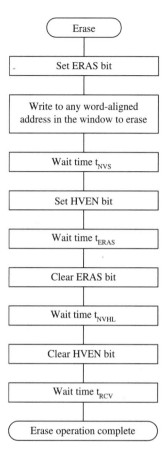

Figure 13.9b ■ Flash memory erase procedure for
912D60, 912DG128/DT128

13.4.3 Programming the Flash Memory

The procedure for programming the flash memory of the B family members is as follows (the programming procedure for the 912D60, the 912DG128, and the 912DT128 can be found on the enclosed CD-ROM):

Step 1
Apply program/erase voltage to the V_{FP} pin.

Step 2
Clear the ERAS bit and set the LAT bit in the FEECTL register to establish program mode and enable programming address and data latches.

Step 3
Write data to a valid address. The address and data latched. If BOOTP is asserted, an attempt to program an address in the boot block will be ignored.

Step 4
Apply programming voltage by setting ENPE.

Step 5
Wait for one programming pulse (t_{PPULSE}). One programming pulse is between 20 to 25 μs.

Step 6
Remove programming voltage by clearing the ENPE bit of the FEECTL register.

Step 7
Delay while high voltage is turning off (t_{VPROG}). This step takes about 10 μs.

Step 8
Read the addressed location to verify that it has been programmed.

1. If the location is not programmed, repeat steps 4 through 7 until the location is programmed or until the specified maximum number of program pulses has been reached (n_{PP}).

2. If the location is programmed, repeat the same number of pulses as required to program the location. This provides a 100% program margin.

Step 9
Read the addressed location to verify that it remains programmed.

Step 10
Clear the LAT bit.

Step 11
If there are more locations to program, repeat steps 2 through 10.

Step 12
Turn off V_{FP} (reduce voltage on V_{FP} pin to V_{DD}).

The flowchart for programming the flash memory is shown in Figures 13.10a and 13.10b.

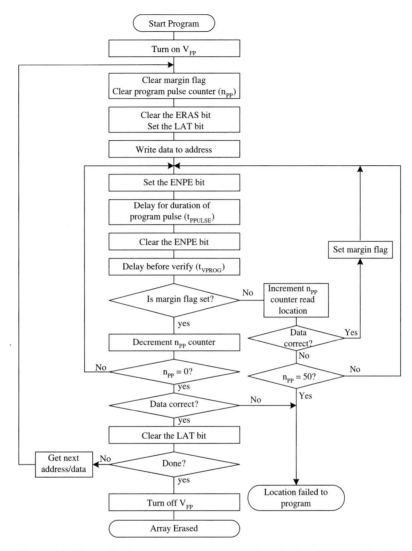

Figure 13.10a ■ Flash memory program procedure for the 68HC12 B family

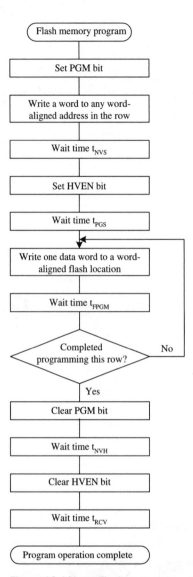

Figure 13.10b ■ Flash memory program procedure for the 912D60, 912DG128, and 912DT128

The purpose of programming the flash memory is mainly to put our application programs or lookup table into the flash memory. It can be done by using an external programmer or by using the BDM pod. We normally do not attempt to erase and program flash memory from application programs. The BDM-12 from Axiom Manufacturing can erase and program the internal flash memory and external EEPROM of the CME-12BC.

13.4.4 Memory Paging in Programming

When paging is enabled, the 68HC12 CPU uses the PPAGE register to view the expanded memory through a 16K window located from $8000 to $BFFF. Since PPAGE is 8-bit wide, the maximum amount of expanded memory is 256 × 16K, or 4 MB. The term CPU address refers to the pair (*bank, address*) that lies within an extended program window. The "bank" must be between 0 and 255 and the "address" must be between $8000 and $BFFF. Due to the way paging is implemented on the 68HC12, it is common to map the last 16KB page of a paged device to also be the last page of the unpaged memory map. For example, a 128K flash device (the 912DG128 or 912DT128) is divided into eight 16KB pages. The last page would have the CPU address (7, $8000–$BFFF) and the unpaged address $C000–$FFFF.

Motorola recommends that the extended program address in the S record output file should be specified as a linear address using the S2 records. For example, for a 128K flash device, the address range is from $0000 to $1FFFF.

The ICC12 professional version supports an expanded program address in the S record output file. To use expanded memory, you would use the physical address to specify the address range of the program memory since it is a linear address for most systems. To enable program paging, you must select a target device (for example, the 912DG128) with expanded memory and then check the "Project -> Options -> Target -> Extended Code Paging" checkbox (see Figure 13.11).

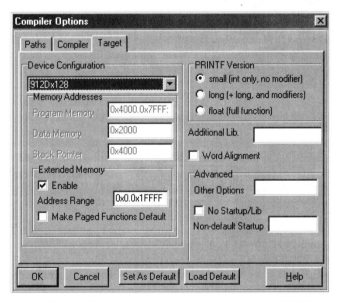

Figure 13.11 ■ Enabling extended memory support in ICC12

When this option is selected, all functions are by default compiled either as paged or non-paged functions, unless otherwise specified by a *pragma* statement. To eliminate confusion, the generated symbol name of a paged function in the assembly file is prefixed by $_ instead of a single underscore _. The library is provided in both forms so that the user can get the benefit of putting the library code in paged memory automatically.

When writing programs in C, it is best to let the C compiler take care of assigning functions to expanded memory pages. It would be most productive for us to focus on program logic. However, in case you prefer to assign some functions to paged memory, use the following statement:

```
#pragma paged_function func_1 func_2 ... func_n
```

where *func_1*, *func_2*, ..., *func_n* are functions to be placed in paged memory.

You can also inform ICC12 not to place a function in a paged memory by using the following statement:

```
#pragma nonpaged_function func_x
```

where *func_x* is a function not to be placed in paged memory.

An outline of a C program that assigns functions to paged and non-paged memory is as follows:

```
#include <hc912dg128.h>
#pragma paged_function random interest
#pragma nonpaged_function disp_time

unsigned int random (unsigned int seed);
float interest (float captial, float rate, int years);
void disp_time (void);

void main (void)
{
        unsigned int ran_1;
        float xy;

        ran_1 = random (23);
        disp_time();
        xy = interest (1000.0, 0.06, 15);
}
unsigned int random (unsigned int seed)
{
...
        return ...;
}

void disp_time (void)
{
...
        printf("\n Current time is hh:mm:ss\n");
}

float interest (float x, float y, int years)
{
...
        return (...);
}
```

Both *random* and *interest* are declared as paged functions and *disp_time* is declared as a non-paged function.

To write assembly programs that utilize expanded memory, you need an assembler that supports expanded memory. Neither the *as12* freeware nor the *MiniIDE* supports expanded memory.

13.5 EEPROM Memory

The on-chip EEPROM of the 68HC12 is often used to store frequently accessed static data or program code for its small size. EEPROM is arranged in a 16-bit configuration. The EEPROM array may be read as bytes, aligned words, or misaligned words. Access time is one bus cycle for byte and aligned word accesses and two bus cycles for misaligned word operations.

Programming is by byte or aligned word. Attempts to program or erase misaligned words will fail. Only the lower byte will be latched and programmed or erased. Programming and erasing of the EEPROM can be done in all operating modes.

Each EEPROM byte or aligned word must be erased before programming. The EEPROM module supports byte, aligned word, row (32 bytes) or bulk erase, all using the internal charge pump. Bulk erasure of odd and even rows is also possible in test modes; the erased state of any byte is $FF. The EEPROM module has hardware interlocks, which protect stored data from corruption by accidentally enabling the program/erase voltage. Programming voltage is derived from the internal V_{DD} supply with an internal charge pump. The EEPROM has a minimum program/erase life of 10,000 cycles over the complete operating temperature range.

13.5.1 EEPROM Programmer's Model

The EEPROM module consists of two separately addressable sections. The first is a 4- or 8-byte (912D60 only) memory mapped control register block used for control, testing, and configuration of the EEPROM array. The second section is the EEPROM array itself.

At reset, the 4-byte (8-byte) register section starts at address $00F0 ($00EC). The size and range of the EEPROM array for each 68HC12 member is shown in Table 13.9.

Member	Size	Address range
812A4	4K	$1000-$1FFF
912B family	768 bytes	$0D00-$0FFF
912D60	1KB	$0C00-$0FFF
912DG128	2KB	$0800-$0FFF
912DT128	2KB	$0800-$0FFF

Table 13.9 ■ Starting address and size of EEPROM

Read accesses to the EEPROM can be enabled or disabled by the EEON bit in the INITEE register. This feature allows the access of memory-mapped resources that have lower priority than EEPROM memory array (when they map to the same area). EEPROM control registers can be accessed and EEPROM locations can be programmed or erased regardless of the state of the EEON bit.

It is possible to continue program/erase operations during the WAIT mode. However, stopping program/erase during WAIT mode can save more power.

If the STOP mode is entered during programming or erasing (the EEPGM bit of the EEPROG register is set), program/erase voltage will be automatically turned off and the RC clock (if enabled) is stopped. However, the EEPGM control bit will remain set. When STOP mode is terminated, the program/erase voltage will automatically be turned back on if EEPGM is set.

13.5.2 EEPROM Control Registers

The programming, erasure, and protection of EEPROM are controlled by four (eight for 912D60) registers. The *EEMCR* register determines whether the EEPROM clock is stopped during wait mode, controls the locking of EEPROM, and selects the charge pump clock source. The contents of this register are shown in Figure 13.12. The EEPROM needs a clock signal to program or erase. The RC oscillator should be selected as the clock source when the system clock is lower than f_{PROG} (1 MHz for the B family, 250 KHz for other members). The user can protect the contents of EEPROM by using block protection. Once the block protection has been set up, the contents of the block protection register can be locked by setting the PROTLCK bit in the EEMCR register.

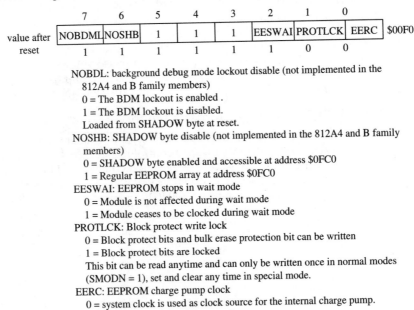

NOBDL: background debug mode lockout disable (not implemented in the
    812A4 and B family members)
    0 = The BDM lockout is enabled .
    1 = The BDM lockout is disabled.
    Loaded from SHADOW byte at reset.
NOSHB: SHADOW byte disable (not implemented in the 812A4 and B family
    members)
    0 = SHADOW byte enabled and accessible at address $0FC0
    1 = Regular EEPROM array at address $0FC0
EESWAI: EEPROM stops in wait mode
    0 = Module is not affected during wait mode
    1 = Module ceases to be clocked during wait mode
PROTLCK: Block protect write lock
    0 = Block protect bits and bulk erase protection bit can be written
    1 = Block protect bits are locked
    This bit can be read anytime and can only be written once in normal modes
    (SMODN = 1), set and clear any time in special mode.
EERC: EEPROM charge pump clock
    0 = system clock is used as clock source for the internal charge pump.
        InternalRC oscillator is stopped.
    1 = Internal RC oscillator drives the charge pump. The RC oscillator is
        required when the system bus clock is lower than f_{PROG}.

Figure 13.12 ■ The EEMCR register

The *EEPROT* register controls the protection of the EEPROM array. The contents of this register are shown in Figure 13.13. An EEPROM block can be protected by setting the corresponding block protection bit as shown in Tables 13.10 through 13.13.

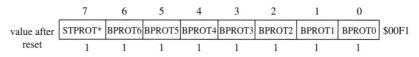

	7	6	5	4	3	2	1	0	
value after	STPROT*	BPROT6	BPROT5	BPROT4	BPROT3	BPROT2	BPROT1	BPROT0	$00F1
reset	1	1	1	1	1	1	1	1	

STPROT: Shadow and test row protection
  0 = Shadow and test rows can be programmed and erased
  1 = Shadow and test rows are protected from being programmed and erased
  This bit is 1 for 812A4.
BPROT[6:0]: EEPROM block protection
  812A4: all 7 bits are used. See Table 13.10.
  B family: BPROT4-BPROT0 are used. See Table 13.11.
  912D60: BPROT4-BPROT0 are used. See Table 13.12.
  912DG128/DT128: BPROT6-BPROT0 are used. See Table 13.13.
  Unused bits are set to 1.

Figure 13.13 ■ The EEPROT register

Bit name	Block protected	Block size
BPROT6	$1000 to $17FF	2048 bytes
BPROT5	$1800 to $1BFF	1024 bytes
BPROT4	$1C00 to $1DFF	512 bytes
BPROT3	$1E00 to $1EFF	256 bytes
BPROT2	$1F00 to $1F7F	128 bytes
BPROT1	$1F80 to $1FBF	64 bytes
BPROT0	$1FC0 to $1FFF	64 bytes

Table 13.10 ■ 4KB EEPROM block protection (812A4)

Bit name	Block protected	Block size
BPROT4	$0D00 to $0DFF	256 bytes
BPROT3	$0E00 to $0EFF	256 bytes
BPROT2	$0F00 to $0F7F	128 bytes
BPROT1	$0F80 to $0FBF	64 bytes
BPROT0	$0FC0 to $0FFF	64 bytes

Table 13.11 ■ 768 byte EEPROM block protection (B family)

Bit name	Block protected	Block size
BPROT4	$0C00 to $0D00	512 bytes
BPROT3	$0E00 to $0EFF	256 bytes
BPROT2	$0F00 to $0F7F	128 bytes
BPROT1	$0F80 to $0FBF	64 bytes
BPROT0	$0FC0 to $0FFF	64 bytes

Table 13.12 ■ 1KB EEPROM block protection (912D60)

Bit name	Block protected	Block size
BPROT5	$0800 to $0BFF	1024 bytes
BPROT4	$0C00 to $0DFF	512 bytes
BPROT3	$0E00 to $0EFF	256 bytes
BPROT2	$0F00 to $0F7F	128 bytes
BPROT1	$0F80 to $0FBF	64 bytes
BPROT0	$0FC0 to $0FFF	64 bytes

Table 13.13 ■ 2KB EEPROM block protection (912DG128/DT128)

The EEPROM module contains an extra word (byte for B family members and the 812A4) called the SHADOW word (byte), which is loaded at reset into the EEMCR, EEDIVH, and EED-IVL registers. The EEDIVH and EEDIVL registers are available only in the 912D60, the 912DG128, and the 912DT128. To program the SHADOW byte, when in special modes (SMODN = 0), the NOSHB bit in the EEMCR register must be cleared. Normal programming routines are used to program the SHADOW byte which becomes accessible at address $0FC0 when the NOSHB bit is cleared.

Example 13.3

▼

Write an initialization subroutine to set up the EEPROM of the B family microcontrollers to allow erasure and programming, and select the system clock as the source for the internal charge pump.

Solution: The initialization routine is as follows:

```
init_eeprom    movb    #$00,EEPROT    ; set the whole EEPROM to be erasable
               movb    #$FC,EEMCR     ; choose system clock as clock source
```

The *EETST* register is provided for testing EEPROM. The contents of this register are shown in Figure 13.14.

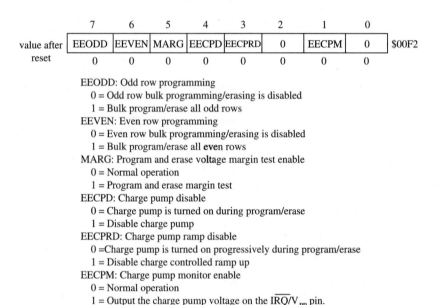

	7	6	5	4	3	2	1	0	
value after	EEODD	EEVEN	MARG	EECPD	EECPRD	0	EECPM	0	$00F2
reset	0	0	0	0	0	0	0	0	

EEODD: Odd row programming
    0 = Odd row bulk programming/erasing is disabled
    1 = Bulk program/erase all odd rows
EEVEN: Even row programming
    0 = Even row bulk programming/erasing is disabled
    1 = Bulk program/erase all **even** rows
MARG: Program and erase voltage margin test enable
    0 = Normal operation
    1 = Program and erase margin test
EECPD: Charge pump disable
    0 = Charge pump is turned on during program/erase
    1 = Disable charge pump
EECPRD: Charge pump ramp disable
    0 = Charge pump is turned on progressively during program/erase
    1 = Disable charge controlled ramp up
EECPM: Charge pump monitor enable
    0 = Normal operation
    1 = Output the charge pump voltage on the $\overline{\text{IRQ}}$/V$_{PP}$ pin.

Figure 13.14 ■ The EETST register

The EEPROG register controls the EEPROM programming and erasing operations. The contents of this register are shown in Figure 13.15. The BULKP, BYTE, ROW, and EELAT bits cannot be changed when EEPGM is set. To complete a program or erase, two successive writes to clear the EEPGM and EELAT bits are required before reading the programmed data. A write to an EEPROM location has no effect when EEPGM is set. Latched address and data cannot be modified during program and erase.

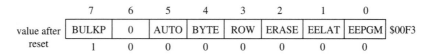

7	6	5	4	3	2	1	0	
BULKP	0	AUTO	BYTE	ROW	ERASE	EELAT	EEPGM	$00F3
1	0	0	0	0	0	0	0	

value after reset

BULKP: bulk erase protection
   0 = EEPROM can be bulk erased
   1 = EEPROM is protected from being bulk or row erased.
AUTO: Automatic shutdown of program/erase operation
   0 = Automatic clear of EEPGM is disabled
   1 = Automatic clear of EEPGM is enabled
BYTE: byte and aligned word erase
   0 = Bulk or row erase is enabled.
   1 = One byte or one aligned word erase only.
ROW: row or bulk erase (when BYTE = 0)
   0 = Erase entire array
   1 = Erase only one 32-byte row
   The erasure selection is shown in Table 13.14.
ERASE: Erase control
   0 = EEPROM configuration for programming. At this value, the BYTE and
      ROW bits have no effect.
   1 = EEPROM configuration for erasure
EELAT: EEPROM latch control
   0 = EEPROM sets up for normal reads
   1 = EEPROM address and data bus latches set up for programming or erasing.
EEPGM: program and erase enable
   0 = Disables program/erase voltage to EEPROM
   1 = Applies program/erase voltage to EEPROM
   The EEPGM bit can be set only after EELAT has been set. When EELAT and
   EEPGM are set simultaneouly, EEPGM remains clear but EELAT is set.

Figure 13.15 ■ The EEPROG register

BULKP	BYTE	ROW	Block size
0	0	0	Bulk erase entire EEPROM array
0	0	1	Row erase 32 bytes
X	1	0	Byte or aligned word erase
X	1	1	Byte or aligned word erase

Table 13.14 ■ Erase selection

To function properly, the EEPROMs of the 912D60 and the 912DG128/912DT128 require a steady internal clock of 35 μs (± 2μs). This clock is divided down from the oscillator clock (EXTAL) by the value in the EEPROM modulus divider registers, EEDIVH and EEDIVL.

The proper divide value is determined by the following formula:

$$\text{EEDIV} = \text{INT} [\text{EXTAL (Hz)} \times (35 \times 10^{-6}) + 0.5]$$

where INT[A] denotes the round-down integer value of A.

The EEPROM contains a special word location called the SHADOW word ($0FC0–$0FC1) that can be used to set up the required time base. At reset, the value programmed into the SHADOW word is automatically loaded into the EEDIVH:EEDIVL registers. Once the time base has been set up in the SHADOW word, the user will be able to use the EEPROM without considering clocking.

To set up the time base, follow these steps:

Step 1
Calculate the EEDIV value.

Step 2
Write the EEDIV value into the EEDIVH:EEDIVL registers.

Step 3
Program the SHADOW word to reflect the EEDIV value and the desired values for the EEPROM module configuration register, EEMCR.

The SHADOW word maps into the upper 4 bits of the EEMCR register and 10 bits of the EEDIV registers. The SHADOW word mapping is shown in Table 13.15.

High byte ($0FC0)							
Register Bit	EEMCR NOBDML	EEMCR NOSHW	EEMCR Bit 5	EEMCR Bit 4	NA NA	NA NA	EEDIVH EEDIV9	EEDIVH EEDIV8
Low byte ($0FC1)							
Register Bit	EEDIVL EEDIV7	EEDIVL EEDIV6	EEDIVL EEDIV5	EEDIVL EEDIV4	EEDIVL EEDIV3	EEDIVL EEDIV2	EEDIVL EEDIV1	EEDIVL EEDIV0

Table 13.15 ■ SHADOW word mapping

Once the time base and module information has been defined and programmed using the SHADOW word, this location can be protected from unintended programming or erasing. This feature is controlled by the SHPROT bit in the EEPROM block protect register, EEPROT. ▲

13.5.3 EEPROM Standard Mode Erasing Algorithm

The EEPROM of all 68HC12 members can be erased using the standard-mode erasing algorithm:

Step 1
Write BULKP, BYTE, and ROW bits in the EEPROM control register (EEPROG) to specify the erase size. Set the ERASE bit to specify the erasing operation. Set the EELAT bit to control erasing latches.

Step 2
Write a byte of data to an EEPROM address or write a word to a word-aligned EEPROM address. If the erase operation is not erasing the entire array or a full row, then this write determines whether a single byte or a word will be erased. Therefore, the address written to must be within the desired erase block.

Step 3
Set the EEPGM bit to one. Apply erasing voltage to the EEPROM.

Step 4
Wait for erase delay time (t_{ERASE}).

Step 5
Clear EEPGM to zero. Disable the erasing voltage from the array.

Step 6
Clear the EELAT bit to zero. This sets the EEPROM into normal mode.

By jumping from Step 5 to Step 2, it is possible to erase more bytes or words without intermediate EEPROM reads. The flowchart of the erasing algorithm for normal mode is shown in Figure 13.16a.

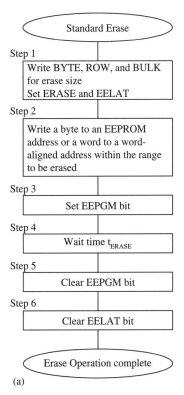

Step 1
Write BYTE, ROW, and BULK for erase size
Set ERASE and EELAT

Step 2
Write a byte to an EEPROM address or a word to a word-aligned address within the range to be erased

Step 3
Set EEPGM bit

Step 4
Wait time t_{ERASE}

Step 5
Clear EEPGM bit

Step 6
Clear EELAT bit

Standard Erase

Erase Operation complete

(a)

Figure 13.16a ■ EEPROM erasing algorithm flowcharts

13.5.4 EEPROM AUTO Mode Erasing Algorithm

The 912D60, the 912DG128, and the 912DT128 can also use the AUTO mode to achieve faster erasure for the EEPROM. The algorithm steps are as follows:

Step 1
Write the BULK, BYTE, and ROW bits in the EEPROM control register (EEPROG) to specify the erase size. Set the ERASE bit to specify an erasing operation. Set the EELAT bit to control erasing latches. Set the AUTO bit for automatic erasing time termination.

Step 2
Write a byte of data to an EEPROM address or write a word of data to a word-aligned EEPROM address.

Step 3

Set the EEPGM bit. This applies the erasing voltage to the EEPROM. If the value stored in the EEDIV registers is a zero, then the EEPGM bit will not be set.

Step 4

Poll the EEPGM bit until it is cleared by the internal timer.

Step 5

Clear the EELAT bit. This sets the EEPROM into the normal mode.

The flowchart of the auto-mode erasing algorithm is shown in Figure 13.16b.

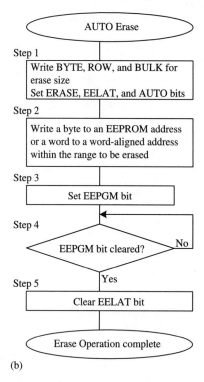

(b)

Figure 13.16b ■ EEPROM erasing algorithm flowcharts

Example 13.4

Write subroutines that use a standard algorithm to erase an EEPROM byte, word, row, and the whole EEPROM array. The address of the byte, word, or row is pointed to by the contents of index register X.

Solution: We need subroutines to erase a byte, a word, a row, and the whole EEPROM. These subroutines are a direct translation of the algorithm illustrated in Figure 13.16a.

```
#include "d:\miniide\hc12.inc"
; **************************************************************
; The following subroutine erases the EEPROM byte specified in
; index register X.
; **************************************************************
```

```
erase_byte    movb      #$16,EEPROG      ; set BYTE, ERASE, and EELAT
              movb      #0,0,X           ; write to the location pointed to by X
              bset      EEPROG,$01       ; set the EEPGM bit
              bsr       delay_10ms       ; wait for 10 ms
              bclr      EEPROG,$01       ; clear the EEPGM bit
              bclr      EEPROG,$02       ; clear the EELAT bit
              rts
;**********************************************************************
; The following subroutine erases the EEPROM word pointed to by
; index register X.
;**********************************************************************
erase_word    movb      #$16,EEPROG      ; set BYTE, ERASE, and EELAT
              movw      #0,0,X           ; write to the location pointed to by X
              bset      EEPROG,$01       ; set the EEPGM bit
              bsr       delay_10ms       ; wait for 10 ms
              bclr      EEPROG,$01       ; clear the EEPGM bit
              bclr      EEPROG,$02       ; clear the EELAT bit
              rts
;**********************************************************************
; The following routine erases the EEPROM row pointed to by X.
;**********************************************************************
row_erase     movb      #$0E,EEPROG      ; set ROW, ERASE, EELAT
              movb      #0,0,X           ; write to the location pointed to by X
              bset      EEPROG,$01       ; set the EEPGM bit
              bsr       delay_10ms       ; wait for 10 ms
              bclr      EEPROG,$01       ; clear the EEPGM bit
              bclr      EEPROG,$02       ; clear the EELAT bit
              rts
;**********************************************************************
; The following subroutine erases the whole EEPROM array.
;**********************************************************************
bulk_erase    movb      #$06,EEPROG      ; set ERASE, EELAT
              movb      #0,0,X           ; write to the location pointed to by X
              bset      EEPROG,$01       ; set the EEPGM bit
              bsr       delay_10ms       ; wait for 10 ms
              bclr      EEPROG,$01       ; clear the EEPGM bit
              bclr      EEPROG,$02       ; clear the EELAT bit
              rts


;**********************************************************************
; The following routine creates a delay of 10 ms.
;**********************************************************************
delay_10ms    pshx
              movb      #$90,TSCR        ; enable TCNT & fast flags clear
              movb      #$03,TMSK2       ; configure prescale factor to 8
              movb      #$01,TIOS        ; enable OC0
              ldd       TCNT
again         addd      #10000           ; start an output compare operation
              std       TC0              ; with 10 ms time delay
              brclr     TFLG1,$01,*      ; wait for OCOF to be set
              pulx
              rts
```

Example 13.5

Write subroutines that use the AUTO mode algorithm to erase an EEPROM byte, word, row, and the whole EEPROM array. The address of the byte, word, or row is pointed to by the contents of index register X.

Solution: The following subroutines use the AUTO mode to erase a byte, a word, a row, or the whole EEPROM array:

```
#include "d:\miniide\hc912dg128.inc"
; ***************************************************************
; The following subroutine erases the EEPROM byte specified in
; index register X.
; ***************************************************************
erase_byte   movb    #$36,EEPROG      ; set AUTO, BYTE, ERASE, and EELAT
             movb    #0,0,X           ; write to the location pointed to by X
             bset    EEPROG,$01       ; set the EEPGM bit
             brset   EEPROG,$01,*     ; wait while the EEPGM bit is still set
             bclr    EEPROG,$02       ; clear the EELAT bit
             rts
; ***************************************************************
; The following subroutine erases the EEPROM word pointed to by
; index register X.
; ***************************************************************
erase_word   movb    #$36,EEPROG      ; set AUTO, BYTE, ERASE, and EELAT
             movw    #0,0,X           ; write to the location pointed to by X
             bset    EEPROG,$01       ; set the EEPGM bit
             brset   EEPROG,$01,*     ; wait while the EEPGM bit is still set
             bclr    EEPROG,$02       ; clear the EELAT bit
             rts
; ***************************************************************
; The following routine erases the EEPROM row pointed to by X.
; ***************************************************************
row_erase    movb    #$2E,EEPROG      ; set ROW, ERASE, EELAT
             movb    #0,0,X           ; write to the location pointed to by X
             bset    EEPROG,$01       ; set the EEPGM bit
             brset   EEPROG,$01,*     ; wait while the EEPGM bit is still set
             bclr    EEPROG,$02       ; clear the EELAT bit
             rts
; ***************************************************************
; The following subroutine erases the whole EEPROM array.
; ***************************************************************
bulk_erase   movb    #$26,EEPROG      ; set ERASE, EELAT
             movb    #0,0,X           ; write to the location pointed to by X
             bset    EEPROG,$01       ; set the EEPGM bit
             brset   EEPROG,$01,*     ; wait while the EEPGM bit is still set
             bclr    EEPROG,$02       ; clear the EELAT bit
             rts
```

These subroutines can only run on the 912D60, the 912DG128, and the 912DT128.

13.5.5 EEPROM Program Operation

EEPROM can be programmed on a per-byte or per-aligned-word (2 bytes) basis. If some protected locations are included in the program area, those bytes will not be affected and only the unprotected locations will be altered.

There are two ways to program the EEPROM: the standard method and the auto mode method. In auto mode programming, the EEPGM bit is polled. When the bit is cleared, the programming has been completed.

13.5.6 Standard Mode Programming Algorithm

The EEPROM of all 68HC12 members can be programmed using the standard-mode programming algorithm:

Step 1
Clear the ERASE bit to specify a programming operation. Set the EELAT bit to control programming latches.

Step 2
Write a byte of data to an EEPROM address or write a word of data to a word-aligned EEPROM address.

Step 3
Set the EEPGM bit to apply programming voltage to the EEPROM.

Step 4
Wait t_{PROG} (t_{PROG} is the high voltage hold time for programming).

Step 5
Clear the EEPGM bit to disable the programming voltage from the EEPROM array.

Step 6
Clear the EELAT bit to set the EEPROM into the normal mode.

This algorithm is also shown in Figure 13.17a.

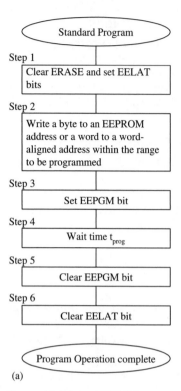

Figure 13.17a ■ EEPROM programming algorithm flowcharts

13.5.7 EEPROM AUTO Mode Programming Algorithm

The 68HC12 members 912D60, 912DG128, and 912DT128 can also utilize the AUTO mode algorithm to program the EEPROM and achieve better performance:

Step 1
Clear the ERASE bit to specify a programming operation. Set the EELAT bit to control programming latches. Set the AUTO bit for automatic programming time termination.

Step 2
Write a byte of data to an EEPROM address or write a word of data to a word-aligned EEPROM address.

Step 3
Set the EEPGM bit, which will apply programming voltage to the EEPROM.

Step 4
Poll the EEPGM bit until it is cleared by the internal timer.

Step 5
Clear the EELAT bit.

In AUTO mode, if the programming block is a protected area, the programming will not be successful and the EEPGM bit will never clear. The user may include a step to verify that the addresses in question are not protected, or include a timer to ensure that the software does not get trapped in that step. This algorithm is also shown in Figure 13.17b.

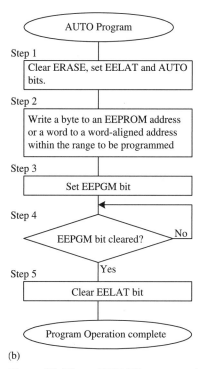

Figure 13.17b ■ EEPROM programming algorithm flowcharts

Example 13.6

Write subroutines to program a byte, a word, and a string of characters into the EEPROM memory location(s) pointed to by index register X using the standard mode algorithm.

Solution: The EEPROM programming subroutines using the standard algorithm are as follows:

```
#include "d:\miniide\hc12.inc"
; **************************************************************
; This subroutine programs the contents of accumulator A into the EEPROM location
; pointed to by index register X.
; **************************************************************
writebyte_eeprom
        bclr        EEPROG,$04      ; clear the ERASE bit
        bset        EEPROG,$02      ; set the EELAT bit
        staa        0,X             ; write the byte into EEPROM
        bset        EEPROG,$01      ; set the EEPGM bit
        bsr         delay_10ms      ; delay for program time
        bclr        EEPROG,$01      ; clear the EEPGM bit
        bclr        EEPROG,$02      ; clear the EELAT bit
        rts
; **************************************************************
; This subroutine programs the contents of accumulator D into the EEPROM location
; pointed to by index register X.
; **************************************************************
```

Inside flowchart (Figure 13.17b):

AUTO Program

Step 1 — Clear ERASE, set EELAT and AUTO bits.

Step 2 — Write a byte to an EEPROM address or a word to a word-aligned address within the range to be programmed

Step 3 — Set EEPGM bit

Step 4 — EEPGM bit cleared? No

Yes

Step 5 — Clear EELAT bit

Program Operation complete

(b)

```
writeword_eeprom
           bclr         EEPROG,$04          ; clear the ERASE bit
           bset         EEPROG,$02          ; set the EELAT bit
           std          0,X                 ; write the word into EEPROM
           bset         EEPROG,$01          ; set the EEPGM bit
           bsr          delay_10ms          ; delay for program time
           bclr         EEPROG,$01          ; clear the EEPGM bit
           bclr         EEPROG,$02          ; clear the EELAT bit
           rts
;*********************************************************************
; The following subroutine program the string pointed to by index register
; Y into EEPROM locations pointed to by index register X.
;*********************************************************************
write_ee_string
           ldaa         1,y+                ; get one character and move the pointer
           beq          done                ; reach the end of the string
           inx                              ; increment the EEPROM location
           bsr          writebyte_eeprom    ; program the character
           bra          write_ee_string     ; continue
done       rts
```

Example 13.7

Write subroutines to program a byte, a word, and a string of characters into the EEPROM memory location(s) pointed to by index register X using the auto mode algorithm.

Solution: The EEPROM programming subroutines using the auto mode algorithm are as follows:

```
#include "d:\miniide\hc912dg128.inc"
;*********************************************************************
; This subroutine programs the contents of accumulator A into the EEPROM location
; pointed to by index register X using auto mode.
;*********************************************************************
writebyte_eeprom
           bclr         EEPROG,$04          ; clear the ERASE bit
           bset         EEPROG,$22          ; set the EELAT and auto bit
           staa         0,X                 ; write the byte into EEPROM
           bset         EEPROG,$01          ; set the EEPGM bit
           brset        EEPROG,$01,*        ; wait until the EEPGM bit is cleared
           bclr         EEPROG,$02          ; clear the EELAT bit
           rts
;*********************************************************************
; This subroutine programs the contents of accumulator D into the EEPROM location
; pointed to by index register X using auto mode.
;*********************************************************************
writeword_eeprom
           bclr         EEPROG,$04          ; clear the ERASE bit
           bset         EEPROG,$22          ; set the EELAT and auto bit
           std          0,X                 ; write the word into EEPROM
           bset         EEPROG,$01          ; set the EEPGM bit
           brset        EEPROG,$01,*        ; wait until the EEPGM bit is cleared
           bclr         EEPROG,$02          ; clear the EELAT bit
           rts
```

```
; ****************************************************************
; The following subroutine program the string pointed to by index register
; Y into EEPROM locations pointed to by index register X.
; ****************************************************************
write_ee_string
            ldaa        1,y+                        ; get one character and move the pointer
            beq         done                        ; reach the end of the string
            inx                                     ; increment the EEPROM pointer
            bsr         writebyte_eeprom            ; program the character
            bra         write_ee_string            ; continue
```

The ICC12 compiler provides two library functions to support the access of EEPROM. The function:

```
unsigned char EEPROMread(int location)
```

reads a byte from the specified EEPROM location.

The function:

```
int EEPROMwrite(int location, unsigned char byte)
```

writes a byte to the specified EEPROM location. A 0 is returned when the write operation is successful.

▲

13.6 External Memory Expansion

There are some embedded systems that require the use of external memory. A microcontroller evaluation (or demo) board is one such example. Many 68HC12 demo boards from Axiom come with external SRAM and EEPROM. To add external memory to the 68HC12 microcontroller, we must configure the microcontroller to operate in the expanded mode. External EEPROMs are used to hold debug monitor such as the D-Bug12. External SRAMs are used to hold user programs to be tested.

To access external memory devices, the microcontroller must provide address signals to specify the memory location to be accessed. A path for carrying data between the microcontroller and the memory device is also needed. The set of conductors that carry the address signals is called an *address bus*, whereas the set of conductors that carry the data is called a *data bus*.

All 68HC12 members except the 812A4 have 16 address pins and 16 data pins to allow the access of 64KB of external memory space. Due to the existence of internal memory resources, not all of this 64KB memory space is available for external memory devices. The 16-bit data bus allows the transfer of 16 bits of data in one *bus cycle*.

The term *memory organization* refers to the number of bits that are accessible in a memory chip at a time (by applying one address). The most common organizations are × 1, × 4, × 8, and × 16. The × 8 organization is used most often for 8-bit and 16-bit microcontrollers.

Adding memory or peripheral memory-mapped devices to a microprocessor or microcontroller involves three issues:

- Memory space assignment
- Address decoder and control circuitry design
- Timing verification

13.6.1 Memory Space Assignment

Assigning memory space using equal block sizes makes the address decoder design simple. However, the advent of *programmable logic devices* (PLD) makes this advantage insignificant.

Before we make a memory space assignment, we need to know the unassigned memory space in the expanded mode. In expanded mode, the on-chip flash memories of all 68HC12 members are turned off. The memory maps for 68HC12 members in expanded mode are shown in Figures 13.18a through 13.18d. All shaded areas have been assigned to internal memory resources. Only unshaded areas are available for external memory use. However, we should keep in mind that it is permissible to overlap a part of the external memory space with that of the internal memory resource. Internal memory resources have higher precedence than external memory devices, which will make the external memory space overlapped with internal memory resources inaccessible.

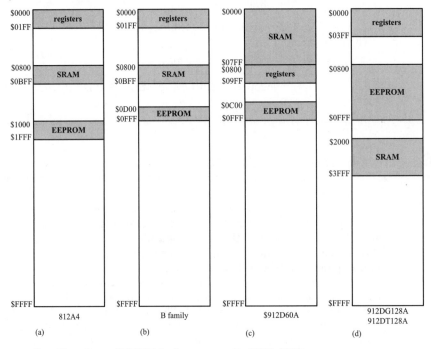

Note: The registers of 912D60A has been remapped to $0800~$09FF

Figure 13.18a-d ■ 68HC12 memory map in expanded mode

When configuring the 68HC12 in expanded mode, we will need to add nonvolatile memory to hold the interrupt vectors and startup routines because the interrupt vector table is located at $FFC0~$FFFF ($FF80~$FFFF for the 912D60A, the 912DG128A, and the 912DT128A).

The capacities of today's memory devices are very large, in many cases, much larger than many applications can take advantage of. Low capacity memory chips are harder to find these days. There are two approaches in memory space assignment:

1. *Equal size assignment.* In this approach, the available memory space is divided into blocks of equal size and then each block is assigned to a memory device without regard for the actual size of each memory-mapped device. A memory-mapped device

could be a memory chip or a peripheral chip. Memory space tends to be wasted using this approach because most memory-mapped peripheral chips (for example, the Intel i8255) need only a few bytes to be assigned to their internal registers.

2. *Demand assignment.* In this approach, we assign the memory space according to the size of memory devices.

Example 13.8

▼

Assign the 68HC12 memory space using a block size of 8KB.

Solution: The 64KB memory space of the 68HC12 can be divided into eight 8KB blocks. The address ranges of the blocks are shown in Table 13.16.

Block number	Address range
0	$0000 - $1FFF
1	$2000 - $3FFF
2	$4000 - $5FFF
3	$6000 - $7FFF
4	$8000 - $9FFF
5	$A000 - $BFFF
6	$C000 - $DFFF
7	$E000 - $FFFF

Table 13.16 ■ 64KB memory space assignment

▲

Example 13.9

▼

Make a memory space assignment so that there is at least 16KB of space assigned to external nonvolatile memory and 32KB of SRAM for holding dynamic data and programs for the 912BC32.

Solution: Since the space for interrupt and reset vectors must be in nonvolatile memory, we will assign the memory space from $C000 to $FFFF to the nonvolatile memory such as EPROM, EEPROM, or flash memory. We prefer not to overlap external memory space with internal memory resources. One viable assignment for the 32KB SRAM is from $4000 to $BFFF.

▲

13.6.2 Address Decoder Design

The function of an address decoder is to make sure that there is no more than one memory device enabled to drive the data bus. If there are two or more memory devices driving the same bus lines, *bus contention* occurs and damage could be resulted. All memory devices have control signals such as *chip enable* (CE), *chip select* (CS), or *output enable* (OE) to control their read and write operations. The outputs of an address decoder will be used as the chip select or chip enable signals of external memory devices.

Two address-decoding schemes have been used: *full* and *partial address decoding*. A memory device is said to be *fully* decoded when each of its addressable locations responds to only a single address on the system bus. A memory component is said to be *partially* decoded when each of its addressable locations responds to more than one address on the system bus. Memory components such as DRAM, SRAM, EPROM, EEPROM, and flash memory chips use the full address-decoding scheme more often, whereas peripheral chips use the partial address-decoding scheme more often.

Address decoder design is closely related to memory space assignment. For the memory space assignment discussed in Section 13.6.1, upper address signals are used as inputs to the decoder while lower address signals are applied to the address inputs of memory device. Address decoder outputs select a memory device to respond to the memory access request, whereas lower address signals specify a location within the selected memory component to be accessed.

The *transistor-transistor logic* (TTL) chip 74138 is a 3-to-8 decoder, and the 74139 is a dual 2-to-4 decoder. Both chips can be used to implement the address decoder. The pin assignments of these two chips are shown in Figure 13.19.

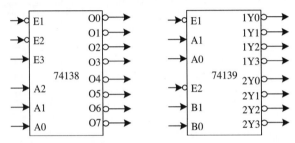

Figure 13.19 ■ The 74138 and 74139 decoder chips

Example 13.10

▼

Use a full decoding scheme to design an address decoder for a 912BC32-based computer that has the following address space assignments:

$2000-$3FFF: SRAM1
$4000-$5FFF: EEPROM1
$6000-$7FFF: Flash memory
$8000-$9FFF: SRAM2
$E000-$FFFF: EEPROM

Solution: Because all of these memory components are 8KB, it would be very easy to use the full address decoding scheme to implement the address decoder. The 3-to-8 decoder 74F138 (F stands for Fast Schottky logic) can be used. The highest three address signals of the 912BC32 are used to select the memory modules:

SRAM1:	001
EEPROM1:	010
Flash memory:	011
SRAM2:	100
EEPROM2:	111

The lower 13 address signals are the address inputs to the memory device and can be either 0 or 1, and hence the address value:

001x xxxx xxxx xxxx

covers the range from 0010 0000 0000 0000 ($2000) to 0011 1111 1111 1111 ($3FFF). Therefore the address input 001 to the address decoder would select the SRAM1 chip. Other address ranges select other memory components for a similar reason.

The address decoder for this computer system is shown in Figure 13.20.

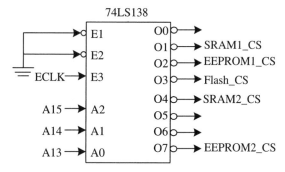

Figure 13.20 ■ Address decoder design for Example 13.10

The internal design of the 74138 and the 74139 requires the use of equal-size memory space assignment. This requirement makes it difficult to achieve the simultaneous use of demand assignment and full decoding for a computer that has external memory components and multiple peripheral chips.

It is easy to perform demand memory assignment and full address decoding simultaneously by using a low-density programmable logic device (PLD) such as GAL16V8, GAL18V10, and GAL22V10 from Lattice Semiconductor or 16V8, 20V8, and 22V10 from Atmel (Atmel names her product as *SPLD*). The functional block diagram of the GAL22V10 is shown in Figure 13.21. In Figure 13.21, the acronym *OLMC* stands for *output logic macro cell*. OLMC can be programmed to be a D flip-flop or combinational logic gate and hence can be used to implement sequential circuit such as registers, counters, and finite state machine. The SPLDs from Atmel have similar architecture.

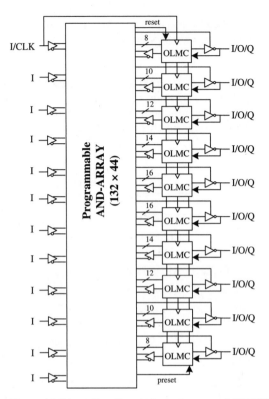

Figure 13.21 ■ Functional block diagram of GAL22V10

To implement the address decoder using a GAL device, we would need to use a hardware description language (HDL) such as ABEL to describe the equations of chip enable (or chip select) signals for all memory devices. To design with the SPLD product from Atmel, we could use the CUPL language. Free compilers for both languages are available from these two companies and can be downloaded from the Internet (www.latticesemi.com and www.atmel.com).

A detail study of these two languages is not in the scope of this text. However, we will use an example to illustrate the use of ABEL language and the GAL22V10 device in the design of an address decoder.

Example 13.11

▼

Design an address decoder to be used in expanded narrow mode for the 912BC32-based computer that has the following address space assignment:

$0200~$0207: LCD
$0208~$020F: PIA
$4000-$7FFF: SRAM1 (one chip)
$8000~$BFFF: SRAM2 (one chip)
$C000-$FFFF: EEPROM (one chip)

Assume all devices are 8-bit wide.

Solution: The highest 13 address bits will be used to select the LCD and PIA, whereas only the highest two address bits will be used to select SRAMs and EEPROM. The equations of the chip enable signals for the above four memory mapped devices are (the exclamation mark ! stands for inversion, the character * stands for *and* operator):

```
CS_LCD = !A15 * !A14 * !A13 * !A12 * !A11 * !A10 * !A10 * A9 * !A8 * !A7
         * !A6 * !A5 * !A4 * !A3
CS_PIA = !A15 * !A14 * !A13 * !A12 * !A11 * !A10 * !A10 * A9 * !A8 * !A7
         * !A6 * !A5 * !A4 * A3
CS_SRAM1 = !A15 * A14
CS_SRAM2 = A15 * !A14
CS_EEPROM = A15 * A14
```

An ABEL program that describes this address decoder is as follows:

```
module Address_decoder
title    'Address decoder for 68HC12'
decoder  DEVICE GAL22V10;
         A15, A14, A13, A12, A11, A10          pin istype 'com';
         A9, A8, A7, A6, A5, A4, A3            pin istype 'com';
"The following are active low signals
         !CS_SRAM1, !CS_SRAM2                  pin istype 'com';
         !CS_EEPROM, !CS_LCD, !CS_PIA          pin istype 'com';
equations
         CS_SRAM1  = !A15 & A14;
         CS_SRAM2  = A15 & !A14;
         CS_LCD = !A15 & !A14 & !A13 & !A12 & !A11 & !A10 & A9 & !A8 & !A7 & !A6 & !A5
                  & !A4 & !A3;
         CS_PIA = !A15 & !A14 & !A13 & !A12 & !A11 & !A10 & A9 & !A8 & !A7 & !A6 & !A5
                  & !A4 & A3;
         CS_EEPROM = A15 & A14;

test_vectors ([A15,A14,A13,A12,A11,A10,A9,A8,A7,A6,A5,A4,A3] -> [CS_SRAM1, CS_SRAM2, CS_LCD, CS_PIA,
             CS_EEPROM])
         [0,0,0,0,0,0,1,0,0,0,0,0,0] -> [0, 0, 1, 0, 0];
         [0,0,0,0,0,0,1,0,0,0,0,0,1] -> [0, 0, 0, 1, 0];
         [0,1,0,0,0,0,0,0,0,0,0,0,0] -> [1, 0, 0, 0, 0];
         [1,0,0,0,0,0,0,0,0,0,0,0,0] -> [0, 1, 0, 0, 0];
         [1,1,0,0,0,0,0,0,0,0,0,0,0] -> [0, 0, 0, 0, 1];
end
```

An ABEL program is self-explanatory and easy to understand.

13.7 Basics of Bus Signals

A bus line is simply a conductor. The voltage level of a bus line is determined by the device that drives it. For this reason, a bus line is often called *passive*. A bus line can be made active by adding a *pull-up* device. As shown in Figure 13.22, a simple pull-up device could be a resistor, a PNP transistor, or a PMOS transistor.

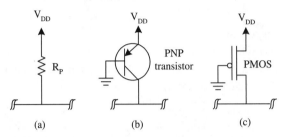

Figure 13.22 ■ Pull-up devices for bus line

In an active bus, the bus voltage will be low only when one or more devices attached to the bus apply a low voltage to the bus. When no device drives the bus, the bus voltage will be pulled to high (V_{DD}).

A passive bus requires the device attached to it to drive the bus to high or low. To drive the bus, a *bus driver* is needed. To receive data from the bus, a *bus receiver* is needed. The bus driver and bus receiver are often combined to form a *bus transceiver*. Usually a bus driver and receiver have an enable signal to control its electrical connection to the bus. When the enable signal is asserted, the transceiver can drive and receive data from the bus. As shown in Figure 13.23, five devices are attached to the bus via bus transceivers.

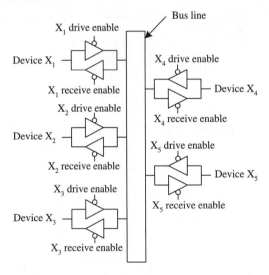

Figure 13.23 ■ Multiple devices attached to the bus line

13.7.1 Waveforms of Bus Signals

The waveform of a typical digital signal is shown in Figure 13.24. A bus signal cannot rise from low to high or drop from high to low instantaneously. The time needed for a signal to rise from 10% of the power supply voltage to 90% of the power supply voltage is called the *rise time* (t_r). The time needed for a signal to drop from 90% of the power supply voltage to 10% of the power supply voltage is called the *fall time* (t_f).

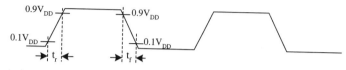

Figure 13.24 ■ A typical digital waveform

A single bus signal is often represented as a set of line segments (see Figure 13.25). The horizontal axis and vertical axis represent the time and the magnitude (in volts) of the signal, respectively. Multiple signals of the same nature, such as address and data, are often grouped together and represented as parallel lines with crossovers, as illustrated in Figure 13.26. A crossover represents the point at which one or multiple signals change value.

Figure 13.25 ■ Single signal waveform

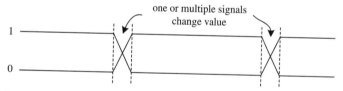

Figure 13.26 ■ Multiple signal waveform

Sometimes a signal value is unknown because the signal is changing. Hatched areas in the timing diagram, shown in Figure 13.27, represent single and multiple unknown signals. Sometimes one or multiple signals are not driven (because their drivers are not enabled) and hence cannot be received. An unknown signal is said to be *floating*. Single and multiple floating signals are represented by a value between the high and the low levels, as shown in Figure 13.28. A floating signal is in a *high-impedance* state.

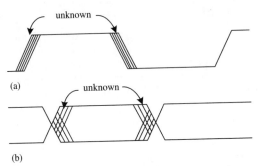

(a)

(b)

Figure 13.27 ■ Unknown signals; (a) single signal, (b) multiple signal

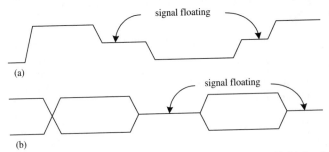

(a)

(b)

Figure 13.28 ■ Floating signals; (a) Single signal, (b) Multiple signals

In a microcontroller or microprocessor system, a bus signal falls into one of three categories: *address, data,* or *control.*

13.7.2 Bus Transactions

A bus transaction includes sending the address and receiving or sending the data. A *read* transaction transfers data from memory to either the CPU or I/O device, and a *write* transaction writes data to memory. In a read transaction, the address is first sent down the bus to the memory, together with the appropriate control signals indicating a read. In Figure 13.29, this means pulling the read signal to high. The memory responds by placing the data on the bus and driving the ready signal to low. The *ready* signal (asserted low) indicates that the data on the data bus is valid.

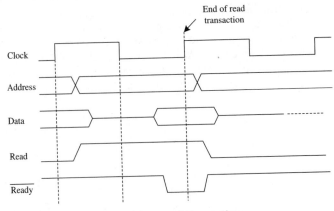

Figure 13.29 ■ A typical bus read transaction

In Figure 13.29, a read bus transaction takes one clock cycle to complete. The ready signal allows the bus transaction to extend to more than one clock cycle and hence can accommodate slower memory components. All 32-bit or 64-bit microprocessors from Intel, Motorola, and other companies have this feature. A write bus transaction requires that the CPU or I/O device sends both address and data and requires no return of data.

In a bus transaction, there must be a device that can initiate a read or a write transaction. The device that can initiate a bus transaction is called a *bus master*. A microcontroller or microprocessor is always a bus master. A device such as a memory chip that cannot initiate a bus transaction is called a *bus slave*.

In a bus transaction, there must be a signal to synchronize the data transfer. The signal that is used most often is the clock signal. The bus is *synchronous* when a clock signal is used to synchronize the data transfer. In a synchronous bus, the timing parameters of all signals use the clock signal as a reference. As long as all timing requirements are satisfied, the bus transaction will be successful.

An asynchronous bus, on the other hand, is not clocked. Instead, self-timed, handshaking protocols are used between the bus sender and receiver. Figure 13.30 shows the steps of a master performing a read on an asynchronous bus.

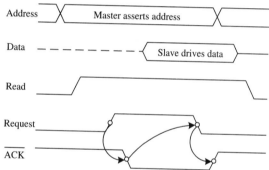

Figure 13.30 ■ Asynchronous read bus transaction

A synchronous bus is often used between the CPU and the memory system, whereas an asynchronous bus is often used to handle different types of devices. A synchronous bus is usually faster because it avoids the overhead of synchronizing the bus for each transaction.

13.7.3 Bus Multiplexing

Designers of microcontrollers prefer to minimize the number of signal pins. By multiplexing the address bus and the data bus, many signal pins can be saved. The drawback of multiplexing bus signals is that the achievable bus transaction performance is compromised. Most 8-bit and many 16-bit microcontrollers (including most 68HC12 members) multiplex their address and data buses.

For any bus transaction, address signals must be stable throughout the whole bus transaction. We need to use a certain logic circuit to latch the address signals so that they stay stable throughout the bus transaction. In a microcontroller that multiplexes the address and data buses, the address signals are placed on the multiplexed bus first, along with certain control signals to indicate that address signals are valid. After the address signals are on the bus long enough so that the external logic have time to latch the address signals, the microcontroller stops driving address signals, and either waits for the memory devices to place data on the multiplexed bus (in a read transaction) or place data on the multiplexed bus (in a write transaction).

13.8 The 68HC12 Bus Timing Diagram

Bus transactions can only be performed in expanded mode. In addition to address and data signals, several control signals are also involved in a bus transaction. These signals are R/$\overline{\text{W}}$, ECLK, LSTRB, and DBE. The R/$\overline{\text{W}}$ (pin PE2) signal specifies the type (read or write) of bus transaction that is performed. The ECLK (pin PE4) signal is used as a timing reference and its *rising edge* will be used to latch address signals from the multiplexed bus pins. The LSTRB (pin PE3) signal is used to indicate 8-bit or 16-bit data. The DBE (pin PE7) signal is used to indicate when the multiplexed bus is used as the data bus. The 812A4 does not need the DBE signal because address and data signals are not multiplexed.

All 68HC12 members except the 812A4 use Port A to carry address signals A15 to A8 and data signals D15~D8 (in expanded wide mode) or data signals D7~D0 (in expanded narrow mode). Port B is used to carry address signals A7~A0 and data signals D7~D0 (in expanded wide mode). The 812A4 uses Port A and Port B to carry address signals A15~A8 and A7~A0, respectively. Port G is used to carry address signals A21~A16. Port C and Port D are used to carry data signals D15~D8 and D7~D0, respectively.

A 68HC12 read bus cycle timing diagram is shown in Figure 13.31. The values of timing parameters are given in Table 13.17.

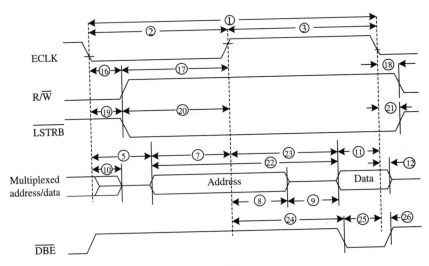

Note: Measurement points are 20% and 70% of V$_{\text{DD}}$

Figure 13.31 ■ 68HC12 read bus cycle timing diagram

Num	Characteristic [1] [2] [3] [4] [5]		Delay	Symbol	8MHz		2MHz		Unit
					Min	Max	Min	Max	
-	Frequency of operation (E-clock frequency)		-	f_o	dc	8.0	dc	8.0	MHz
1	Cycle time	$t_{cyc} = 1/f_o$	-	t_{cyc}	125	-	500	-	ns
2	Pulse width, E low	$PW_{EL} = t_{cyc}/2 + delay$	-4	PW_{EL}	59	-	246	-	ns
3	Pulse width, E high [6]	$PW_{EH} = t_{cyc}/2 + delay$	-2	PW_{EH}	59	-	248	-	ns
5	Address delay time	$t_{AD} = t_{cyc}/4 + delay$	27	t_{AD}	-	67.5	-	152	ns
7	Address valid time to ECLK rise	$t_{AV} = PW_{EL} - t_{AD}$	-	t_{AV}	-6.2	-	94	-	ns
8	Multiplexed address hold time	$t_{MAH} = t_{cyc}/4 + delay$	-18	t_{MAH}	13	-	107	-	ns
9	Address hold to data valid		-	t_{AHDS}	30	-	20	-	ns
10	Data hold to high impedance	$t_{DHZ} = t_{AD} + 20$	-	t_{DHZ}	-	45.2	-	132	ns
11	Read data setup time		-	t_{DSR}	31.2	-	25	-	ns
12	Read data hold time		-	t_{DHR}	0	-	0	-	ns
13	Write data delay time		-	t_{DDW}	-	62.5	-	165	ns
14	Write data hold time		-	t_{DHW}	25	-	20	-	ns
15	Write data setup time [6]	$t_{DSW} = PW_{EH} - t_{DDW}$	-	t_{DSW}	5.8	-	83	-	ns
16	Read/Write delay time	$t_{RWD} = t_{cyc}/4 + delay$	18	t_{RWD}	-	57.5	-	143	ns
17	Read/Write valid time to E rise	$t_{RWV} = PW_{EL} - t_{RWD}$	-	t_{RWV}	3.8	-	103	-	ns
18	Read/Write hold time		-	t_{RWH}	25	-	20	-	ns
19	Low strobe [7] delay time	$t_{LSD} = t_{cyc}/4 + delay$	18	t_{LSD}	-	57.5	-	143	ns
20	Low strobe valid time to E rise	$t_{LSV} = PW_{EL} = t_{LSD}$	-	t_{LSL}	3.8	-	103	-	ns
21	Low strobe [7] hold time		-	t_{LSH}	25	-	20	-	ns
22	Address access time [6]	$t_{ACCA} = t_{cyc} - t_{AD} - t_{DSR}$	-	t_{ACCA}	-	27.6	-	323	ns
23	Access time from E rise [6]	$t_{ACCE} = PW_{EH} - t_{DSR}$	-	t_{ACCE}	-	27.8	-	223	ns
24	$\overline{DBE}$ delay from ECLK rise[6]	$t_{DBED} = t_{cyc}/4 + delay$	8	t_{DBED}	-	57.5	-	133	ns
25	$\overline{DBE}$ valid time	$t_{DBE} = PW_{EH} - t_{DBED}$	-	t_{DBE}	11.8	-	115	-	ns
26	DBE hold time from ECLK fall		-	t_{DBEH}	-3	10	-3	10	ns

1. $V_{DD} = 5.0V \pm 10\%$, $V_{SS} = 0V$, $T_A = T_L$ to T_H, unless otherwise noted.
2. All timings are calculated for normal port drives.
3. Crystal input is required to be within 45% to 55% duty cycle.
4. Reduced drive must be off to meet these timings.
5. Unequalled loading of pins will affect relative timing numbers.
6. This characteristic is affected by clock stretch.
   Add $N \times t_{cyc}$ where N = 0, 1, 2, or 3, depending on the number of clock stretches.
7. Without TAG enabled.

Table 13.17 ■ 68HC12 expanded bus timing parameters

As shown in Figure 13.31, a normal read cycle takes one E clock cycle to complete. If slower memory is used, the read cycle can be stretched by one to three E clock cycles. The stretching of the bus cycle is done by programming the bit 3 and bit 2 of the MISC register.

The R/$\overline{W}$ signal will go high t_{RWV} ns (16) before the rising edge of the E clock to indicate that this is a read cycle. The CPU will drive the address signals t_{AD} ns (5) after the falling edge of the E clock. Address signals will become valid t_{AV} ns (7) before the rising edge of the E clock and remain valid until t_{MAH} ns (t_8) after the rising edge of the E clock signal. After that, the CPU stops driving the address bus and waits for data to arrive. In order for the CPU to correctly latch the data, a memory device must place data on the data bus t_{DSR} ns (11) before the falling edge of the E clock signal, and keep the data valid for t_{DHR} ns (12) after the falling edge.

The write bus cycle timing diagram is shown in Figure 13.32. A shortest write bus cycle takes one E clock cycle to complete. Like a read bus cycle, a write bus cycle can be stretched by one to three E clock cycles to accommodate slow memory devices. A write bus cycle timing diagram is similar to a read bus cycle timing diagram with two differences:

1. The R/$\overline{\text{W}}$ signal is asserted low to indicate that it is a write bus cycle.

2. Data is driven by the CPU rather than by memory devices. The CPU drives data on the data bus t_{DSW} ns (15) before the falling edge of the E clock signal, and keeps it valid until t_{DHW} ns (14) after the falling edge of the E clock signal.

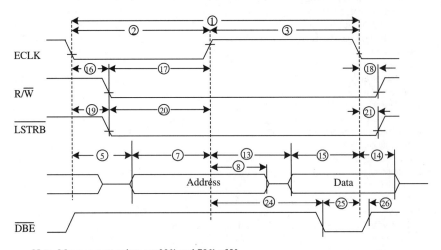

Note: Measurement points are 20% and 70% of V_{DD}.

Figure 13.32 ■ 68HC12 write bus cycle timing diagram

13.9 Adding External Memory to the 68HC12

When adding external memory to the 68HC12, we have several possible memory technologies to choose from: DRAM, SRAM, EPROM, EEPROM, and flash memory. DRAM and SRAM are suitable for storing dynamic data and programs, whereas other memory technologies are suitable for storing static data and programs. Because the 68HC12 has a relatively small memory space, most external memory needs can be satisfied by one or two memory chips. DRAM has the lowest per bit cost. However, the following two reasons make it the undesirable choice for external memory expansion for the 68HC12:

1. *Periodic refresh requirement.* DRAM chips use capacitors to store information, but without periodic refreshes the information stored in the capacitors will leak away. A timer is needed to generate a periodic refresh request. However, the 68HC12 may need to access the DRAM when the refresh operation is being performed. This would require the bus cycle to be dynamically frozen. Dynamic bus freezing is not easy to implement.

2. *Address multiplexing.* DRAM chips use an address multiplexing technique to reduce the number of address pins. This requires the memory system to include an address multiplexor, further increasing the cost of the memory system.

The control circuit designs for interfacing SRAM, EPROM, EEPROM, and flash memory to a microcontroller are quite similar. In the following sections, we will use two 8K × 8 SRAM chips, two 8K × 8 EEPROM chips, and an LCD module as an example to illustrate adding external memory devices to the 912BC32.

13.9.1 The IDT7164 SRAM

The IDT7164 is an 8K × 8 SRAM that operates with a 5V power supply. The IDT7164 offers a range of access times from as slow as 100 ns to as fast as 15 ns. The IDT7164 has three-state outputs that can be put in a high-impedance state when the chip is not selected for access. The pin assignment of the IDT7164 is shown in Figure 13.33.

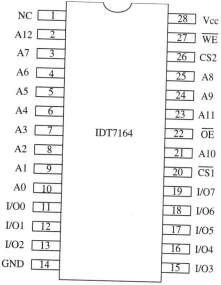

Figure 13.33 ■ The IDT7164 pin assignment

The signals A12-A0 select one of the 8192 locations within the chip to be read from or written into. Pins I/O7–I/O0 carry the data to be transferred between the chip and the CPU. There are two chip-select signals: $\overline{CS1}$ is active low and $CS2$ is active high. The active low signal $\overline{OE}$ controls the read access from the chip, whereas the active low signal $\overline{WE}$ controls the write access to the chip. All control signals must be asserted in order to read from or write into the IDT7164. Table 13.18 gives the function of I/O pins as a combination of these four control signals.

The $\overline{OE}$ signal can be tied to low permanently. Either $\overline{CS1}$ or $CS2$ can be asserted permanently. With two control signals tied to fixed voltage levels, we only need to provide control to the remaining two control signals.

Depending on the assertion times of control signals, there are three timing diagrams for the read cycle and two timing diagrams for the write cycle (shown in Figures 13.34 and 13.35). The values of related timing parameters for the read and write cycles are listed in Table 13.19.

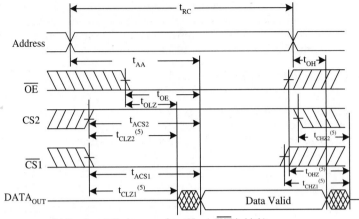

(a) Read cycle timing waveform No. 1 ($\overline{WE}$ is high)

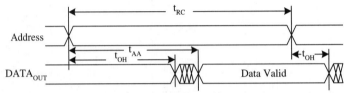

(b) Read cycle timing waveform No. 2 (device permanently selected)

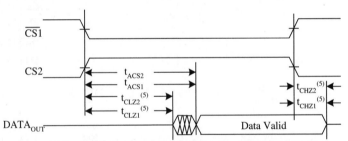

(c) Read cycle timing waveform 3 (OE is low, address valid before or coincident with $\overline{CS1}$ transition low and CS2 transition high)

Notes.
1. $\overline{WE}$ is high for read cycle.
2. Device is continuously selected, $\overline{CS1}$ is low and CS2 is high.
3. Address valid prior to or coincident with $\overline{CS1}$ transition low and CS2 transition high.
4. $\overline{OE}$ is low.
5. Transition is measured ±200 mV from steady state.

Figure 13.34 ■ IDT7164 read cycle timing diagrams

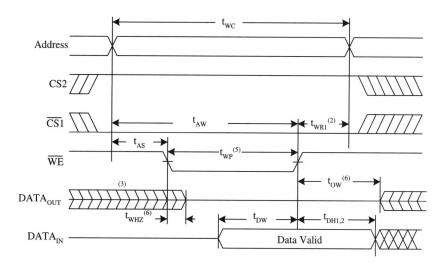

(a) Write cycle timing waveform No. 1 ($\overline{WE}$ controlled timing)

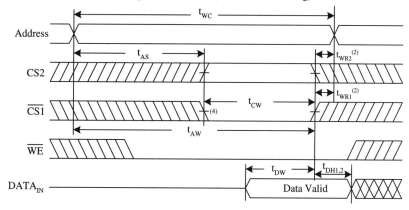

(b) Write cycle timing waveform No. 2 ($\overline{CS}$ controlled timing)

Notes:
1. A write occurs during the overlap of a low $\overline{WE}$, a low $\overline{CS1}$ and a high CS2.
2. $t_{WR1,2}$ is measured from the earlier of $\overline{CS1}$ or $\overline{WE}$ going high or CS2 going low to the end of the write cycle.
3. During this period, I/O pins are in the output state so that the input signals must not be applied.
4. If the $\overline{CS1}$ low transition or CS2 high transition occurs simultaneously with or after the $\overline{WE}$ low transition, the outputs remain in a high-impedance state.
5. $\overline{OE}$ is continuously high. If $\overline{OE}$ is low during a $\overline{WE}$ controlled write cycle, the write pulse width must be the larger of t_{WP} or ($t_{WHZ} + t_{DW}$) to allow the I/O drivers to turn off and data to be placed on the bus for the required t_{DW}. If $\overline{OE}$ is high during a $\overline{WE}$ controlled write cycle, this requirement does not apply and the minimum write pulse width is as short as specified t_{WP}.
6. Transition is measured ±200mV from steady state.

Figure 13.35 ■ IDT7164 write cycle timing diagrams

WE	CS1	CS2	OE	I/O	Function
X	H	X	X	High-Z	Deselected -Standby
X	X	L	X	High-Z	Deselected -Standby
X	V_{HC}	V_{HC} or V_{LC}	X	High-Z	Deselected -Standby
X	X	V_{LC}	X	High-Z	Deselected -Standby
H	L	H	H	High-Z	Output Disabled
H	L	H	L	Data$_{OUT}$	Read Data
L	L	H	X	Data$_{IN}$	Write Data

Table 13.18 ■ I/O pin functions as a combination of control signals

The IDT7164 has eight versions that differ in their access times. Only six of them are included in Table 13.19. The last two digits (for example, 25 in 7164L25) stand for the access time of the device in *ns* (10^{-9} seconds). The IDT7164 has four read access times:

Address access time (t_{AA}). This is the delay time from the moment that the address signals become valid until valid data appear on the data pins (I/O7–I/O0), if all other control signals ($\overline{CS1}$, CS2, and $\overline{OE}$) are asserted.

$\overline{CS1}$ access time (t_{ACS1}). This is the delay time from the moment that the $\overline{CS1}$ signal becomes active until valid data become available at the data pins (I/O7–I/O0), if address signals and all other control signals (CS2 and $\overline{OE}$) are asserted.

CS2 access time (t_{ACS2}). This is the delay time from the moment that the CS2 signal becomes active until valid data are present at the data pins (I/O7–I/O0), if address signals and all other control signals ($\overline{CS1}$ and $\overline{OE}$) are asserted.

$\overline{OE}$ access time (t_{OE}). This is the delay time from the moment that the $\overline{OE}$ signal becomes asserted until IDT7164 drives valid data on the data pins (I/O7–I/O0), if address signals and all other control signals ($\overline{CS1}$ and CS2) are asserted.

The data on the data pins will remain valid for t_{OH} ns after the address changes. The read cycle time is the required minimum separation between two consecutive read accesses to the memory.

Symbol	Parameter	7164L20(2)		7164L25		7164L35		unit
		Min.	Max.	Min.	Max.	Min.	Max.	
t_{RC}	Read cycle time	20	-	25	-	35		ns
t_{AA}	Address access time	-	19	-	25	-	35	ns
t_{ACS1} [3]	Chip select-1 access time	-	20	-	25	-	35	ns
t_{ACS2} [3]	Chip select-2 access time	-	25	-	30	-	40	ns
$t_{CLZ1,2}$ [4]	Chip select-1, 2 to output in low Z	5	-	5	-	5	-	ns
t_{OE}	Output enable to output valid	-	8	-	12	-	18	ns
t_{OLZ} [4]	Output enable to output in low Z	0	-	0	-	0	-	ns
$t_{CHZ1,2}$ [4]	Chip select-1, 2 to output in high Z	-	9	-	13	-	15	ns
t_{OHZ} [4]	Output disable to output in high Z	-	8	-	10	-	15	ns
t_{OH}	Output hold from address change	5	-	5	-	5	-	ns
t_{PU} [4]	Chip select to power up time	0	-	0	-	0	-	ns
t_{PD} [4]	Chip deselect to power down time	-	20	-	25	-	35	ns
								ns
t_{WC}	Write cycle time	20	-	25	-	35	-	ns
$t_{CW1,2}$	Chip select to end-of-write	15	-	18	-	25	-	ns
t_{AW}	Address valid to end-of-write	15	-	18	-	25	-	ns
t_{AS}	Address setup time	0	-	0	-	0	-	ns
t_{WP}	Write pulse width	15	-	21	-	25	-	ns
t_{WR1}	Write recovery time ($\overline{CS1}$, $\overline{WE}$)	0	-	0	-	0	-	ns
t_{WR2}	Write recovery time (CS2)	5	-	5	-	5	-	ns
t_{WHZ} [4]	Write enable to output in high-Z	-	8	-	10	-	14	ns
t_{DW}	Data to write time overlap	10	-	13	-	15	-	ns
t_{DH1}	Data hold from write time ($\overline{CS1}$,$\overline{WE}$)	0	-	0	-	0	-	ns
t_{DH2}	Data hold from write time (CS2)	5	-	5	-	5	-	ns
t_{OW} [4]	Output active from end-of-write	4	-	4	-	4	-	ns

Notes.
1. 0° to +70°C temperature range only.
2. 0° to +70°C and -55°C to 125°C temperature ranges only.
3. Both chip selects must be active for the device to be selected.
4. This parameter is guaranteed by device characterization, but is not production tested.

Table 13.19 ■ IDT7164 timing parameter values

Symbol	Parameter	7164L45		7164L55		7164L70		unit
		Min.	Max.	Min.	Max.	Min.	Max.	
t_{RC}	Read cycle time	45	-	55	-	70	-	ns
t_{AA}	Address access time	-	45	-	55	-	70	ns
t_{ACS1} [1]	Chip select-1 access time	-	45	-	55	-	70	ns
t_{ACS2} [1]	Chip select-2 access time	-	45	-	55	-	70	ns
$t_{CLZ1, 2}$ [2]	Chip select-1, 2 to output in low Z	5	-	5	-	5	-	ns
t_{OE}	Output enable to output valid	-	25	-	30	-	35	ns
t_{OLZ} [2]	Output enable to output in low Z	0	-	0	-	0	-	ns
$t_{CHZ1, 2}$ [2]	Chip select-1, 2 to output in high Z	-	20	-	25	-	30	ns
t_{OHZ} [2]	Output disable to output in high Z	-	20	-	25	-	30	ns
t_{OH}	Output hold from address change	5	-	5	-	5	-	ns
t_{PU} [2]	Chip select to power up time	0	-	0	-	0	-	ns
t_{PD} [2]	Chip deselect to power down time	-	45	-	55	-	70	ns
								ns
t_{WC}	Write cycle time	45	-	55	-	70	-	ns
$t_{CW1, 2}$	Chip select to end-of-write	33	-	50	-	60	-	ns
t_{AW}	Address valid to end-of-write	33	-	50	-	60	-	ns
t_{AS}	Address setup time	0	-	0	-	0	-	ns
t_{WP}	Write pulse width	25	-	50	-	60	-	ns
t_{WR1}	Write recovery time ($\overline{CS1}$, $\overline{WE}$)	0	-	0	-	0	-	ns
t_{WR2}	Write recovery time (CS2)	5	-	5	-	5	-	ns
t_{WHZ} [2]	Write enable to output in high-Z	-	18	-	25	-	30	ns
t_{DW}	Data to write time overlap	20	-	25	-	30	-	ns
t_{DH1}	Data hold from write time ($\overline{CS1}$,$\overline{WE}$)	0	-	0	-	0	-	ns
t_{DH2}	Data hold from write time (CS2)	5	-	5	-	5	-	ns
t_{OW} [2]	Output active from end-of-write	4	-	4	-	4	-	ns

Notes.
1. Both chip selects must be active for the device to be selected.
2. This parameter is guaranteed by device characterization, but is not production tested.

Table 13.19 ■ IDT7164 timing parameter values (continued)

There are two write cycle timing diagrams. In the first write timing diagram, both chip select signals are active before the $\overline{WE}$ signal becomes asserted. In this situation, the $\overline{WE}$ pulse must be at least t_{WP} ns wide. The write data must be valid for t_{DW} ns before the $\overline{WE}$ signal goes high and must remain valid for $t_{DH1, 2}$ ns after the $\overline{WE}$ signal goes high. The address signals must be valid for t_{AS} before the $\overline{WE}$ signal goes low. This parameter is also called address setup time relative to the $\overline{WE}$ signal.

In the second situation, the $\overline{WE}$ signal is asserted before two chip-select signals. In this condition, the chip-select signals must be active at least for $t_{CW1, 2}$ ns. The write data must be valid for t_{DW} ns before the chip-select signal become inactive and must remain valid for $t_{DH1, 2}$ ns after the chip-select signals become inactive ($\overline{CS1}$ goes high and CS2 goes low). The address signals must be valid for t_{AS} before chip-select signals become active.

13.9.2 The Atmel AT28HC64B EEPROM

The AT28HC64B is an 8K × 8 EEPROM, fabricated with Atmel's floating-gate CMOS technology. The AT28HC64B needs only a single 5V power supply to operate and achieves an access time of 55 ns.

The AT28HC64B supports a 1- to 64-byte page-write operation. The AT28HC64B also features *data polling* and *toggle bit polling* that enables early end-of-write detection. The device contains a 64-byte page register to allow the CPU to write up to 64 bytes of data without waiting for the EEPROM to complete the internal write operation. The end of a write cycle can be detected by data polling of the I/O$_7$ pin. Once the end of a write cycle has been detected, a new access for a read or write can begin. An optional software data protection mechanism is available to guard against inadvertent writes. The device also includes extra 64 bytes of EEPROM for device identification or tracking. The pin assignment of AT28HC64B is shown in Figure 13.36.

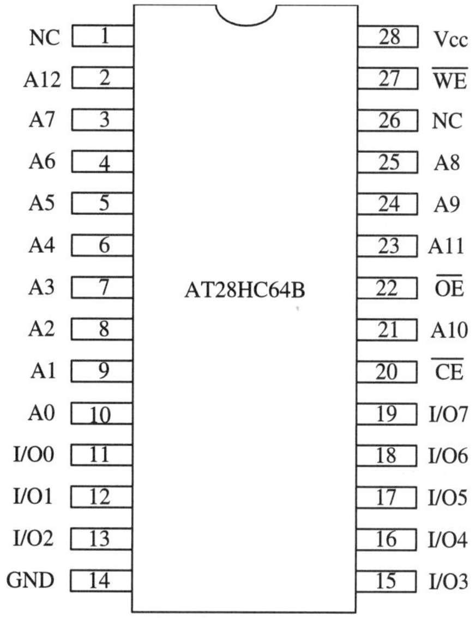

Figure 13.36 ■ The AT28HC64B pin assignment

The AT28HC64B is accessed (read and write) like a SRAM. When the $\overline{CE}$ and $\overline{OE}$ signals are low and the $\overline{WE}$ signal is high, the data stored at the memory location determined by the address pins is driven out to the I/O pins. The I/O pins are put in a high-impedance state when either $\overline{CE}$ or $\overline{OE}$ signal is high. This dual line control gives designers flexibility in preventing bus contention in their system.

A low pulse on the $\overline{WE}$ (or $\overline{CE}$) input with $\overline{CE}$ (or $\overline{WE}$) low and $\overline{OE}$ high initiates a write cycle. The address is latched on the falling edge of $\overline{CE}$ or $\overline{WE}$, whichever occurs first. The data is latched by the first rising edge of $\overline{CE}$ or $\overline{WE}$. Once a byte write has been started, it will automatically time itself to completion. Once a programming operation has been initiated and for the duration of t_{WC}, a read operation will effectively be a polling operation.

The *page-write* operation of the AT28HC64B allows 1 to 64 bytes of data to be written into the device during a single internal programming period. A page write operation is initiated in the same manner as a byte write; after the first byte is written, it can then be followed by 1 to 63 additional bytes. Each successive byte must be loaded within 150 μs (t_{BLC}) of the previous byte. If the t_{BLC} limit is exceeded, the AT28HC64B will cease accepting data and commence the internal programming operation. All bytes during a page write operation must reside on the same page as defined by the state of the A6 to A12 inputs. For each WE high-to-low transition during the page write operation, A6 to A12 must be the same. The A0 to A5 inputs specify which bytes within the page are to be written. The bytes may be loaded in any order and may be altered within the same load period. Only bytes that are specified for writing will be written; unnecessary cycling or other bytes within the page does not occur.

The AT28HC64B features data polling to indicate the end of a write cycle. During a byte or page write cycle, an attempted read of the last byte written will result in the complement of the written data to be presented on I/O_7. Once the write cycle has been completed, true data is valid on all outputs, and the next write cycle may begin. Data polling may begin at any time during the write cycle.

In addition to data polling, the AT28HC64B provides another method for determining the end of the write cycle. During the write operation, successive attempts to read data from the device will result in I/O_6 toggling between one and zero. Once the write has completed, I/O_6 will stop toggling, and valid data will be read. Toggle bit reading may begin at any time during the write cycle.

If precautions are not taken, inadvertent writes may occur during transitions of the host system power supply. Atmel has incorporated both hardware and software features that will protect the memory against inadvertent writes.

The AT28HC64B has a circuit that senses the level of V_{CC}. If V_{CC} is lower than 3.8V, the write function is inhibited. When the device is first powered on and the V_{CC} has reached 3.8V, the device will automatically time out for 5 ms before allowing a new write operation. The AT28HC64B also has a feature that disallows a write cycle to start if pulses on the $\overline{WE}$ or $\overline{CE}$ input are shorter than 15 ns.

The AT28HC64B also implements the *software data protection* (SDP) feature. This feature is disabled by default and needs to be enabled if desired. SDP is enabled by the user by issuing a series of three write commands in which three specific bytes of data are written to three specific addresses. After writing the 3-byte command sequence and waiting for t_{WC}, the entire AT28HC64B will be protected against inadvertent writes. It should be noted that even after SDP is enabled, the user may still perform a byte or page write to the AT28HC64B. This is done by preceding the data to be written by the same 3-byte command sequence used to enable SDP.

Once set, SDP remains active unless the disable command sequence is issued. Power transitions do not disable SDP, and SDP protects the AT28HC64B during power-up and power-down conditions. All command sequences must conform to the page-write timing specifications. The data in the enable and disable command sequence are not actually written into the device; their addresses may still be written with user data in either a byte or page-write operation.

After setting SDP, any attempt to write to the device without the 3-byte command sequence will start the internal write timers. No data will be written to the device, however. For the duration of t_{WC}, read operations will effectively be polling operations.

The algorithm for enabling software data protection is as follows:

Step 1
Write $AA to the memory location at $1555.

Step 2
Write $55 to the memory location at $0AAA.

Step 3
Write $A0 to the memory location at $1555. At the end of the write, the write protect state will be activated. After this step, the write operation is also enabled.

Step 4
Write any value to any location (1 to 64 bytes of data are written).

Software data protection can be disabled any time when it is undesirable. The algorithm for disabling software data protection is as follows:

Step 1
Write $AA to the memory location at $1555.

Step 2
Write $55 to the memory location at $0AAA.

Step 3
Write $80 to the memory location at $1555.

Step 4
Write $AA to the memory location at $1555.

Step 5
Write $55 to the memory location at $0AAA.

Step 6
Write $20 to the memory location at $1555. After this step, software data protection is exited.

Step 7
Write xx to any location.

Step 8
Write the last byte to the last address.

An extra 64 bytes of EEPROM memory are available to the user for device identification. By raising A9 to 12V ± 0.5V and using address locations 1FC0H to 1FFFH, the additional bytes may be written to or read from in the same manner as the regular memory.

The read, write, and page-write timing diagrams are shown in Figures 13.37, 13.38, and 13.39. The timing parameters for read, write, and page-write are listed in Tables 13.20, 13.21, and 13.22.

The AT28HC64B has three read access times:

- Address access time t_{ACC}.

- CE access time t_{CE}.

- OE access time t_{OE}.

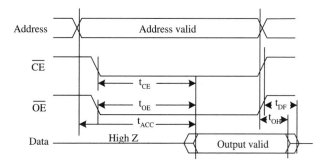

Figure 13.37 ■ AT28HC64B read timing diagram

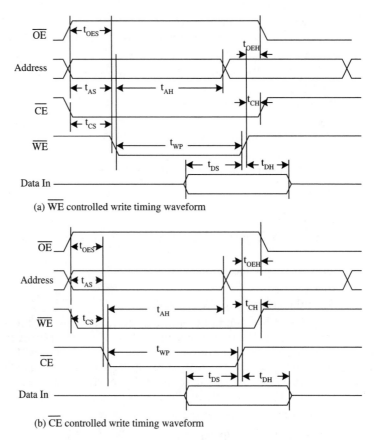

(a) $\overline{\text{WE}}$ controlled write timing waveform

(b) $\overline{\text{CE}}$ controlled write timing waveform

Figure 13.38 ■ AT28HC64B write cycle timing waveform

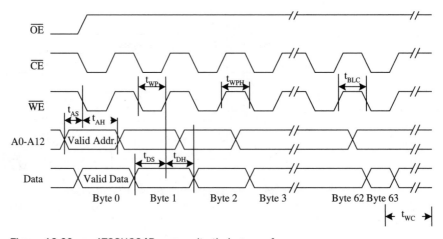

Figure 13.39 ■ AT28HC64B page write timing waveform

Symbol	Parameter	AT28HC64B-55 Min.	Max.	AT28HC64B-70 Min.	Max.	AT28HC64B-90 Min.	Max.	AT28HC64B-120 Min.	Max.	Unit
t_{ACC}	Address to output delay		55		70		90		120	ns
$t_{CE}^{(1)}$	CE to output delay		55		70		90		120	ns
$t_{OE}^{(2)}$	OE to output delay	0	30	0	35	0	40	0	50	ns
$t_{DF}^{(3)(4)}$	OE to output float	0	30	0	35	0	40	0	50	ns
t_{DH}	Output hold	0		0		0		0		ns

Notes.
1. $\overline{CE}$ may be delayed up to t_{ACC} - t_{CE} after the address transition without impact on t_{ACC}.
2. $\overline{OE}$ may be delayed up to t_{CE} - t_{OE} after the falling edge of $\overline{CE}$ without impact on t_{CE} or by t_{ACC} - t_{OE} after an address change without impact on t_{ACC}.
3. t_{DF} is specified from $\overline{OE}$ or $\overline{CE}$ whichever occurs first (C_L = 5pF).
4. This parameter is characterized and is not 100% tested.

Table 13.20 ■ AT28HC64B read characteristics

Symbol	Parameter	Min.	Max.	Unit
t_{AS}, t_{OES}	Address, $\overline{OE}$ setup time	0		ns
t_{AH}	Address hold time	50		ns
t_{CS}	Chip select setup time	0		ns
t_{CH}	Chip select hold time	0		ns
t_{WP}	Write pulse width ($\overline{WE}$ or $\overline{CE}$)	100		ns
t_{DS}	Data setup time	50		ns
t_{DH}, t_{OEH}	Data, $\overline{OE}$ hold time	0		ns

Table 13.21 ■ AT28HC64B write characteristics

Symbol	Parameter	Min.	Max.	Unit
t_{WC}	Write cycle time		10	ms
t_{WC}	Write cycle time (option available; contact Atmel)	0	2	ms
t_{AS}	Address setup time	0		ns
t_{AH}	Address hold time	50		ns
t_{DS}	Data setup time	50		ns
t_{DH}	Data hold time	0		ns
t_{WP}	Write pulse width	100		ns
t_{BLC}	Byte load cycle time		150	μs
t_{WPH}	Write pulse width high	50		ns

Table 13.22 ■ AT28HC64B page-write characteristics

The meanings of these three access times are identical to their equivalents in the IDT7164. Data on the I/O pins will remain valid for at least t_{DH} ns after address changes. The I/O pins will go to a high-impedance state t_{DF} ns after the $\overline{CE}$ or $\overline{OE}$ signal goes high. The AT28HC64B has versions with access times ranging from 55 ns to 120 ns.

There are two byte-write cycle timing waveforms. In the $\overline{WE}$-controlled write timing waveform, the $\overline{CE}$ signal is asserted (goes low) before the $\overline{WE}$ signal and becomes inactive (goes high) after the $\overline{WE}$ signal. In the $\overline{CE}$-controlled write timing waveform, the $\overline{WE}$ signal becomes active earlier and becomes inactive later than the $\overline{CE}$ signal. In both cases, the write data must be valid for t_{DS} ns before the controlling signal ($\overline{WE}$ or $\overline{CE}$) becomes inactive, and remain valid for t_{DH} ns after the controlling signal goes high. The write timing waveform only illustrates how the CPU writes data into the EEPROM. The EEPROM still needs to initiate an internal programming process to actually write data into the specified location. The CPU can find out whether the internal programming process has been completed by using the data polling or toggle bit polling method.

13.9.3 The Optrex DMC-20434 LCD Kit

The function and programming of the Optrex DMC-20434 LCD kit has been illustrated in Chapter 7. The block diagram of the DMC-20434 is shown in Figure 7.16. This LCD kit uses the Hitachi HD44780 as the controller. The HC44780 is the most widely used LCD controller today. For easy reference, Figure 7.16 is repeated here.

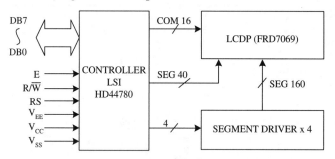

Figure 7.16 ■ Block diagram of the DMC-20434 LCD kit

Data and commands from the microcontroller are written into the HC44780 via the data bus pins DB7-DB0. The status of the LCD is also read from the data bus. The E signal is used to enable the operation of the LCD kit and should be controlled by the chip enable signal generated by the address decoder. The read and write timing diagrams of the HD44780 are shown in Figures 13.40 and 13.41. The values of timing parameters are listed in Table 13.23.

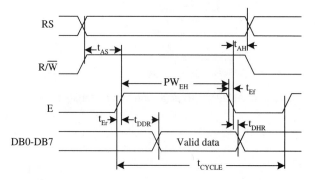

Figure 13.40 ■ HD44780 LCD controller read timing diagram

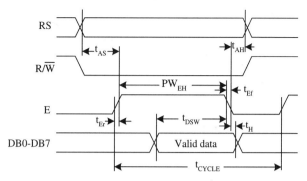

Figure 13.41 ■ HD44780 LCD controller write timing diagram

Symbol	Meaning	Min	Typ	Max.	Unit
t_{CYCLE}	Enable cycle time	500	-	-	ns
PW_{EH}	Enable pulse width (high level)	230	-	-	ns
t_{Er}, t_{Ef}	Enable rise and decay time	-	-	20	ns
t_{AS}	Address setup time, RS, R/W, E	40	-	-	ns
t_{DDR}	Data delay time	-	-	160	ns
t_{DSW}	Data setup time	80	-	-	ns
t_H	Data hold time (write)	10	-	-	ns
t_{DHR}	Data hold time (read)	5	-	-	ns
t_{AH}	Address hold time	10	-	-	ns

Table 13.23 ■ HD44780 bus cycle timing parameters

In Figure 13.40, the address signal (RS) should be valid for at least 40 ns (t_{AS}) before the E clock goes high. The LCD data will become valid 160 ns (t_{DDR}) after the rising edge of the E signal and remain valid for 5 ns (t_{DHR}) after the falling edge of the E signal. The pulse width of the E signal must be at least 230 ns (PW_{EH}).

When writing data to the LCD, the RS signal must be valid for at least 40 ns (t_{AS}) before the rising edge of the E signal. Data to be written into the LCD must be valid for at least 80 ns (t_{DSW}) before the falling edge of the E signal and must remain valid for 10 ns after the falling edge of the E signal.

13.10 The Memory Expansion Issue

There are infinite alternatives in external memory expansion. In this section, we will choose one alternative and carry it through the whole process.

13.10.1 Memory Space Assignment

Suppose we are designing a 912BC32-based demo board that has 16KB external EEPROM, 16KB external SRAM, and a DMC-20434 LCD kit. This demo board would use an 8 MHz E clock signal to control all the operations. Both the EEPROM and SRAM are to be configured in expanded wide mode to achieve the best memory performance. The LCD kit will be assigned to the register-following map and configured to operate in expanded narrow mode because it has an 8-bit wide data bus.

The memory space assignment is as follows:

- LCD kit: $0200-$0201
- SRAM: $4000-$7FFF
- EEPROM: $C000-$FFFF

13.10.2 Control Signal Requirements

In this section, we will discuss the timing requirements for all control signals and how to generate them.

ADDRESS SIGNALS

Because address signals must remain stable and valid throughout the read and write bus cycles, we need to use latches (for example, the 74HC573) to hold the upper (A15-A8) and lower (A7-A0) address signals. By examining the 68HC12 read cycle and write cycle timing diagrams (Figures 13.10 and 13.31), we conclude that we should use the rising edge of the ECLK signal to latch A15-A0 into two 74HC573s. The block diagram of the 74HC573 is shown in Figure 13.42. Inputs D0-D7 are latched into Q0-Q7 on the falling edge of the $\overline{LE}$ signal. D0-D7 must be stable for at least 15 ns (t_{su}) before the rising edge of the $\overline{LE}$ signal in order to be correctly latched. Outputs Q0-Q7 will be in a high-impedance state if the $\overline{OE}$ signal is high. The 74HC573 has a propagation delay of 35 ns for a 4.5V power supply. We will use this value for timing calculation purpose because the propagation delay for a 5V power supply will be very close.

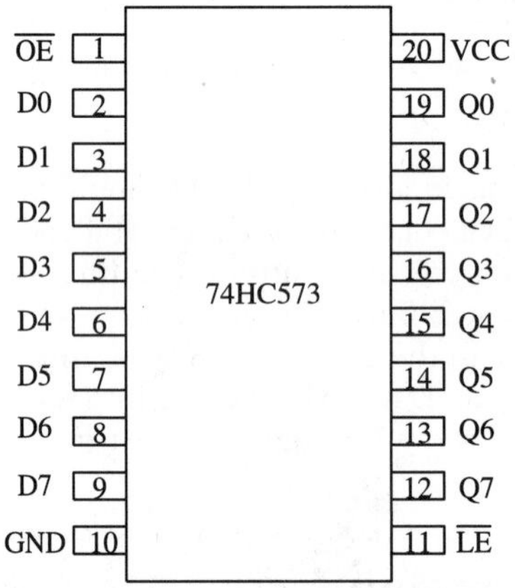

Figure 13.42 ■ Pin assignment of the 74HC573

At 8 MHz ECLK, address signals A15-A0 become valid 6.2 ns after the rising edge of the ECLK signal. Therefore we need to delay and invert the ECLK signal before we can use it to latch address signals. One possibility is to invert it once and buffer it (delay it by cascading a buffer to it) twice using the GAL18V10B-7LP (with 7.5 ns propagation delay). This should provide enough time for the 74HC573 to latch address signals (3×7.5 ns > 6.2 ns $+ 15$ ns). The resultant address-latching signal (call it $\overline{AS}$) will become valid 22.5 ns after the rising edge of ECLK, and the latched address signals A15-A0 will be valid 57.5 ns (22.5 ns + 35 ns) after the rising edge of ECLK. The equations for generating the $\overline{AS}$ signal are as follows:

t1 = ECLK;
t2 = t1
$\overline{AS}$ = !t2;

The circuit for generating the $\overline{AS}$ signal and latching address signals is shown in Figure 13.43.

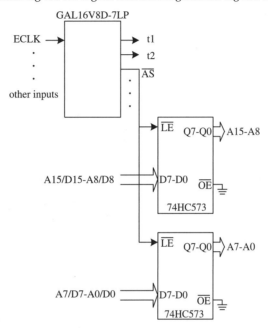

Figure 13.43 ■ Address latching circuit

CONTROL SIGNALS FOR THE LCD KIT DMC-20434

Since the LCD kit data bus is 8-bit, it must be connected to port A pins in expanded narrow mode. According to Table 13.23, DMC-20434 has a cycle time of 500 ns. This is equal to 4 ECLK cycles at 8 MHz. Therefore, we need to extend the high interval of the ECLK signal by three cycles.

We will use the full decoding method to generate the chip select signal (CS_LCD) for the LCD. Figure 13.41 shows that the LCD needs an E clock to operate. This signal should be a delayed version of ECLK. The minimum delay of E from the RS input (t_{AS}) is 40 ns. We can use an 8-bit addressable latch 1-of-8 decoder 74HC259 to generate the E signal for the LCD kit by delaying the ECLK output from the 68HC12.

The pin assignment of the 74HC259 is shown in Figure 13.44.

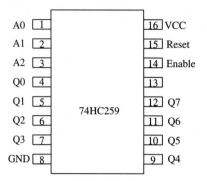

Figure 13.44 ■ Pin assignment of the 74HC259

The 74HC259 has four operation modes. We will use the 8-line demultiplexing mode only. In this mode, the inputs Enable and Reset must be tied to low. Suppose we use the GAL22V10 (with 7.5 ns propagation delay) to generate the active low chip select signal (CS_LCD) for LCD and make the following connection to the 74HC259:

- connect *CS_LCD* to A0 pin of 74HC259
- connect A1 (from 74HC573) to A1 pin of 74HC259
- connect A2 (from 74HC573) to A2 pin of 74HC259
- connect ECLK to Data In pin of 74HC259

Then the Q0 output will be the delayed version of ECLK with a delay equal to:

AS delay + address latch delay + address decoder delay + delay of 74HC259
= 22.5 ns + 35 ns + 7.5 ns + 43 ns = 108 ns

Q0 will be used as the E input of the LCD kit. For the LCD kit, the RS input (connected to A0) is valid 57.5 ns after the rising edge of ECLK. Therefore, the E input for the LCD kit is valid 50.5 ns (= 108 ns – 57.5 ns) after RS becomes valid and hence satisfies the timing requirement. The R/W input of the LCD should come directly from the 68HC12's R/W signal. Among all the GAL devices that we mentioned, only the GAL22V10 provides product terms with more than 14 variables and therefore must be selected if we decide to use the full-decoding scheme to implement our address decoder.

A sketch of this drawing is shown in Figure 13.45. In Figure 13.45, the logic equation for the *CS_LCD* signal is:

!CS_LCD = (!A15 * !A14 * !A13 * !A12 * !A11 * !A10 * A9 * !A8 *
            !A7 * !A6 * !A5 * !A4 * !A3 * ECLK)

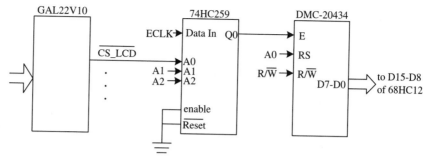

Figure 13.45 ■ Partial circuit for the DMC-20434 LCD kit

Example 13.12

For the circuit shown in Figure 13.45, verify that the read and write timing requirements for the 68HC12 are satisfied.

Solution:

Read timing analysis:

The 68HC12 requires data to be valid 31.5 ns before the falling edge of the ECLK signal in a read cycle. The data from the LCD becomes valid t_{DDR} ns after the rising edge of the E signal and is 268 ns (160 ns + 108 ns) after the rising edge of the ECLK signal. This is equivalent to 169.5 ns (500 ns – 62.5 ns – 268 ns) before the falling edge of the ECLK signal. Therefore, the read data setup time is satisfied.

The RS (A0) signal will not change until a new value is latched into the 74HC573, which won't occur until 120 ns (0.5 t_{cyc} + 22.5 ns + 35 ns) into the next ECLK cycle. The E signal will fall 43 ns after the start of the next ECLK cycle. Therefore the LCD data will remain valid 48 ns (= 43 ns + t_{DHR} ns) after the start of the next ECLK cycle. The 68HC12 requires read data to remain valid 0 ns after the falling edge of the ECLK signal. The data hold time requirement is satisfied. The LCD read cycle timing is illustrated in Figure 13.46.

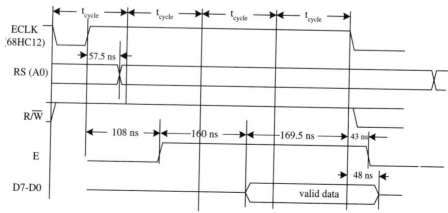

Figure 13.46 ■ DMC-20434 LCD read cycle timing diagram

Write Timing Analysis:

The 68HC12 drives the write data to the data bus 62.5 ns after the rising edge of the ECLK signal or 375 ns $(3.5 \times t_{cyc} - t_{DDW})$ before the falling edge of the ECLK signal. Write data will be held valid for 25 ns after the falling edge of the ECLK signal. The DMC-20434 requires write data to be valid 80 ns before the falling edge of the E signal and remain valid for 10 ns. Since E goes low 43 ns after ECLK goes low, the write data setup time is 418 ns (= 375 ns + 43 ns) and is satisfied. Write data hold time is − 18 ns (= 25 ns − 43 ns) and violates the requirement. However, the 68HC12 does not drive the multiplex address/data bus for 68.7 ns (62.5 ns + 6.2 ns) after the end of a bus cycle. The capacitor of the printed circuit board will keep the write data during this period and satisfies the write data hold time requirement. The LCD write cycle timing is illustrated in Figure 13.47.

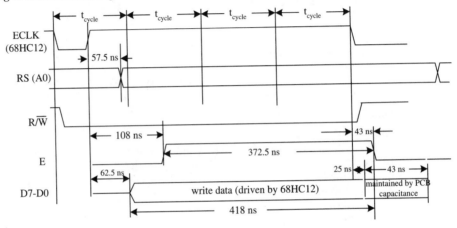

Figure 13.47 ■ 68HC12 to DMC-20434 LCD write cycle timing diagram

Example 13.13

Provide an analytical proof that the capacitance of the printed circuit board can hold the data voltage for enough time to satisfy the data hold time requirement of the LCD.

Solution:

The voltage of the data input to the LCD is degraded by the following leakage currents:

1. Input current into the 68HC12 data pin (on the order of 2.5 µA).
2. Input leakage current into the LCD kit (not given by the LCD data sheet, but we can assume it is three times as large as the leakage current into the 68HC12 data pin).
3. Other leakage paths on the printed circuit board, depending on the system configuration.

The capacitance of the printed circuit board is estimated to be 20 pF per foot. Let C, I, ΔV, and Δt be printed circuit board capacitance of one data line, total leakage current, voltage change due to leakage current, and the time it takes for voltage to degraded by ΔV, respectively.

The elapsed time before the data bus signal degrades to an invalid level can be estimated by the following equation:

$$\Delta t \approx C \Delta V \div I$$

The voltage degradation of 2.5V is considered enough to cause the data input to the LCD kit to change the logic value from one to zero. The data bus line on the printed circuit board is normally not longer than one foot for a single board computer. We can use one foot as its length. The elapsed time before the data bus signal degrades to an invalid level is:

$$\Delta t = 20 \text{ pF} \times 2.5\text{V} \div 10 \text{ μA} \approx 5 \text{ μs}$$

Although the above equation is oversimplified, it does give us some idea about the order of the time over which the charge across the data bus capacitor will hold after the CPU stops driving the data bus. Suppose the leakage current of other path is 10 times larger, the Δt value will be 500 ns and is still longer than the minimum hold time requirement.

In Figure 13.47, the pulse width of the E signal is 372.5 ns (= 3.5 t_{cyc} – 108 ns + 43 ns) and exceeds the minimum requirement (230 ns).

▲

CONTROL SIGNALS FOR THE AT28HC64B

We need two AT28HC64B chips to construct a 16-bit EEPROM for the 68HC12. The upper byte is connected to Port A (D15-D8) and the lower byte is connected to Port B (D7-D0). The $\overline{OE}$ signal would be permanently grounded to save a control signal.

These two chips could use the same select input: CS_EEPROM. The address signals A13-A1 will be connected to the address inputs A12-A0 of the AT28HC64B. The logic equation of the chip-select for the AT28HC64B is as follows:

!CS_EEPROM = (A15 * A14 * ECLK)

The chip select signal for EEPROM would become valid 65 ns after the rising edge of the ECLK:

AS delay + address latch delay + address decoder delay =
22.5 ns + 35 ns + 7.5 ns = 65 ns

The access time of the EEPROM is 55 ns for read. There is no possibility for completing a read or write cycle of the EEPROM in one ECLK cycle. We need to stretch each bus cycle by one ECLK period to meet the access time requirement. After stretching one ECLK cycle, the data will become valid 67.5 ns before the falling edge of the ECLK signal and is longer than the data setup time requirement (t_{su} = 31.5 ns):

$0.5 \times t_{cyc} + t_{cyc}$ – EEPROM chip select valid time – EEPROM access time =
62.5 ns + 125 ns – 65 ns – 55 ns = 67.5 ns

The EEPROM data hold time is 0 ns. The EEPROM read data would become invalid (7.5 ns after the current bus cycle) at the earlier one of the following two times:

1. t_{OH} ns after address changes. This won't happen until 120 ns (62.5 ns + 57.5 ns) after the current bus cycle.

2. t_{DF} ns after $\overline{CE}$ becomes invalid. The $\overline{CE}$ signal becomes invalid 7.5 ns after the falling edge of ECLK.

This analysis verifies that the 68HC12 read data setup and hold times are satisfied. The read cycle timing for the AT28HC64B is shown in Figure 13.48.

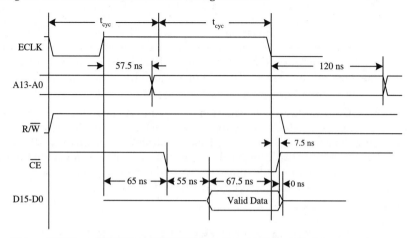

Figure 13.48 ■ 68HC12 to AT28HC64 read cycle timing diagram

Depending on the application, we may or may not want to allow a write operation to the EEPROM during the normal operation. If we don't want the EEPROM to be written into, we can simply pull the $\overline{WE}$ input of the EEPROM to high. This is done in the CME-12BC. Since we want to store the monitor program in EEPROM, this is a legitimate decision.

A partial circuit for the AT28HC64Bs is shown in Figure 13.49.

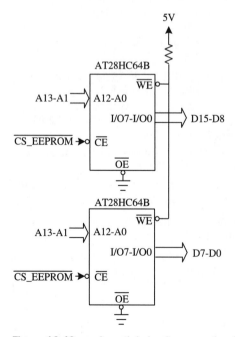

Figure 13.49 ■ A partial circuit connection for the AT28HC64B

CONTROL SIGNALS FOR IDT7164

All external memory chips (except those in the register-following map) must use the same amount of clock stretch. Since the bus cycle of the AT28HC64B has been stretched by one ECLK cycle, the IDT7164 must also stretch its bus cycle by one ECLK cycle. For this reason we don't need to use the fastest IDT7164 available, we will choose the version that has access time equal to 55 ns and verify that it meets all the timing requirements.

The 68HC12 is designed to follow the *Big-Endian* convention in its data bus order. The high order byte in a word is always stored at the lower address. Depending on whether a word is aligned with an even address, a 16-bit data access may take one or two bus cycles. The even byte is always aligned with the upper byte of the data bus (D15-D8), whereas the odd byte is always aligned with the lower byte of the data bus (D7-D0). When the lower data byte is accessed, the 68HC12 will assert the $\overline{LSTRB}$ signal as an indication.

For a read access, we can allow the SRAM to send out data on both bytes and the 68HC12 will figure out how many bytes and which byte(s) to accept. However, we need to control the write access to the upper and lower bytes of SRAM to avoid any undesirable write operation during a write bus cycle. Let $\overline{WE1}$ and $\overline{WE0}$ be the write enable signals of the upper and the lower bytes of SRAM. The generation of $\overline{WE1}$ and $\overline{WE0}$ are illustrated in Table 13.24.

A0	LSTRB	R/W	WE1	WE0	Description
x	x	high	high	high	68HC12 is not performing a write
0	0	low	low	low	68HC12 is writing an aligned word
0	1	low	low	high	68HC12 is writing the upper data byte
1	0	low	high	low	68HC12 is writing the lower data byte
1	1	low	high	high	this combination won't happen

Table 13.24 ■ Generation of WE1 and WE0

Therefore:

!WE1 = (!R/W * !A0 * ECLK)
!WE0 = (!R/W * A0 * !LSTRB * ECLK)

The logic equation of the chip select signal for the IDT7164 is as follows:

!CS_SRAM = (!A15 * A14 * ECLK)

The data from SRAM should be allowed to drive the data bus when SRAM is selected and the operation is a read. The logic equation of the output enable (OE) signal for SRAM is as follows:

!OE = (!A15 * A14 * R/W * ECLK)

▲

Example 13.14

▼

Verify that the timing requirements for the read and the write bus cycles to the IDT7164 SRAM are satisfied.

Solution: The chip select and output enable signals become valid 65 ns after the rising edge of the ECLK:

AS delay + address latch delay + address decoder delay =
22.5 ns + 35 ns + 7.5 ns = 65 ns

Since the IDT7164 and AT28HC64B have the same access time, the data setup time for the 68HC12 is satisfied by the same argument. Since address signals change much later, the read data hold time is determined by the $\overline{OE}$ or the $\overline{CS}$ signal. The IDT7164 output data signals go into high impedance 25 ns after either the $\overline{OE}$ or the $\overline{CS}$ signal becomes invalid. Therefore, data will hold for at least 32.5 ns after the falling edge of ECLK and hence the data hold time requirement is satisfied.

The read bus cycle timing for the IDT7164 is illustrated in Figure 13.50.

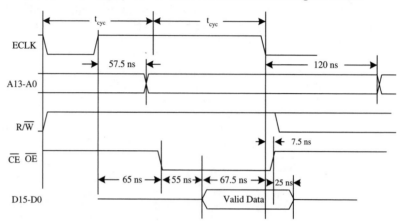

Figure 13.50 ■ 68HC12 to IDT7164 read cycle timing diagram

During a write bus cycle, the 68HC12 drives the data 62.5 ns after the rising edge of ECLK and keeps the write data for 25 ns after the falling edge of ECLK. The $\overline{CS}$, $\overline{WEI}$, and $\overline{WE0}$ go to high 7.5 ns after the falling edge of ECLK. The write data becomes valid 132.5 ns before the rising edge of $\overline{WE}$ signal and remains valid 17.5 ns after the rising edge of $\overline{WE}$ and satisfy the requirements of the IDT7164. The write pulse width is 130 ns and also exceeds the minimum requirement.

The write bus cycle timing is illustrated in Figure 13.51.

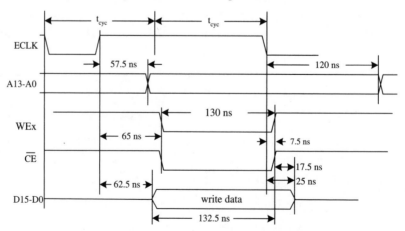

Figure 13.51 ■ 68HC12 to IDT7164 write cycle timing diagram

The circuit connection for the IDT7164 is shown in Figure 13.52.

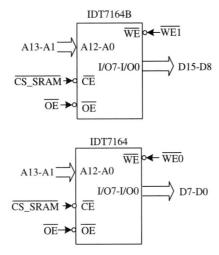

Figure 13.52 ■ A partial circuit connection for the IDT7164

ADDRESS DECODER & CONTROL DESIGN

The address decoder and all control signals can be implemented in one GAL22V10 and one GAL18V10. The GAL18V10 contains 10 OLMCs that are identical to those contained in the GAL22V10. The GAL22V10 supports product terms with more variables (8 to 16) than the GAL18V10 does (8 to 10). The pin assignment for these signals is shown in Figure 13.53.

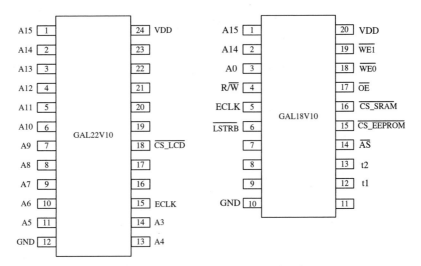

Figure 13.53 ■ Pin assignment for memory control signals

To make this design to work, we need to program the MISC register to set up the following configuration:

- Register-following map space acts like an 8-bit external data bus
- Stretch E clocks by three cycles on accesses to the 512-byte register-following map
- Stretch E clocks by one cycle on accesses to the external address space
- Disable internal flash memory

This configuration can be achieved by programming the MISC register in the reset handling routine:

```
movb  #$74,MISC
```

The PEAR register must be configured as follows:

- Enables external ECLK (is enabled by default after reset)
- Enables a *low byte strobe signal* ($\overline{\text{LSTRB}}$)
- Enables external memory read and write

This configuration can be achieved by executing the following instruction in the reset handling routine:

```
movb  #$0C,PEAR
```

Overall, the external memory system of this demo board consists of:

- One 74HC259 demultiplexor
- One DMC-20434 LCD kit
- Two AT28HC64B EEPROM chips
- Two IDT7164 SRAM chips
- Two 74HC573 latch chips
- One GAL22V10 and one GAL18V10

The complete circuit diagram of this memory system can be obtained by combining Figures 13.43, 13.45, 13.49, 13.52, and 13.53.

We will need two ABEL programs for this memory control circuit—one for the GAL22V10 and one for the GAL18V10. Creating these programs will be left as an exercise problem.

13.11 Summary

The 68HC12 microcontroller has on-chip memory resources including SRAM, EEPROM, and flash memory. Each of these memory resources has its own default assigned space but can be remapped to a different block. It is possible to have two or more memory resources overlapped. The 68HC12 assigns precedence to each memory resource. The memory resource with the highest precedence is selected when two or more memory chips overlap in their spaces. The remapping of a memory resource is done by programming the associated initialization register.

All 68HC12 members except the 812HCA4 have on-chip flash memory. The B family members have 32KB on-chip flash memory. The 912D60A have 60KB flash memory. The 912DG128 and the 912DT128 have 128KB of on-chip flash memory. The 912DG128 and the 912DT128 use the paging method to access more than 64KB of memory. These two microcontrollers use the PPAGE register to perform on-chip memory expansion. They have a user's program space window, a register space window for flash module registers, and a test program space window.

The user's program page window consists of 16KB flash EEPROM bytes. One of eight pages is viewed through this window for a total of 128KB accessible flash memory. The program space window is located from $8000 to $BFFF.

The register space window consists of a 4-byte register block. One of four pages is viewed through this window for each of the 32K flash module register blocks.

The test mode program page window consists of 32KB flash memory. One of the four 32 KB arrays is viewed through this window. This window is only available in special mode for test purposes and replaces the user's program page window.

External memory expansion involves three issues:

- Memory space assignment
- Address decoder and control circuitry design
- Timing verification

The 512 bytes after the register file is called the *register-following* map. This space can be configured to expanded narrow mode when the microcontroller is configured to expanded wide mode so that it can accommodate 8-bit peripheral devices. LCD and peripheral devices are often assigned to the register-following map.

Assigning memory space into blocks of equal size is easy but has the drawback of inefficient memory space utilization. Programmable logic devices (PLDs) such as PAL or GAL have been used in designing address decoders in many products. PLDs support product terms of many variables and hence can support the *best-fit* memory space assignment. A best-fit memory assignment assigns as many bytes to a memory-mapped device as required.

The 68HC12 supports bus cycle stretching to accommodate slower memory components. A normal 68HC12 bus cycle takes exactly one E clock cycle to complete. However, a bus cycle can be stretched by up to three E clock cycles. Bus cycle stretching is implemented by programming the MISC register and should be done in the reset handling routine.

When adding external memory to a microcontroller, we need to make sure all timing requirements of the microcontroller and the memory devices are satisfied. Any timing violation may prevent the memory system to operate correctly. The last section of this chapter provides a detailed timing verification to a design example.

13.12 Exercises

E13.1 Write an instruction sequence to remap the register block so that it starts at $1000.

E13.2 Write an instruction sequence to remap the on-chip SRAM of the 912BC32 so that it starts at $5000.

E13.3 Write an instruction sequence to remap the on-chip EEPROM of the 912BC32 so that it starts at $2000.

E13.4 Suppose we replace the AT28HC64B in Figure 13.49 with a version that has 70 ns access time. Would timing requirements still be satisfied? Verify your answer.

E13.5 Suppose we replace the AT28HC64B in Figure 13.49 with a version that has 90 ns access time. Would timing requirements still be satisfied? Verify your answer.

E13.6 Suppose we replace the IDT7164 in Figure 13.52 with a version that has 70 ns access time. Would timing requirements still be satisfied? Verify your answer.

13.13 Lab Exercises & Assignments

L13.1 *Memory test for the CME-12BC.* Write a program that tests the memory locations immediately after your program until the end of the external SRAM ($7FFF). Your program is written to test stuck-at-1 and stuck-at-0 faults, and proper data connections.

1. To test stuck-at-0 faults, write $FF to each memory location and read it back to check.

2. To test stuck-at-1 faults, write $00 to each memory location and read it back to check.

3. To test for proper connection, write 10101010 and 01010101 to each memory location and read it back to check.

Output a message to indicate the test result at the end of the program.

L13.2 *EEPROM programming.* Write two programs. The first program is to output a message "Hello world!" to the terminal port. The second program is to copy the first program to the EEPROM area and call it as a subroutine (this program needs to perform EEPROM programming).

L13.3 *Flash memory programming.* Write a program to be programmed into the on-chip flash memory (to be run on the CME-12BC demo board). This program would perform two operations. The first operation is to keep track of the current time-of-day and the second operation is to generate a 1 KHz square wave with duty cycle specified by the end user interactively. When the program starts, it outputs a prompt to ask the user to enter the current time-of-day in the format of *hhmmss* and waits for the user to enter the time. After the user enters the time, the program outputs another prompt to ask the user to specify the duty cycle for the square wave. After the user enters the duty cycle, the square wave would be generated using the output compare function and be displayed on the oscilloscope. You also need to write subroutines to perform input/output to/from the SCI port.

To make this program work, you need to store the starting address of your program at $FFFE-$FFFF. You can set the stack pointer at $C00 in your program. Enter and assemble this program using the AxIDE software distributed by Axiom. Select "CME12BC" under AxIDE.

Select the "Program" option and when prompted for a file name, enter your program's file name with *.s19* as the filename extension.

Set the CONFIG SWITCH positions 1, 2, 3, 4, and 6 to the ON position on the demo board. Press the **reset** button on the demo board, then select **continue** or hit **enter**. A new utilities menu should be displayed.

Set the CONFIG SWITCH position 5 to the ON position. The red VPP light should come on. When prompted to "Erase" choose **yes**. When finished with the flash memory programming, set the CONFIG SWITCH positions 1, 2, 3, 4, and 5 to the OFF position. The VPP light should turn off.

Press the RESET button on the demo board. Your program should start automatically and the operations will be performed.

To return to the D-Bug12 monitor prompt, set the CONFIG SWITCH positions 1, 2, 3, and 4 back to ON and then press RESET.

Appendix A

Instruction Reference

A.3 Instruction Set Summary

Table A-1 is a quick reference to the CPU12 instruction set. The table shows source form, describes the operation performed, lists the addressing modes used, gives machine encoding in hexadecimal form, and describes the effect of execution on the condition code bits.

A.3.1 Notation Used in Instruction Set Summary

A.3.1.1 Explanation of Italic Expressions in Source Form Column

abc — A or B or CCR

abcdxys — A or B or CCR or D or X or Y or SP. Some assemblers also allow T2 or T3.

abd — A or B or D

abdxys — A or B or D or X or Y or SP

dxys — D or X or Y or SP

msk8 — 8-bit mask, some assemblers require # symbol before value

opr8i — 8-bit immediate value

opr16i — 16-bit immediate value

opr8a — 8-bit address used with direct address mode

opr16a — 16-bit address value

oprx0_xysp — Indexed addressing postbyte code:
  oprx3,–xys Predecrement X or Y or SP by 1 . . . 8
  oprx3,+xys Preincrement X or Y or SP by 1 . . . 8
  oprx3,xys– Postdecrement X or Y or SP by 1 . . . 8
  oprx3,xys+ Postincrement X or Y or SP by 1 . . . 8
  oprx5,xysp 5-bit constant offset from X or Y or SP or PC
  abd,xysp   Accumulator A or B or D offset from X or Y or SP or PC

oprx3 — Any positive integer 1 . . . 8 for pre/post increment/decrement

oprx5 — Any value in the range –16 . . . +15

oprx9 — Any value in the range –256 . . . +255

oprx16 — Any value in the range –32,768 . . . 65,535

page — 8-bit value for PPAGE, some assemblers require # symbol before this value

rel8 — Label of branch destination within –256 to +255 locations

rel9 — Label of branch destination within –512 to +511 locations

rel16 — Any label within 64-Kbyte memory space

trapnum — Any 8-bit value in the range $30–$39 or $40–$FF

xys — X or Y or SP

xysp — X or Y or SP or PC

A.3.1.2 Address Modes

IMM	—	Immediate
IDX	—	Indexed (no extension bytes) includes:
		5-bit constant offset
		Pre/post increment/decrement by 1 . . . 8
		Accumulator A, B, or D offset
IDX1	—	9-bit signed offset (one extension byte)
IDX2	—	16-bit signed offset (two extension bytes)
[D, IDX]	—	Indexed indirect (accumulator D offset)
[IDX2]	—	Indexed indirect (16-bit offset)
INH	—	Inherent (no operands in object code)
REL	—	Two's complement relative offset (branches)

A.3.1.3 Machine Coding

dd — 8-bit direct address $0000 to $00FF high byte assumed to be $00

ee — High-order byte of a 16-bit constant offset for indexed addressing

eb — Exchange/Transfer post-byte.

ff — Low-order eight bits of a 9-bit signed constant offset for indexed addressing or low-order byte of a 16-bit constant offset for indexed addressing

hh — High-order byte of a 16-bit extended address

ii — 8-bit immediate data value

jj — High-order byte of a 16-bit immediate data value

kk — Low-order byte of a 16-bit immediate data value

lb — Loop primitive (DBNE) post-byte.

ll — Low-order byte of a 16-bit extended address

mm — 8-bit immediate mask value for bit manipulation instructions Set bits indicate bits to be affected.

pg — Program page (bank) number used in CALL instruction.

qq — High-order byte of a 16-bit relative offset for long branches

tn — Trap number $30–$39 or $40–$FF

rr — Signed relative offset $80 (–128) to $7F (+127) Offset relative to the byte following the relative offset byte or low-order byte of a 16-bit relative offset for long branches

xb — Indexed addressing post-byte.

A.3.1.4 Access Detail

Each code letter equals one CPU cycle. Uppercase = 16-bit operation and lowercase = 8-bit operation.

f — Free cycle, CPU doesn't use bus

g — Read PPAGE internally

I — Read indirect pointer (indexed indirect)

i — Read indirect PPAGE value (call indirect)

n — Write PPAGE internally

O — Optional program word fetch (P) if instruction is misaligned and has an odd number of bytes of object code — otherwise, appears as a free cycle (f)

P — Program word fetch (always an aligned word read)

r — 8-bit data read

R — 16-bit data read

s — 8-bit stack write

S — 16-bit stack write

w — 8-bit data write

W — 16-bit data write

u — 8-bit stack read

U — 16-bit stack read

V — 16-bit vector fetch

t — 8-bit conditional read (or free cycle)

T — 16-bit conditional read (or free cycle)

x — 8-bit conditional write

Special Cases

PPP/P — Short branch, PPP if branch taken, P if not

OPPP/OPO — Long branch, OPPP if branch taken, OPO if not

A.3.1.5 Condition Codes Columns

– — Status bit is not affected by operation.

0 — Status bit is cleared by operation.

1 — Status bit is set by operation.

Δ — Status bit is affected by operation.

⇓ — Status bit may be cleared or remain set, but is not set by operation.

⇑ — Status bit may be set or remain cleared, but is not cleared by operation.

? — Status bit may be changed by operation but the final state is not defined.

! — Status bit is used for a special purpose.

Source Form	Operation	Addr. Mode	Machine Coding (hex)	Access Detail	S	X	H	I	N	Z	V	C
ABA	(A) + (B) ⇒ A Add Accumulators A and B	INH	18 06	OO	–	–	Δ	–	Δ	Δ	Δ	Δ
ABX	(B) + (X) ⇒ X *Translates to* LEAX B,X	IDX	1A E5	PP[1]	–	–	–	–	–	–	–	–
ABY	(B) + (Y) ⇒ Y *Translates to* LEAY B,Y	IDX	19 ED	PP[1]	–	–	–	–	–	–	–	–
ADCA #opr8i ADCA opr8a ADCA opr16a ADCA oprx0_xysp ADCA oprx9,xysp ADCA oprx16,xysp ADCA [D,xysp] ADCA [oprx16,xysp]	(A) + (M) + C ⇒ A Add with Carry to A	IMM DIR EXT IDX IDX1 IDX2 [D,IDX] [IDX2]	89 ii 99 dd B9 hh ll A9 xb A9 xb ff A9 xb ee ff A9 xb A9 xb ee ff	P rfP rOP rfP rPO frPP fIfrfP fIPrfP	–	–	Δ	–	Δ	Δ	Δ	Δ
ADCB #opr8i ADCB opr8a ADCB opr16a ADCB oprx0_xysp ADCB oprx9,xysp ADCB oprx16,xysp ADCB [D,xysp] ADCB [oprx16,xysp]	(B) + (M) + C ⇒ B Add with Carry to B	IMM DIR EXT IDX IDX1 IDX2 [D,IDX] [IDX2]	C9 ii D9 dd F9 hh ll E9 xb E9 xb ff E9 xb ee ff E9 xb E9 xb ee ff	P rfP rOP rfP rPO frPP fIfrfP fIPrfP	–	–	Δ	–	Δ	Δ	Δ	Δ
ADDA #opr8i ADDA opr8a ADDA opr16a ADDA oprx0_xysp ADDA oprx9,xysp ADDA oprx16,xysp ADDA [D,xysp] ADDA [oprx16,xysp]	(A) + (M) ⇒ A Add without Carry to A	IMM DIR EXT IDX IDX1 IDX2 [D,IDX] [IDX2]	8B ii 9B dd BB hh ll AB xb AB xb ff AB xb ee ff AB xb AB xb ee ff	P rfP rOP rfP rPO frPP fIfrfP fIPrfP	–	–	Δ	–	Δ	Δ	Δ	Δ
ADDB #opr8i ADDB opr8a ADDB opr16a ADDB oprx0_xysp ADDB oprx9,xysp ADDB oprx16,xysp ADDB [D,xysp] ADDB [oprx16,xysp]	(B) + (M) ⇒ B Add without Carry to B	IMM DIR EXT IDX IDX1 IDX2 [D,IDX] [IDX2]	CB ii DB dd FB hh ll EB xb EB xb ff EB xb ee ff EB xb EB xb ee ff	P rfP rOP rfP rPO frPP fIfrfP fIPrfP	–	–	Δ	–	Δ	Δ	Δ	Δ
ADDD #opr16i ADDD opr8a ADDD opr16a ADDD oprx0_xysp ADDD oprx9,xysp ADDD oprx16,xysp ADDD [D,xysp] ADDD [oprx16,xysp]	(A:B) + (M:M+1) ⇒ A:B Add 16-Bit to D (A:B)	IMM DIR EXT IDX IDX1 IDX2 [D,IDX] [IDX2]	C3 jj kk D3 dd F3 hh ll E3 xb E3 xb ff E3 xb ee ff E3 xb E3 xb ee ff	OP RfP ROP RfP RPO fRPP fIfRfP fIPRfP	–	–	–	–	Δ	Δ	Δ	Δ
ANDA #opr8i ANDA opr8a ANDA opr16a ANDA oprx0_xysp ANDA oprx9,xysp ANDA oprx16,xysp ANDA [D,xysp] ANDA [oprx16,xysp]	(A) • (M) ⇒ A Logical And A with Memory	IMM DIR EXT IDX IDX1 IDX2 [D,IDX] [IDX2]	84 ii 94 dd B4 hh ll A4 xb A4 xb ff A4 xb ee ff A4 xb A4 xb ee ff	P rfP rOP rfP rPO frPP fIfrfP fIPrfP	–	–	–	–	Δ	Δ	0	–
ANDB #opr8i ANDB opr8a ANDB opr16a ANDB oprx0_xysp ANDB oprx9,xysp ANDB oprx16,xysp ANDB [D,xysp] ANDB [oprx16,xysp]	(B) • (M) ⇒ B Logical And B with Memory	IMM DIR EXT IDX IDX1 IDX2 [D,IDX] [IDX2]	C4 ii D4 dd F4 hh ll E4 xb E4 xb ff E4 xb ee ff E4 xb E4 xb ee ff	P rfP rOP rfP rPO frPP fIfrfP fIPrfP	–	–	–	–	Δ	Δ	0	–

Note 1. Due to internal CPU requirements, the program word fetch is performed twice to the same address during this instruction.

Reprinted with permission of Motorola

Source Form	Operation	Addr. Mode	Machine Coding (hex)	Access Detail	S	X	H	I	N	Z	V	C
ANDCC #opr8i	(CCR) • (M) ⇒ CCR Logical And CCR with Memory	IMM	10 ii	P	⇓	⇓	⇓	⇓	⇓	⇓	⇓	⇓
ASL opr16a ASL oprx0_xysp ASL oprx9,xysp ASL oprx16,xysp ASL [D,xysp] ASL [oprx16,xysp] ASLA ASLB	 Arithmetic Shift Left Arithmetic Shift Left Accumulator A Arithmetic Shift Left Accumulator B	EXT IDX IDX1 IDX2 [D,IDX] [IDX2] INH INH	78 hh ll 68 xb 68 xb ff 68 xb ee ff 68 xb 68 xb ee ff 48 58	rOPw rPw rPOw frPPw fIfrPw fIPrPw O O	–	–	–	–	Δ	Δ	Δ	Δ
ASLD	 Arithmetic Shift Left Double	INH	59	O	–	–	–	–	Δ	Δ	Δ	Δ
ASR opr16a ASR oprx0_xysp ASR oprx9,xysp ASR oprx16,xysp ASR [D,xysp] ASR [oprx16,xysp] ASRA ASRB	 Arithmetic Shift Right Arithmetic Shift Right Accumulator A Arithmetic Shift Right Accumulator B	EXT IDX IDX1 IDX2 [D,IDX] [IDX2] INH INH	77 hh ll 67 xb 67 xb ff 67 xb ee ff 67 xb 67 xb ee ff 47 57	rOPw rPw rPOw frPPw fIfrPw fIPrPw O O	–	–	–	–	Δ	Δ	Δ	Δ
BCC rel8	Branch if Carry Clear (if C = 0)	REL	24 rr	PPP/P[1]	–	–	–	–	–	–	–	–
BCLR opr8a, msk8 BCLR opr16a, msk8 BCLR oprx0_xysp, msk8 BCLR oprx9,xysp, msk8 BCLR oprx16,xysp, msk8	(M) • (mm) ⇒ M Clear Bit(s) in Memory	DIR EXT IDX IDX1 IDX2	4D dd mm 1D hh ll mm 0D xb mm 0D xb ff mm 0D xb ee ff mm	rPOw rPPw rPOw rPwP frPwOP	–	–	–	–	Δ	Δ	0	–
BCS rel8	Branch if Carry Set (if C = 1)	REL	25 rr	PPP/P[1]	–	–	–	–	–	–	–	–
BEQ rel8	Branch if Equal (if Z = 1)	REL	27 rr	PPP/P[1]	–	–	–	–	–	–	–	–
BGE rel8	Branch if Greater Than or Equal (if N ⊕ V = 0) (signed)	REL	2C rr	PPP/P[1]	–	–	–	–	–	–	–	–
BGND	Place CPU in Background Mode	INH	00	VfPPP	–	–	–	–	–	–	–	–
BGT rel8	Branch if Greater Than (if Z ; (N ⊕ V) = 0) (signed)	REL	2E rr	PPP/P[1]	–	–	–	–	–	–	–	–
BHI rel8	Branch if Higher (if C ; Z = 0) (unsigned)	REL	22 rr	PPP/P[1]	–	–	–	–	–	–	–	–
BHS rel8	Branch if Higher or Same (if C = 0) (unsigned) same function as BCC	REL	24 rr	PPP/P[1]	–	–	–	–	–	–	–	–
BITA #opr8i BITA opr8a BITA opr16a BITA oprx0_xysp BITA oprx9,xysp BITA oprx16,xysp BITA [D,xysp] BITA [oprx16,xysp]	(A) • (M) Logical And A with Memory	IMM DIR EXT IDX IDX1 IDX2 [D,IDX] [IDX2]	85 ii 95 dd B5 hh ll A5 xb A5 xb ff A5 xb ee ff A5 xb A5 xb ee ff	P rfP rOP rfP rPO frPP fIfrfP fIPrfP	–	–	–	–	Δ	Δ	0	–
BITB #opr8i BITB opr8a BITB opr16a BITB oprx0_xysp BITB oprx9,xysp BITB oprx16,xysp BITB [D,xysp] BITB [oprx16,xysp]	(B) • (M) Logical And B with Memory	IMM DIR EXT IDX IDX1 IDX2 [D,IDX] [IDX2]	C5 ii D5 dd F5 hh ll E5 xb E5 xb ff E5 xb ee ff E5 xb E5 xb ee ff	P rfP rOP rfP rPO frPP fIfrfP fIPrfP	–	–	–	–	Δ	Δ	0	–

Note 1. PPP/P indicates this instruction takes three cycles to refill the instruction queue if the branch is taken and one program fetch cycle if the branch is not taken.

Source Form	Operation	Addr. Mode	Machine Coding (hex)	Access Detail	S	X	H	I	N	Z	V	C
BLE rel8	Branch if Less Than or Equal (if Z ; (N ⊕ V) = 1) (signed)	REL	2F rr	PPP/P[1]	–	–	–	–	–	–	–	–
BLO rel8	Branch if Lower (if C = 1) (unsigned) same function as BCS	REL	25 rr	PPP/P[1]	–	–	–	–	–	–	–	–
BLS rel8	Branch if Lower or Same (if C ; Z = 1) (unsigned)	REL	23 rr	PPP/P[1]	–	–	–	–	–	–	–	–
BLT rel8	Branch if Less Than (if N ⊕ V = 1) (signed)	REL	2D rr	PPP/P[1]	–	–	–	–	–	–	–	–
BMI rel8	Branch if Minus (if N = 1)	REL	2B rr	PPP/P[1]	–	–	–	–	–	–	–	–
BNE rel8	Branch if Not Equal (if Z = 0)	REL	26 rr	PPP/P[1]	–	–	–	–	–	–	–	–
BPL rel8	Branch if Plus (if N = 0)	REL	2A rr	PPP/P[1]	–	–	–	–	–	–	–	–
BRA rel8	Branch Always (if 1 = 1)	REL	20 rr	PPP	–	–	–	–	–	–	–	–
BRCLR opr8a, msk8, rel8 BRCLR opr16a, msk8, rel8 BRCLR oprx0_xysp, msk8, rel8 BRCLR oprx9,xysp, msk8, rel8 BRCLR oprx16,xysp, msk8, rel8	Branch if (M) • (mm) = 0 (if All Selected Bit(s) Clear)	DIR EXT IDX IDX1 IDX2	4F dd mm rr 1F hh ll mm rr 0F xb mm rr 0F xb ff mm rr 0F xb ee ff mm rr	rPPP rfPPP rPPP rffPPP frPffPPP	–	–	–	–	–	–	–	–
BRN rel8	Branch Never (if 1 = 0)	REL	21 rr	P	–	–	–	–	–	–	–	–
BRSET opr8, msk8, rel8 BRSET opr16a, msk8, rel8 BRSET oprx0_xysp, msk8, rel8 BRSET oprx9,xysp, msk8, rel8 BRSET oprx16,xysp, msk8, rel8	Branch if ($\overline{M}$) • (mm) = 0 (if All Selected Bit(s) Set)	DIR EXT IDX IDX1 IDX2	4E dd mm rr 1E hh ll mm rr 0E xb mm rr 0E xb ff mm rr 0E xb ee ff mm rr	rPPP rfPPP rPPP rffPPP frPffPPP	–	–	–	–	–	–	–	–
BSET opr8, msk8 BSET opr16a, msk8 BSET oprx0_xysp, msk8 BSET oprx9,xysp, msk8 BSET oprx16,xysp, msk8	(M) ; (mm) ⇒ M Set Bit(s) in Memory	DIR EXT IDX IDX1 IDX2	4C dd mm 1C hh ll mm 0C xb mm 0C xb ff mm 0C xb ee ff mm	rPOw rPPw rPOw rPwP frPwOP	–	–	–	–	Δ	Δ	0	–
BSR rel8	(SP) − 2 ⇒ SP; RTN$_H$:RTN$_L$ ⇒ M$_{(SP)}$:M$_{(SP+1)}$ Subroutine address ⇒ PC Branch to Subroutine	REL	07 rr	PPPS	–	–	–	–	–	–	–	–
BVC rel8	Branch if Overflow Bit Clear (if V = 0)	REL	28 rr	PPP/P[1]	–	–	–	–	–	–	–	–
BVS rel8	Branch if Overflow Bit Set (if V = 1)	REL	29 rr	PPP/P[1]	–	–	–	–	–	–	–	–
CALL opr16a, page CALL oprx0_xysp, page CALL oprx9,xysp, page CALL oprx16,xysp, page CALL [D,xysp] CALL [oprx16, xysp]	(SP) − 2 ⇒ SP; RTN$_H$:RTN$_L$ ⇒ M$_{(SP)}$:M$_{(SP+1)}$ (SP) − 1 ⇒ SP; (PPG) ⇒ M$_{(SP)}$; pg ⇒ PPAGE register; Program address ⇒ PC Call subroutine in extended memory (Program may be located on another expansion memory page.) Indirect modes get program address and new pg value based on pointer.	EXT IDX IDX1 IDX2 [D,IDX] [IDX2]	4A hh ll pg 4B xb pg 4B xb ff pg 4B xb ee ff pg 4B xb 4B xb ee ff	gnfSsPPP gnfSsPPP gnfSsPPP fgnfSsPPP fIignSsPPP fIignSsPPP	–	–	–	–	–	–	–	–
CBA	(A) − (B) Compare 8-Bit Accumulators	INH	18 17	OO	–	–	–	–	Δ	Δ	Δ	Δ
CLC	0 ⇒ C *Translates to* ANDCC #$FE	IMM	10 FE	P	–	–	–	–	–	–	–	0
CLI	0 ⇒ I *Translates to* ANDCC #$EF (enables I-bit interrupts)	IMM	10 EF	P	–	–	–	0	–	–	–	–

Note 1. PPP/P indicates this instruction takes three cycles to refill the instruction queue if the branch is taken and one program fetch cycle if the branch is not taken.

Source Form	Operation	Addr. Mode	Machine Coding (hex)	Access Detail	S	X	H	I	N	Z	V	C
CLR opr16a CLR oprx0,xysp CLR oprx9,xysp CLR oprx16,xysp CLR [D,xysp] CLR [oprx16,xysp] CLRA CLRB	0 ⇒ M Clear Memory Location 0 ⇒ A Clear Accumulator A 0 ⇒ B Clear Accumulator B	EXT IDX IDX1 IDX2 [D,IDX] [IDX2] INH INH	79 hh ll 69 xb 69 xb ff 69 xb ee ff 69 xb 69 xb ee ff 87 C7	wOP Pw PwO PwP PIfPw PIPPw O O	–	–	–	–	0	1	0	0
CLV	0 ⇒ V *Translates to* ANDCC #$FD	IMM	10 FD	P	–	–	–	–	–	–	0	–
CMPA #opr8i CMPA opr8a CMPA opr16a CMPA oprx0_xysp CMPA oprx9,xysp CMPA oprx16,xysp CMPA [D,xysp] CMPA [oprx16,xysp]	(A) – (M) Compare Accumulator A with Memory	IMM DIR EXT IDX IDX1 IDX2 [D,IDX] [IDX2]	81 ii 91 dd B1 hh ll A1 xb A1 xb ff A1 xb ee ff A1 xb A1 xb ee ff	P rfP rOP rfP rPO frPP fIfrfP fIPrfP	–	–	–	–	Δ	Δ	Δ	Δ
CMPB #opr8i CMPB opr8a CMPB opr16a CMPB oprx0_xysp CMPB oprx9,xysp CMPB oprx16,xysp CMPB [D,xysp] CMPB [oprx16,xysp]	(B) – (M) Compare Accumulator B with Memory	IMM DIR EXT IDX IDX1 IDX2 [D,IDX] [IDX2]	C1 ii D1 dd F1 hh ll E1 xb E1 xb ff E1 xb ee ff E1 xb E1 xb ee ff	P rfP rOP rfP rPO frPP fIfrfP fIPrfP	–	–	–	–	Δ	Δ	Δ	Δ
COM opr16a COM oprx0_xysp COM oprx9,xysp COM oprx16,xysp COM [D,xysp] COM [oprx16,xysp] COMA COMB	$\overline{(M)}$ ⇒ M *equivalent to* $FF – (M) ⇒ M 1's Complement Memory Location $\overline{(A)}$ ⇒ A Complement Accumulator A $\overline{(B)}$ ⇒ B Complement Accumulator B	EXT IDX IDX1 IDX2 [D,IDX] [IDX2] INH INH	71 hh ll 61 xb 61 xb ff 61 xb ee ff 61 xb 61 xb ee ff 41 51	rOPw rPw rPOw frPPw fIfrPw fIPrPw O O	–	–	–	–	Δ	Δ	0	1
CPD #opr16i CPD opr8a CPD opr16a CPD oprx0_xysp CPD oprx9,xysp CPD oprx16,xysp CPD [D,xysp] CPD [oprx16,xysp]	(A:B) – (M:M+1) Compare D to Memory (16-Bit)	IMM DIR EXT IDX IDX1 IDX2 [D,IDX] [IDX2]	8C jj kk 9C dd BC hh ll AC xb AC xb ff AC xb ee ff AC xb AC xb ee ff	OP RfP ROP RfP RPO fRPP fIfRfP fIPRfP	–	–	–	–	Δ	Δ	Δ	Δ
CPS #opr16i CPS opr8a CPS opr16a CPS oprx0_xysp CPS oprx9,xysp CPS oprx16,xysp CPS [D,xysp] CPS [oprx16,xysp]	(SP) – (M:M+1) Compare SP to Memory (16-Bit)	IMM DIR EXT IDX IDX1 IDX2 [D,IDX] [IDX2]	8F jj kk 9F dd BF hh ll AF xb AF xb ff AF xb ee ff AF xb AF xb ee ff	OP RfP ROP RfP RPO fRPP fIfRfP fIPRfP	–	–	–	–	Δ	Δ	Δ	Δ
CPX #opr16i CPX opr8a CPX opr16a CPX oprx0_xysp CPX oprx9,xysp CPX oprx16,xysp CPX [D,xysp] CPX [oprx16,xysp]	(X) – (M:M+1) Compare X to Memory (16-Bit)	IMM DIR EXT IDX IDX1 IDX2 [D,IDX] [IDX2]	8E jj kk 9E dd BE hh ll AE xb AE xb ff AE xb ee ff AE xb AE xb ee ff	OP RfP ROP RfP RPO fRPP fIfRfP fIPRfP	–	–	–	–	Δ	Δ	Δ	Δ
CPY #opr16i CPY opr8a CPY opr16a CPY oprx0_xysp CPY oprx9,xysp CPY oprx16,xysp CPY [D,xysp] CPY [oprx16,xysp]	(Y) – (M:M+1) Compare Y to Memory (16-Bit)	IMM DIR EXT IDX IDX1 IDX2 [D,IDX] [IDX2]	8D jj kk 9D dd BD hh ll AD xb AD xb ff AD xb ee ff AD xb AD xb ee ff	OP RfP ROP RfP RPO fRPP fIfRfP fIPRfP	–	–	–	–	Δ	Δ	Δ	Δ

Reprinted with permission of Motorola

Source Form	Operation	Addr. Mode	Machine Coding (hex)	Access Detail	S	X	H	I	N	Z	V	C
DAA	Adjust Sum to BCD Decimal Adjust Accumulator A	INH	18 07	OfO	–	–	–	–	Δ	Δ	?	Δ
DBEQ abdxys, rel9	(cntr) – 1 ⇒ cntr if (cntr) = 0, then Branch else Continue to next instruction Decrement Counter and Branch if = 0 (cntr = A, B, D, X, Y, or SP)	REL (9-bit)	04 lb rr	PPP	–	–	–	–	–	–	–	–
DBNE abdxys, rel9	(cntr) – 1 ⇒ cntr If (cntr) not = 0, then Branch; else Continue to next instruction Decrement Counter and Branch if ≠ 0 (cntr = A, B, D, X, Y, or SP)	REL (9-bit)	04 lb rr	PPP	–	–	–	–	–	–	–	–
DEC opr16a DEC oprx0_xysp DEC oprx9,xysp DEC oprx16,xysp DEC [D,xysp] DEC [oprx16,xysp] DECA DECB	(M) – $01 ⇒ M Decrement Memory Location (A) – $01 ⇒ A Decrement A (B) – $01 ⇒ B Decrement B	EXT IDX IDX1 IDX2 [D,IDX] [IDX2] INH INH	73 hh ll 63 xb 63 xb ff 63 xb ee ff 63 xb 63 xb ee ff 43 53	rOPw rPw rPOw frPPw fIfrPw fIPrPw O O	–	–	–	–	Δ	Δ	Δ	–
DES	(SP) – $0001 ⇒ SP *Translates to* LEAS –1,SP	IDX	1B 9F	PP[1]	–	–	–	–	–	–	–	–
DEX	(X) – $0001 ⇒ X Decrement Index Register X	INH	09	O	–	–	–	–	–	Δ	–	–
DEY	(Y) – $0001 ⇒ Y Decrement Index Register Y	INH	03	O	–	–	–	–	–	Δ	–	–
EDIV	(Y:D) ÷ (X) ⇒ Y Remainder ⇒ D 32 × 16 Bit Divide (unsigned)	INH	11	ffffffffffO	–	–	–	–	Δ	Δ	Δ	Δ
EDIVS	(Y:D) ÷ (X) ⇒ Y Remainder ⇒ D 32 × 16 Bit Divide (signed)	INH	18 14	OffffffffffO	–	–	–	–	Δ	Δ	Δ	Δ
EMACS opr16a [2]	$(M_{(X)}:M_{(X+1)}) \times (M_{(Y)}:M_{(Y+1)}) + (M\sim M+3) \Rightarrow M\sim M+3$ 16 × 16 Bit ⇒ 32 Bit Multiply and Accumulate (signed)	Special	18 12 hh ll	ORROfffRRfWWP	–	–	–	–	Δ	Δ	Δ	Δ
EMAXD oprx0_xysp EMAXD oprx9,xysp EMAXD oprx16,xysp EMAXD [D,xysp] EMAXD [oprx16,xysp]	MAX((D), (M:M+1)) ⇒ D MAX of 2 Unsigned 16-Bit Values N, Z, V, and C status bits reflect result of internal compare ((D) – (M:M+1))	IDX IDX1 IDX2 [D,IDX] [IDX2]	18 1A xb 18 1A xb ff 18 1A xb ee ff 18 1A xb 18 1A xb ee ff	ORfP ORPO OfRPP OfIfRfP OfIPRfP	–	–	–	–	Δ	Δ	Δ	Δ
EMAXM oprx0_xysp EMAXM oprx9,xysp EMAXM oprx16,xysp EMAXM [D,xysp] EMAXM [oprx16,xysp]	MAX((D), (M:M+1)) ⇒ M:M+1 MAX of 2 Unsigned 16-Bit Values N, Z, V, and C status bits reflect result of internal compare ((D) – (M:M+1))	IDX IDX1 IDX2 [D,IDX] [IDX2]	18 1E xb 18 1E xb ff 18 1E xb ee ff 18 1E xb 18 1E xb ee ff	ORPW ORPWO OfRPWP OfIfRPW OfIPRPW	–	–	–	–	Δ	Δ	Δ	Δ
EMIND oprx0_xysp EMIND oprx9,xysp EMIND oprx16,xysp EMIND [D,xysp] EMIND [oprx16,xysp]	MIN((D), (M:M+1)) ⇒ D MIN of 2 Unsigned 16-Bit Values N, Z, V, and C status bits reflect result of internal compare ((D) – (M:M+1))	IDX IDX1 IDX2 [D,IDX] [IDX2]	18 1B xb 18 1B xb ff 18 1B xb ee ff 18 1B xb 18 1B xb ee ff	ORfP ORPO OfRPP OfIfRfP OfIPRfP	–	–	–	–	Δ	Δ	Δ	Δ
EMINM oprx0_xysp EMINM oprx9,xysp EMINM oprx16,xysp EMINM [D,xysp] EMINM [oprx16,xysp]	MIN((D), (M:M+1)) ⇒ M:M+1 MIN of 2 Unsigned 16-Bit Values N, Z, V, and C status bits reflect result of internal compare ((D) – (M:M+1))	IDX IDX1 IDX2 [D,IDX] [IDX2]	18 1F xb 18 1F xb ff 18 1F xb ee ff 18 1F xb 18 1F xb ee ff	ORPW ORPWO OfRPWP OfIfRPW OfIPRPW	–	–	–	–	Δ	Δ	Δ	Δ
EMUL	(D) × (Y) ⇒ Y:D 16 × 16 Bit Multiply (unsigned)	INH	13	ffO	–	–	–	–	Δ	Δ	–	Δ

Notes:
1. Due to internal CPU requirements, the program word fetch is performed twice to the same address during this instruction.
2. *opr16a* is an extended address specification. Both X and Y point to source operands.

Reprinted with permission of Motorola

Source Form	Operation	Addr. Mode	Machine Coding (hex)	Access Detail	S	X	H	I	N	Z	V	C
EMULS	(D) × (Y) ⇒ Y:D 16 × 16 Bit Multiply (signed)	INH	18 13	OfO	–	–	–	–	Δ	Δ	–	Δ
EORA #opr8i EORA opr8a EORA opr16a EORA oprx0_xysp EORA oprx9,xysp EORA oprx16,xysp EORA [D,xysp] EORA [oprx16,xysp]	(A) ⊕ (M) ⇒ A Exclusive-OR A with Memory	IMM DIR EXT IDX IDX1 IDX2 [D,IDX] [IDX2]	88 ii 98 dd B8 hh ll A8 xb A8 xb ff A8 xb ee ff A8 xb A8 xb ee ff	P rfP rOP rfP rPO frPP fIfrfP fIPrfP	–	–	–	–	Δ	Δ	0	–
EORB #opr8i EORB opr8a EORB opr16a EORB oprx0_xysp EORB oprx9,xysp EORB oprx16,xysp EORB [D,IDX] EORB [oprx16,xysp]	(B) ⊕ (M) ⇒ B Exclusive-OR B with Memory	IMM DIR EXT IDX IDX1 IDX2 [D,IDX] [IDX2]	C8 ii D8 dd F8 hh ll E8 xb E8 xb ff E8 xb ee ff E8 xb E8 xb ee ff	P rfP rOP rfP rPO frPP fIfrfP fIPrfP	–	–	–	–	Δ	Δ	0	–
ETBL oprx0_xysp	(M:M+1)+ [(B)×((M+2:M+3) − (M:M+1))] ⇒ D 16-Bit Table Lookup and Interpolate Initialize B, and index before ETBL. <ea> points at first table entry (M:M+1) and B is fractional part of lookup value (no indirect address modes or extensions allowed)	IDX	18 3F xb	ORRfffffP	–	–	–	–	Δ	Δ	–	⇒
EXG abcdxys,abcdxys	(r1) ⇔ (r2) (if r1 and r2 same size) or $00:(r1) ⇒ r2 (if r1 = 8-bit; r2 = 16-bit) or (r1$_{low}$) ⇔ (r2) (if r1 = 16-bit; r2 = 8-bit) r1 and r2 may be A, B, CCR, D, X, Y, or SP	INH	B7 eb	P	–	–	–	–	–	–	–	–
FDIV	(D) ÷ (X) ⇒ X; Remainder ⇒ D 16 × 16 Bit Fractional Divide	INH	18 11	OfffffffffO	–	–	–	–	–	Δ	Δ	Δ
IBEQ abdxys, rel9	(cntr) + 1 ⇒ cntr If (cntr) = 0, then Branch else Continue to next instruction Increment Counter and Branch if = 0 (cntr = A, B, D, X, Y, or SP)	REL (9-bit)	04 lb rr	PPP	–	–	–	–	–	–	–	–
IBNE abdxys, rel9	(cntr) + 1 ⇒ cntr if (cntr) not = 0, then Branch; else Continue to next instruction Increment Counter and Branch if ≠ 0 (cntr = A, B, D, X, Y, or SP)	REL (9-bit)	04 lb rr	PPP	–	–	–	–	–	–	–	–
IDIV	(D) ÷ (X) ⇒ X; Remainder ⇒ D 16 × 16 Bit Integer Divide (unsigned)	INH	18 10	OfffffffffO	–	–	–	–	–	Δ	0	Δ
IDIVS	(D) ÷ (X) ⇒ X; Remainder ⇒ D 16 × 16 Bit Integer Divide (signed)	INH	18 15	OfffffffffO	–	–	–	–	Δ	Δ	Δ	Δ
INC opr16a INC oprx0_xysp INC oprx9,xysp INC oprx16,xysp INC [D,xysp] INC [oprx16,xysp] INCA INCB	(M) + $01 ⇒ M Increment Memory Byte (A) + $01 ⇒ A Increment Accumulator A (B) + $01 ⇒ B Increment Accumulator B	EXT IDX IDX1 IDX2 [D,IDX] [IDX2] INH INH	72 hh ll 62 xb 62 xb ff 62 xb ee ff 62 xb 62 xb ee ff 42 52	rOPw rPw rPOw frPPw fIfrPw fIPrPw O O	–	–	–	–	Δ	Δ	Δ	–
INS	(SP) + $0001 ⇒ SP Translates to LEAS 1,SP	IDX	1B 81	PP[1]	–	–	–	–	–	–	–	–
INX	(X) + $0001 ⇒ X Increment Index Register X	INH	08	O	–	–	–	–	–	Δ	–	–

Note 1. Due to internal CPU requirements, the program word fetch is performed twice to the same address during this instruction.

Source Form	Operation	Addr. Mode	Machine Coding (hex)	Access Detail	S	X	H	I	N	Z	V	C
INY	$(Y) + \$0001 \Rightarrow Y$ Increment Index Register Y	INH	02	O	–	–	–	–	–	Δ	–	–
JMP opr16a JMP oprx0_xysp JMP oprx9,xysp JMP oprx16,xysp JMP [D,xysp] JMP [oprx16,xysp]	Subroutine address $\Rightarrow$ PC Jump	EXT IDX IDX1 IDX2 [D,IDX] [IDX2]	06 hh ll 05 xb 05 xb ff 05 xb ee ff 05 xb 05 xb ee ff	PPP PPP PPP fPPP fIfPPP fIfPPP	–	–	–	–	–	–	–	–
JSR opr8a JSR opr16a JSR oprx0_xysp JSR oprx9,xysp JSR oprx16,xysp JSR [D,xysp] JSR [oprx16,xysp]	$(SP) - 2 \Rightarrow SP$; $RTN_H{:}RTN_L \Rightarrow M_{(SP)}{:}M_{(SP+1)}$; Subroutine address $\Rightarrow$ PC Jump to Subroutine	DIR EXT IDX IDX1 IDX2 [D,IDX] [IDX2]	17 dd 16 hh ll 15 xb 15 xb ff 15 xb ee ff 15 xb 15 xb ee ff	PPPS PPPS PPPS PPPS fPPPS fIfPPPS fIfPPPS	–	–	–	–	–	–	–	–
LBCC rel16	Long Branch if Carry Clear (if C = 0)	REL	18 24 qq rr	OPPP/OPO[1]	–	–	–	–	–	–	–	–
LBCS rel16	Long Branch if Carry Set (if C = 1)	REL	18 25 qq rr	OPPP/OPO[1]	–	–	–	–	–	–	–	–
LBEQ rel16	Long Branch if Equal (if Z = 1)	REL	18 27 qq rr	OPPP/OPO[1]	–	–	–	–	–	–	–	–
LBGE rel16	Long Branch Greater Than or Equal (if N $\oplus$ V = 0) (signed)	REL	18 2C qq rr	OPPP/OPO[1]	–	–	–	–	–	–	–	–
LBGT rel16	Long Branch if Greater Than (if Z ; (N $\oplus$ V) = 0) (signed)	REL	18 2E qq rr	OPPP/OPO[1]	–	–	–	–	–	–	–	–
LBHI rel16	Long Branch if Higher (if C ; Z = 0) (unsigned)	REL	18 22 qq rr	OPPP/OPO[1]	–	–	–	–	–	–	–	–
LBHS rel16	Long Branch if Higher or Same (if C = 0) (unsigned) same function as LBCC	REL	18 24 qq rr	OPPP/OPO[1]	–	–	–	–	–	–	–	–
LBLE rel16	Long Branch if Less Than or Equal (if Z ; (N $\oplus$ V) = 1) (signed)	REL	18 2F qq rr	OPPP/OPO[1]	–	–	–	–	–	–	–	–
LBLO rel16	Long Branch if Lower (if C = 1) (unsigned) same function as LBCS	REL	18 25 qq rr	OPPP/OPO[1]	–	–	–	–	–	–	–	–
LBLS rel16	Long Branch if Lower or Same (if C ; Z = 1) (unsigned)	REL	18 23 qq rr	OPPP/OPO[1]	–	–	–	–	–	–	–	–
LBLT rel16	Long Branch if Less Than (if N $\oplus$ V = 1) (signed)	REL	18 2D qq rr	OPPP/OPO[1]	–	–	–	–	–	–	–	–
LBMI rel16	Long Branch if Minus (if N = 1)	REL	18 2B qq rr	OPPP/OPO[1]	–	–	–	–	–	–	–	–
LBNE rel16	Long Branch if Not Equal (if Z = 0)	REL	18 26 qq rr	OPPP/OPO[1]	–	–	–	–	–	–	–	–
LBPL rel16	Long Branch if Plus (if N = 0)	REL	18 2A qq rr	OPPP/OPO[1]	–	–	–	–	–	–	–	–
LBRA rel16	Long Branch Always (if 1 = 1)	REL	18 20 qq rr	OPPP	–	–	–	–	–	–	–	–
LBRN rel16	Long Branch Never (if 1 = 0)	REL	18 21 qq rr	OPO	–	–	–	–	–	–	–	–
LBVC rel16	Long Branch if Overflow Bit Clear (if V = 0)	REL	18 28 qq rr	OPPP/OPO[1]	–	–	–	–	–	–	–	–
LBVS rel16	Long Branch if Overflow Bit Set (if V = 1)	REL	18 29 qq rr	OPPP/OPO[1]	–	–	–	–	–	–	–	–
LDAA #opr8i LDAA opr8a LDAA opr16a LDAA oprx0_xysp LDAA oprx9,xysp LDAA oprx16,xysp LDAA [D,xysp] LDAA [oprx16,xysp]	$(M) \Rightarrow A$ Load Accumulator A	IMM DIR EXT IDX IDX1 IDX2 [D,IDX] [IDX2]	86 ii 96 dd B6 hh ll A6 xb A6 xb ff A6 xb ee ff A6 xb A6 xb ee ff	P rfP rOP rfP rPO frPP fIfrfP fIPrfP	–	–	–	–	Δ	Δ	0	–

Note 1. OPPP/OPO indicates this instruction takes four cycles to refill the instruction queue if the branch is taken and three cycles if the branch is not taken.

Source Form	Operation	Addr. Mode	Machine Coding (hex)	Access Detail	S	X	H	I	N	Z	V	C
LDAB #opr8i LDAB opr8a LDAB opr16a LDAB oprx0_xysp LDAB oprx9,xysp LDAB oprx16,xysp LDAB [D,xysp] LDAB [oprx16,xysp]	(M) ⇒ B Load Accumulator B	IMM DIR EXT IDX IDX1 IDX2 [D,IDX] [IDX2]	C6 ii D6 dd F6 hh ll E6 xb E6 xb ff E6 xb ee ff E6 xb E6 xb ee ff	P rfP rOP rfP rPO frPP fIfrfP fIPrfP	–	–	–	–	Δ	Δ	0	–
LDD #opr16i LDD opr8a LDD opr16a LDD oprx0_xysp LDD oprx9,xysp LDD oprx16,xysp LDD [D,xysp] LDD [oprx16,xysp]	(M:M+1) ⇒ A:B Load Double Accumulator D (A:B)	IMM DIR EXT IDX IDX1 IDX2 [D,IDX] [IDX2]	CC jj kk DC dd FC hh ll EC xb EC xb ff EC xb ee ff EC xb EC xb ee ff	OP RfP ROP RfP RPO fRPP fIfRfP fIPRfP	–	–	–	–	Δ	Δ	0	–
LDS #opr16i LDS opr8a LDS opr16a LDS oprx0_xysp LDS oprx9,xysp LDS oprx16,xysp LDS [D,xysp] LDS [oprx16,xysp]	(M:M+1) ⇒ SP Load Stack Pointer	IMM DIR EXT IDX IDX1 IDX2 [D,IDX] [IDX2]	CF jj kk DF dd FF hh ll EF xb EF xb ff EF xb ee ff EF xb EF xb ee ff	OP RfP ROP RfP RPO fRPP fIfRfP fIPRfP	–	–	–	–	Δ	Δ	0	–
LDX #opr16i LDX opr8a LDX opr16a LDX oprx0_xysp LDX oprx9,xysp LDX oprx16,xysp LDX [D,xysp] LDX [oprx16,xysp]	(M:M+1) ⇒ X Load Index Register X	IMM DIR EXT IDX IDX1 IDX2 [D,IDX] [IDX2]	CE jj kk DE dd FE hh ll EE xb EE xb ff EE xb ee ff EE xb EE xb ee ff	OP RfP ROP RfP RPO fRPP fIfRfP fIPRfP	–	–	–	–	Δ	Δ	0	–
LDY #opr16i LDY opr8a LDY opr16a LDY oprx0_xysp LDY oprx9,xysp LDY oprx16,xysp LDY [D,xysp] LDY [oprx16,xysp]	(M:M+1) ⇒ Y Load Index Register Y	IMM DIR EXT IDX IDX1 IDX2 [D,IDX] [IDX2]	CD jj kk DD dd FD hh ll ED xb ED xb ff ED xb ee ff ED xb ED xb ee ff	OP RfP ROP RfP RPO fRPP fIfRfP fIPRfP	–	–	–	–	Δ	Δ	0	–
LEAS oprx0_xysp LEAS oprx9,xysp LEAS oprx16,xysp	Effective Address ⇒ SP Load Effective Address into SP	IDX IDX1 IDX2	1B xb 1B xb ff 1B xb ee ff	PP[1] PO PP	–	–	–	–	–	–	–	–
LEAX oprx0_xysp LEAX oprx9,xysp LEAX oprx16,xysp	Effective Address ⇒ X Load Effective Address into X	IDX IDX1 IDX2	1A xb 1A xb ff 1A xb ee ff	PP[1] PO PP	–	–	–	–	–	–	–	–
LEAY oprx0_xysp LEAY oprx9,xysp LEAY oprx16,xysp	Effective Address ⇒ Y Load Effective Address into Y	IDX IDX1 IDX2	19 xb 19 xb ff 19 xb ee ff	PP[1] PO PP	–	–	–	–	–	–	–	–
LSL opr16a LSL oprx0_xysp LSL oprx9,xysp LSL oprx16,xysp LSL [D,xysp] LSL [oprx16,xysp] LSLA LSLB	 Logical Shift Left same function as ASL Logical Shift Accumulator A to Left Logical Shift Accumulator B to Left	EXT IDX IDX1 IDX2 [D,IDX] [IDX2] INH INH	78 hh ll 68 xb 68 xb ff 68 xb ee ff 68 xb 68 xb ee ff 48 58	rOPw rPw rPOw frPPw fIfrPw fIPrPw O O	–	–	–	–	Δ	Δ	Δ	Δ
LSLD	 Logical Shift Left D Accumulator same function as ASLD	INH	59	O	–	–	–	–	Δ	Δ	Δ	Δ

Note 1. Due to internal CPU requirements, the program word fetch is performed twice to the same address during this instruction.

Source Form	Operation	Addr. Mode	Machine Coding (hex)	Access Detail	S	X	H	I	N	Z	V	C
LSR opr16a LSR oprx0_xysp LSR oprx9,xysp LSR oprx16,xysp LSR [D,xysp] LSR [oprx16,xysp]	$0 \rightarrow$ b7 ... b0 $\rightarrow$ C Logical Shift Right	EXT IDX IDX1 IDX2 [D,IDX] [IDX2]	74 hh ll 64 xb 64 xb ff 64 xb ee ff 64 xb 64 xb ee ff	rOPw rPw rPOw frPPw fIfrPw fIPrPw	–	–	–	–	0	Δ	Δ	Δ
LSRA LSRB	Logical Shift Accumulator A to Right Logical Shift Accumulator B to Right	INH INH	44 54	O O								
LSRD	$0 \rightarrow$ b7 A b0 b7 B b0 $\rightarrow$ C Logical Shift Right D Accumulator	INH	49	O	–	–	–	–	0	Δ	Δ	Δ
MAXA oprx0_xysp MAXA oprx9,xysp MAXA oprx16,xysp MAXA [D,xysp] MAXA [oprx16,xysp]	MAX((A), (M)) $\Rightarrow$ A MAX of 2 Unsigned 8-Bit Values N, Z, V, and C status bits reflect result of internal compare ((A) – (M)).	IDX IDX1 IDX2 [D,IDX] [IDX2]	18 18 xb 18 18 xb ff 18 18 xb ee ff 18 18 xb 18 18 xb ee ff	OrfP OrPO OfrPP OfIfrfP OfIPrfP	–	–	–	–	Δ	Δ	Δ	Δ
MAXM oprx0_xysp MAXM oprx9,xysp MAXM oprx16,xysp MAXM [D,xysp] MAXM [oprx16,xysp]	MAX((A), (M)) $\Rightarrow$ M MAX of 2 Unsigned 8-Bit Values N, Z, V, and C status bits reflect result of internal compare ((A) – (M)).	IDX IDX1 IDX2 [D,IDX] [IDX2]	18 1C xb 18 1C xb ff 18 1C xb ee ff 18 1C xb 18 1C xb ee ff	OrPw OrPwO OfrPwP OfIfrPw OfIPrPw	–	–	–	–	Δ	Δ	Δ	Δ
MEM	μ (grade) $\Rightarrow M_{(Y)}$; (X) + 4 $\Rightarrow$ X; (Y) + 1 $\Rightarrow$ Y; A unchanged if (A) < P1 or (A) > P2 then m = 0, else m = MIN[((A) – P1)×S1, (P2 – (A))×S2, \$FF] where: A = current crisp input value; X points at 4-byte data structure that describes a trapezoidal membership function (P1, P2, S1, S2); Y points at fuzzy input (RAM location).	Special	01	RRfOw	–	–	?	–	?	?	?	?
MINA oprx0_xysp MINA oprx9,xysp MINA oprx16,xysp MINA [D,xysp] MINA [oprx16,xysp]	MIN((A), (M)) $\Rightarrow$ A MIN of 2 Unsigned 8-Bit Values N, Z, V, and C status bits reflect result of internal compare ((A) – (M)).	IDX IDX1 IDX2 [D,IDX] [IDX2]	18 19 xb 18 19 xb ff 18 19 xb ee ff 18 19 xb 18 19 xb ee ff	OrfP OrPO OfrPP OfIfrfP OfIPrfP	–	–	–	–	Δ	Δ	Δ	Δ
MINM oprx0_xysp MINM oprx9,xysp MINM oprx16,xysp MINM [D,xysp] MINM [oprx16,xysp]	MIN((A), (M)) $\Rightarrow$ M MIN of 2 Unsigned 8-Bit Values N, Z, V, and C status bits reflect result of internal compare ((A) – (M)).	IDX IDX1 IDX2 [D,IDX] [IDX2]	18 1D xb 18 1D xb ff 18 1D xb ee ff 18 1D xb 18 1D xb ee ff	OrPw OrPwO OfrPwP OfIfrPw OfIPrPw	–	–	–	–	Δ	Δ	Δ	Δ
MOVB #opr8, opr16a[1] MOVB #opr8i, oprx0_xysp[1] MOVB opr16a, opr16a[1] MOVB opr16a, oprx0_xysp[1] MOVB oprx0_xysp, opr16a[1] MOVB oprx0_xysp, oprx0_xysp[1]	$(M_1) \Rightarrow M_2$ Memory to Memory Byte-Move (8-Bit)	IMM-EXT IMM-IDX EXT-EXT EXT-IDX IDX-EXT IDX-IDX	18 0B ii hh ll 18 08 xb ii 18 0C hh ll hh ll 18 09 xb hh ll 18 0D xb hh ll 18 0A xb xb	OPwP OPwO OPrwPO OPrPw OrPwP OrPwO	–	–	–	–	–	–	–	–
MOVW #oprx16, opr16a[1] MOVW #opr16i, oprx0_xysp[1] MOVW opr16a, opr16a[1] MOVW opr16a, oprx0_xysp[1] MOVW oprx0_xysp, opr16a[1] MOVW oprx0_xysp, oprx0_xysp[1]	$(M:M+1_1) \Rightarrow M:M+1_2$ Memory to Memory Word-Move (16-Bit)	IMM-EXT IMM-IDX EXT-EXT EXT-IDX IDX-EXT IDX-IDX	18 03 jj kk hh ll 18 00 xb jj kk 18 04 hh ll hh ll 18 01 xb hh ll 18 05 xb hh ll 18 02 xb xb	OPwPO OPPW OrPwPO OPrPw OrPwP OrPwO	–	–	–	–	–	–	–	–
MUL	$(A) \times (B) \Rightarrow A:B$ 8×8 Unsigned Multiply	INH	12	ffO	–	–	–	–	–	–	–	Δ

Note 1. The first operand in the source code statement specifies the source for the move.

Source Form	Operation	Addr. Mode	Machine Coding (hex)	Access Detail	S	X	H	I	N	Z	V	C
NEG opr16a NEG oprx0_xysp NEG oprx9,xysp NEG oprx16,xysp NEG [D,xysp] NEG [oprx16,xysp]	$0 - (M) \Rightarrow M\ or\ (\overline{M}) + 1 \Rightarrow M$ Two's Complement Negate	EXT IDX IDX1 IDX2 [D,IDX] [IDX2]	70 hh ll 60 xb 60 xb ff 60 xb ee ff 60 xb 60 xb ee ff	rOPw rPw rPOw frPPw fIfrPw fIPrPw	–	–	–	–	Δ	Δ	Δ	Δ
NEGA	$0 - (A) \Rightarrow A$ equivalent to $(\overline{A}) + 1 \Rightarrow A$ Negate Accumulator A	INH	40	O								
NEGB	$0 - (B) \Rightarrow B$ equivalent to $(\overline{B}) + 1 \Rightarrow B$ Negate Accumulator B	INH	50	O								
NOP	No Operation	INH	A7	O	–	–	–	–	–	–	–	–
ORAA #opr8i ORAA opr8a ORAA opr16a ORAA oprx0_xysp ORAA oprx9,xysp ORAA oprx16,xysp ORAA [D,xysp] ORAA [oprx16,xysp]	$(A) ; (M) \Rightarrow A$ Logical OR A with Memory	IMM DIR EXT IDX IDX1 IDX2 [D,IDX] [IDX2]	8A ii 9A dd BA hh ll AA xb AA xb ff AA xb ee ff AA xb AA xb ee ff	P rfP rOP rfP rPO frPP fIfrfP fIPrfP	–	–	–	–	Δ	Δ	0	–
ORAB #opr8i ORAB opr8a ORAB opr16a ORAB oprx0_xysp ORAB oprx9,xysp ORAB oprx16,xysp ORAB [D,xysp] ORAB [oprx16,xysp]	$(B) ; (M) \Rightarrow B$ Logical OR B with Memory	IMM DIR EXT IDX IDX1 IDX2 [D,IDX] [IDX2]	CA ii DA dd FA hh ll EA xb EA xb ff EA xb ee ff EA xb EA xb ee ff	P rfP rOP rfP rPO frPP fIfrfP fIPrfP	–	–	–	–	Δ	Δ	0	–
ORCC #opr8i	$(CCR) ; M \Rightarrow CCR$ Logical OR CCR with Memory	IMM	14 ii	P	⇑	–	⇑	⇑	⇑	⇑	⇑	⇑
PSHA	$(SP) - 1 \Rightarrow SP; (A) \Rightarrow M_{(SP)}$ Push Accumulator A onto Stack	INH	36	Os	–	–	–	–	–	–	–	–
PSHB	$(SP) - 1 \Rightarrow SP; (B) \Rightarrow M_{(SP)}$ Push Accumulator B onto Stack	INH	37	Os	–	–	–	–	–	–	–	–
PSHC	$(SP) - 1 \Rightarrow SP; (CCR) \Rightarrow M_{(SP)}$ Push CCR onto Stack	INH	39	Os	–	–	–	–	–	–	–	–
PSHD	$(SP) - 2 \Rightarrow SP; (A:B) \Rightarrow M_{(SP)}:M_{(SP+1)}$ Push D Accumulator onto Stack	INH	3B	OS	–	–	–	–	–	–	–	–
PSHX	$(SP) - 2 \Rightarrow SP; (X_H:X_L) \Rightarrow M_{(SP)}:M_{(SP+1)}$ Push Index Register X onto Stack	INH	34	OS	–	–	–	–	–	–	–	–
PSHY	$(SP) - 2 \Rightarrow SP; (Y_H:Y_L) \Rightarrow M_{(SP)}:M_{(SP+1)}$ Push Index Register Y onto Stack	INH	35	OS	–	–	–	–	–	–	–	–
PULA	$(M_{(SP)}) \Rightarrow A; (SP) + 1 \Rightarrow SP$ Pull Accumulator A from Stack	INH	32	ufO	–	–	–	–	–	–	–	–
PULB	$(M_{(SP)}) \Rightarrow B; (SP) + 1 \Rightarrow SP$ Pull Accumulator B from Stack	INH	33	ufO	–	–	–	–	–	–	–	–
PULC	$(M_{(SP)}) \Rightarrow CCR; (SP) + 1 \Rightarrow SP$ Pull CCR from Stack	INH	38	ufO	Δ	⇓	Δ	Δ	Δ	Δ	Δ	Δ
PULD	$(M_{(SP)}:M_{(SP+1)}) \Rightarrow A:B; (SP) + 2 \Rightarrow SP$ Pull D from Stack	INH	3A	UfO	–	–	–	–	–	–	–	–

Source Form	Operation	Addr. Mode	Machine Coding (hex)	Access Detail	S	X	H	I	N	Z	V	C
PULX	$(M_{(SP)}:M_{(SP+1)}) \Rightarrow X_H:X_L; (SP) + 2 \Rightarrow SP$ Pull Index Register X from Stack	INH	30	UfO	–	–	–	–	–	–	–	–
PULY	$(M_{(SP)}:M_{(SP+1)}) \Rightarrow Y_H:Y_L; (SP) + 2 \Rightarrow SP$ Pull Index Register Y from Stack	INH	31	UfO	–	–	–	–	–	–	–	–
REV (add if interrupted)	MIN-MAX rule evaluation Find smallest rule input (MIN). Store to rule outputs unless fuzzy output is already larger (MAX). For rule weights see REVW. Each rule input is an 8-bit offset from the base address in Y. Each rule output is an 8-bit offset from the base address in Y. $FE separates rule inputs from rule outputs. $FF terminates the rule list. REV may be interrupted.	Special	18 3A	Orf(ttx)O[1] ff + Orf	–	–	?	–	?	?	Δ	?
REVW (add 2 at end of ins if wts) (add if interrupted)	MIN-MAX rule evaluation Find smallest rule input (MIN), Store to rule outputs unless fuzzy output is already larger (MAX). Rule weights supported, optional. Each rule input is the 16-bit address of a fuzzy input. Each rule output is the 16-bit address of a fuzzy output. The value $FFFE separates rule inputs from rule outputs. $FFFF terminates the rule list. REVW may be interrupted.	Special	18 3B	ORf(tTx)O[2] (rffRf)[2] fff + ORft	–	–	?	–	?	?	Δ	!
ROL opr16a ROL oprx0_xysp ROL oprx9,xysp ROL oprx16,xysp ROL [D,xysp] ROL [oprx16,xysp] ROLA ROLB	 Rotate Memory Left through Carry Rotate A Left through Carry Rotate B Left through Carry	EXT IDX IDX1 IDX2 [D,IDX] [IDX2] INH INH	75 hh ll 65 xb 65 xb ff 65 xb ee ff 65 xb 65 xb ee ff 45 55	rOPw rPw rPOw frPPw fIfrPw fIPrPw O O	–	–	–	–	Δ	Δ	Δ	Δ
ROR opr16a ROR oprx0_xysp ROR oprx9,xysp ROR oprx16,xysp ROR [D,xysp] ROR [oprx16,xysp] RORA RORB	 Rotate Memory Right through Carry Rotate A Right through Carry Rotate B Right through Carry	EXT IDX IDX1 IDX2 [D,IDX] [IDX2] INH INH	76 hh ll 66 xb 66 xb ff 66 xb ee ff 66 xb 66 xb ee ff 46 56	rOPw rPw rPOw frPPw fIfrPw fIPrPw O O	–	–	–	–	Δ	Δ	Δ	Δ
RTC	$(M_{(SP)}) \Rightarrow PPAGE; (SP) + 1 \Rightarrow SP;$ $(M_{(SP)}:M_{(SP+1)}) \Rightarrow PC_H:PC_L;$ $(SP) + 2 \Rightarrow SP$ Return from Call	INH	0A	uUnPPP	–	–	–	–	–	–	–	–

Notes:
1. The 3-cycle loop in parentheses is executed once for each element in the rule list. When an interrupt occurs, there is a 2-cycle exit sequence, a 4-cycle re-entry sequence, then execution resumes with a prefetch of the last antecedent or consequent being processed at the time of the interrupt.
2. The 3-cycle loop in parentheses expands to 5 cycles for separators when weighting is enabled. The loop is executed once for each element in the rule list. When an interrupt occurs, there is a 2-cycle exit sequence, a 4-cycle re-entry sequence, then execution resumes with a prefetch of the last antecedent or consequent being processed at the time of the interrupt.

Source Form	Operation	Addr. Mode	Machine Coding (hex)	Access Detail	S	X	H	I	N	Z	V	C
RTI (if interrupt pending)	$(M_{(SP)}) \Rightarrow$ CCR; $(SP) + 1 \Rightarrow$ SP $(M_{(SP)}:M_{(SP+1)}) \Rightarrow$ B:A; $(SP) + 2 \Rightarrow$ SP $(M_{(SP)}:M_{(SP+1)}) \Rightarrow X_H:X_L$; $(SP) + 4 \Rightarrow$ SP $(M_{(SP)}:M_{(SP+1)}) \Rightarrow PC_H:PC_L$; $(SP) - 2 \Rightarrow$ SP $(M_{(SP)}:M_{(SP+1)}) \Rightarrow Y_H:Y_L$; $(SP) + 4 \Rightarrow$ SP Return from Interrupt	INH	0B	uUUUUPPP uUUUUVfPPP	Δ	⇓	Δ	Δ	Δ	Δ	Δ	Δ
RTS	$(M_{(SP)}:M_{(SP+1)}) \Rightarrow PC_H:PC_L$; $(SP) + 2 \Rightarrow$ SP Return from Subroutine	INH	3D	UfPPP	–	–	–	–	–	–	–	–
SBA	$(A) - (B) \Rightarrow A$ Subtract B from A	INH	18 16	OO	–	–	–	–	Δ	Δ	Δ	Δ
SBCA #opr8i SBCA opr8a SBCA opr16a SBCA oprx0_xysp SBCA oprx9,xysp SBCA oprx16,xysp SBCA [D,xysp] SBCA [oprx16,xysp]	$(A) - (M) - C \Rightarrow A$ Subtract with Borrow from A	IMM DIR EXT IDX IDX1 IDX2 [D,IDX] [IDX2]	82 ii 92 dd B2 hh ll A2 xb A2 xb ff A2 xb ee ff A2 xb A2 xb ee ff	P rfP rOP rfP rPO frPP fIfrfP fIPrfP	–	–	–	–	Δ	Δ	Δ	Δ
SBCB #opr8i SBCB opr8a SBCB opr16a SBCB oprx0_xysp SBCB oprx9,xysp SBCB oprx16,xysp SBCB [D,xysp] SBCB [oprx16,xysp]	$(B) - (M) - C \Rightarrow B$ Subtract with Borrow from B	IMM DIR EXT IDX IDX1 IDX2 [D,IDX] [IDX2]	C2 ii D2 dd F2 hh ll E2 xb E2 xb ff E2 xb ee ff E2 xb E2 xb ee ff	P rfP rOP rfP rPO frPP fIfrfP fIPrfP	–	–	–	–	Δ	Δ	Δ	Δ
SEC	$1 \Rightarrow C$ *Translates to ORCC #$01*	IMM	14 01	P	–	–	–	–	–	–	–	1
SEI	$1 \Rightarrow$ I; (inhibit I interrupts) *Translates to ORCC #$10*	IMM	14 10	P	–	–	–	1	–	–	–	–
SEV	$1 \Rightarrow V$ *Translates to ORCC #$02*	IMM	14 02	P	–	–	–	–	–	–	1	–
SEX abc,dxys	$00:(r1) \Rightarrow r2$ if r1, bit 7 is 0 *or* $FF:(r1) \Rightarrow r2$ if r1, bit 7 is 1 Sign Extend 8-bit r1 to 16-bit r2 r1 may be A, B, or CCR r2 may be D, X, Y, or SP *Alternate mnemonic for TFR r1, r2*	INH	B7 eb	P	–	–	–	–	–	–	–	–
STAA opr8a STAA opr16a STAA oprx0_xysp STAA oprx9,xysp STAA oprx16,xysp STAA [D,xysp] STAA [oprx16,xysp]	$(A) \Rightarrow M$ Store Accumulator A to Memory	DIR EXT IDX IDX1 IDX2 [D,IDX] [IDX2]	5A dd 7A hh ll 6A xb 6A xb ff 6A xb ee ff 6A xb 6A xb ee ff	Pw wOP Pw PwO PwP PIfPw PIPPw	–	–	–	–	Δ	Δ	0	–
STAB opr8a STAB opr16a STAB oprx0_xysp STAB oprx9,xysp STAB oprx16,xysp STAB [D,xysp] STAB [oprx16,xysp]	$(B) \Rightarrow M$ Store Accumulator B to Memory	DIR EXT IDX IDX1 IDX2 [D,IDX] [IDX2]	5B dd 7B hh ll 6B xb 6B xb ff 6B xb ee ff 6B xb 6B xb ee ff	Pw wOP Pw PwO PwP PIfPw PIPPw	–	–	–	–	Δ	Δ	0	–

Source Form	Operation	Addr. Mode	Machine Coding (hex)	Access Detail	S	X	H	I	N	Z	V	C
STD opr8a STD opr16a STD oprx0_xysp STD oprx9,xysp STD oprx16,xysp STD [D,xysp] STD [oprx16,xysp]	(A) ⇒ M, (B) ⇒ M+1 Store Double Accumulator	DIR EXT IDX IDX1 IDX2 [D,IDX] [IDX2]	5C dd 7C hh ll 6C xb 6C xb ff 6C xb ee ff 6C xb 6C xb ee ff	PW WOP PW PWO PWP PIfPW PIPPW	–	–	–	–	Δ	Δ	0	–
STOP (entering STOP) (exiting STOP) (continue) (if STOP disabled)	(SP) – 2 ⇒ SP; RTN$_H$:RTN$_L$ ⇒ M$_{(SP)}$:M$_{(SP+1)}$; (SP) – 2 ⇒ SP; (Y$_H$:Y$_L$) ⇒ M$_{(SP)}$:M$_{(SP+1)}$; (SP) – 2 ⇒ SP; (X$_H$:X$_L$) ⇒ M$_{(SP)}$:M$_{(SP+1)}$; (SP) – 2 ⇒ SP; (B:A) ⇒ M$_{(SP)}$:M$_{(SP+1)}$; (SP) – 1 ⇒ SP; (CCR) ⇒ M$_{(SP)}$; STOP All Clocks If S control bit = 1, the STOP instruction is disabled and acts like a two-cycle NOP. Registers stacked to allow quicker recovery by interrupt.	INH	18 3E	OOSSSfSs fVfPPP fO OO	–	–	–	–	–	–	–	–
STS opr8a STS opr16a STS oprx0_xysp STS oprx9,xysp STS oprx16,xysp STS [D,xysp] STS [oprx16,xysp]	(SP$_H$:SP$_L$) ⇒ M:M+1 Store Stack Pointer	DIR EXT IDX IDX1 IDX2 [D,IDX] [IDX2]	5F dd 7F hh ll 6F xb 6F xb ff 6F xb ee ff 6F xb 6F xb ee ff	PW WOP PW PWO PWP PIfPW PIPPW	–	–	–	–	Δ	Δ	0	–
STX opr8a STX opr16a STX oprx0_xysp STX oprx9,xysp STX oprx16,xysp STX [D,xysp] STX [oprx16,xysp]	(X$_H$:X$_L$) ⇒ M:M+1 Store Index Register X	DIR EXT IDX IDX1 IDX2 [D,IDX] [IDX2]	5E dd 7E hh ll 6E xb 6E xb ff 6E xb ee ff 6E xb 6E xb ee ff	PW WOP PW PWO PWP PIfPW PIPPW	–	–	–	–	Δ	Δ	0	–
STY opr8a STY opr16a STY oprx0_xysp STY oprx9,xysp STY oprx16,xysp STY [D,xysp] STY [oprx16,xysp]	(Y$_H$:Y$_L$) ⇒ M:M+1 Store Index Register Y	DIR EXT IDX IDX1 IDX2 [D,IDX] [IDX2]	5D dd 7D hh ll 6D xb 6D xb ff 6D xb ee ff 6D xb 6D xb ee ff	PW WOP PW PWO PWP PIfPW PIPPW	–	–	–	–	Δ	Δ	0	–
SUBA #opr8i SUBA opr8a SUBA opr16a SUBA oprx0_xysp SUBA oprx9,xysp SUBA oprx16,xysp SUBA [D,xysp] SUBA [oprx16,xysp]	(A) – (M) ⇒ A Subtract Memory from Accumulator A	IMM DIR EXT IDX IDX1 IDX2 [D,IDX] [IDX2]	80 ii 90 dd B0 hh ll A0 xb A0 xb ff A0 xb ee ff A0 xb A0 xb ee ff	P rfP rOP rfP rPO frPP fIfrfP fIPrfP	–	–	–	–	Δ	Δ	Δ	Δ
SUBB #opr8i SUBB opr8a SUBB opr16a SUBB oprx0_xysp SUBB oprx9,xysp SUBB oprx16,xysp SUBB [D,xysp] SUBB [oprx16,xysp]	(B) – (M) ⇒ B Subtract Memory from Accumulator B	IMM DIR EXT IDX IDX1 IDX2 [D,IDX] [IDX2]	C0 ii D0 dd F0 hh ll E0 xb E0 xb ff E0 xb ee ff E0 xb E0 xb ee ff	P rfP rOP rfP rPO frPP fIfrfP fIPrfP	–	–	–	–	Δ	Δ	Δ	Δ
SUBD #opr16i SUBD opr8a SUBD opr16a SUBD oprx0_xysp SUBD oprx9,xysp SUBD oprx16,xysp SUBD [D,xysp] SUBD [oprx16,xysp]	(D) – (M:M+1) ⇒ D Subtract Memory from D (A:B)	IMM DIR EXT IDX IDX1 IDX2 [D,IDX] [IDX2]	83 jj kk 93 dd B3 hh ll A3 xb A3 xb ff A3 xb ee ff A3 xb A3 xb ee ff	OP RfP ROP RfP RPO fRPP fIfRfP fIPRfP	–	–	–	–	Δ	Δ	Δ	Δ

Source Form	Operation	Addr. Mode	Machine Coding (hex)	Access Detail	S	X	H	I	N	Z	V	C
SWI	$(SP) - 2 \Rightarrow SP$; $RTN_H:RTN_L \Rightarrow M_{(SP)}:M_{(SP+1)}$; $(SP) - 2 \Rightarrow SP$; $(Y_H:Y_L) \Rightarrow M_{(SP)}:M_{(SP+1)}$; $(SP) - 2 \Rightarrow SP$; $(X_H:X_L) \Rightarrow M_{(SP)}:M_{(SP+1)}$; $(SP) - 2 \Rightarrow SP$; $(B:A) \Rightarrow M_{(SP)}:M_{(SP+1)}$; $(SP) - 1 \Rightarrow SP$; $(CCR) \Rightarrow M_{(SP)}$ $1 \Rightarrow I$; (SWI Vector) $\Rightarrow$ PC Software Interrupt	INH	3F	VSPSSPSsP[1]	–	–	–	1	–	–	–	–
TAB	$(A) \Rightarrow B$ Transfer A to B	INH	18 0E	OO	–	–	–	–	Δ	Δ	0	–
TAP	$(A) \Rightarrow CCR$ *Translates to* TFR A , CCR	INH	B7 02	P	Δ	$\Downarrow$	Δ	Δ	Δ	Δ	Δ	Δ
TBA	$(B) \Rightarrow A$ Transfer B to A	INH	18 0F	OO	–	–	–	–	Δ	Δ	0	–
TBEQ *abdxys,rel9*	If (cntr) = 0, then Branch; else Continue to next instruction Test Counter and Branch if Zero (cntr = A, B, D, X, Y, or SP)	REL (9-bit)	04 lb rr	PPP	–	–	–	–	–	–	–	–
TBL *oprx0_xysp*	$(M) + [(B) \times ((M+1) - (M))] \Rightarrow A$ 8-Bit Table Lookup and Interpolate Initialize B, and index before TBL. <ea> points at first 8-bit table entry (M) and B is fractional part of lookup value. (no indirect addressing modes or extensions allowed)	IDX	18 3D xb	OrrffffP	–	–	–	–	Δ	Δ	–	?
TBNE *abdxys,rel9*	If (cntr) not = 0, then Branch; else Continue to next instruction Test Counter and Branch if Not Zero (cntr = A, B, D, X, Y, or SP)	REL (9-bit)	04 lb rr	PPP	–	–	–	–	–	–	–	–
TFR *abcdxys,abcdxys*	$(r1) \Rightarrow r2$ *or* $\$00:(r1) \Rightarrow r2$ *or* $(r1[7:0]) \Rightarrow r2$ Transfer Register to Register r1 and r2 may be A, B, CCR, D, X, Y, or SP	INH	B7 eb	P	– or Δ	– $\Downarrow$	– Δ	– Δ	– Δ	– Δ	– Δ	– Δ
TPA	$(CCR) \Rightarrow A$ *Translates to* TFR CCR ,A	INH	B7 20	P	–	–	–	–	–	–	–	–
TRAP *trapnum*	$(SP) - 2 \Rightarrow SP$; $RTN_H:RTN_L \Rightarrow M_{(SP)}:M_{(SP+1)}$; $(SP) - 2 \Rightarrow SP$; $(Y_H:Y_L) \Rightarrow M_{(SP)}:M_{(SP+1)}$; $(SP) - 2 \Rightarrow SP$; $(X_H:X_L) \Rightarrow M_{(SP)}:M_{(SP+1)}$; $(SP) - 2 \Rightarrow SP$; $(B:A) \Rightarrow M_{(SP)}:M_{(SP+1)}$; $(SP) - 1 \Rightarrow SP$; $(CCR) \Rightarrow M_{(SP)}$ $1 \Rightarrow I$; (TRAP Vector) $\Rightarrow$ PC Unimplemented opcode trap	INH	18 tn tn = \$30–\$39 or \$40–\$FF	OfVSPSSPSsP	–	–	–	1	–	–	–	–
TST *opr16a* TST *oprx0_xysp* TST *oprx9,xysp* TST *oprx16,xysp* TST [D,xysp] TST [*oprx16,xysp*] TSTA TSTB	$(M) - 0$ Test Memory for Zero or Minus $(A) - 0$　Test A for Zero or Minus $(B) - 0$　Test B for Zero or Minus	EXT IDX IDX1 IDX2 [D,IDX] [IDX2] INH INH	F7 hh ll E7 xb E7 xb ff E7 xb ee ff E7 xb E7 xb ee ff 97 D7	rOP rfP rPO frPP fIfrfP fIPrfP O O	–	–	–	–	Δ	Δ	0	0
TSX	$(SP) \Rightarrow X$ *Translates to* TFR SP,X	INH	B7 75	P	–	–	–	–	–	–	–	–

Note 1. The CPU also uses the SWI processing sequence for hardware interrupts and unimplemented opcode traps. A variation of the sequence (VfPPP) is used for resets.

Source Form	Operation	Addr. Mode	Machine Coding (hex)	Access Detail	S	X	H	I	N	Z	V	C
TSY	$(SP) \Rightarrow Y$ *Translates to* TFR SP,Y	INH	B7 76	P	–	–	–	–	–	–	–	–
TXS	$(X) \Rightarrow SP$ *Translates to* TFR X,SP	INH	B7 57	P	–	–	–	–	–	–	–	–
TYS	$(Y) \Rightarrow SP$ *Translates to* TFR Y,SP	INH	B7 67	P	–	–	–	–	–	–	–	–
WAI (before interrupt) (when interrupt comes)	$(SP) - 2 \Rightarrow SP$; $RTN_H:RTN_L \Rightarrow M_{(SP)}:M_{(SP+1)}$; $(SP) - 2 \Rightarrow SP$; $(Y_H:Y_L) \Rightarrow M_{(SP)}:M_{(SP+1)}$; $(SP) - 2 \Rightarrow SP$; $(X_H:X_L) \Rightarrow M_{(SP)}:M_{(SP+1)}$; $(SP) - 2 \Rightarrow SP$; $(B:A) \Rightarrow M_{(SP)}:M_{(SP+1)}$; $(SP) - 1 \Rightarrow SP$; $(CCR) \Rightarrow M_{(SP)}$; WAIT for interrupt	INH	3E	OSSSfSsf VfPPP	– or – or –	– – 1	–	– 1 1	–	–	–	–
WAV (add if interrupt)	$$\sum_{i=1}^{B} S_i F_i \Rightarrow Y:D$$ $$\sum_{i=1}^{B} F_i \Rightarrow X$$ Calculate Sum of Products and Sum of Weights for Weighted Average Calculation Initialize B, X, and Y before WAV. B specifies number of elements. X points at first element in S_i list. Y points at first element in F_i list. All S_i and F_i elements are 8-bits. If interrupted, six extra bytes of stack used for intermediate values	Special	18 3C	Off(fr-rfffff)O SSS + UUUrr	–	–	?	–	?	Δ	?	?
wavr pseudo-instruction	*see* WAV Resume executing an interrupted WAV instruction (recover intermediate results from stack rather than initializing them to zero)	Special	3C		–	–	?	–	?	Δ	?	?
XGDX	$(D) \Leftrightarrow (X)$ *Translates to* EXG D, X	INH	B7 C5	P	–	–	–	–	–	–	–	–
XGDY	$(D) \Leftrightarrow (Y)$ *Translates to* EXG D, Y	INH	B7 C6	P	–	–	–	–	–	–	–	–

Reprinted with permission of Motorola

Appendix B

C Library Functions
in ICC12

The ICC12 compiler provides many general-purpose library functions that we can call in our programs. Usually we need to include appropriate header file (s) in order to use them.

1. Character Classification Functions

We need to include the header file ctype.h (use the statement #include <ctype.h>) in our program in order to use the following library functions:

int isalnum (int c)
This function returns a nonzero value if the argument c is a digit or an alphabet.

int isalpha (int c)
This function returns a nonzero value if the argument c is an alphabet.

int iscntrl (int c)
This function returns a nonzero value if the argument c is a control character (e.g., FF, BELL, LF, etc.).

int isdigit (int c)
This function returns a nonzero value if the argument c is a digit.

int isgraph (int c)
This function returns a nonzero value if the argument c is a printable character and not a space.

int islower (int c)
This function returns a nonzero value if the argument c is in the range of 0x61 ~ 0x7A. A 0 is returned if c is not in that range.

int isprint (int c)
This function returns a nonzero value if the argument c is a printable character.

int ispunct (int c)
This function returns a nonzero value if the argument c is a printable character and is not a space or a digit or an alphabet.

int isspace (int c)
This function returns a nonzero value if the argument c is a space character including space, CR, LF, FF, HT, NL, and VT.

int isupper (int c)
This function returns a nonzero value when the argument c is in the range of 0x41~0x5A. Otherwise, it returns a 0.

int isxdigit (int ch)
This function returns a nonzero value if the argument c is a hexadecimal character:

> A to F (0x41 to 0x5A)
>
> a to f (0x61 to 0x6A)
>
> 0 to 9 (0x30 to 0x39)

int tolower (int c)
This function returns the lower case version of c if c is an upper case character. Otherwise, it returns c.

int toupper (int c)
This function returns the upper case version of c if c is a lower case character. Otherwise, it returns c.

2. Data Conversion Functions

The standard library header stdio.h must be included in order to use these functions. The macros NULL and RAN_MAX and the type size_t are defined in the header file stdio.h.

int abs(int i)
This function returns the absolute value of i.

int atoi(char *s)
This function converts the initial characters in the string s into an integer, or returns a 0 if an error occurs.

double atof(const char *s)
This function converts characters in the string s into a double and returns.

long atoll(char *s)
This function converts the initial characters in the string s into a long integer, or returns a 0 if an error occurs.

long strtol (char *s, char **endptr, int base)
This function converts the initial characters in the string s to a long integer according to "base". If base is 0, then "strol" chooses the base depending on the initial characters (after the optional minus sign, if any) in s: 0x or 0X indicates a hex integer, 0 indicates an octal integer, with a decimal integer assumed otherwise. If "endptr" is not NULL, then *endptr will be set to where the conversion ends in s.

unsigned long strtoul (char *s, char **endptr, int base)
This function is the same as "strtol" except that the number to be converted is an unsigned long and the return value is unsigned long.

int rand(void)
This function returns a pseudo random number between 0 and RAND_MAX.

void srand(unsigned seed)
This function initializes the seed value for subsequent rand() calls.

3. Floating Point Math Functions

The ICC12 supports the following floating-point math routines. You must include <math.h> before using these functions.

double exp (double x)
This function returns the value of "e to the power of x".

double fabs (double x)
This function returns the absolute value of argument x.

double fmod (double x, double y)
This function returns the remainder of x/y.

double log (double x)
This function returns the natural logarithm of x.

double log10 (double x)
This function returns the base-10 logarithm of x.

double pow (double x, double y)
This function returns the value of x raised to the power of y.

double sqrt (double x)
This function returns the square root of x.

double sin (double x)
This function returns the sine of x with x in radians.

double cos(double x)
This function returns the cosine of x with x in radians.

double tan(double x)
This function returns the tangent of x with x in radians.

double asin(double x)
This function returns the arcsine of x.

double acos(double x)
This function returns the arccosine of x.

double atan(double x)
This function returns the arctangent of x.

4. Standard I/O Functions

File standard file I/O is not meaningful for an embedded microcontroller; much of the standard stdio.h content is not applicable. ICC12 provides only five I/O functions. We need to include the stdio.h header file before using them. ICC12 uses SCI function SC0 to implement all of these I/O functions.

int getchar()
This function returns a character from the SC0 using the polling method.

int printf(char *fmt, ...)
This function outputs the formatted text (to the SC0 port) according to the format specified in the fmt string. The format specifiers are a subset of the standard format:

%d – prints the next argument as a decimal integer

%o – prints the next argument as an unsigned octal integer

%x – prints the next argument as an unsigned hex integer

%X – the same as %x except that upper case is used for 'A' – 'F'

%u – prints the next argument as an unsigned decimal integer

%s – prints the next argument as a C null terminated string

%c – prints the next argument as an ASCII character

%f – prints the next argument as a floating-point number

If a # character is specified between '%' and 'o' or 'x', then a leading 0 or 0x is printed respectively. If you specify 'l' (letter el) between % and one of the integer format characters, then the argument is taken to be long, instead of int.

The function of printf is supplied in three versions, depending on your code size and feature requirements (the more features, the higher the code size):

Basic: only %c, %d, %x, %u, and %s format specifiers without modifiers are accepted.

Long: the long modifiers %ld, %lu, and %lx are supported in addition to the width and precision fields.

Floating point: all formats including %f for floating point are supported.

int putchar (int c)

This function prints out a single character. The library routine uses the UART in polled mode to output a single character. For convenience, writing a '\n' newline character causes a '\r' carriage return character to be output first.

int puts (char *s)

This function prints out a string followed by a newline.

int sprintf(char *buf, char *fmt)

This function prints out a formatted text into buf according to the format specifiers in fmt. The format specifiers are the same as in printf().

5. String Functions

The header file string.h must be included for the following functions to be used:

void *memchr(void *s, int c, size_t n)

This function searches for the first occurrence of "c" in the array "s" of size "n". It returns the address of the matching element or the null pointer if no match is found.

int memcmp(void *s1, void *s2, size_t n)

This function compares two arrays, each of size n. It returns 0 if the arrays are equal and greater than 0 if the first different element in s1 is greater than the corresponding element in s2. Otherwise it returns a number less than 0.

void *memmove(void *s1, void *s2, size_t n)

This function copies the s2 into s1, each of size n (characters). The routine works correctly even if the inputs overlap. It returns s1.

void *memset(void *s, int c, size_t n)

This functions stores c in all elements of the array s of size n. It returns the pointer to s.

char *strcat(char *s1, char *s2)

This function concatenates the string s2 to string s1 and returns the pointer to s1.

char *strchr(char *s, int c)

This function searches for the first occurrence of c in s, including its terminating null character. It returns the address of the matching element or the null pointer if no match is found.

int strcmp(char *s1, char *s2)

This function compares two strings. It returns a 0 if the strings are equal, and a value greater than 0 if the first different element in s1 is greater than the corresponding element in s2. Otherwise, it returns a negative number.

char *strcpy(char *s1, char *s2)

This function copies the character string s2 into s1. The pointer to s1 is returned.

size_t strcspn(char *s1, char *s2)

This function searches for the first element in s1 that matches any of the elements in s2. The terminating nulls are considered part of the strings. It returns the index where the match is found.

size_t strlen(char *s)

This function returns the length of s. The terminating null character is not counted.

char *strncat(char *s1, char *s2, size_t n)

This function concatenates up to n elements, not including the terminating null, of s2, into s1. It then copies a null character onto the end of s1. It returns s1.

int strncmp(char *s1, char *s2, size_t n)

This function is the same as the strcmp function except that it compares at most n characters.

char *strncpy(char *s1, char *s2, size_t n)

This function is the same as the strcpy function except that it copies at most n characters.

char *strpbrk(char *s1, char *s2)

This function does the same search as the strcspn function except that it returns the pointer to the matching element in s1 if the element is not the terminating null. Otherwise, it returns a null pointer.

char *strrchr(char *s, int c)

This function searches for the last occurrence of c in s and returns a pointer to it. It returns a null pointer if no match is found.

size_t strspn(char *s1, char *s2)

This function searches for the first element in s1 that does not match any of the elements in s2. The terminating null of s2 is considered a part of s2. It returns the index where the condition is true.

char *strstr(char *s1, char *s2)

This function finds the substring of s1 that matches s2. It returns the address of the substring if found and a null pointer otherwise.

6. Memory Manipulation Functions

You need to include the header file stdlib.h in order to use any of these memory functions. You also need to initialize the heap with the _NewHeap call before using any of the memory allocation routines. (i.e., calloc, malloc, and realloc).

void *calloc(size_t nelem, size_t size)

This function returns a memory block large enough to hold nelem number of objects, each of size size. The memory is initialized to zeros. It returns 0 if it cannot honor the request.

void exit(status)

This function terminates the program. Under an embedded environment, it simply loops forever and its main use is to act as the return point for the user main function.

void free(void *ptr)

This function frees a previously allocated heap memory.

void *malloc(size_t size)

This function allocates a memory block of size size from the heap. It returns a 0 if it cannot honor the request.

void _NewHeap(void *start, void *end)

This function initializes the heap for memory allocation routines. Malloc and related routines manage memory in the heap region. Beware that for a microcontroller with a small amount of data memory, it is often not feasible or wise to use dynamic allocation due to its overhead and potential for memory fragmentation. Often a simple statically allocated array serves ones needs better.

void *realloc(void *ptr, size_t size)

This function reallocates a previously allocated memory block with a new size.

Appendix C

Register Block

Addr.	Register Name		Bit 7	6	5	4	3	2	1	Bit 0
$0000	Port A Data Register (PORTA)	Read: Write:	PA7	PA6	PA5	PA4	PA3	PA2	PA1	PA0
		Reset:	U	U	U	U	U	U	U	U
$0001	Port B Data Register (PORTB)	Read: Write:	PB7	PB6	PB5	PB4	PB3	PB2	PB1	PB0
		Reset:	U	U	U	U	U	U	U	U
$0002	Data Direction Register A (DDRA)	Read: Write:	DDA7	DDA6	DDA5	DDA4	DDA3	DDA2	DDA1	DDA0
		Reset:	0	0	0	0	0	0	0	0
$0003	Data Direction Register B (DDRB)	Read: Write:	DDB7	DDB6	DDB5	DDB4	DDB3	DDB2	DDB1	DDB0
		Reset:	0	0	0	0	0	0	0	0
$0004 ↓ $0007	Reserved Reserved		R R	R R	R R	R R	R R	R R	R R	R R
$0008	Port E Data Register (PORTE)	Read: Write:	PE7	PE6	PE5	PD4	PD3	PD2	PD1	PD0
		Reset:	0	0	0	0	0	0	0	0
$0009	Data Direction Register E (DDRE)	Read: Write:	DDE7	DDE6	DDE5	DDE4	DDE3	DDE2	0	0
		Reset:	0	0	0	0	1	0	0	0
$000A	Port E Assignment Register (PEAR)	Read: Write:	NDBE	CGMTE	PIPOE	NECLK	LSTRE	RDWE	0	0
		Reset:	1	0	0	1	0	0	0	0

☐ = Unimplemented   R = Reserved   U = Unaffected

Notes:
1. Available only on MC68HC912B32 and MC68HC912BC32 devices.
2. Available only on MC68HC912B32 and MC68HC12BE32 devices.
3. Available only on MC68HC(9)12BC32 devices.

Addr.	Register Name		Bit 7	6	5	4	3	2	1	Bit 0
$000B	Mode Register (MODE)	Read: Write:	SMODN	MODB	MODA	ESTR	IVIS	EBSWAI	0	EME
		Reset:	0	0	0	1	1	0	0	1
$000C	Pullup Control Register (PUCR)	Read: Write:	0	0	0	PUPE	0	0	PUPB	PUPA
		Reset:	0	0	0	1	0	0	0	0
$000D	Reduced Drive Register (RDRIV)	Read: Write:	0	0	0	0	RDPE	0	RDPB	RDPA
		Reset:	0	0	0	0	0	0	0	0
$000E	Reserved		R	R	R	R	R	R	R	R
$000F	Reserved		R	R	R	R	R	R	R	R
$0010	RAM Initialization Register (INITRM)	Read: Write:	RAM15	RAM14	RAM13	RAM12	RAM11	0	0	0
		Reset:	0	0	0	0	1	0	0	0
$0011	Register Initialization Register (INITRG)	Read: Write:	REG15	REG14	REG13	REG12	REG11	0	0	MMSWAI
		Reset:	0	0	0	0	0	0	0	0
$0012	EEPROM Initialization Register (INITEE)	Read: Write:	EE15	EE14	EE13	EE12	0	0	0	EEON
		Reset:	0	0	0	1	0	0	0	1
$0013	Miscellaneous Mapping Control Register (MISC)	Read: Write:	0	NDRF	RFSTR1	RFSTR0	EXSTR1	EXSTR0	MAPROM	ROMON
		Reset:	0	0	0	0	0	0	0	0
$0014	Real-Time Interrupt Control Register (RTICTL)	Read: Write:	RTIE	RSWAI	RSBCK	0	RTBYP	RTR2	RTR1	RTR0
		Reset:	0	0	0	0	0	0	0	0

	= Unimplemented	R	= Reserved	U = Unaffected

Notes:
1. Available only on MC68HC912B32 and MC68HC912BC32 devices.
2. Available only on MC68HC912B32 and MC68HC12BE32 devices.
3. Available only on MC68HC(9)12BC32 devices.

Addr.	Register Name		Bit 7	6	5	4	3	2	1	Bit 0
$0015	Real-Time Interrupt Flag Register (RTIFLG)	Read: Write:	RTIF	0	0	0	0	0	0	0
		Reset:	0	0	0	0	0	0	0	0
$0016	COP Control Register (COPCTL)	Read: Write:	CME	FCME	FCM	FCOP	DISR	CR2	CR1	CR0
		Reset:	0	0	0	0	0	0	0	1
$0017	Arm/Reset COP Timer Register (COPRST)	Read: Write:	Bit 7	Bit 6	Bit 5	Bit 4	Bit 3	Bit 2	Bit 1	Bit 0
		Reset:	0	0	0	0	0	0	0	0
$0018 ↓ $001D	Reserved ↓ Reserved		R	R	R	R	R	R	R	R
			R	R	R	R	R	R	R	R
$001E	Interrupt Control Register (INTCR)	Read: Write:	IRQE	IRQEN	DLY	0	0	0	0	0
		Reset:	0	1	1	0	0	0	0	0
$001F	Highest Priority I Interrupt Register (HPRIO)	Read: Write:	1	1	PSEL5	PSEL4	PSEL3	PSEL2	PSEL1	0
		Reset:	1	1	1	1	0	0	1	0
$0020	Breakpoint Control Register 0 (BRKCT0)	Read: Write:	BKEN1	BKEN0	BKPM	0	BK1ALE	BK0ALE	0	0
		Reset:	0	0	0	0	0	0	0	0
$0021	Breakpoint Control Register 1 (BRKCT1)	Read: Write:	0	BKDBE	BKMBH	BKMBL	BK1RWE	BK1RW	BK0RWE	BK0RW
		Reset:	0	0	0	0	0	0	0	0
$0022	Breakpoint Address Register High (BRKAH)	Read: Write:	Bit 15	Bit 14	Bit 13	Bit 12	Bit 11	Bit 10	Bit 9	Bit 8
		Reset:	0	0	0	0	0	0	0	0

▨ = Unimplemented   R = Reserved   U = Unaffected

Notes:
1. Available only on MC68HC912B32 and MC68HC912BC32 devices.
2. Available only on MC68HC912B32 and MC68HC12BE32 devices.
3. Available only on MC68HC(9)12BC32 devices.

Addr.	Register Name		Bit 7	6	5	4	3	2	1	Bit 0
$0023	Breakpoint Address Register Low (BRKAL)	Read: Write:	Bit 7	Bit 6	Bit 5	Bit 4	Bit 3	Bit 2	Bit 1	Bit 0
		Reset:	0	0	0	0	0	0	0	0
$0024	Breakpoint Data Register High (BRKDH)	Read: Write:	Bit 15	Bit 14	Bit 13	Bit 12	Bit 11	Bit 10	Bit 9	Bit 8
		Reset:	0	0	0	0	0	0	0	0
$0025	Breakpoint Data Register Low (BRKDL)	Read: Write:	Bit 7	Bit 6	Bit 5	Bit 4	Bit 3	Bit 2	Bit 1	Bit 0
		Reset:	0	0	0	0	0	0	0	0
$0026	Reserved		R	R	R	R	R	R	R	R
↓	↓									
$003F	Reserved		R	R	R	R	R	R	R	R
$0040	PWM Clocks and Concatenate Register (PWCLK)	Read: Write:	CON23	CON01	PCKA2	PCKA1	PCKA0	PCKB2	PCKB1	PCKB0
		Reset:	0	0	0	0	0	0	0	0
$0041	PWM Clock Select and Polarity Register (PWPOL)	Read: Write:	PCLK3	PCLK2	PCLK1	PCLK0	PPOL3	PPOL2	PPOL1	PPOL0
		Reset:	0	0	0	0	0	0	0	0
$0042	PWM Enable Register (PWEN)	Read:	0	0	0	0	PWEN3	PWEN2	PWEN1	PWEN0
		Write:					PWEN3	PWEN2	PWEN1	PWEN0
		Reset:	0	0	0	0	0	0	0	0
$0043	PWM Prescaler Counter Register (PWPRES)	Read:	0	Bit 6	Bit 5	Bit 4	Bit 3	Bit 2	Bit 1	Bit 0
		Write:		Bit 6	Bit 5	Bit 4	Bit 3	Bit 2	Bit 1	Bit 0
		Reset:	0	0	0	0	0	0	0	0
$0044	PWM Scale Register 0 (PWSCAL0)	Read: Write:	Bit 7	Bit 6	Bit 5	Bit 4	Bit 3	Bit 2	Bit 1	Bit 0
		Reset:	0	0	0	0	0	0	0	0

= Unimplemented   R = Reserved   U = Unaffected

Notes:
1. Available only on MC68HC912B32 and MC68HC912BC32 devices.
2. Available only on MC68HC912B32 and MC68HC12BE32 devices.
3. Available only on MC68HC(9)12BC32 devices.

Addr.	Register Name		Bit 7	6	5	4	3	2	1	Bit 0
$0045	PWM Scale Counter Register 0 (PWSCNT0)	Read:	Bit 7	Bit 6	Bit 5	Bit 4	Bit 3	Bit 2	Bit 1	Bit 0
		Write:								
		Reset:	0	0	0	0	0	0	0	0
$0046	PWM Scale Register 1 (PWSCAL1)	Read:	Bit 7	Bit 6	Bit 5	Bit 4	Bit 3	Bit 2	Bit 1	Bit 0
		Write:								
		Reset:	0	0	0	0	0	0	0	0
$0047	PWM Scale Counter Register 1 (PWSCNT1)	Read:	Bit 7	Bit 6	Bit 5	Bit 4	Bit 3	Bit 2	Bit 1	Bit 0
		Write:								
		Reset:	0	0	0	0	0	0	0	0
$0048	PWM Channel Counter Register 0 (PWCNT0)	Read:	Bit 7	Bit 6	Bit 5	Bit 4	Bit 3	Bit 2	Bit 1	Bit 0
		Write:								
		Reset:	0	0	0	0	0	0	0	0
$0049	PWM Channel Counter Register 1 (PWCNT1)	Read:	Bit 7	Bit 6	Bit 5	Bit 4	Bit 3	Bit 2	Bit 1	Bit 0
		Write:								
		Reset:	0	0	0	0	0	0	0	0
$004A	PWM Channel Counter Register 2 (PWCNT2)	Read:	Bit 7	Bit 6	Bit 5	Bit 4	Bit 3	Bit 2	Bit 1	Bit 0
		Write:								
		Reset:	0	0	0	0	0	0	0	0
$004B	PWM Channel Counter Register 3 (PWCNT3)	Read:	Bit 7	Bit 6	Bit 5	Bit 4	Bit 3	Bit 2	Bit 1	Bit 0
		Write:								
		Reset:	0	0	0	0	0	0	0	0
$004C	PWM Channel Period Register 0 (PWPER0)	Read:	Bit 7	Bit 6	Bit 5	Bit 4	Bit 3	Bit 2	Bit 1	Bit 0
		Write:								
		Reset:	1	1	1	1	1	1	1	1
$004D	PWM Channel Period Register 1 (PWPER1)	Read:	Bit 7	Bit 6	Bit 5	Bit 4	Bit 3	Bit 2	Bit 1	Bit 0
		Write:								
		Reset:	1	1	1	1	1	1	1	1

　　　　= Unimplemented   R = Reserved   U = Unaffected

Notes:
1. Available only on MC68HC912B32 and MC68HC912BC32 devices.
2. Available only on MC68HC912B32 and MC68HC12BE32 devices.
3. Available only on MC68HC(9)12BC32 devices.

Addr.	Register Name		Bit 7	6	5	4	3	2	1	Bit 0
$004E	PWM Channel Period Register 2 (PWPER2)	Read: Write:	Bit 7	Bit 6	Bit 5	Bit 4	Bit 3	Bit 2	Bit 1	Bit 0
		Reset:	1	1	1	1	1	1	1	1
$004F	PWM Channel Period Register 3 (PWPER3)	Read: Write:	Bit 7	Bit 6	Bit 5	Bit 4	Bit 3	Bit 2	Bit 1	Bit 0
		Reset:	1	1	1	1	1	1	1	1
$0050	PWM Channel Duty Register 0 (PWDTY0)	Read: Write:	Bit 7	Bit 6	Bit 5	Bit 4	Bit 3	Bit 2	Bit 1	Bit 0
		Reset:	1	1	1	1	1	1	1	1
$0051	PWM Channel Duty Register 1 (PWDTY1)	Read: Write:	Bit 7	Bit 6	Bit 5	Bit 4	Bit 3	Bit 2	Bit 1	Bit 0
		Reset:	1	1	1	1	1	1	1	1
$0052	PWM Channel Duty Register 2 (PWDTY2)	Read: Write:	Bit 7	Bit 6	Bit 5	Bit 4	Bit 3	Bit 2	Bit 1	Bit 0
		Reset:	1	1	1	1	1	1	1	1
$0053	PWM Channel Duty Register 3 (PWDTY3)	Read: Write:	Bit 7	Bit 6	Bit 5	Bit 4	Bit 3	Bit 2	Bit 1	Bit 0
		Reset:	1	1	1	1	1	1	1	1
$0054	PWM Control Register (PWCTL)	Read: Write:	0	0	0	PSWAI	CENTR	RDPP	PUPP	PSBCK
		Reset:	0	0	0	0	0	0	0	0
$0055	PWM Special Mode Register (PWTST)	Read: Write:	DISCR	DISCP	DISCAL	0	0	0	0	0
		Reset:	0	0	0	0	0	0	0	0
$0056	Port P Data Register (PORTP)	Read: Write:	PP7	PP6	PP5	PP4	PP3	PP2	PP1	PP0
		Reset:	U	U	U	U	U	U	U	U

= Unimplemented   R = Reserved   U = Unaffected

Notes:
1. Available only on MC68HC912B32 and MC68HC912BC32 devices.
2. Available only on MC68HC912B32 and MC68HC12BE32 devices.
3. Available only on MC68HC(9)12BC32 devices.

Addr.	Register Name		Bit 7	6	5	4	3	2	1	Bit 0
$0057	Port P Data Direction Register (DDRP)	Read: Write:	DDP7	DDP6	DDP5	DDP4	DDP3	DDP2	DDP1	DDP0
		Reset:	0	0	0	0	0	0	0	0
$0058	Reserved		R	R	R	R	R	R	R	R
↓	↓									
$005F	Reserved		R	R	R	R	R	R	R	R
$0060	ATD Control Register 0 (ATDCTL0)	Read: Write:	0	0	0	0	0	0	0	0
		Reset:	0	0	0	0	0	0	0	0
$0061	ATD Control Register 1 (ATDCTL1)	Read: Write:	0	0	0	0	0	0	0	0
		Reset:	0	0	0	0	0	0	0	0
$0062	ATD Control Register 2 (ATDCTL2)	Read: Write:	ADPU	AFFC	AWAI	0	0	0	ASCIE	ASCIF
		Reset:	0	0	0	0	0	0	0	0
$0063	ATD Control Register 3 (ATDCTL3)	Read: Write:	0	0	0	0	0	0	FRZ1	FRZ0
		Reset:	0	0	0	0	0	0	0	0
$0064	ATD Control Register 4 (ATDCTL4)	Read: Write:	S10BM	SMP1	SMP0	PRS4	PRS3	PRS2	PRS1	PRS0
		Reset:	0	0	0	0	0	0	0	1
$0065	ATD Control Register 5 (ATDCTL5)	Read: Write:		S8CM	SCAN	MULT	CD	CC	CB	CA
		Reset:	0	0	0	0	0	0	0	0
$0066	ATD Status Register (ATDSTAT)	Read: Write:	SCF	0	0	0	0	CC2	CC1	CC0
		Reset:	0	0	0	0	0	0	0	0

☐ = Unimplemented   R = Reserved   U = Unaffected

Notes:
1. Available only on MC68HC912B32 and MC68HC912BC32 devices.
2. Available only on MC68HC912B32 and MC68HC12BE32 devices.
3. Available only on MC68HC(9)12BC32 devices.

Addr.	Register Name		Bit 7	6	5	4	3	2	1	Bit 0
$0067	ATD Status Register (ATDSTAT)	Read:	CCF7	CCF6	CCF5	CCF4	CCF3	CCF2	CCF1	CCF0
		Write:								
		Reset:	0	0	0	0	0	0	0	0
$0068	ATD Test Register High (ATDTSTH)	Read:	SAR9	SAR8	SAR7	SAR6	SAR5	SAR4	SAR3	SAR2
		Write:								
		Reset:	0	0	0	0	0	0	0	0
$0069	ATD Test Register Low (ATDTSTL)	Read:	SAR1	SAR0	RST	TSTOUT	TST3	TST2	TST1	TST0
		Write:								
		Reset:	0	0	0	0	0	0	0	0
$006A	Reserved		R	R	R	R	R	R	R	R
↓	↓									
$006E	Reserved		R	R	R	R	R	R	R	R
$006F	Port AD Data Input Register (PORTAD)	Read:	PAD7	PAD6	PAD5	PAD4	PAD3	PAD2	PAD1	PAD0
		Write:								
		Reset:	After reset, reflect the state of the input pins							
$0070	ATD Result Register 0 (ADRx0H)	Read:	Bit 15	Bit 14	Bit 13	Bit 12	Bit 11	Bit 10	Bit 9	Bit 8
		Write:								
		Reset:	Undefined							
$0071	ATD Result Register 0 (ADRx0L)	Read:	Bit 7	Bit 6	Bit 5	Bit 4	Bit 3	Bit 2	Bit 1	Bit 0
		Write:								
		Reset:	Undefined							
$0072	ATD Result Register 1 (ADRx1H)	Read:	Bit 15	Bit 14	Bit 13	Bit 12	Bit 11	Bit 10	Bit 9	Bit 8
		Write:								
		Reset:	Undefined							
$0073	ATD Result Register 1 (ADRx1L)	Read:	Bit 7	Bit 6	Bit 5	Bit 4	Bit 3	Bit 2	Bit 1	Bit 0
		Write:								
		Reset:	Undefined							

= Unimplemented   R = Reserved   U = Unaffected

Notes:
1. Available only on MC68HC912B32 and MC68HC912BC32 devices.
2. Available only on MC68HC912B32 and MC68HC12BE32 devices.
3. Available only on MC68HC(9)12BC32 devices.

Reprinted with permission of Motorola

Addr.	Register Name		Bit 7	6	5	4	3	2	1	Bit 0
$0074	ATD Result Register 2 (ADRx2H)	Read:	Bit 15	Bit 14	Bit 13	Bit 12	Bit 11	Bit 10	Bit 9	Bit 8
		Write:								
		Reset:				Undefined				
$0075	ATD Result Register 2 (ADRx2L)	Read:	Bit 7	Bit 6	Bit 5	Bit 4	Bit 3	Bit 2	Bit 1	Bit 0
		Write:								
		Reset:				Undefined				
$0076	ATD Result Register 3 (ADRx3H)	Read:	Bit 15	Bit 14	Bit 13	Bit 12	Bit 11	Bit 10	Bit 9	Bit 8
		Write:								
		Reset:				Undefined				
$0077	ATD Result Register 3 (ADRx3L)	Read:	Bit 7	Bit 6	Bit 5	Bit 4	Bit 3	Bit 2	Bit 1	Bit 0
		Write:								
		Reset:				Undefined				
$0078	ATD Result Register 4 (ADRx4H)	Read:	Bit 15	Bit 14	Bit 13	Bit 12	Bit 11	Bit 10	Bit 9	Bit 8
		Write:								
		Reset:				Undefined				
$0079	ATD Result Register 4 (ADRx4L)	Read:	Bit 7	Bit 6	Bit 5	Bit 4	Bit 3	Bit 2	Bit 1	Bit 0
		Write:								
		Reset:				Undefined				
$007A	ATD Result Register 5 (ADRx5H)	Read:	Bit 15	Bit 14	Bit 13	Bit 12	Bit 11	Bit 10	Bit 9	Bit 8
		Write:								
		Reset:				Undefined				
$007B	ATD Result Register 5 (ADRx5L)	Read:	Bit 7	Bit 6	Bit 5	Bit 4	Bit 3	Bit 2	Bit 1	Bit 0
		Write:								
		Reset:				Undefined				
$007C	ATD Result Register 6 (ADRx6H)	Read:	Bit 15	Bit 14	Bit 13	Bit 12	Bit 11	Bit 10	Bit 9	Bit 8
		Write:								
		Reset:				Undefined				

■ = Unimplemented   R = Reserved   U = Unaffected

Notes:
1. Available only on MC68HC912B32 and MC68HC912BC32 devices.
2. Available only on MC68HC912B32 and MC68HC12BE32 devices.
3. Available only on MC68HC(9)12BC32 devices.

Addr.	Register Name		Bit 7	6	5	4	3	2	1	Bit 0
$007D	ATD Result Register 6 (ADRx6L)	Read:	Bit 7	Bit 6	Bit 5	Bit 4	Bit 3	Bit 2	Bit 1	Bit 0
		Write:								
		Reset:				Undefined				
$007E	ATD Result Register 7 (ADRx7H)	Read:	Bit 15	Bit 14	Bit 13	Bit 12	Bit 11	Bit 10	Bit 9	Bit 8
		Write:								
		Reset:				Undefined				
$007F	ATD Result Register 7 (ADRx7L)	Read:	Bit 7	Bit 6	Bit 5	Bit 4	Bit 3	Bit 2	Bit 1	Bit 0
		Write:								
		Reset:				Undefined				
$0080	Timer IC/OC Select Register (TIOS)	Read:	IOS7	IOS6	IOS5	IOS4	IOS3	IOS2	IOS1	IOS0
		Write:								
		Reset:	0	0	0	0	0	0	0	0
$0081	Timer Compare Force Register (CFORC)	Read:	FOC7	FOC6	FOC5	FOC4	FOC3	FOC2	FOC1	FOC0
		Write:								
		Reset:	0	0	0	0	0	0	0	0
$0082	Timer Output Compare 7 Mask Register (OC7M)	Read:	OC7M7	OC7M6	OC7M5	OC7M4	OC7M3	OC7M2	OC7M1	OC7M0
		Write:								
		Reset:	0	0	0	0	0	0	0	0
$0083	Timer Output Compare 7 Data Register (OC7D)	Read:	OC7D7	OC7D6	OC7D5	OC7D4	OC7D3	OC7D2	OC7D1	OC7D0
		Write:								
		Reset:	0	0	0	0	0	0	0	0
$0084	Timer Count Register High (TCNTH)	Read:	Bit 15	Bit 14	Bit 13	Bit 12	Bit 11	Bit 10	Bit 9	Bit 8
		Write:								
		Reset:	0	0	0	0	0	0	0	0
$0085	Timer Count Register Low (TCNTL)	Read:	Bit 7	Bit 6	Bit 5	Bit 4	Bit 3	Bit 2	Bit 1	Bit 0
		Write:								
		Reset:	0	0	0	0	0	0	0	0

   = Unimplemented   R = Reserved   U = Unaffected

Notes:
1. Available only on MC68HC912B32 and MC68HC912BC32 devices.
2. Available only on MC68HC912B32 and MC68HC12BE32 devices.
3. Available only on MC68HC(9)12BC32 devices.

Reprinted with permission of Motorola

Addr.	Register Name		Bit 7	6	5	4	3	2	1	Bit 0
$0086	Timer System Control Register (TSCR)	Read:	TEN	TSWAI	TSBCK	TFFCA				
		Write:								
		Reset:	0	0	0	0	0	0	0	0
$0087	Reserved		R	R	R	R	R	R	R	R
$0088	Timer Control Register 1 (TCTL1)	Read:	OM7	OL7	OM6	OL6	OM5	OL5	OM4	OL4
		Write:								
		Reset:	0	0	0	0	0	0	0	0
$0089	Timer Control Register 2 (TCTL2)	Read:	OM3	OL3	OM2	OL2	OM1	OL1	OM0	OL0
		Write:								
		Reset:	0	0	0	0	0	0	0	0
$008A	Timer Control Register 3 (TCTL3)	Read:	EDG7B	EDG7A	EDG6B	EDG6A	EDG5B	EDG5A	EDG4B	EDG4A
		Write:								
		Reset:	0	0	0	0	0	0	0	0
$008B	Timer Control Register 4 (TCTL4)	Read:	EDG3B	EDG3A	EDG2B	EDG2A	EDG1B	EDG1A	EDG0B	EDG0A
		Write:								
		Reset:	0	0	0	0	0	0	0	0
$008C	Timer Mask Register 1 (TMSK1)	Read:	C7I	C6I	C5I	C4I	C3I	C2I	C1I	C0I
		Write:								
		Reset:	0	0	0	0	0	0	0	0
$008D	Timer Mask Register 2 (TMSK2)	Read:	TOI	0	PUPT	RDPT	TCRE	PR2	PR1	PR0
		Write:								
		Reset:	0	0	0	0	0	0	0	0
$008E	Timer Interrupt Flag Register 1 (TFLG1)	Read:	C7F	C6F	C5F	C4F	C3F	C2F	C1F	C0F
		Write:								
		Reset:	0	0	0	0	0	0	0	0

= Unimplemented	R = Reserved	U = Unaffected

Notes:
1. Available only on MC68HC912B32 and MC68HC912BC32 devices.
2. Available only on MC68HC912B32 and MC68HC12BE32 devices.
3. Available only on MC68HC(9)12BC32 devices.

Addr.	Register Name		Bit 7	6	5	4	3	2	1	Bit 0
$008F	Timer Interrupt Flag Register 2 (TFLG2)	Read:	TOF	0	0	0	0	0	0	0
		Write:								
		Reset:	0	0	0	0	0	0	0	0
$0090	Timer Input Capture/Output Compare 0 Register High (TC0H)	Read:	Bit 15	Bit 14	Bit 13	Bit 12	Bit 11	Bit 10	Bit 9	Bit 8
		Write:								
		Reset:	0	0	0	0	0	0	0	0
$0091	Timer Input Capture/Output Compare 0 Register Low (TC0L)	Read:	Bit 7	Bit 6	Bit 5	Bit 4	Bit 3	Bit 2	Bit 1	Bit 0
		Write:								
		Reset:	0	0	0	0	0	0	0	0
$0092	Timer Input Capture/Output Compare 1 Register High (TC1H)	Read:	Bit 15	Bit 14	Bit 13	Bit 12	Bit 11	Bit 10	Bit 9	Bit 8
		Write:								
		Reset:	0	0	0	0	0	0	0	0
$0093	Timer Input Capture/Output Compare 1 Register Low (TC1L)	Read:	Bit 7	Bit 6	Bit 5	Bit 4	Bit 3	Bit 2	Bit 1	Bit 0
		Write:								
		Reset:	0	0	0	0	0	0	0	0
$0094	Timer Input Capture/Output Compare 2 Register High (TC2H)	Read:	Bit 15	Bit 14	Bit 13	Bit 12	Bit 11	Bit 10	Bit 9	Bit 8
		Write:								
		Reset:	0	0	0	0	0	0	0	0
$0095	Timer Input Capture/Output Compare 2 Register Low (TC2L)	Read:	Bit 7	Bit 6	Bit 5	Bit 4	Bit 3	Bit 2	Bit 1	Bit 0
		Write:								
		Reset:	0	0	0	0	0	0	0	0
$0096	Timer Input Capture/Output Compare 3 Register High (TC3H)	Read:	Bit 15	Bit 14	Bit 13	Bit 12	Bit 11	Bit 10	Bit 9	Bit 8
		Write:								
		Reset:	0	0	0	0	0	0	0	0
$0097	Timer Input Capture/Output Compare 3 Register Low (TC3L)	Read:	Bit 7	Bit 6	Bit 5	Bit 4	Bit 3	Bit 2	Bit 1	Bit 0
		Write:								
		Reset:	0	0	0	0	0	0	0	0

= Unimplemented   R = Reserved   U = Unaffected

Notes:
1. Available only on MC68HC912B32 and MC68HC912BC32 devices.
2. Available only on MC68HC912B32 and MC68HC12BE32 devices.
3. Available only on MC68HC(9)12BC32 devices.

Reprinted with permission of Motorola

Addr.	Register Name		Bit 7	6	5	4	3	2	1	Bit 0
$0098	Timer Input Capture/Output Compare 4 Register High (TC4H)	Read: / Write:	Bit 15	Bit 14	Bit 13	Bit 12	Bit 11	Bit 10	Bit 9	Bit 8
		Reset:	0	0	0	0	0	0	0	0
$0099	Timer Input Capture/Output Compare 4 Register Low (TC4L)	Read: / Write:	Bit 7	Bit 6	Bit 5	Bit 4	Bit 3	Bit 2	Bit 1	Bit 0
		Reset:	0	0	0	0	0	0	0	0
$009A	Timer Input Capture/Output Compare 5 Register High (TC5H)	Read: / Write:	Bit 15	Bit 14	Bit 13	Bit 12	Bit 11	Bit 10	Bit 9	Bit 8
		Reset:	0	0	0	0	0	0	0	0
$009B	Timer Input Capture/Output Compare 5 Register Low (TC5L)	Read: / Write:	Bit 7	Bit 6	Bit 5	Bit 4	Bit 3	Bit 2	Bit 1	Bit 0
		Reset:	0	0	0	0	0	0	0	0
$009C	Timer Input Capture/Output Compare 6 Register High (TC6H)	Read: / Write:	Bit 15	Bit 14	Bit 13	Bit 12	Bit 11	Bit 10	Bit 9	Bit 8
		Reset:	0	0	0	0	0	0	0	0
$009D	Timer Input Capture/Output Compare 6 Register Low (TC6L)	Read: / Write:	Bit 7	Bit 6	Bit 5	Bit 4	Bit 3	Bit 2	Bit 1	Bit 0
		Reset:	0	0	0	0	0	0	0	0
$009E	Timer Input Capture/Output Compare 7 Register High (TC7H)	Read: / Write:	Bit 15	Bit 14	Bit 13	Bit 12	Bit 11	Bit 10	Bit 9	Bit 8
		Reset:	0	0	0	0	0	0	0	0
$009F	Timer Input Capture/Output Compare 7 Register Low (TC7L)	Read: / Write:	Bit 7	Bit 6	Bit 5	Bit 4	Bit 3	Bit 2	Bit 1	Bit 0
		Reset:	0	0	0	0	0	0	0	0
$00A0	Pulse Accumulator Control Register (PACTL)	Read: / Write:	0	PAEN	PAMOD	PEDGE	CLK1	CLK0	PAOVI	PAI
		Reset:	0	0	0	0	0	0	0	0

= Unimplemented   R = Reserved   U = Unaffected

Notes:
1. Available only on MC68HC912B32 and MC68HC912BC32 devices.
2. Available only on MC68HC912B32 and MC68HC12BE32 devices.
3. Available only on MC68HC(9)12BC32 devices.

Addr.	Register Name		Bit 7	6	5	4	3	2	1	Bit 0
$00A1	Pulse Accumulator Flag Register (PAFLG)	Read:	0	0	0	0	0	0	PAOVF	PAIF
		Write:								
		Reset:	0	0	0	0	0	0	0	0
$00A2	Pulse Accumulator Count Register 3 (PACN3)	Read:	Bit 7	Bit 6	Bit 5	Bit 4	Bit 3	Bit 2	Bit 1	Bit 0
		Write:								
		Reset:	0	0	0	0	0	0	0	0
$00A3	Pulse Accumulator Count Register 2 (PACN2)	Read:	Bit 7	Bit 6	Bit 5	Bit 4	Bit 3	Bit 2	Bit 1	Bit 0
		Write:								
		Reset:	0	0	0	0	0	0	0	0
$00A4	Pulse Accumulator Count Register 1 (PACN1)	Read:	Bit 7	Bit 6	Bit 5	Bit 4	Bit 3	Bit 2	Bit 1	Bit 0
		Write:								
		Reset:	0	0	0	0	0	0	0	0
$00A5	Pulse Accumulator Count Register 0 (PACN0)	Read:	Bit 7	Bit 6	Bit 5	Bit 4	Bit 3	Bit 2	Bit 1	Bit 0
		Write:								
		Reset:	0	0	0	0	0	0	0	0
$00A6	16-Bit Modulus Down-Counter Control Regster (MCCTL)	Read:	MCZI	MODMC	RDMCL	ICLAT	FLMC	MCEN	MCPR1	MCPR0
		Write:								
		Reset:	0	0	0	0	0	0	0	0
$00A7	16-Bit Modulus Down-Counter Flag Regster (MCFLG)	Read:	MCZF	0	0	0	POLF3	POLF2	POLF1	POLF0
		Write:								
		Reset:	0	0	0	0	0	0	0	0
$00A8	Input Control Pulse Accumulators Control Register (ICPACR)	Read:	0	0	0	0	PA3EN	PA2EN	PA1EN	PA0EN
		Write:								
		Reset:	0	0	0	0	0	0	0	0
$00A9	Delay Counter Control Register (DLYCT)	Read:	0	0	0	0	0	0	DLY1	DLY0
		Write:								
		Reset:	0	0	0	0	0	0	0	0

▨ = Unimplemented   R = Reserved   U = Unaffected

Notes:
1. Available only on MC68HC912B32 and MC68HC912BC32 devices.
2. Available only on MC68HC912B32 and MC68HC12BE32 devices.
3. Available only on MC68HC(9)12BC32 devices.

Addr.	Register Name		Bit 7	6	5	4	3	2	1	Bit 0
$00AA	Input Control Overwrite Register (ICOVW)	Read: Write:	NOVW7	NOVW6	NOVW5	NOVW4	NOVW3	NOVW2	NOVW1	NOVW0
		Reset:	0	0	0	0	0	0	0	0
$00AB	Input Control System Control Register (ICSYS)	Read: Write:	SH37	SH26	SH15	HS04	TFMOD	PACMX	BUFEN	LATQ
		Reset:	0	0	0	0	0	0	0	0
$00AC	Reserved		R	R	R	R	R	R	R	R
$00AD	Timer Test Register (TIMTST)	Read:	0	0	0	0	0	0	TCBYP	PCBYP[1]
		Write:								
		Reset:	0	0	0	0	0	0	0	0
$00AE	Timer Port Data Register (PORTT)	Read: Write:	PT7	PT6	PT5	PT4	PT3	PT2	PT1	PT0
		Reset:	0	0	0	0	0	0	0	0
$00AF	Data Direction Register for Timer Port (DDRT)	Read: Write:	DDT7	DDT6	DDT5	DDT4	DDT3	DDT2	DDT1	DDT0
		Reset:	0	0	0	0	0	0	0	0
$00B0	16-Bit Pulse Accumulator B Control Register (PBCTL)	Read: Write:	0	PBEN	0	0	0	0	PBOV	0
		Reset:	0	0	0	0	0	0	0	0
$00B1	Pulse Accumulator B Flag Register (PBFLG)	Read: Write:	0	0	0	0	0	0	PBOV	0
		Reset:	0	0	0	0	0	0	0	0
$00B2	8-Bit Pulse Accumulator Holding Register 3 (PA3H)	Read: Write:	Bit 7	Bit 6	Bit 5	Bit 4	Bit 3	Bit 2	Bit 1	Bit 0
		Reset:	0	0	0	0	0	0	0	0

☐ = Unimplemented   R = Reserved   U = Unaffected

Notes:
1. Available only on MC68HC912B32 and MC68HC912BC32 devices.
2. Available only on MC68HC912B32 and MC68HC12BE32 devices.
3. Available only on MC68HC(9)12BC32 devices.

Addr.	Register Name		Bit 7	6	5	4	3	2	1	Bit 0
$00B3	8-Bit Pulse Accumulator Holding Register 2 (PA2H)	Read: Write:	Bit 7	Bit 6	Bit 5	Bit 4	Bit 3	Bit 2	Bit 1	Bit 0
		Reset:	0	0	0	0	0	0	0	0
$00B4	8-Bit Pulse Accumulator Holding Register 1 (PA1H)	Read: Write:	Bit 7	Bit 6	Bit 5	Bit 4	Bit 3	Bit 2	Bit 1	Bit 0
		Reset:	0	0	0	0	0	0	0	0
$00B5	8-Bit Pulse Accumulator Holding Register 0 (PA0H)	Read: Write:	Bit 7	Bit 6	Bit 5	Bit 4	Bit 3	Bit 2	Bit 1	Bit 0
		Reset:	0	0	0	0	0	0	0	0
$00B6	Modulus Down-Counter Count Register (MCCNT)	Read: Write:	Bit 15	Bit 14	Bit 13	Bit 12	Bit 11	Bit 10	Bit 9	Bit 8
		Reset:	1	1	1	1	1	1	1	1
$00B7	Modulus Down-Counter Count Register (MCCNT)	Read: Write:	Bit 7	Bit 6	Bit 5	Bit 4	Bit 3	Bit 2	Bit 1	Bit 0
		Reset:	1	1	1	1	1	1	1	1
$00B8	Timer Input Capture Holding Register 0 (TC0H)	Read: Write:	Bit 15	Bit 14	Bit 13	Bit 12	Bit 11	Bit 10	Bit 9	Bit 8
		Reset:	0	0	1	0	0	0	0	0
$00B9	Timer Input Capture Holding Register 0 (TC0H)	Read: Write:	Bit 7	Bit 6	Bit 5	Bit 4	Bit 3	Bit 2	Bit 1	Bit 0
		Reset:	0	0	0	0	0	0	0	0
$00BA	Timer Input Capture Holding Register 1 (TC1H)	Read: Write:	Bit 15	Bit 14	Bit 13	Bit 12	Bit 11	Bit 10	Bit 9	Bit 8
		Reset:	0	0	0	0	0	0	0	0
$00BB	Timer Input Capture Holding Register 1 (TC1H)	Read: Write:	Bit 7	Bit 6	Bit 5	Bit 4	Bit 3	Bit 2	Bit 1	Bit 0
		Reset:	0	0	0	0	0	0	0	0

　　　　　= Unimplemented　　R = Reserved　　U = Unaffected

Notes:
1. Available only on MC68HC912B32 and MC68HC912BC32 devices.
2. Available only on MC68HC912B32 and MC68HC12BE32 devices.
3. Available only on MC68HC(9)12BC32 devices.

Addr.	Register Name		Bit 7	6	5	4	3	2	1	Bit 0
$00BC	Timer Input Capture Holding Register 2 (TC2H)	Read: Write:	Bit 15	Bit 14	Bit 13	Bit 12	Bit 11	Bit 10	Bit 9	Bit 8
		Reset:	0	0	0	0	0	0	0	0
$00BD	Timer Input Capture Holding Register 2 (TC2H)	Read: Write:	Bit 7	Bit 6	Bit 5	Bit 4	Bit 3	Bit 2	Bit 1	Bit 0
		Reset:	0	0	0	0	0	0	0	0
$00BE	Timer Input Capture Holding Register 3 (TC3H)	Read: Write:	Bit 15	Bit 14	Bit 13	Bit 12	Bit 11	Bit 10	Bit 9	Bit 8
		Reset:	0	0	0	0	0	0	0	0
$00BF	Timer Input Capture Holding Register 3 (TC3H)	Read: Write:	Bit 7	Bit 6	Bit 5	Bit 4	Bit 3	Bit 2	Bit 1	Bit 0
		Reset:	0	0	0	0	0	0	0	0
$00C0	SCI 0 Baud Rate Control Register High (SC0BDH)	Read: Write:	BTST	BSPL	BRLD	SBR12	SBR11	SBR10	SBR9	SBR8
		Reset:	0	0	0	0	0	0	0	0
$00C1	SCI 0 Baud Rate Control Register Low (SC0BDL)	Read: Write:	SBR7	SBR6	SBR5	SBR4	SBR3	SBR2	SBR1	SBR0
		Reset:	0	0	0	0	0	1	0	0
$00C2	SCI Control Register 1 (SC0CR1)	Read: Write:	LOOPS	WOMS	RSRC	M	WAKE	ILT	PE	PT
		Reset:	0	0	0	0	0	0	0	0
$00C3	SCI Control Register 2 (SC0CR2)	Read: Write:	TIE	TCIE	RIE	ILIE	TE	RE	RWU	SBK
		Reset:	0	0	0	0	0	0	0	0
$00C4	SCI Status Register 1 (SC0SR1)	Read: Write:	TDRE	TC	RDRF	IDLE	OR	NF	FE	PF
		Reset:	1	1	0	0	0	0	0	0

▨ = Unimplemented   R = Reserved   U = Unaffected

Notes:
1. Available only on MC68HC912B32 and MC68HC912BC32 devices.
2. Available only on MC68HC912B32 and MC68HC12BE32 devices.
3. Available only on MC68HC(9)12BC32 devices.

Addr.	Register Name		Bit 7	6	5	4	3	2	1	Bit 0
$00C5	SCI Status Register 2 (SC0SR2)	Read:	0	0	0	0	0	0	0	RAF
		Write:								
		Reset:	0	0	0	0	0	0	0	0
$00C6	SCI Data Register High (SC0DRH)	Read:	R8	T8	0	0	0	0	0	0
		Write:		T8						
		Reset:	U	U	0	0	0	0	0	0
$00C7	SCI Data Register Low (SC0DRL)	Read:	R7T7	R6T6	R5T5	R4T4	R3T3	R2T2	R1T1	R0T0
		Write:								
		Reset:				Unaffected by reset				
$00C8	Reserved		R	R	R	R	R	R	R	R
↓	↓									
$00CF	Reserved		R	R	R	R	R	R	R	R
$00D0	SPI Control Register 1 (SP0CR1)	Read:	SPIE	SPE	SWOM	MSTR	CPOL	CPHA	SSOE	LSBF
		Write:								
		Reset:	0	0	0	0	0	0	0	0
$00D1	SPI Control Register 2 (SP0CR2)	Read:	0	0	0	0	PUPS	RDS	0	SPC0
		Write:					PUPS	RDS		SPC0
		Reset:	0	0	0	0	1	0	0	0
$00D2	SPI Baud Rate Register (SP0BR)	Read:	0	0	0	0	0	SPR2	SPR1	SPR0
		Write:						SPR2	SPR1	SPR0
		Reset:	0	0	0	0	0	0	0	0
$00D3	SPI Status Register (SP0SR)	Read:	SPIF	WCOL	0	MODF	0	0	0	0
		Write:								
		Reset:	0	0	0	0	0	0	0	0
$00D4	Reserved		R	R	R	R	R	R	R	R

　　　　　　　　　　　= Unimplemented　　　R　= Reserved　　　U = Unaffected

Notes:
1. Available only on MC68HC912B32 and MC68HC912BC32 devices.
2. Available only on MC68HC912B32 and MC68HC12BE32 devices.
3. Available only on MC68HC(9)12BC32 devices.

Reprinted with permission of Motorola

Addr.	Register Name		Bit 7	6	5	4	3	2	1	Bit 0
$00D5	SPI Data Register (SP0DR)	Read:	Bit 7	Bit 6	Bit 5	Bit 4	Bit 3	Bit 2	Bit 1	Bit 0
		Write:								
		Reset:	Unaffected by reset							
$00D6	Port S Data Register (PORTS)	Read:	PS7	PS6	PS5	PS4	PS3	PS2	PS1	PS0
		Write:								
		Reset:	After reset all bits configured as general-purpose inputs							
$00D7	Port S Data Direction Register (DDRS)	Read:	DDS7	DDS6	DDS5	DDS4	DDS3	DDS2	DDS1	DDS0
		Write:								
		Reset:	After reset all bits configured as general-purpose inputs							
$00D8 ↓ $00DA	Reserved		R	R	R	R	R	R	R	R
	Reserved		R	R	R	R	R	R	R	R
$00DB	Port S Pullup/Reduced Drive Register (PURDS)	Read:	0	RDPS2	RDPS1	RDPS0	0	PUPS2	PUPS1	PUPS0
		Write:								
		Reset:	0	0	0	0	0	0	0	0
$00DC ↓ $00DF	Reserved		R	R	R	R	R	R	R	R
	Reserved		R	R	R	R	R	R	R	R
$00E0	Slow Mode Divider Register (SLOW)	Read:	0	0	0	0	0	SLDV2	SLDV1	SLDV0
		Write:								
		Reset:	0	0	0	0	0	0	0	0
$00E1 ↓ $00EF	Reserved		R	R	R	R	R	R	R	R
	Reserved		R	R	R	R	R	R	R	R

```
[shaded] = Unimplemented     R = Reserved     U = Unaffected
```

Notes:
1. Available only on MC68HC912B32 and MC68HC912BC32 devices.
2. Available only on MC68HC912B32 and MC68HC12BE32 devices.
3. Available only on MC68HC(9)12BC32 devices.

Addr.	Register Name		Bit 7	6	5	4	3	2	1	Bit 0
$00F0	EEPROM Configuration Register (EEMCR)	Read:	1	1	1	1	1	EESWAI	PROTLCK	EERC
		Write:								
		Reset:	1	1	1	1	1	1	0	0
$00F1	EEPROM Block Protect Register (EEPROT)	Read:	1	1	1	BRPROT4	BRPROT3	BRPROT2	BRPROT1	BRPROT0
		Write:								
		Reset:	1	1	1	1	1	1	1	1
$00F2	EEPROM Test Register (EETST)	Read:	EEODD	EEVEN	MARG	EECPD	EECPRD	0	EECPM	0
		Write:								
		Reset:	0	0	0	0	0	0	0	0
$00F3	EEPROM Control Register (EEPROG)	Read:	BULKP	0	0	BYTE	ROW	ERASE	EELAT	EEPGM
		Write:								
		Reset:	1	0	0	0	0	0	0	0
$00F4	FLASH EEPROM Lock Control Register (FEELCK)[1]	Read:	0	0	0	0	0	0	0	LOCK
		Write:								
		Reset:	0	0	0	0	0	0	0	0
$00F5	FLASH EEPROM Configuration Register (FEEMCR)[1]	Read:	0	0	0	0	0	0	0	BOOTP
		Write:								
		Reset:	0	0	0	0	0	0	0	1
$00F6	FLASH EEPROM Test Register (FEETST)[1]	Read:	FSTE	GADR	HVT	FENLV	FDISVFP	VTCK	STRE	MWPR
		Write:								
		Reset:	0	0	0	0	0	0	0	0
$00F7	FLASH EEPROM Control Register (FEECTL)[1]	Read:	0	0	0	FEESWAI	SVFP	ERAS	LAT	ENPE
		Write:								
		Reset:	0	0	0	0	0	0	0	0
$00F8	BDLC Control Register 1 (BCR1)[2]	Read:	IMSG	CLKS	R1	R0	0	0	IE	WCM
		Write:						R	R		
		Reset:	1	1	1	0	0	0	0	0

☐ = Unimplemented   R = Reserved   U = Unaffected

Notes:
1. Available only on MC68HC912B32 and MC68HC912BC32 devices.
2. Available only on MC68HC912B32 and MC68HC12BE32 devices.
3. Available only on MC68HC(9)12BC32 devices.

Addr.	Register Name		Bit 7	6	5	4	3	2	1	Bit 0
	BDLC State Vector Register (BSVR)[2]	Read:	0	0	I3	I2	I1	I0	0	0
$00F9		Write:								
		Reset:	0	0	0	0	0	0	0	0
	BDLC Control Register 2 (BCR2)[2]	Read:	ALOOP	DLOOP	RX4XE	NBFS	TEOD	TSIFR	TMIFR1	TMIFR0
$00FA		Write:								
		Reset:	1	1	0	0	0	0	0	0
	BDLC Data Register (BDR)[2]	Read:	BD7	BD6	BD5	BD4	BD3	BD2	BD1	BD0
$00FB		Write:								
		Reset:				Indeterminate after reset				
	BDLC Analog Roundtrip Delay Register (BARD)[2]	Read:	ATE	RXPOL	0	0	BO3	BO2	BO1	BO0
$00FC		Write:								
		Reset:	1	1	0	0	0	1	1	1
	Port DLC Control Register (DLCSCR)[2]	Read:	0	0	0	0	0	BDLCEN	PUPDLC	RDPDLC
$00FD		Write:								
		Reset:	0	0	0	0	0	0	0	0
	Port DLC Data Register (PORTDLC)[2]	Read:	0	Bit 6	Bit 5	Bit 4	Bit 3	Bit 2	Bit 1	Bit 0
$00FE		Write:								
		Reset:	0	U	U	U	U	U	U	U
	Port DLC Data Direction Register (DDRDLC)[2]	Read:	0	DDDLC6	DDDLC5	DDDLC4	DDDLC3	DDDLC2	DDDLC1	DDDLC0
$00FF		Write:								
		Reset:	0	0	0	0	0	0	0	0
	msCAN12 Module Control Register 0 (CMCR0)[3]	Read:	0	0	CSWAI	SYNCH	TLNKEN	SLPAK	SLPRQ	SFTRES
$0100		Write:								
		Reset:	0	0	1	0	0	0	0	1
	msCAN12 Module Control Register 1 (CMCR1)[3]	Read:	0	0	0	0	0	LOOPB	WUPM	CLKSRC
$0101		Write:								
		Reset:	0	0	0	0	0	0	0	0

= Unimplemented   R = Reserved   U = Unaffected

Notes:
1. Available only on MC68HC912B32 and MC68HC912BC32 devices.
2. Available only on MC68HC912B32 and MC68HC12BE32 devices.
3. Available only on MC68HC(9)12BC32 devices.

Addr.	Register Name		Bit 7	6	5	4	3	2	1	Bit 0
$0102	msCAN12 Bus Timing Register 0 (CBTR0)[3]	Read: Write:	SJW1	SJW0	BRP5	BRP4	BPR3	BPR2	BPR1	BPR0
		Reset:	0	0	0	0	0	0	0	0
$0103	msCAN12 Bus Timing Register 1 (CBTR1)[3]	Read: Write:	SAMP	TSEG22	TSEG21	TSEG20	TSEG13	TSEG12	TSEG11	TSEG10
		Reset:	0	0	0	0	0	0	0	0
$0104	msCAN12 Receiver Flag Register (CRFLG)[3]	Read: Write:	WUPIF	RWRNIF	TWRNIF	RERRIF	TERRIF	BOFFIF	OVRIF	RXF
		Reset:	0	0	0	0	0	0	0	0
$0105	msCAN12 Receiver Interrupt Enable Register (CRIER)[3]	Read: Write:	WUPIE	RWRNIE	TWRNIE	RERRIE	TERRIE	BOFFIE	OVRIE	RXFIE
		Reset:	0	0	0	0	0	0	0	0
$0106	msCAN12 Transmitter Flag Register (CTFLG)[3]	Read: Write:	0	ABTAK2	ABTAK1	ABTAK0	0	TXE2	TXE1	TXE0
		Reset:	0	0	0	0	0	1	1	1
$0107	msCAN12 Transmitter Control Register (CTCR)[3]	Read: Write:	0	ABTRQ2	ABTRQ1	ABTRQ0	0	TXEIE2	TXEIE1	TXEIE0
		Reset:	0	0	0	0	0	0	0	0
$0108	msCAN12 Identifier Acceptance Control Register (CIDAC)[3]	Read: Write:	0	0	IDAM1	IDAM0	0	IDHIT2	IDHIT1	IDHIT0
		Reset:	0	0	0	0	0	0	0	0
$0109 ↓ $010D	Reserved		R	R	R	R	R	R	R	R
	Reserved		R	R	R	R	R	R	R	R
$010E	msCAN12 Receive Error Counter (CRXERR)[3]	Read: Write:	RXERR7	RXERR6	RXERR5	RXERR4	RXERR3	RXERR2	RXERR1	RXERR0
		Reset:	0	0	0	0	0	0	0	0

= Unimplemented   R = Reserved   U = Unaffected

Notes:
1. Available only on MC68HC912B32 and MC68HC912BC32 devices.
2. Available only on MC68HC912B32 and MC68HC12BE32 devices.
3. Available only on MC68HC(9)12BC32 devices.

Reprinted with permission of Motorola

Addr.	Register Name		Bit 7	6	5	4	3	2	1	Bit 0
$010F	msCAN12 Transmit Error Counter (CTXERR)[3]	Read:	TXERR7	TXERR6	TXERR5	TXERR4	TXERR3	TXERR2	TXERR1	TXERR0
		Write:								
		Reset:	0	0	0	0	0	0	0	0
$0110	msCAN12 Identifier Acceptance Register 0 (CIDAR0)[3]	Read:	AC7	AC6	AC5	AC4	AC3	AC2	AC1	AC0
		Write:								
		Reset:				Unaffected by reset				
$0111	msCAN12 Identifier Acceptance Register 1 (CIDAR1)[3]	Read:	AC7	AC6	AC5	AC4	AC3	AC2	AC1	AC0
		Write:								
		Reset:				Unaffected by reset				
$0112	msCAN12 Identifier Acceptance Register 2 (CIDAR2)[3]	Read:	AC7	AC6	AC5	AC4	AC3	AC2	AC1	AC0
		Write:								
		Reset:				Unaffected by reset				
$0113	msCAN12 Identifier Acceptance Register 3 (CIDAR3)[3]	Read:	AC7	AC6	AC5	AC4	AC3	AC2	AC1	AC0
		Write:								
		Reset:				Unaffected by reset				
$0114	msCAN12 Identifier Mask Register 0 (CIDMR0)[3]	Read:	AM7	AM6	AM5	AM4	AM3	AM2	AM1	AM0
		Write:								
		Reset:				Unaffected by reset				
$0115	msCAN12 Identifier Mask Register 1 (CIDMR1)[3]	Read:	AM7	AM6	AM5	AM4	AM3	AM2	AM1	AM0
		Write:								
		Reset:				Unaffected by reset				
$0116	msCAN12 Identifier Mask Register 2 (CIDMR2)[3]	Read:	AM7	AM6	AM5	AM4	AM3	AM2	AM1	AM0
		Write:								
		Reset:				Unaffected by reset				
$0117	msCAN12 Identifier Mask Register 3 (CIDMR3)[3]	Read:	AM7	AM6	AM5	AM4	AM3	AM2	AM1	AM0
		Write:								
		Reset:				Unaffected by reset				

[shaded] = Unimplemented   R = Reserved   U = Unaffected

Notes:
1. Available only on MC68HC912B32 and MC68HC912BC32 devices.
2. Available only on MC68HC912B32 and MC68HC12BE32 devices.
3. Available only on MC68HC(9)12BC32 devices.

Addr.	Register Name		Bit 7	6	5	4	3	2	1	Bit 0
$0118	msCAN12 Identifier Acceptance Register 4 (CIDAR4)[3]	Read:	AC7	AC6	AC5	AC4	AC3	AC2	AC1	AC0
		Write:								
		Reset:	Unaffected by reset							
$0119	msCAN12 Identifier Acceptance Register 5 (CIDAR5)[3]	Read:	AC7	AC6	AC5	AC4	AC3	AC2	AC1	AC0
		Write:								
		Reset:	Unaffected by reset							
$011A	msCAN12 Identifier Acceptance Register 6 (CIDAR6)[3]	Read:	AC7	AC6	AC5	AC4	AC3	AC2	AC1	AC0
		Write:								
		Reset:	Unaffected by reset							
$011B	msCAN12 Identifier Acceptance Register 7 (CIDAR7)[3]	Read:	AC7	AC6	AC5	AC4	AC3	AC2	AC1	AC0
		Write:								
		Reset:	Unaffected by reset							
$011C	msCAN12 Identifier Mask Register 4 (CIDMR4)[3]	Read:	AM7	AM6	AM5	AM4	AM3	AM2	AM1	AM0
		Write:								
		Reset:	Unaffected by reset							
$011D	msCAN12 Identifier Mask Register 5 (CIDMR5)[3]	Read:	AM7	AM6	AM5	AM4	AM3	AM2	AM1	AM0
		Write:								
		Reset:	Unaffected by reset							
$011E	msCAN12 Identifier Mask Register 6 (CIDMR6)[3]	Read:	AM7	AM6	AM5	AM4	AM3	AM2	AM1	AM0
		Write:								
		Reset:	Unaffected by reset							
$011F	msCAN12 Identifier Mask Register 7 (CIDMR7)[3]	Read:	AM7	AM6	AM5	AM4	AM3	AM2	AM1	AM0
		Write:								
		Reset:	Unaffected by reset							
$0120	Reserved		R	R	R	R	R	R	R	R
↓	↓									
$013C	Reserved		R	R	R	R	R	R	R	R

 = Unimplemented   R = Reserved   U = Unaffected

Notes:
1. Available only on MC68HC912B32 and MC68HC912BC32 devices.
2. Available only on MC68HC912B32 and MC68HC12BE32 devices.
3. Available only on MC68HC(9)12BC32 devices.

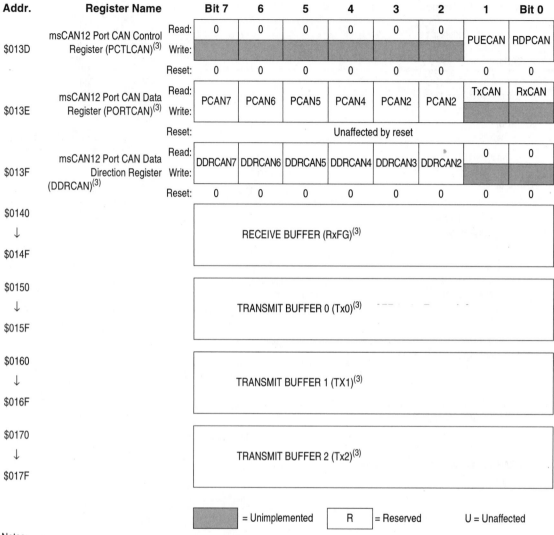

Addr.	Register Name		Bit 7	6	5	4	3	2	1	Bit 0
$013D	msCAN12 Port CAN Control Register (PCTLCAN)[3]	Read:	0	0	0	0	0	0	PUECAN	RDPCAN
		Write:								
		Reset:	0	0	0	0	0	0	0	0
$013E	msCAN12 Port CAN Data Register (PORTCAN)[3]	Read:	PCAN7	PCAN6	PCAN5	PCAN4	PCAN2	PCAN2	TxCAN	RxCAN
		Write:								
		Reset:				Unaffected by reset				
$013F	msCAN12 Port CAN Data Direction Register (DDRCAN)[3]	Read:	DDRCAN7	DDRCAN6	DDRCAN5	DDRCAN4	DDRCAN3	DDRCAN2	0	0
		Write:								
		Reset:	0	0	0	0	0	0	0	0

| $0140 ↓ $014F | RECEIVE BUFFER (RxFG)[3] |

| $0150 ↓ $015F | TRANSMIT BUFFER 0 (Tx0)[3] |

| $0160 ↓ $016F | TRANSMIT BUFFER 1 (TX1)[3] |

| $0170 ↓ $017F | TRANSMIT BUFFER 2 (Tx2)[3] |

 = Unimplemented   R = Reserved   U = Unaffected

Notes:
 1. Available only on MC68HC912B32 and MC68HC912BC32 devices.
 2. Available only on MC68HC912B32 and MC68HC12BE32 devices.
 3. Available only on MC68HC(9)12BC32 devices.

Appendix D

Interrupt Vectors

Vector address	Interrupt source	CCR mask	Local enable	HPRIO value to elevate to highest I bit
$FFFE	Reset	none	none	-
$FFFC	Clock monitor reset	none	COPCTL(CME,FCME)	-
$FFFA	COP failure reset	none	COP rate selected	-
$FFF8	Unimplemented instruction trap	none	none	-
$FFF6	SWI	none	none	-
$FFF4	XIRQ	X bit	none	-
$FFF2	IRQ	I bit	INTCR(IRQEN)	$F2
$FFF0	Real time interrupt	I bit	RTICTL(RTIE)	$F0
$FFEE	Timer channel 0	I bit	TMSK1(C0I)	$EE
$FFEC	Timer channel 1	I bit	TMSK1(C1I)	$EC
$FFEA	Timer channel 2	I bit	TMSK1(C2I)	$EA
$FFE8	Timer channel 3	I bit	TMSK1(C3I)	$E8
$FFE6	Timer channel 4	I bit	TMSK1(C4I)	$E6
$FFE4	Timer channel 5	I bit	TMSK1(C5I)	$E4
$FFE2	Timer channel 6	I bit	TMSK1(C6I)	$E2
$FFE0	Timer channel 7	I bit	TMSK1(C7I)	$E0
$FFDE	Timer overflow	I bit	TMSK2(TOI)	$DE
$FFDC	Pulse accumulator overflow	I bit	PACTL(PAOVI)	$DC
$FFDA	Pulse accumulator input edge	I bit	PACTL(PAI)	$DA
$FFD8	SPI serial transfer complete	I bit	SP0CR1(SPIE)	$D8
$FFD6	SCI0	I bit	SC0CR2(TIE,TCIE,RIE,ILIE)	$D6
$FFD4	SCI1	I bit	SC1CR2(TIE,TCIE,RIE,ILIE)	$D4 (1,3,4)
$FFD2	ATD0 or ATD1	I bit	ATDxCTL2(ASCIE)	$D2
$FFD0	MSCAN 0 wakeup	I bit	C0RIER(WUPIE)	$D0 (1*,2,2*)
$FFCE	Key wakeup J or H	I bit	KWIEJ[7:0] and KWIEH[7:0]	$CE (1,3,4)
$FFCC	Modulus down counter underflow	I bit	MCCTL(MCZI)	$CC
$FFCA	Pulse accumulator B overflow	I bit	PBCTL(PBOVI)	$CA
$FFC8	MSCAN 0 errors	I bit	C0RIER(RWRNIE,TWRNIE, RERRIE,TERRIE, BOFFIE,OVRIE)	$C8 (2*,3,4)
$FFC6	MSCAN 0 receive	I bit	C0RIER(RXFIE)	$C6 (2*,3,4)
$FFC4	MSCAN 0 transmit	I bit	C0TCR(TXEIE[2:0])	$C4 (2*,3,4)
$FFC2	CGK lock and limp home	I bit	PLLCR(LOCKIE, LHIE)	$C2 (3,4)
$FFC0	IIC Bus	I bit	IBCR(IBIE)	$C0 (4)
$FFBE	MSCAN 1 wakeup	I bit	C1RIER(WUPIE)	$BE (4)
$FFBC	MSCAN 1 errors	I bit	C1RIER(RWRNIE,TWRNIE, RERRIE,TERRIE, BOFFIE,OVRIE)	$BC (4)
$FFBA	MSCAN 1 receive	I bit	C1RIER(RXFIE)	$BA (4)
$FFB8	MSCAN 1 transmit	I bit	C1TCR(TXEIE[2:0])	$B8 (4)
$FFB6	Reserved	I bit		$B6
$FF80-$FFB5	Reserved	I bit		$80-$B4

Note: 1. Available in 812 A4
     2. Used as BDLC interrupt vector for 912B32, 912BE32.
     3. Available in D60
     4. Available in DG128 (DT128)

1*. Used as wake up key J in 812A4
2*. Available in 912BC32

Vector number	D-Bug12 version 1.xxx	D-Bug12 version 2.xxx
7	Port H key Wakeup	Port H key Wakeup
8	Port J key Wakeup	Port J key Wakeup
9	Analog-to-Digital converter	Analog-to-Digital converter
10	Serial Communication Interface 1	Serial Communication Interface 1
11	Serial Communication Interface 0	Serial Communication Interface 0
12	Serial Peripheral Interface 0	Serial Peripheral Interface 0
13	Timer Channel 0	Pulse Accumulator Edge
14	Timer Channel 1	Pulse Accumulator Overflow
15	Timer Channel 2	Timer Overflow
16	Timer Channel 3	Timer Channel 7
17	Timer Channel 4	Timer Channel 6
18	Timer Channel 5	Timer Channel 5
19	Timer Channel 6	Timer Channel 4
20	Timer Channel 7	Timer Channel 3
21	Pulse Accumulator Overflow	Timer Channel 2
22	Pulse Accumulator Edge	Timer Channel 1
23	Timer Overflow	Timer Channel 0
24	Real Time Interrupt	Real Time Interrupt
25	$\overline{\text{IRQ}}$ interrupt	$\overline{\text{IRQ}}$ interrupt
26	$\overline{\text{XIRQ}}$ interrupt	$\overline{\text{XIRQ}}$ interrupt
27	SWI instruction	SWI instruction
28	Unimplemented Instruction Trap	Unimplemented Instruction Trap
-1	Return to the starting address of the RAM vector table	Return to the starting address of the RAM vector table

Appendix E

SPI-Compatible Chips

The SPI (acronym of serial peripheral interface) is a protocol defined by Motorola that allows the microprocessor to exchange data with a peripheral chip using the serial format of data transfer. This protocol classes devices into two types: master and slave. There may be multiple slaves in a system. However, only one master device is allowed in a system. The master device initiates the data transfer and also provides the clock signal to synchronize the data transfer. When there are multiple slaves, an enable signal is required to select a slave device for data transfer. The master device is usually the MPU. However, an MPU can also de configured to be a slave device. The following signals are defined in the SPI protocol:

MISO: *master-in-slave-out.* This signal allows a slave to shift data out to the master device.

MOSI: *master-out-slave-in.* This signal allows the master device to shift data out to a slave device.

SCK: *shift clock.* This clock is generated by the master device to synchronize data shifting.

Microcontrollers from vendors other than Motorola may also implement the SPI protocol. For example, some of the PIC microcontrollers from Microchip and some 8051 variants from Atmel have incorporated the SPI interface.

Peripheral chips that may incorporate the SPI interface include: display (LED and LCD) driver chips, A/D and D/A converters, phase-lock loop devices, serial EEPROM, serial SRAM, timer/clock chips, etc. The peripheral chip that implements the SPI protocol is called SPI compatible. The list of SPI-compatible chips is getting longer quickly. A partial list of these devices are in the following:

Vendor	Part Number	Description
Motorola	MC28HC14	256-byte EEPROM in 8-pin DIP
	MC14021	8-bit input port
	MC74HC165	8-bit input port
	MC74LS165	8-bit input port
	MC74HC589	8-bit parallel-in/serial-out shift register
	MC74HC595	8-bit serial-in/parallel-out shift register
	MC74LS673	16-bit output port
	MC144115	16-segment non-muxed LCD driver
	MC144117	4-digit LCD driver
	MC14549	A/D converter successive approximation register
	MC14559	A/D converter successive approximation register
	MC14489	5-digit 7-segment LED display decoder/driver
	MC14499	4-digit 7-segment LED display decoder/driver
	MC145000	serial input multiplexed LCD driver (master)
	MC145001	serial input multiplexed LCD driver (slave)
	MC145453	LCD driver with serial interface
	MC144110	6 six-bit D/A converter
	MC144111	4 six-bit D/A converter
	MC144040	8-bit A/D converter
	MC144051	8-bit A/D converter
	MC145050	10-bit A/D converter
	MC145051	10-bit A/D converter
	MC145053	10-bit A/D converter
	MC68HC68T1	real-time clock
	MCCS1850	serial real-time clock
	MC145155	serial input PLL FS
	MC145156	serial input PLL FS
	MC145157	serial input PLL FS
	MC145158	serial input PLL FS
National Semiconductor	ADC0811	8-bit 11-channel A/D converter
	ADC0819	8-bit 19-channel A/D converter
	ADC0831	8-bit serial I/O A/D converter with mux option
	ADC0832	"
	ADC0834	"
	ADC0838	"
	ADC0833	8-bit 4-channel serial I/O A/D converter
	ADC08031	8-bit high-speed serial I/O A/D converter
	ADC08032	"
	ADC08034	"
	ADC08038	"
	ADC08131	"
	ADC08132	"
	ADC08134	"
	ADC08138	"
	ADC08231	"
	ADC08234	"
	ADC08238	"

Vendor	Part Number	Description
National Semiconductor	ADC0831	8-bit A/D converter with analog mux
	ADC0832	"
	ADC0834	"
	ADC0838	"
	ADC08831	"
	ADC08832	"
	ADC10731	10-bit plus sign A/D converter with mux, sample & hold
	ADC10732	"
	ADC10738	"
	ADC12030	12-bit self-calibrating plus sign A/D converter
	ADC12032	"
	ADC12034	"
	ADC12038	"
	ADC12130	"
	ADC12132	"
	ADC12138	"
	ADC12H030	"
	ADC12H032	12-bit self-calibrating plus sign with mux A/D converter
	ADC12H034	"
	ADC12H038	"
	ADC12L030	"
	ADC12L032	"
	ADC12L038	"
	LM70	SPI 10-bit plus sign digital temperature sensor
	LM74	SPI digital temperature sensor
	LM80	Serial interface ACPI-compatible mp system hardware monitor
	LM81	Serial interface ACPI-compatible mp system hardware monitor
	LM87	Serial interface system hardware monitor with remote diode temperature sensor
Signetics	PCx2100	40-segment LCD duplex driver
	PCx2110	60-segmeng LCD duplex driver
	PCx2111	64-segment LCD duplex driver
	PCx2112	32-segment LCD duplex driver
	PCx3311	DTMF generator with parallel inputs
	PCx3312	DTMF generator with I2C bus inputs
	PCx8570A	256×8 SRAM
	PCx8571	128×8 SRAM
	PCx8573	clock/timer
	PCx8474	8-bit remote I/O expander
	PCx8476	1:4 mux LCD driver
	PCx8477	32/64 segment LCD driver
	PCx8491	8-bit, 4-channel A/D converter and D/A converter
	*SAA1057	PLL tuning circuit: 512 KHz to 120 MHz
	*SAA1060	32-segment LED driver
	*SAA1061	16-segment LED driver
	*SAA1062A	20-segment LED driver
	*SAA1063	fluorescent display driver
	SAA1300	switching circuit

Vendor	Part Number	Description
Signetics	*SAA3019	clock/timer
	PCx2100	40-segment LCD duplex driver
	PCx2110	60-segmeng LCD duplex driver
	PCx2111	64-segment LCD duplex driver
	PCx2112	32-segment LCD duplex driver
	PCx3311	DTMF generator with parallel inputs
	PCx3312	DTMF generator with I2C bus inputs
	PCx8570A	256×8 SRAM
	PCx8571	128×8 SRAM
	PCx8573	clock/timer
	PCx8474	8-bit remote I/O expander
	PCx8476	1:4 mux LCD driver
	PCx8477	32/64 segment LCD driver
	PCx8491	8-bit, 4-channel A/D converter and D/A converter
	*SAA1057	PLL tuning circuit: 512 KHz to 120 MHz
	*SAA1060	32-segment LED driver
	*SAA1061	16-segment LED driver
	*SAA1062A	20-segment LED driver
	*SAA1063	fluorescent display driver
	SAA1300	switching circuit
	*SAA3019	clock/timer
	SAA3028	I/R ranscoder
	SAB3013	hex 6-bit D/A converter
	SAB3035	tuning circuit with 8 D/A converters
	SAB3036	PLL digital tuning circuit
	SAB3037	PLL digital tuning circuit with 4 D/A converters
	SAA5240	teletext controller chip—625 line system
	TDA3820	digital stereo sound control IC
	TEA6000	MUSTI: FM/RF system
	TDA1534A	14-bit D/A converter—serial output
	TDA1540P,D	14-bit D/A converter—serial output
	NE5036	6-bit A/D converter—serial output
RCA	CDP68HC68A1	10-bit A/D converter
	CDP68HC68R1	128×8-bit SRAM
	CDP68HC68R2	256×8-bit SRAM
	CDP68HC68T1	real-time clock
XICOR	X5001	CPU supervisor with 0K memory, single voltage monitor
	X5043	CPU supervisor with 4K memory, single voltage monitor
	X5045	CPU supervisor with 4K memory, single voltage monitor
	X5083	CPU supervisor with 8K memory, single voltage monitor
	X5163	CPU supervisor with 16K memory, single voltage monitor
	X5165	CPU supervisor with 16K memory, single voltage monitor
	X51638	CPU supervisor with 16K memory, single voltage monitor
	X5168	CPU supervisor with 16K memory, single voltage monitor
	X5169	CPU supervisor with 16K memory, single voltage monitor
	X5323	CPU supervisor with 32K memory, single voltage monitor
	X5325	CPU supervisor with 32K memory, single voltage monitor

Vendor	Part Number	Description
XICOR	X5328	CPU supervisor with 32K memory, single voltage monitor
	X5329	CPU supervisor with 32K memory, single voltage monitor
	X5643	CPU supervisor with 64K memory, single voltage monitor
	X5645	CPU supervisor with 644K memory, single voltage monitor
	X5648	CPU supervisor with 64K memory, single voltage monitor
	X5649	CPU supervisor with 64K memory, single voltage monitor
	X5563	CPU supervisor with 256K memory, single voltage monitor
	X55020	CPU supervisor with 4K memory, dual voltage monitor
	X55040	CPU supervisor with 16K memory, dual voltage monitor
	X55060	CPU supervisor with 64K memory, dual voltage monitor
	X55620	CPU supervisor with 256K memory, dual voltage monitor
	X55621	CPU supervisor with 256K memory, dual voltage monitor
ATMEL	AT93C46/56/66	1/2/4 K, 3-wire bus serial EEPROM
	AT93C46A	1K, 3-wire bus serial EEPROM ($\times$ 16 organization)
	AT93C46C	1K, 3-wire bus serial EEPROM ($\times$ 16 organization)
	AT93C86	16K, 3-wire bus serial EEPROM ($\times$ 8 or $\times$ 16 organization)
	AT25010/020/040	1/2/4K SPI bus serial EEPROM
	AT25080/160/320/640	8/16/32/64K SPI bus serial EEPROM
	AT25P1024	1M, SPI bus serial EEPROM
	AT25128/256	256K SPI bus serial EEPOM
	AT25HP256/512	256/512K-bit SPI bus serial EEPROM
	AT25F512/1024	1 M-bit SPI bus serial flash
MAXIM	MAX1080, MAX1081	10-bit, 300ksps/400ksps, 8-channel, ADC with internal reference
	MAX1082, MAX1083	10-bit, 300ksps/400ksps, 4-channel, ADC with internal reference
	MAX1084, MAX1085	10-bit, 300ksps/400ksps, low-power, ADC with internal reference
	MAX1098, MAX1099	10-bit serial-output temperature sensor with five channel ADC
	MAX110, MAX111	$\pm$14-bit, low-cost, 2-channel, serial ADCs
	MAX1106, MAX1107	8-bit, single-supply, low-power serial ADCs
	MAX1108, MAX1109	8-bit single-supply, low-power, serial 2-channel, ADCs
	MAX1110, MAX1111	8-bit, +2.7V, low-power, multichannel, serial ADCs
	MAX1112, MAX1113	8-bit, +5V, low-power, multichannel, serial ADCs
	MAX1115, MAX1116	8-bit, single-supply, low-power, serial ADCs
	MAX1117, MAX1118	8-bit, single-supply, low-power, 2-channel, serial ADCs
	MAX1119	"
	MAX1204	10-bit, 5V, 8-channel, serial ADC
	MAX1240, MAX1241	12-bit, +2.7V, low-power, serial ADCs in 8-pin SO
	MAX1242, MAX1243	10-bit, +2.7V to 5.5V, low-power, serial ADCs
	MAX1245	12-bit, +2.375V, low-power, 8-channel, serial ADC
	MAX1246, MAX1247	12-bit, +2.7V, low-power, 4-channel, serial ADCs in QSOP
	MAX1270, MAX1271	12-bit, multirange, +5V, 8-channel, serial ADCs
	MAX1282, MAX1283	12-bit, 300ksps/400ksps, single supply, 4-channel, serial ADC
	MAX1284, MAX1285	12-bit, 300ksps/400ksps, single supply, low-power,4-channel, serial ADC, with internal reference
	MAX1298, MAX1299	12-bit serial-output temperature sensors with 5-channel ADC
	MAX144, MAX145	12-bit, +2.7V, low-power, 2-channel, 108ksps, serial ADCs
	MAX146, MAX147	12-bit, +2.7V, low-power, 8-channel, serial ADCs
	MAX148, MAX149	10-bit, +2.7V to 5.5V, low-power, 8-channel serial ADCs

Vendor	Part Number	Description
MAXIM	MAX157, MAX159	10-bit, +2.7V, low-power, 2-channel, 108ksps, serial ADCs
	MAX170	12-bit, serial output, 5.6 ms, ADC with internal reference
	MAX176	12-bit, serial output, 250 ksps, ADC with internal reference
	MAX186, MAX188	12-bit, low-power, 8-channel, serial ADCs
	MAX187, MAX189	12-bit, +5V, low-power, serial ADCs
	MAX192	10-bit, low-power, 8-channel, serial ADCs
	MAX509, MAX510	8-bit, quad, serial, DACs with rail-to-rail voltage outputs
	MAX5104	12-bit, dual, voltage-output, DAC with serial interface
	MAX512, MAX513	8-bit, tripple, voltage output DACs with serial interface
	MAX5120, MAX5121	12-bit, +3V/+5V, serial, voltage output DAC
	MAX5122, MAX5123	12-bit, +3V/+5V, serial, force-sense DACs
	MAX5130, MAX5131	13-bit, +3V/+5V, serial, voltage output DACs with internal ref.
	MAX5132, MAX5133	13-bit, +3V/+5V, serial, force-sense DACs with internal ref.
	MAX5141..MAX5144	14-bit, +3V/+5V, serial, voltage output DACs
	MAX5150, MAX5151	13-bit, +3V/+5V, dual, serial, voltage output DACs
	MAX5154, MAX5155	12-bit, +3V/+5V, dual, serial, voltage output DACs
	MAX5158, MAX5159	10-bit, +3V/+5V, dual, serial, voltage output DACs
	MAX5170..MAX5172	14-bit, low-power, serial, DACs with voltage output
	MAX5171, MAX5173	14-bit, low-power, serial, DACs with force-sense voltage output
	MAX5174, MAX5176	12-bit, low-power, serial, DACs with voltage output
	MAX5175, MAX5177	12-bit, low-power, serial, DACs with force-sense voltage output
	MAX522	8-bit, dual, voltage-ouptut, serial DAC in 8-pin SO package
	MAX5222	8-bit, dual, voltage-output, serial DAC in 8-pin SOT23
	MAX5223	8-bit, dual, low-power, voltage output serial DAC
	MAX525	12-bit, low-power, quad, serial voltage-output DAC
	MAX5250	10-bit, low-power, quad, serial voltage-output DAC
	MAX5251	10-bit, +3V, quad, serial voltage-output DAC
	MAX5253	12-bit, +3V, quad, serial voltage-output DAC
	MAX5302	12-bit, low-power, quad, serial voltage-output DAC
	MAX531, MAX538	12-bit, +5V, low-power, voltage-output, serial DAC
	MAX539	12-bit, +5V, low-power, voltage-output, serial DAC
	MAX532	12-bit, dual, serial input, voltage output, multiplying, DAC
	MAX535, MAX5351	13-bit, low-power, voltage-output DACs with serial interface
	MAX5352, MAX5353	12-bit, low-power, voltage output DACs with serial interface
	MAX536, MAX537	12-bit, quad, voltage output DACs with serial interface
	MAX5363..MAX5365	6-bit DACs with 3-wire serial interface in SOT23 package
	MAX5383..MAX5385	8-bit DACs with 3-wire serial interface in SOT23 package
	MAX541, MAX542	16-bit, +5V, serial-input, voltage-output, DACs
	MAX5441..MAX5444	16-bit, +3V/+5V, serial input, voltage-output DACs
	MAX5541	16-bit, +5V, serial input, voltage-output, DAC
	MAX5544	14-bit, +5V, serial-input, voltage-output, DAC
	MAX5711	10-bit, serial, voltage-output, rail-to-rail, DAC
	MAX5712	12-bit, serial, voltage-output, rail-to-rail, DAC
	MAX5721	10-bit, dual, serial, voltage-output, rail-to-rail, DAC
	MAX5722	12-bit, dual, serial, voltage-output, rail-to-rail, DAC
	MAX5741	10-bit, quad, serial, voltage-output, rail-to-rail, DAC
	MAX5742	12-bit, quad, serial, voltage-output, rail-to-rail, DAC

Vendor	Part Number	Description
MAXIM	MAX7543	12-bit, CMOS, serial-input DAC
	MAX132	±18-bit serial DAC
	MAX7219, MAX7221	8-digit, seven-segment display driver with serial interface
	DS1305, DS1306	serial alarm real-time clock (RTC)
	DS1307	64×8 SRAM, serial alarm real-time clock (RTC)
	DS1337	serial real-time clock
	MAX6902	SPI-compatible serial real-time clock
	DS1722	digital thermometerwith SPI interface
	MAX1098, MAX1099	10-bit serial-output temperature sensor with 5-channel ADC
	MAX6629..MAX6632	12-bit + sign digital temperature sensors with serial interface
	MAX6662	12-bit + sign digital temperature sensors with SPI interface
Analog Devices	AD7898	5V, 12-bit serial 220ksps ADC
	AD7827	3/5V, 1 Msps, 8-bit, serial ADC
	AD7812	10-bit, 8-channel, 350 Ksps, serial ADC
	AD7811	10-bit, 4-channel, 350 Ksps, serial ADC
	AD977	16-bit, 100 Ksps, serial ADC
	AD977A	16-bit, 200 Ksps, serial ADC
	AD7851	14-bit, 333 Ksps, serial ADC
	AD7895	12-bit, true bipolar input, 5V supply, serial ADC
	AD7858	12-bit, 8-channel, 3 to 5V supply, 200 Ksps, serial ADC
	AD7853	12-bit, 3 to 5V supply, 200 Ksps, serial ADC
	AD7890	12-bit, true bipolar, serial ADC
	AD7893	12-bit, true bipolar, 8 ms, serial ADC
	AD677	16-bit, 100 Ksps, serial ADC
	AD7872	14-bit, CMOS, serial ADC
	AD7943	12-bit 3.3V/5.0V, multiplying serial DAC
	DAC8043	12-bit serial input multiplying CMOS DAC
	AD5531	14-bit, serial input DAC
	ADT7316	12-bit, SPI interface digital temperature sensor, quad output
	ADT7317	10-bit, SPI interface, digital temperature sensor, quad output
	ADT7318	8-bit, SPI interface, digital temperature sensor, quad output
	AD5532HS	14-bit, 32-channel, serial DAC
	AD7399	10-bit, quad, serial input DAC
	AD5551	14-bit, 5V, serial input DAC
	AD7394	12-bit, dual, +3V/+5V, serial input DAC
	AD7395	10-bit, dual, +3V/+5V, serial input DAC
	AD7303	8-bit, dual, +2.7V to 5.5V, serial input DAC
	AD7390	12-bit, +3V/+5V, serial input DAC
	AD7391	10-bit, +3V/+5V, serial input DAC

Appendix F

68HC12 Development Tool Vendor

Company	Products	Contact Address
Software tools		
Avocet	macro assembler, BDM debugger	www.hmi.com
COSMIC Software Inc.	IDE, C compiler, source-level debugger, assembler	www.cosmic-software.com
Hiware Inc. (acquired by Metrowerk)	C compiler, IDE, assembler, source-level debugger	www.hiware.com
IAR System	C compiler, assembler, IDE, source-level debugger, simulator	www.iar.com
Imagecraft	C compiler, IDE	www.imagecraft.com
P&E	assembler, IDE	www.pemicro.com
Hardware tools		
Axiom Manufacturing	demo boards, BDM-12 kit (with software)	www.axman.com
Electronik Laden	demo board, BDM kit	www.elektronikladen.de/en_starprog.html
Motorola	demo boards	www.mot.com
Nohau	in-circuit emulator	www.nohau.com
Techi Inc.	in-circuit emulator	www.tech-I.com
Technological Arts	demo boards, BDM pod	www.technologicalarts.com

Glossary

Accumulator A register in a computer that contains an operand to be used in an arithmetic operation.

Acknowledgement Error In CAN bus protocol, an acknowledgement error is detected whenever the transmitter does not monitor a dominant bit in the ACK slot.

Activation record Another term for stack frame.

Address access time The amount of time it takes for a memory component to send out valid data to the external data pins after address signals have been applied (assuming that all other control signals have been asserted).

Addressing The application of a unique combination of high and low logic levels to select a corresponding unique memory location.

Address multiplexing A technique that allows the same address pin to carry different signals at different time; used mainly by DRAM technology. Address multiplexing can dramatically reduce the number of address pins required by DRAM chips and reduce the size of the memory chip package.

Algorithm A set of procedure steps represented in pseudo code that is designed to solve certain computation issue.

ALU (arithmetic logic unit) The part of the processor in which all arithmetic and logical operations are performed.

Array An ordered set of elements of the same type. The elements of the array are arranged so that there is a zeroth, first, second, third, and so forth. An array may be one-, two-, or multi-dimensional.

ASCII (American Standard Code for Information Interchange) code. A code that uses seven bits to encode all printable and control characters.

Assembler A program that converts a program in assembly language into machine instructions so that it can be executed by a computer.

Assembler directive A command to the assembler for defining data and symbols, setting assembler conditions, and specifying output format. Assembler directives do not produce machine code.

Assembly instruction A mnemonic representation of a machine instruction.

Assembly program A program written in assembly language.

Automatic variable A variable defined inside a function that comes into existence when the function is entered and disappears when the function returns.

Background receive buffer The receive buffer where the incoming frame is placed before it is forwarded to the foreground receive buffer.

Barometric pressure The air pressure existing at any point within the earth's atmosphere.

BDM mode A mode provided by the 68HC12 and some other Motorola microcontrollers to facilitate the design of debugging tools. This mode has been utilized to implement source-level debugger.

Binary Coded Decimal (BCD) A coding method that uses four binary digits to represent one decimal digit. The binary codes 00002-10012 correspond to the decimal digits 0 to 9.

Bit error   In the CAN bus protocol, a node that is sending a bit on the bus also monitors the bus. When the bit value monitored is different from the bit value being sent, the node interprets the situation as a bit error.

Branch instruction   An instruction that causes the program flow to change.

Break   The transmission or reception of a low for at least one complete character time.

Breakpoint   A memory location in a program where the user program execution will be stopped and the monitor program will take over the CPU control and display the contents of CPU registers.

Bubble sort   A simple sorting method in which an array or a file to be sorted is gone through sequentially several times. Each iteration consists of comparing each element in the array or file with its successor (x[i] with x[i+1]) and interchanging the two elements if they are not in proper order (either ascending or descending)

Bus   A set of signal lines through which the processor of a computer communicates with memory and I/O devices.

Bus cycle timing diagram   A diagram that describes the transitions of all the involved signals during a read or write operation.

Bus multiplexing   A technique that allows more than one set of signals to share the same group of bus lines.

Bus off   The situation that a CAN node having transmit error count above 256.

CAN transceiver   A chip used to interface a CAN controller to the CAN bus.

Controller Area Network (CAN)   A serial communication protocol initially proposed to be used in automotive applications. In this protocol, data is transferred frame by frame. Each frame can carry up to 8 bytes of data.

Central processing unit (CPU)   The combination of the register file, the ALU, and the control unit.

Charge pump   A circuit technique that can raise a low voltage to a level above the power supply. A charge pump is often used in A/D converter, in EEPROM and EPROM programming, etc.

Clock monitor reset   A mechanism provided to detect whether the frequency of the system clock signal inside a CPU is lower than certain value.

Column address strobe (CAS)   The signal used by DRAM chips to indicate that column address logic levels are applied to the address input pins.

Comment   A statement that explains the function of a single instruction or directive or a group of instructions or directives. Comments make a program more readable.

Communication program   A program that allows a PC to communicate with another computer.

Computer   A computer consists of hardware and software. The hardware includes four major parts: the central processing unit, the memory unit, the input unit, and the output unit. Software is a sequence of instructions that control the operations of the hardware.

Computer operate properly (COP) watchdog timer   A special timer circuit designed to detect software processing errors. If software is written correctly, then it should complete all operations within certain amount of time. Software problems can be detected by enabling a watchdog timer so that the software resets the watchdog timer before it times out.

Contact bounce   A phenomenon in which a mechanical key switch will go up and down several times before it settles down when it is pressed.

Control unit   The part of the processor that decodes and monitors the execution of instructions. It arbitrates the use of computer resources and makes sure that all computer operations are performed in proper order.

CRC error   Cyclic redundancy check error. In data communications, the CRC sequence consists of the result of the CRC calculation by the transmitter. The receiver calculates the CRC in the same way as the transmitter. A CRC error is detected if the calculated result is not the same as that received in the CRC sequence.

Cross assembler   An assembler that runs on one computer but generates machine instructions that will be executed by another computer that has a different instruction set.

Cross compiler   A compiler that runs on one computer but generates machine instructions that will be executed by another computer that has a different instruction set.

D/A converter   A circuit that can convert a digital value into an analog voltage.

Data hold time   The length of time over which the data must remain stable after the edge of the control signal that latches the data.

Datapath   The part of processor that consists of a register file and the ALU.

Data setup time   The amount of time over which the data must become valid before the edge of the control signal that latches the data.

D-Bug12 Monitor   A monitor program designed for the 68HC12 microcontroller.

DCE   The acronym of data communication equipment. DCE usually refers to equipment such as a modem, concentrator, router, etc.

Demo board   A single board computer that contains the target microcontroller as the CPU and a monitor to help the user to perform embedded product development.

Direct mode   An addressing mode that uses an 8-bit value to represent the address of a memory location.

Dominant level   A voltage level in a CAN bus that will prevail when a voltage level at this state and a different level (recessive level) are applied to the CAN bus at the same time.

DTE   The acronym for data terminal equipment. DTE usually refers to a computer or terminal.

Dynamic memories Memory devices that require periodic refreshing of the stored information, even when power is on.

EBCDIC (Extended Binary Coded Decimal Interchange Code)   A code used mainly in IBM mainframe computers; it uses eight bits to represent each character.

EIA   The acronym of electronic industry association.

Electrically erasable programmable read-only memory (EEPROM)   A type of read-only memory that can be erased and reprogrammed using electrical signals. EEPROM allows each individual location inside the chip to be erased and reprogrammed.

Embedded system   A product that uses a microcontroller as the controller to provide the features. End users are interested in these features rather than the power of the microcontroller. A cell phone, a charge card, and a home security system are examples of the embedded system.

Erasable programmable read only memory (EPROM)   A type of read-only memory that can be erased by subjecting it to strong ultraviolet light. It can be reprogrammed using an EPROM programmer. A quartz window on top of the EPROM chip allows light to be shone directly on the silicon chip inside.

Error active   A CAN node that has both transmit error count and receive error count lower than 127.

Error passive   A CAN node that has either transmit error count or receive error count between 128 and 256.

Exception   A software interrupt such as an illegal opcode, an overflow, division by zero, or an underflow.

Expanded mode   The operation mode in which the 68HC12 can access external memory components by sending out address signals. A 64 KB memory space is available in this mode.

Extended mode   An addressing mode that uses a 16-bit value to represent the address of a memory location.

Fall time   The amount of time a digital signal takes to go from logic high to logic low.

Floating signal   An undriven signal.

Foreground receive buffer   The receive buffer that is accessible to the programmer.

Form error An error detected when a fixed-form bit field contains one or more illegal bits (in CAN bus protocol).

Framing error   A data communication error in which a received character is not properly framed by the start and stop bits.

Frame pointer   A pointer used to facilitate access to parameters in a stack frame.

Framing error   A data communication error in which a received character is not properly framed by the start and stop bits.

Full-duplex link   A four-wire communication link that allows both transmission and reception to proceed simultaneously.

Global memory   Memory that is available to all programs in a computer system.

Half duplex link   A communication link that can be used for either transmission or reception, but only in one direction at a time.

Hard synchronization   The synchronization performed by all CAN nodes at the beginning of a frame.

Hardware breakpoint   A hardware circuit that compares actual address and data values to predetermined data in setup registers. A successful comparison places the CPU in background debug mode or initiates a software interrupt (SWI).

Identifier acceptance filter   A group of registers that can be programmed to select those identifier bits of the incoming frames to be compared for acceptance.

Idle   A continuous logic high on the RxD line for one complete character time.

Immediate mode   An addressing mode that will be used as the operand of the instruction.

Illegal opcode   A binary bit pattern of the opcode byte for which an operation is not defined.

Indexed addressing mode   An addressing mode that uses the sum of the contents of an index register and a value contained in the instruction to specify the address of a memory location. The value to be added to the index register can be a 5-bit, 9-bit, or 16-bit signed value. The contents of an accumulator (A, B, or D) can also be used as the value to be added to the index register to compute the address.

Indexable data structure   A data structure in which each element is associated with an integer that can be used to access it. Arrays and matrices are examples of indexable data structures.

Inline assembly instruction   Assembly instructions that are embedded in a high-level language program.

Input capture   The 68HC12 function that captures the value of the 16-bit free-running main timer into a latch when the falling or rising edge of the signal connected to the input capture pin arrives.

Input handshake A protocol that uses two handshake signals to make sure that peripheral chip receives data correctly from the input device.

Input port The part of the microcontroller that consists of input pins, input data register, and other control circuitry to perform input function.

Instruction queue A circuit in the CPU to hold the instruction bytes prefetched by the CPU. The 68HC12 and some other microprocessors utilize the bus idle time to perform instruction prefetch in the hope to enhance the processor throughput.

Instruction tagging A feature that provides a way of forcing the 68HC12 to stop executing an application program when the tagged instruction reaches the CPU. When the tagged instruction is detected, the 68HC12 CPU enters background debug mode rather than executing the tagged instruction.

Integrated development environment A piece of software that combines a text editor, a terminal program, a cross compiler and/or cross assembler, and/or simulator that allows the user to perform program development activities without quitting any one of the programs.

Interframe space A field in the CAN bus that is used to separate data frames or remote frames from the previous frames.

Interrupt An unusual event that requires the CPU to stop normal program execution and perform some service to the event.

Interrupt overhead The time spent on handling an interrupt. This time consists of the saving and restoring of registers and the execution of instructions contained in the service routine.

Interrupt priority The order in which the CPU will service interrupts when all of them occur at the same time.

Interrupt service The service provided to a pending interrupt by CPU execution of a program called a service routine.

Interrupt vector The starting address of an interrupt service routine.

Interrupt vector table A table that stores all interrupt vectors.

I/O synchronization A mechanism that can make sure that CPU and I/O devices exchange data correctly.

ISO The acronym for the international standard organization.

Keyboard debouncing A process that can eliminate the key-switch bouncing problem so that the computer can detect correctly whether a key has indeed been pressed.

Keyboard scanning A process that is performed to detect whether any key has been pressed by the user.

Key wakeup A mechanism that can generate interrupt requests to wake up a sleeping CPU. The key wakeup ability is associated with I/O ports.

Label field The field in an assembly program statement that represents a memory location.

Linked list A data structure that consists of linked nodes. Each node consists of two fields, an information field and a next address field. The information field holds the actual element on the list, and the next address field contains the address of the next node in the list.

Load cell A transducer that can convert weight into a voltage.

Local variable Temporary variables that exist only when a subroutine is called. They are used as loop indices, working buffers, etc. Local variables are often allocated in the system stack.

Low power mode An operation mode in which less power is consumed. In CMOS technology, the low-power mode is implemented by either slowing down the clock frequency or turning off some circuit modules within a chip.

Machine instruction A set of binary digits that tells the computer what operation to perform.

Mark A term used to indicate a binary 1.

Maskable interrupts Interrupts that can be ignored by the CPU. This type of interrupt can be disabled by setting a mask bit or by clearing an enable bit.

Masked ROM (MROM) A type of ROM that is programmed when it is fabricated.

Matrix A two-dimensional data structure that is organized into rows and columns. The elements of a matrix are of the same length and are accessed using their row and column numbers (i, j), where i is the row number and j is the column number.

Memory Storage for software and information.

Memory capacity The total amount of information that a memory device can store; also called memory density.

Memory organization A description of the number of bits that can be read from or written into a memory chip during a read or write operation.

Microprocessor A CPU packaged in a single integrated circuit.

Microcontroller A computer system implemented on a single, very large-scale integrated circuit. A microcontroller contains everything that is in a microprocessor and may contain memories, an I/O device interface, a timer circuit, an A/D converter, and so on.

Mode fault An SPI error that indicates that there may have been a multimaster conflict for system control. Mode fault is detected when the master SPI device has its SS pin pulled low.

Modem A device that can accept digital bits and change them into a form suitable for analog transmission (modulation) and can also receive a modulated signal and transform it back to its original digital representation (demodulation).

Multi-drop A data communication scheme in which more than two stations share the same data link. One station is designated as the master, and the other stations are designated as slaves. Each station has its own unique address, with the primary station controlling all data transfers over the link.

Multitasking A computing technique in which CPU time is divided into slots that are usually 10 to 20 ms in length. When multiple programs are resident in the main memory waiting for execution, the operating system assigns a program to be executed to one time slot. At the end of a time slot or when a program is waiting for completion of I/O, the operating system takes over and assigns another program to be executed.

Multiprecision arithmetic Arithmetic (add, subtract, multiply, or divide) performed by a computer that deals with operands longer than the computer's word length.

Nibble A group of four-bit information.

Nonmaskable interrupts Interrupts that the CPU cannot ignore.

Nonvolatile memory Memory that retains stored information even when power to the memory is removed.

Null modem A circuit connection between two DTEs in which the leads are interconnected in such a way as to fool both DTEs into thinking that they are connected to modems. A null modem is only used for short distance interconnections.

Object code The sequence of machine instructions that results from the process of assembling and/or compiling a source program.

Output-compare A 68HC12 timer function that allows the user to make a copy of the value of the 16-bit main timer, add a delay to the copy, and then store the sum in a register. The output-compare function compares the sum with the main timer in each of the following E clock cycles. When these two values are equal, the circuit can trigger a signal change on an output-compare pin and may also generate an interrupt request to the 68HC12.

Output handshake A protocol that uses two handshake signals to make sure that output device correctly receives the data driven by the peripheral chip (sent by the CPU).

Output port The part of the circuit in a microcontroller that consists of output pins, data register, and control circuitry to send data to output device.

Overflow A condition that occurs when the result of an arithmetic operation cannot be accommodated by the preset number of bits (say, 8 or 16 bits); it occurs fairly often when numbers are represented by fixed numbers of bits.

Parameter passing The process and mechanism of sending parameters from a caller to a subroutine, where they are used in computations; parameters can be sent to a subroutine using CPU registers, the stack, or global memory.

Parity error An error in which odd number of bits change value; it can be detected by a parity checking circuit.

Phase_seg1 & Phase_seg2 Segments that are used to compensate for edge phase errors. These segments can be lengthened or shortened by synchronization.

Physical layer The lowest layer in the layered network architecture. This layer deals with how signals are actually transmitted, the descriptions of bit timing, bit encoding, and synchronization.

Physical time In the 68HC12 timer system, the time represented by the count in the 16-bit main timer counter.

Point-to-point A data communication scheme in which there are two stations communicate as peers.

Precedence of operators The order in which operators are processed.

Program A set of instructions that the computer hardware can execute.

Program counter (PC) A register that keeps track of the address of the next instruction to be executed.

Program loops A group of instructions or statements that are executed by the processor more than once.

PROM (programmable read-only memory) A type of ROM that allows the end use to program it once and only once using a device called PROM programmer.

Prop_seg The segment within a bit time used to compensate for the physical delay times within the CAN network.

Pseudo code An expressive method that combines the use of plain English and statements similar to certain programming languages to represents an algorithm.

Pull The operation that removes the top element from a stack data structure.

Pulse accumulator A timer function that uses a counter to count the number of events that occur or measure the duration of a single pulse.

Pulse width modulation A timer function that allows the user to specify the frequency and duty cycle of the digital waveform to be generated.

Push The operation that adds a new element to the top of a stack data structure.

Queue A data structure to which elements can be added at only one end and removed only from the other end. The end to which new elements can be added is called the tail of the queue, and the end from which elements can be removed is called the head of the queue.

RAM (random-access memory)   RAM allows read and write access to every location inside the memory chip. Furthermore, read access and write access take the same amount of time for any location within the RAM chip.

Receiver overrun   A data communication error in which a character or a number of characters were received but not read from the buffer before subsequent characters being received.

Refresh   An operation performed on dynamic memories in order to retain the stored information during normal operation.

Refresh period   The time interval within which each location of a DRAM chip must be refreshed at least once in order to retain its stored information.

Register   A storage location in the CPU. It is used to hold data and/or a memory address during the execution of an instruction.

Register following map   The 512 bytes of space immediately following the register block of the hc12 microcontroller.

Relative mode   An addressing mode that uses an 8-bit value to specify the branch distance for branch instructions. If the sign of the value is negative then the branch is a backward branch. Otherwise, the branch is a forward branch.

Remote frame   A frame sent out by a CAN node to request another node to send data frames.

Reset   A signal or operation that sets the flip-flops and registers of a chip or microprocessor to some predefined values or states so that the circuit or microprocessor can start from a known state.

Reset handling routine   The routine that will be executed when the microcontroller or microprocessor gets out of the reset state.

Reset state   The state in which the voltage level of the RESET pin of the 68HC12 is low. In this state, a default value is established for most on-chip registers, including the program counter. The operation mode is established when the 68HC12 exits the reset state.

Resynchronization   All CAN nodes perform resynchronization within a frame whenever a change of bit value from recessive to dominant occurs outside of the expected sync_seg segment after the hard synchronization.

Resynchronization jump width The amount of lengthening in phase_seg1 or shortening in phase_seg2 in order to achieve resynchronization in every bit within a frame. The resynchronization jump width is programmable to between 1 and 4.

Return address   The address of the instruction that immediately follows the subroutine call instruction (either JSR or BSR).

Rise time   The amount of time a digital signal takes to go from logic low to logic high.

ROM (read-only memory)   A type of memory that is nonvolatile in the sense that when power is removed from ROM and then reapplied, the original data are still there. ROM data can only be read—not written—during normal computer operation.

Row address strobe (RAS)   The signal used by DRAM chips to indicate that row address logic levels are applied to the address input pins.

RS232   An interface standard recommended for interfacing between a computer and a modem. This standard was established by EIA in 1960 and has since then been revised several times.

Signal conditioning circuit   A circuit added to the output of a transducer to scale and shift the voltage output from the transducer to a range that can take advantage of the whole dynamic range of the A/D converter being used.

Simulator A program that allows the user to execute microcontroller programs without having the actual hardware.

Simplex link A line is dedicated either for transmission or reception, but not both.

Single-chip mode The operation mode in which the 68HC12 functions without external address and data buses.

Source code A program written in either assembly language or a high-level language; also called a source program.

Source-level debugger A program that allows the user to find problems in user code at the high-level language (such as C) or assembly language level.

Space A term space is used to indicate a binary 0.

Special test mode The 68HC12 operation mode used primarily during Motorola's internal production testing.

SPI Serial Peripheral Interface. A protocol proposed by Motorola that uses three wires to perform data communication between a master device and a slave device.

Stack A last-in-first-out data structure whose elements can be accessed only from one end. A stack structure has a top and a bottom. A new item can be added only to the top, and the stack elements can be removed only from the top.

Stack frame A region in the stack that holds incoming parameters, the subroutine return address, local variables, saved registers, and so on.

Standard timer module The timer module implemented in 812A4, 912B32, and 912BC32. This module consists of a 16-bit main timer, 8 channels of output-compare/input-capture function, and one 16-bit pulse accumulator.

Static memories Memory devices that do not require periodic refreshing in order to retain the stored information as long as power is applied.

Status register A register located in the CPU that keeps track of the status of instruction execution by noting the presence of carries, zeros, negatives, overflows, and so on.

Stepper motor A digital motor that rotates certain degrees clockwise or counterclockwise whenever certain sequence of values is applied to the motor.

String A sequence of characters.

Subroutine A sequence of instructions that can be called from various places in the program and will return to the caller after its execution. When a subroutine is called, the return address will be saved on the stack.

Subroutine call The process of invoking the subroutine to perform the desired operations. The 68HC12 has BSR, JSR, and CALL instructions for making subroutine calls.

Successive approximation method A method for performing A/D conversion that works from the most significant bit toward the least significant bit. For every bit, the algorithm guesses the bit to be 1 and then converts the resultant value into the analog voltage and then compares it with the input voltage. If the converted voltage is smaller than the input voltage, the guess is right. Otherwise, the guess is wrong and the bit is cleared to 0.

Switch Statement A multiway decision based on the value of a control expression.

Sync_seg In the CAN format, the sync_seg segment is the segment within a bit time used to synchronize all CAN nodes.

Temperature sensor A transducer that can convert temperature into a voltage.

Text editor A program that allows the end user to enter and edit text and program files.

Thermocouple A transducer that converts a high temperature into a voltage.

Transducer A device that can convert a nonelectric quantity into a voltage.

Transpose An operation that converts the rows of a matrix into columns and vice versa.

Trap A software interrupt; an exception.

UART The acronym of universal asynchronous receiver and transmitter; an interface chip that allows the microprocessor to perform asynchronous serial data communication.

Union A variable that may hold (at different times) objects of different types and sizes, with the compiler keeping track of size and alignment requirements.

Vector A vector is a unidimensional data structure in which each element is associated with an index i. The elements of a vector are of the same length.

Volatile memory Semiconductor memory that loses its stored information when power is removed.

Volatile variable A variable that has a value that can be changed by something other than user code. A typical example is an input port or a timer register. These variables must be declared as volatile so the compiler makes no assumptions on their values while performing optimizations.

Write collision The SPI error that occurs when an attempt is made to write to the SPDR register while data transfer is taking place.

References

1. BOSCH, *CAN Specification, Version 2.0*, Stuggart Germany, 1991.

2. CANopen, *Cabling and Connector Pin Assignment, Version 1.1*, April 2001.

3. Han-Way Huang, *MC68HC11: An Introduction*, 2nd ed. Clifton Park, NY: Delmar Thompson Learning, 2001.

4. Motorola, *M68HC12B Family Advance Information, Rev 3*. Phoenix, AZ: Motorola, 2001

5. Motorola, *MC68HC912D60A Technical Data, Rev. 1.0*, Phoenix, AZ: Motorola, 2001.

6. Motorola, *MC68HC912DG128A/MC68HC912DT128 Technical Data, Rev. 2.0*, Phoenix, AZ, Motorola, 2001.

7. Motorola, *CAN Bit Timing Requirements, AN1798*, Phoenix, AZ: Motorola, 1999.

8. Philips, *PCA82C250 CAN Controller Interface*, Amsterdam, Netherlands: Philips, January 2000.

Index